ActivEpi Companion Textbook

David G. Kleinbaum
Kevin M. Sullivan
Nancy D. Barker

ActivEpi Companion Textbook

A supplement for use with the *ActivEpi* CD-ROM

David G. Kleinbaum
Department of Epidemiology
Emory University
Atlanta, GA 30322
USA
dkleinb@sph.emory.edu

Kevin M. Sullivan
Department of Epidemiology
Emory University
Atlanta, GA 30322
USA
cdckms@sph.emory.edu

Nancy D. Barker
Consultant
Atlanta, GA
USA
ndbarker@earthlink.net

Library of Congress Cataloging-in-Publication Data
Kleinbaum, David G.
ActivEpi companion textbook : A supplement for use with the ActivEpi CD-ROM / David G. Kleinbaum, Kevin M. Sullivan, Nancy D. Barker.
p. cm.
Includes bibliographical references and index.
ISBN 0-387-95574-7 (softcover : alk. paper)
1. Epidemiology. I. Sullivan, Kevin M. II. Barker, Nancy D. III. Title.
RA651 .K545 2002
614.4—dc21 2002036547

ISBN 0-387-95574-7 Printed on acid-free paper.

Printed in the United States of America.

9 8 7 6 5 4 3 Corrected third printing, 2005. SPIN 11331124

springeronline.com

To

Edna
John and Barbara; Enid, Patrick, and Zoë
Bud

Preface

This text on epidemiology is a companion to the **ActivEpi** CD-ROM. The **ActivEpi** CD-ROM provides a multimedia presentation of epidemiologic concepts commonly taught in an introductory course in epidemiology. **ActivEpi** CD-ROM uses a range of multimedia effects to motivate, explain, visualize, and apply introductory epidemiologic concepts, integrating video, animation, narration, text, and interactive question and answer sessions. Since individuals differ in their learning skills, the **ActivEpi** CD-ROM and **ActivEpi** Companion Textbook offer readers different but nevertheless intertwined options on how to learn epidemiology. The **ActivEpi** CD-ROM provides an exciting way of presenting epidemiologic concepts through use of animation. The **ActivEpi** Companion Textbook can be utilized as a hardcopy reference of the textual materials contained in the CD-ROM, as a resource for the practice exercises, as a general reference, or even a self-contained textbook. The **ActivEpi** CD-ROM and **ActivEpi** Companion Textbook can be used for self-study or for a course in epidemiology, either in a traditional classroom setting or in a distance learning setting.

In general, virtually all of the material on the **ActivEpi** CD-ROM is included in the **ActivEpi** Companion Textbook. Some of the narration on the **ActivEpi** CD-ROM was altered for the Companion Textbook. This difference occurs primarily when the CD-ROM narration refers to an animation on the screen. Another difference between the **ActivEpi** CD-ROM and the Companion Textbook is in the Study Questions and the Quizzes. On the CD-ROM, the answers are provided interactively. For the text, the Study Questions and Quizzes are sequentially numbered throughout each lesson with the answers provided at the end of the lesson. Finally, there are some interactive activities on the **ActivEpi** CD-ROM that cannot be duplicated in the text, such as the exercises using the Data Desk program.

Contents

LESSON 1

ActivEpi

1-1 Getting Started: The Lesson Book

INTRODUCTION

Epidemiology is the study of health and illness in human or other (veterinary) populations. In this course, we consider real-world health and illness problems, and we show how epidemiologic concepts and methods allow us to study, understand and solve such problems. And, most important, we apply each new concept or method as we develop it to help you attain a growing understanding of the subject.

About the Authors of ActivEpi Companion Textbook
David G. Kleinbaum, Kevin M. Sullivan, and Nancy Barker

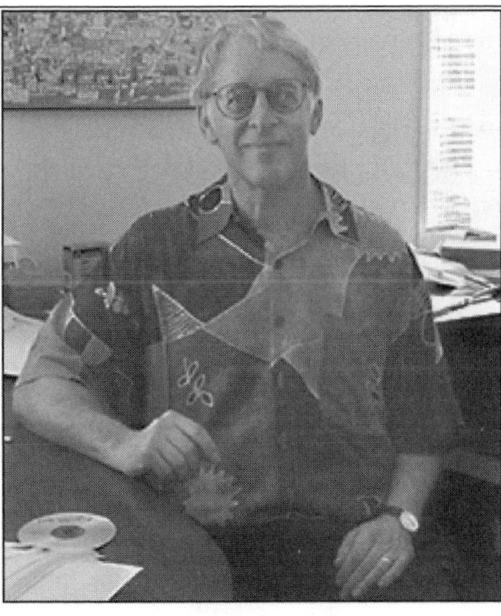

David G. Kleinbaum is a Professor of Epidemiology at Emory University's Rollins School of Public Health in Atlanta, GA, and an internationally recognized expert in teaching biostatistical and epidemiological concepts and methods at all levels. He is the author of several widely acclaimed textbooks including, *Applied Regression Analysis and Other Multivariable Methods*, *Epidemiologic Research: Principles and Quantitative Methods*, *Logistic Regression-A Self-Learning Text*, and *Survival Analysis-A Self-Learning Text*.

Dr. Kleinbaum has more than 25 years of experience teaching over 100 short courses on statistical and epidemiologic methods to a variety of international audiences, and has published widely in both the methodological and applied public health literature. He is also an experienced and sought-after consultant, and is presently an ad-hoc consultant to all research staff at the Centers for Disease Control and Prevention.

On a personal note, Dr. Kleinbaum is an accomplished jazz flutist, and plays weekly in Atlanta with his jazz combo, The Moonlighters Jazz Band.

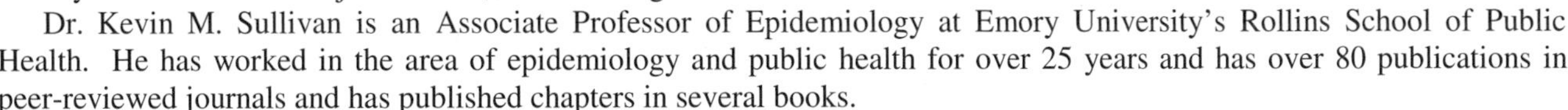

Dr. Kevin M. Sullivan is an Associate Professor of Epidemiology at Emory University's Rollins School of Public Health. He has worked in the area of epidemiology and public health for over 25 years and has over 80 publications in peer-reviewed journals and has published chapters in several books.

Ms. Nancy Barker is a statistical consultant who formerly worked at the Centers for Disease Control and Prevention.

Acknowledgements

There are many people that we wish to thank for their contributions to the ActivEpi CD and Companion Textbook project. First, Dr. Kleinbaum wishes to thank his wife, Edna Kleinbaum, his ActivEpi widow, who for the entire 4 years of this project, including continuous long work days, nights and weekends, had been his primary source of inspiration and support for the worth of this project. Dr. Sullivan would like to thank his wife, Enid, and children Patrick and Zoë for their support and patience. Ms. Barker would like to thank her husband, Bud, for his support.

We also wish to thank Paul Velleman, John Sammis, Tracy Stewart, Matt Clark and the staff at Data Description Inc. in Ithaca, New York: Christine Crane, Sarah Culley, Mike Hollenbeck, Jennie Lavine, Nick Lahr, Aaron Lowenkron, Pete Nix, David Velleman, Brian Young, all of whom were essential to carrying out this project. Paul Velleman is the voice (i.e., narrator) on all expositions in ActivEpi, but moreover, he is the main developer and author of ActivStats, the CD ROM course on Statistics that provided the authoring template used in the development of ActivEpi. John Sammis has been the overall coordinator of all aspects of ActivEpi development, including contract negotiations for financial support and maintenance as well as the primary developer of the Data Desk computer exercises in ActivEpi. Tracy Stewart has steadfastly carried out the unenviable task of coordinating all data management activities for the project, including synchronizing all audio, video and scripting tasks as well as coordinating long distance communication between Data Description staff and Atlanta-based staff. Matt Clark developed the authoring template for ActivStats and expanded its capability for use in developing ActivEpi.

Also essential to this project was Jan Hill, having been involved in this project from the beginning and enthusiastically committed to making ActivEpi a successful product. She had been the primary instructional media consultant in the development of this project. Her tasks have involved finding, producing, and/or transmitting all the visual components of ActivEpi, including video, still photographs, and animation. She also has guided and monitored all artistic aspects of the project, including producing the drawings used for all icons on the lesson pages.

In addition, we thank Richard Dicker of the Center for Disease Control and Prevention (CDC) who reviewed the epidemiological content the expositions in Lessons/Chapters 3 through 6 and also drafted and reviewed quizzes for these same lessons.

We also wish to thank Irene M. Vander Meer and Albert Hoffman of the Departments of Epidemiology and Biostatistics at Erasmus University in Rotterdam, for contributing several activities on epidemiologic studies carried out in the Netherlands that have been incorporated into ActivEpi expositions and/or quizzes.

We thank Stephen Pitts at Emory University's School of Medicine and School of Public Health for contributing three activities on clinical decision-making and its relationship to misclassification bias issues. Furthermore, we thank Mitch Klein of the Departments of Epidemiology and Environmental and Occupational Health at Emory University who provided review and consultation on the epidemiologic content and concepts described in many expositions that covered topics that were conceptually and methodologically complicated.

Also, we thank Kristie Polk of the Department of Biostatistics at Emory University for her assistance in the development of activities and asterisks on the "rare disease assumption" in Chapter 5.

Dr. Kleinbaum thanks the University of Maastricht, Netherlands Departments of General Practice, Epidemiology and Methodology and Statistics, for their interest as well as financial support for this project in the form of an adjunct professorial appointment at the University of Maastricht.

And, finally Dr. Kleinbaum wishes to thank the Centers of Disease Control and Prevention (CDC), who provided the financial support that made this product possible through contract number 200-2000-10036. He also thank CDC for financial support of my work on this project through IPA and ORISE contracts. Moreover, he thanks Donna Stroup and G. David Williamson of CDC for their continued belief in the value of this project and for playing an important role in setting up and maintaining the funding for the project. Lastly, he thanks Colette Zyrkowski of CDC for her valuable assistance in coordinating and monitoring this project from CDC's end and for overseeing the chapter-by-chapter review of the content by a committee of research scientists from several of the Centers at CDC. The CDC committee members include: Richard Garfein, Diane Bennett, Consuelo Beck-Sague, Rick Waxweiller, Tom Sinks, Stephanie Bock, Stacie Greby, Maya Sternberg, Jose Becera, and Stephanie Sansom.

Credits

We gratefully acknowledge the following sources for images used in ActivEpi:

- New South Wales Environment Protection Authority
- CDC Public Health Image Library
- Sandra Ford, CDC
- James "Mack" McGehee, mayor, City of Bogalusa, Louisiana
- Dr. Daniel Levy, Director, Framingham Heart Study
- Irene M. van der Meer, Erasmus Medical Centre Rotterdam
- Photographer G. van der Molen, The Netherlands Brain Bank, Coordinator Dr. R. Ravid; American Cancer Society (logo)
- Dr. George D. LeMaitre, LeMaitre Vascular
- Rick Sutton, Cartoonist
- WHO - World Health Organization (from National Library of Medicine, History of Medicine Images)

The Lesson Book

Each page of this Lesson Book presents only a few concepts, introduced by brief paragraphs. The actual learning is done by viewing, and interacting with, a launchable ***Activity****. ActivEpi employs three types of activities:*

- ***Narrated Expositions*** *that use animation, text, pictures, and video, synchronized with an audio track, to teach concepts;*
- ***Drag-and-Drop Quizzes*** *that provide feedback so you can determine whether to go back and review or move forward to the next set of activities;*
- ***Data Desk*** *data analysis activities that let you practice applying what you have learned.*

The Lesson Book is the home base for the course. Each page of the Lesson Book focuses on a few concepts, introduced by brief paragraphs. The Lesson Book has three general areas. The elements of each area respond to a single mouse click, opening new windows for each function.

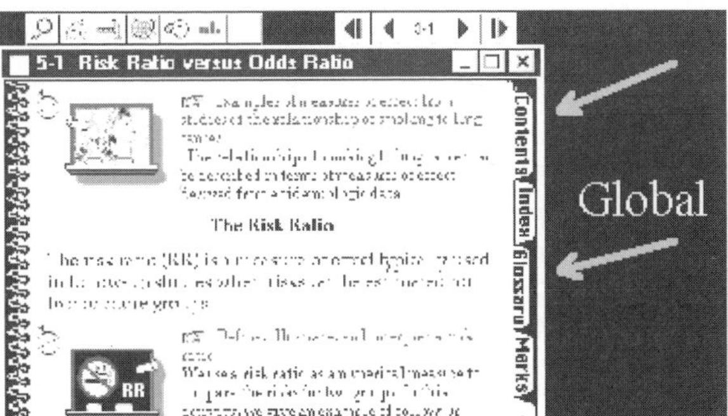

The tabs hold global matters such as the Table of Contents, the Index, and the Glossary.

The control bar holds page-level matters. Icons in the control bar provide access to the statistics environment that accompanies the course, offer a way to move forward and backward in the Lesson Book, and provide projects and exercises appropriate for the page.

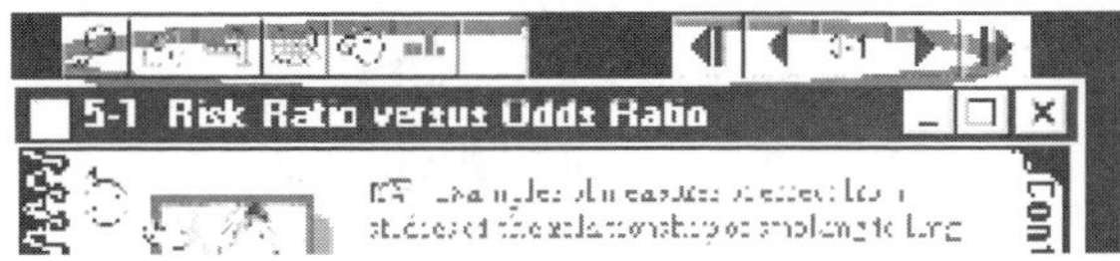

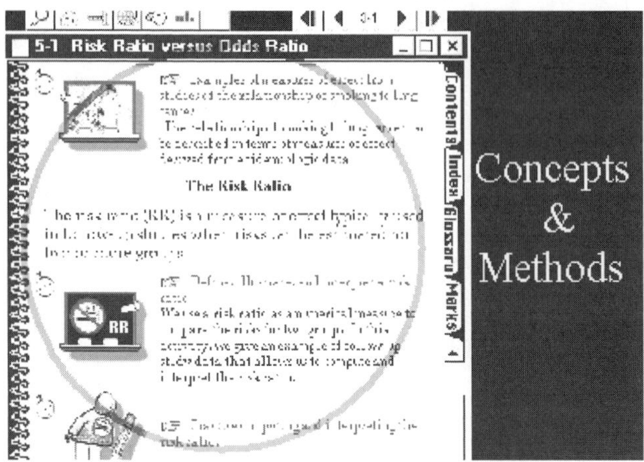

The body of a page holds discussions and examples of the concepts and methods that make up the content of the course. You initiate Activities by clicking once on their icons on the page, as you did to view this discussion. Pages may be too long to fit on the screen. The scroll bar on the right scrolls the page contents up and down. The details of using each of these features are discussed in separate Activities below.

Getting Started

This course uses multimedia to

- *Show real-world examples*
- *Let you apply methods as you learn them*
- *Provide computer-based experiments*
- *Allow you to check your understanding privately, and*
- *Supply a range of exercises, projects, and examples so you can take your learning beyond the computer.*

Epidemiology is the study of health and illness in human populations. For example, a randomized clinical trial conducted by Epidemiologists at the Harvard School of Public Health showed that taking aspirin reduces heart attack risk by 20 to 30 percent. Public health studies in the 1950's demonstrated that smoking cigarettes causes lung cancer. Environmental epidemiologists have been evaluating the evidence that living near power lines may have a high risk for childhood leukemia. Cancer researchers wonder why older women are less likely to be screened for breast cancer than younger women. All of these are examples of epidemiologic research, because they all attempt to describe the relationship between a health outcome and one or more explanations or causes of that outcome. All of these examples share several challenges: they must choose an appropriate study design, they must be careful to avoid bias, and they must use appropriate statistical methods to analyze the data. Epidemiology deals with each of these three challenges.

epidemiology

Health Outcome	Explanation
Heart Attack Status	Aspirin Intake
Lung Cancer	Smoking
Childhood Leukemia	Powerline Exposure
Breast Cancer Screening	Age

Choose Study Design

Be Careful to Avoid Bias

Use Appropriate Statistical Methods

1-2 Preferences and Activities

Preferences

*You can customize the look of the Lesson Book page, choose to hide or show various kinds of Activities, and control how many of the features of the course work. The **Preferences...** command in the **Edit** menu offers many choices.*

Preferences

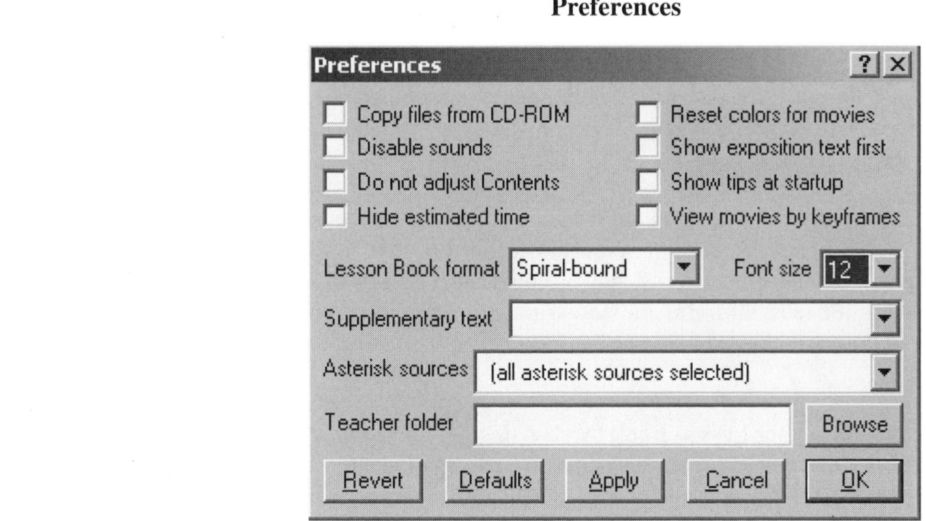

These are notes on the available preference choices. For all checkboxes, the default state is **unchecked**.

Copy files from CD-ROM
Ordinarily, Activities play from the CD. Checking this box asks ActivEpi to copy the files for playing each Activity from the CD to your hard disk (if possible) before the Activity begins. You will have to wait about 15-20 seconds before the Activity can begin while the disk is working. However, the Activity may perform better because your hard disk can transfer data to ActivEpi much faster than your CD drive. This option may help performance when playing from slower (2X) CD drives. It offers fewer advantages when playing from a 4X or faster CD drive.

Disable sounds
A variety of sound effects accompany ActivEpi actions. You may prefer silence. Check this box to silence those sounds. None of the sounds is essential to using ActivEpi or understanding epidemiology.

Do not adjust Contents
The Table of Contents provides convenient navigation through the course. Use the Table of Contents to show the top level of an outline of the course, and to expand that outline for the current Lesson and current page. The Table of Contents adjusts automatically whenever you turn a page or goes directly to another page in the Lesson book, again expanding to display information for the open page. Check this box to have the Table of Contents stay the same even though the Lesson Book page has changed.

Hide control bar (Macintosh only)
The Control Bar runs across the top of each Lesson Book page and provides access to key components of the course. Check this box to hide the control bar when you launch the program. You can click on the triangle on the upper-left comer of the Lesson Book bring back the control bar. On Windows computers, the control bar is displayed from the Tools Menu.

Hide estimated time
Each Activity shows a stopwatch with an estimated completion time for that Activity. For example, a stopwatch with a quarter of its area red estimates that the Activity will take you about 15 minutes. Check this box to hide the stopwatches.

Continued on next page

Reset colors for movies
Each movie has a unique color palette that can improve video playback for some computers, but may cause colors to flicker after the movie plays. Most computers will play movies well without unique palettes. If the colors in a movie seem wrong, try this option.

Show exposition text
All narrated Activities can display the text of the narration in a separate window. This may be especially helpful if English is not your first language. Check this box to open the narrated script text window automatically whenever a narrated Activity is launched.

Show tips at startup
When this box is checked, the Hints & Tips window is automatically opened when the program is launched. This preference can also be set from within the Hints & Tips window.

View movies by keyframes
Movie files are large, and may display poorly over a network. As an alternative for network use, ActivEpi offers a "keyframe only" version of each movie that plays the full sound track, but displays a series of "slides" taken from the movie. This version will usually play better over networks, but of course it does not offer the full-motion video of the complete version. Click this box to use the reduced-frame movies. If movies seem to be only slide shows rather than full-motion movies, open Preferences and un-check this box.

Lesson Book format
This one is just for fun. You can personalize your Lesson Book by choosing how it should look. Click on the black triangle and choose one of the formats in the menu. The selected format shows a check mark next to it.

Font size
You can specify the size of the type in the Lesson Book, in asterisks, and other places. Teachers may prefer a larger type font size when displaying the computer screen on a projector. Click on the black triangle and choose the correct size for you. The selected font size shows a check mark next to it. The default value is **12**.

Teacher folder
The Teacher Folder holds supplementary material that appears as additional Activities on pages of the Lesson book. For this feature to work, you must tell ActivEpi where to find the folder. It can be on any disk that your computer can access, including a network server or a floppy disk. If your teacher has created supplementary files, he or she will tell you how to locate the Teacher Folder.

Activities

Lesson Book pages hold a number of ***Activities****. Click on the Activity icon to launch it. Close the Activity window to return to the Lesson Book.* ***Narrated Exposition*** *activity windows offer controls so that you can pause the discussion, repeat any portion, skip over parts, and close the window. These controls are the same in every window, so you need to learn them only once.*

Using Controls in Activities

Each Activity opens its own window. You can take control over the Activity with the controls along the bottom of the window.

The stopwatch icon to the left of each Activity button shows the approximate time required to complete the Activity. On the CD-ROM, click on it to drop down a menu with other commands relating to the Activity.

Using the Stopwatch Commands

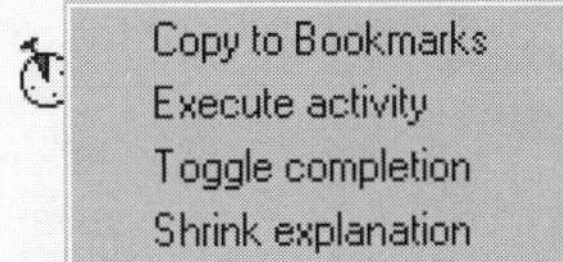

Copy to Bookmarks places a Bookmark to the Activity. Open the **Marks** tab from the lesson page to see a list of Activities that were copied. Clicking on the Activity inside the Bookmark takes you to that Activity.

Execute activity opens the Activity. This is equivalent to clicking on the Activity icon from the lesson page.

Set activity as completed puts a check mark next to the Activity icon on the lesson page to indicate completion of that Activity. This command will change to say **Set activity as uncompleted** if there is a check mark next to the Activity icon.

Shrink explanation hides the Activity icon and the explanation leaving only the goal statement on the lesson page. Click on the goal statement to bring back the Activity explanation and the icon.

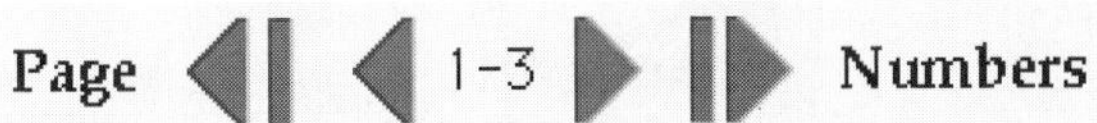

The page number at the right of the Control Bar at the top of the page identifies the current page, offers arrows that turn pages forward or backward and speed arrows that turn to the next or previous lessons. Click the page number itself to turn directly to any other page.

Glossary

Terms that appear in color and underlined on the CD-ROM are in the Glossary. Click on any glossary term to open the glossary to the appropriate definition. Whenever you see a glossary term, ask yourself whether you know what it means. If you are not certain, just click on the term.

Asterisks

Asterisks cover concepts in greater depth and offer additional material such as examples. Asterisks are not optional material, but rather can contain important information or comments. You should generally click on asterisks as you find them. In the Companion Textbook, items that have an asterisk on the CD-ROM will be presented in a box at the end of the Activity.

Using the Asterisks

Clicking on an asterisk on the CD-ROM opens a window. Asterisk windows present new or additional information and often offer links to the Glossary, and are referred to in the Index.

Each Lesson has an initial asterisk (using the symbol on the right below) that provides references for the material covered in that lesson. In the Companion Textbook, references are placed at the end of each lesson and the new or additional information placed in a box at the end of the activity.

★ §

1-4 Global Features

Global Features of the Course

Table of Contents

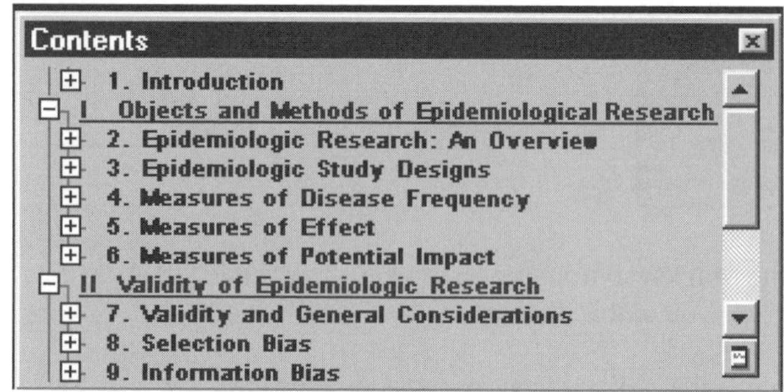

The Table of Contents serves 3 important functions: First, it gives an overview of the course, listing each lesson in order. Second, it provides a quick way to go to any lesson in the course. Just click on any lesson title to turn to the first page of that lesson. Finally, the Table of Contents shows you where you are in the course.

The Table of Contents opens to show each of the pages of the current lesson, and opens the current page to show each of the Activities on that page.

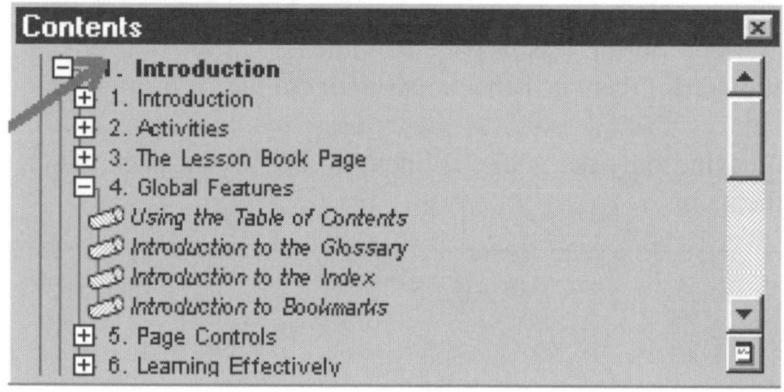

Activities that are checked off as viewed on the page show check marks next to them in the table of contents as well. Click on any line of the table to go to that page or even to that specific Activity.

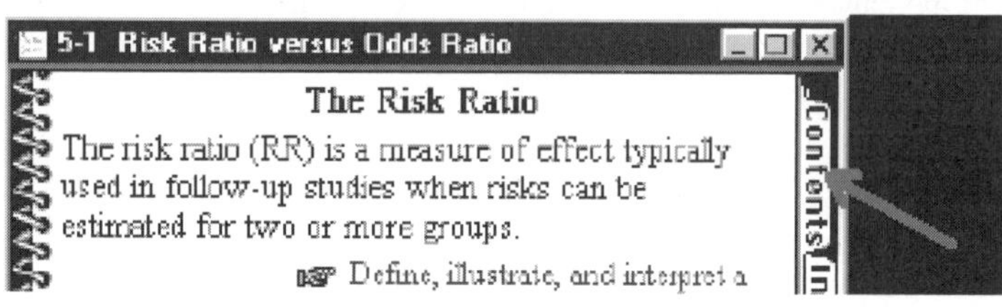

Close the Table of Contents by clicking its close box. Open the Table of Contents by clicking on the Contents Tab of the Lesson Book.

Glossary

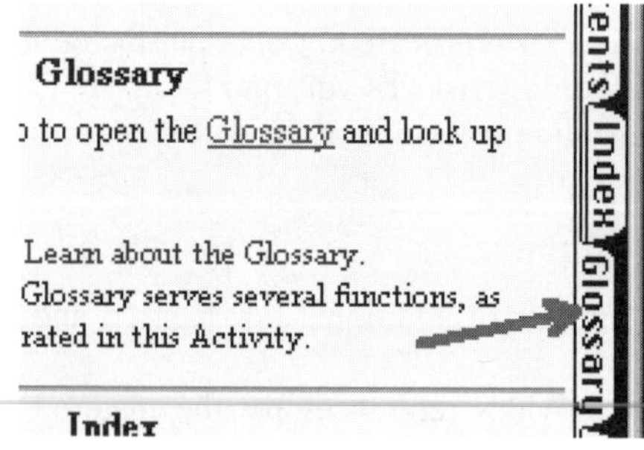

Throughout the Lesson Book, and in supplementary files, you will find words highlighted in color and underlined. These are Glossary terms. Click on any Glossary term to open the glossary to its definition. Alternatively, you can click on the Glossary tab and drag it out:

The Glossary window defines terms discussed in this course. Select the term to define by clicking on the term in the right panel. The definition is displayed in the adjacent panel on the left. You can scroll through the alphabetized list to find a specific term or press any key to find terms beginning with that letter.

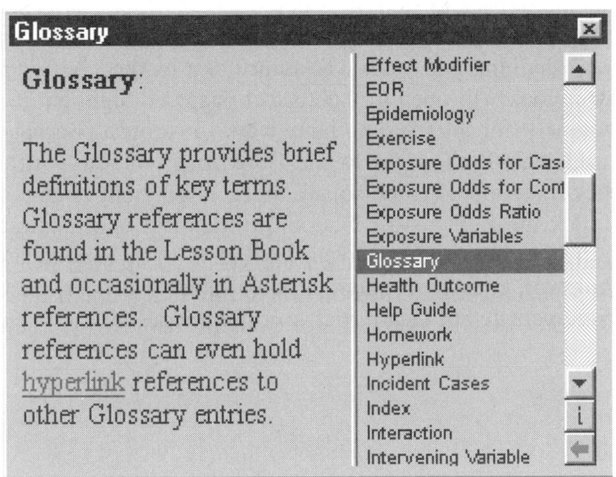

If a definition is too long to fit in the window, scroll or resize the window. Most definitions refer to other definitions. Click on any colored term to see its definition. To return to previous definitions (all the way back to the first you selected), click on the return arrow. To locate where the term is discussed in the course, click the small i button to open the index. To close the glossary, click its close box.

Index

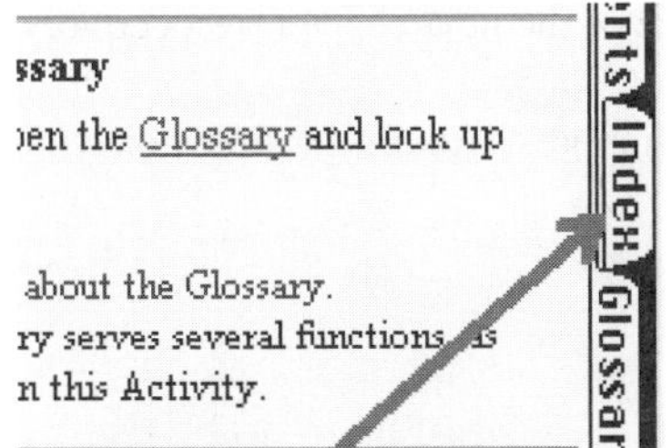

Open the Index by clicking on the Index tab:

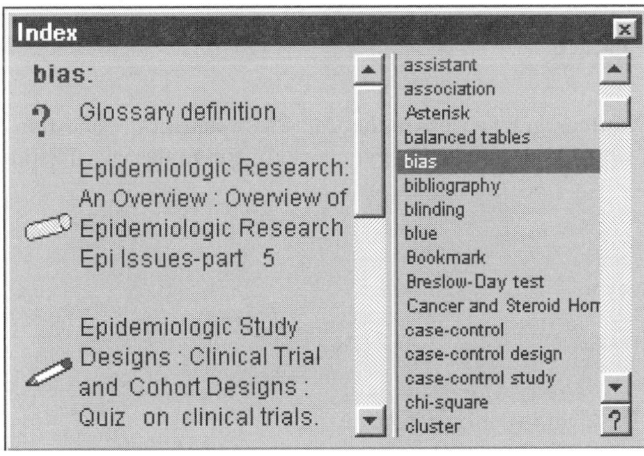

The Index window shows a scrolling alphabetical list of terms. Click on the term to locate. The adjacent panel gives links to references in the course. The icon next to the reference indicates the type of Activity, in the same way as in the Table of Contents and Bookmarks. The reference gives the Lesson name, the page name, and the Activity name. Click on the icon to turn the Lesson Book to the corresponding page. Many indexed terms have Glossary definitions. When there is a glossary definition, it is offered both in the references and at the bottom of the reference scroll bar. You can scroll through the alphabetized list to find a specific term. To close the index, click its close box.

Bookmarks

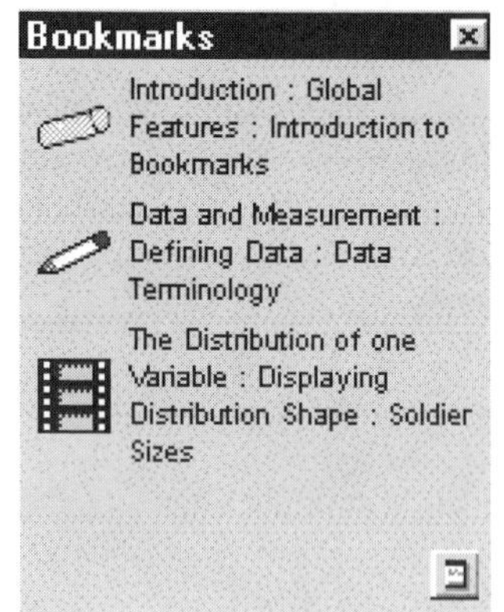

The Bookmark window holds icons that point back to Activities in the Lesson Book. To mark any Activity with a Bookmark, drag the Activity's icon to the Bookmark window. You can edit the text next to a Bookmark to say anything you want. To return to a marked Activity, just click on the Bookmark. The Lesson book will open to the correct page and highlight the Activity. You can even make a Bookmark for an asterisk, homework, or project. Because these windows have no icons, you add their Bookmarks to the Bookmark window with a menu command. On a Macintosh, click on the local menu square in the lower right of any of these windows. On Windows, just click the right mouse button anywhere in one of these windows. A menu will pop down offering a command to add the bookmark to the Bookmark window. You can save separate Bookmark files each under its own name and open them whenever you'd like. This makes it easy to create bookmark files for different needs.

Shortcuts

- To make the **sound** in any Activity **louder**, press the up-arrow key. To make the **sound softer**, press the down-arrow key.
- **Turn pages** from the keyboard on a Macintosh by pressing Command-2 to view a previous page and Command-3 to advance to the next page. On Windows computers, use the Control key in place of the Command key. Command-l and Command-4 take you to the first page of the previous chapter and the first page of the next chapter, respectively.
- Press the space bar or click the mouse to **pause** an Activity.
- To **continue** a paused Activity, double-click in the body of its exposition area or press Return or Enter on your keyboard.

1-5 Page Controls

Page Controls

The Control Bar at the top of the page in the ActivEpi CD-ROM offers direct access to the Data Desk statistics applications, the Homework and Projects for each lesson, the World Wide Web (if your computer is connected), and to the visualization tools used in each lesson.

Use this icon to learn about computing using Data Desk with your own data.

Exercises, Homework, Study Questions, and Projects

Homework exercises appropriate to each lesson are kept in the Homework icon of the control bar. Do the Homework. Click the WORK icon on the control bar to open the Homework

ACE-I. Homework Introduction

Homework exercises typically provide data and ask that you apply the methods or concepts you have been learning to understand something about the data. The most common request is that you write a paragraph or two about your conclusions, possibly illustrated with graphs or tables. If you use Data Desk, graphs and tables can be copied and pasted into any standard word processor. You will probably find that you learn more by doing homework exercises than by working with the tools or following the expositions. You will recognize your progress when you are able to phrase your question in proper terms to your teacher or your teaching assistant.

Note: Some homework exercises that require computations can be completed using Data Desk and Data Desk templates. Data Desk templates are special Data Desk files that extend the capabilities of the program. To use a template with data in a Data Desk datafile, you would typically merge the template into the file holding the data using the Import command. For all of the Data Desk activities launched from the Lesson Book in this course we have already imported any required templates into the Data Desk datafile.

To bring a template into your current Data Desk file, choose the Import command from the File menu. Use the dialog that appears to find the template you wish to import and click the Import button.

Projects usually include the collection or generation of new ideas. Identify projects that apply concepts and methods from the current lesson. Click the PROJ icon to open the Project Browse. Projects provide an opportunity to apply the skills and concepts learned in the Activities to new real-world problems.

The **Web** icon is for linking to the World Wide Web for a wealth of related data, activities, and information. This is where links to Internet resources are made, especially for gathering data for statistical analysis.

The **Guide** icon opens the Help Guide, which indicates the different types of help available.

1-6 Learning Effectively

Learning Effectively with this Course

To work effectively in this course, you must take control of the key parts of your learning. In particular, you should:

Take Notes: There is real mnemonic value in the physical experience of writing notes. We encourage you to take notes on paper in the traditional way. (See how to to copy the text to the clipboard at the end of this lesson). Remember that you can pause an exposition or video at any time by clicking the stop/play button or pressing the space bar. If you miss something or want to see it again, slide the progress bar back to that point.
Control the Expositions: Everyone's mind wanders sometimes. And even if you are paying close attention, some of the material just doesn't make sense the first time you see it. (Frankly, nobody understands this stuff the first time they see it. A drawback of a standard classroom lecture is that you can't pause or rewind most lecturers.) You have full control over the explanation of new material. You can stop at any point just to sit and think for a minute to absorb a new idea, to write some additional notes, to refer to the corresponding section of a text, or to confer with another student. You can review any part as often as you like and work with any tool as often as you like.
Do the Exercises: Nobody is watching, so it is easy to skip the review material. Don't skip it! Some important parts of the course are taught in the exercises. If you skip them, you'll miss some important stuff.
Work Sequentially: Yes, it's multimedia, with hypertext and many options. But epidemiologic methods, as well as statistics, is a sequential subject. You are free to jump around in the course, but you'll find that the material makes much more sense when you learn basic ideas first and then build on them. And that old trick of first trying to do the homework and then looking back to try to find a similar part of the text to copy for the answer just won't work with multimedia. The content is often found inside an Activity or Exercise, so you'll waste much more time looking for it than you ever could have saved.
Accessibility: ActivEpi has several features that make it more accessible to those with hearing or visual impairments, or those who have learned English as a second language. See the box at the end of this lesson to see how to use ActivEpi most effectively.

Copying text

For asterisk, projects, and homework, you can choose **Copy** from the **Edit** menu to copy the entire text. Copy works in the **Lesson Book** either for all the text or just with selected paragraphs. To select a paragraph on Macintosh platforms, hold down the option key and click on the paragraph. To select a continuous paragraph list, hold down the shift-option keys and click. For a non-contiguous grouping, use option-command. To select a paragraph on Windows platforms, hold down the control key and click on the paragraph. To select more than a single paragraph, hold down the shift-control keys and click.

Accessibility

Accessibility means creating products usable and friendly to a wide range of users, including those with disabilities. There are over forty million people in the United States who have some type of disability. ActivEpi has been designed to address accessibility issues.

There are several accessibility tools already available for personal computers. Macintosh users should be familiar with **Close View** and **Easy Access**, and Windows users should have **Accessibility Options** installed.

People With a Physical Disability

People who have a physical disability mainly have difficulty with computer input devices, such as the mouse or keyboard, and with handling storage media. Where possible, both mouse behavior and command-key equivalents are used to perform the same action. For example, a movie or an exposition can be started or stopped by clicking the mouse button on the play/pause button or by using the space bar. For the hands-on activities that require a generated data file from either clicking on a target or guessing fractional parts, sample data files have been provided on the ActivEpi CD-ROM.

People With a Visual Disability

People with a visual disability have the most trouble with the output display, the screen. For these users, it is possible to set the size of text in the Lesson Book, Asterisk, and Homework windows. Choose from the menu **Edit→Preferences** and change the font size setting. For the expositions and tear-off tools, consider either changing the desktop window size (using the **Monitor** control panel on Mac or the **Start→Settings→Control Panel→Display** on Windows) or using a zoom utility (**Close View** on Mac, **ZoomIn** on Windows, **Magnifier** on Windows98). For people with color-vision difficulties, several hands-on activities allow the colors of data items to be changed. For example, click on a holding bin in either the Randomness or Probability activities and select a new color. On Macintosh, switching the monitor setting to black-and-white will display the different data types using patterns instead of colors.

People With a Hearing Disability

Hearing-disabled people cannot hear normal volume levels or at all. With the exception of the spoken text in movies, expositions, and hands-on activities, all sounds are used as assisting mechanisms, for example hearing a page turn when a new Lesson Book page is displayed. For the spoken text in activities, written text is available. After activating an activity, choose **Exposition→View Exposition Text** from the menu to see the narration. If you would like to see this text before every activity, choose **View exposition text** from the Preferences dialog. People who are not fluent in English may also wish to take advantage of this feature. The volume of movies and expositions can be set by clicking on the speaker icon at the lower-left comer of these activity windows. Hold down the **shift** key to set the volume for all activities instead of just the current one. The exposition text is also in the ActivEpi Companion Textbook.

People With a Speech or Language Disability

People who have a speech or language disability may have normal to above-average cognitive ability but no capacity for oral communication. The speech or language disability may be caused by an injury or a stroke, for example. In these cases, we recommend using the **Exposition→View Exposition Text** option whenever possible (see previous section) and the ActivEpi Companion Textbook.

Continued on next page

People With a Seizure Disorder

Some people with a seizure disorder are sensitive to certain flicker frequencies, which may cause them to go into seizure. The most problematic part of this frequency range is from 15 to 30 Hz. ActivEpi simulations are designed to use a graphical technique known as double buffering to reduce the amount of flickering.

Collaborative Computing

Collaborative computing is a shared computing environment or application that facilitates communication and teamwork among groups of people. If you are using ActivEpi in this type of environment, we recommend using shared headphones in a computer laboratory environment (i.e. a classroom with multiple computers).

Further Information

Apple Computer Disability Connection, www.apple.com/disability/
Microsoft on Disabilities and Accessibility, www.microsoftcom/enable/microsoft/

Answers to Study Questions

Q1.1

1. (no question asked)
2. (no question asked)
3. Although it is tempting to just click to see the answer, you will learn much more if you try to answer the question first and just click for the solution as a check afterwards.
4. To change the sound volume from the keyboard, press the up or down arrow keys

LESSON 2

EPIDEMIOLOGIC RESEARCH: AN OVERVIEW

2-1 Important Methodologic Issues

The field of epidemiology was initially concerned with providing a methodological basis for the study and control of population epidemics. Now, however, epidemiology has a much broader scope, including the study of both acute and chronic diseases, the quality of health care, and mental health problems. As the focus of epidemiologic inquiry has broadened, so has the methodology. In this overview lesson, we describe examples of epidemiologic research and introduce several important methodologic issues typically considered in such research.

The Sydney Beach Users Study

***Epidemiology** is primarily concerned with identifying the important factors or variables that influence a health outcome of interest. In the Sydney Beach Users Study, the key question was "Is swimming at the beaches in Sydney associated with an increased risk of acute infectious illness?"*

In Sydney, Australia, throughout the 1980s, complaints were expressed in the local news media that the popular public beaches surrounding the city were becoming more and more unsafe for swimming. Much of the concern focused on the suspicion that the beaches were being increasingly polluted by waste disposal.

In 1989, the New South Wales Department of Health decided to undertake a study to investigate the extent to which swimming and possible pollution at 12 popular Sydney beaches affected the public's health, particularly during the summer months when the beaches were most crowded. The primary research question of interest was: *are persons who swim at Sydney beaches at increased risk for developing an acute infectious illness?*

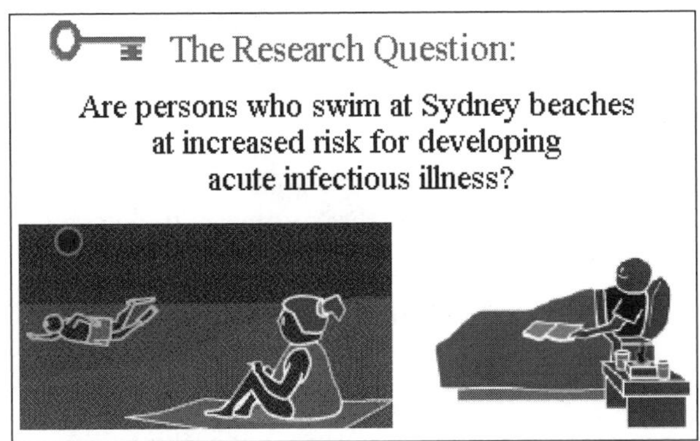
The Research Question:
Are persons who swim at Sydney beaches at increased risk for developing acute infectious illness?

The study was carried out by selecting subjects on the beaches throughout the summer months of 1989-90. Those subjects eligible to participate at this initial interview were then followed-up by phone a week later to determine swimming exposure on the day of the beach interview and subsequent illness status during the week following the interview.

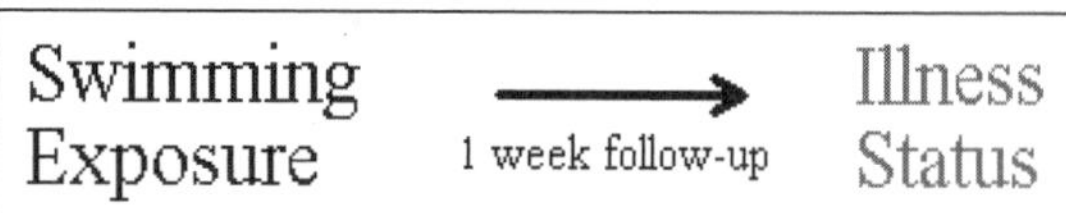

Water quality measurements at the beaches were also taken on each day that subjects were sampled in order to match swimming exposure to pollution levels at the beaches.

Analysis of the study data lead to the overall conclusion that swimming in polluted water carried a statistically significant 33% increased risk for an infectious illness when compared to swimming in non-polluted water. These results were considered by health department officials and the public alike to confirm that swimming in Sydney beaches posed an important health problem. Consequently, the state and local health departments together with other environmental agencies in the Sydney area undertook a program to reduce sources of pollution of beach water that lead to improved water quality at the beaches during the 1990's.

Summary

- The Sydney Beach Users Study is an example of the application of epidemiologic principles and methods to investigate a localized public health issue.
- The key question in the Sydney Beach Users Study was:
  - Does swimming at the beaches in Sydney, Australia (in 1989-90) pose an increased health risk for acute infectious illnesses?
  - The conclusion was yes, a 33% increased risk.

Important Methodologic Issues

We provide a general perspective of epidemiologic research by highlighting several broad issues that arise during the course of most epidemiologic investigations.

There are many issues to worry about when planning an epidemiologic research study (see Box below). In this activity, we will begin to describe a list of broad methodologic issues that need to be addressed. We will illustrate each issue using the previously described Sydney Beach Users Study of 1989.

Issues to consider when planning an epidemiologic research study

Question	Define a question of interest and key variables
Variables	What to measure and how; exposure (**E**), disease (**D**), and control (**C**) variables
Design	What study design and sampling frame?
Frequency	Measures of disease frequency
Effect	Measures of effect
Bias	Flaws in study design, collection, or analysis
Analysis	Perform appropriate analyses

The first is to clearly define the study **question** of interest, including specifying the key variables to be measured. Typically, we ask: *What is the relationship of one or more hypothesized determinants to a disease or health outcome of interest?*

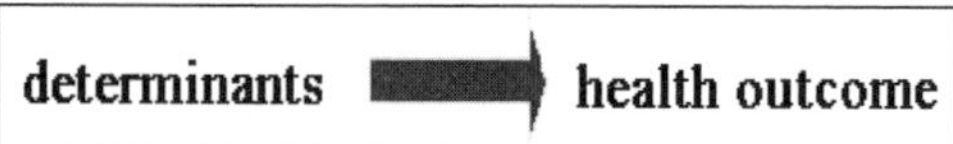

A **determinant** is often called an **exposure variable** and is denoted by the letter **E**. The disease or health outcome is denoted as **D**. Generally, variables other than exposure and disease that are known to predict the health outcome must be taken into account. We often call these variables **control variables** and denote them using the letter **C**.

Next, we must determine how to actually measure these **variables**. This step requires determining the information-gathering instruments and survey questionnaires to be obtained or developed.

Data is Obtained From:

surveys
interviews
samples
laboratory

The next issue is to select an appropriate **study design** and devise a sampling plan for enrolling subjects into the study. The choice of study design and sampling plan depends on feasibility and cost as well as a variety of characteristics of the population being studied and the study purpose.

Terms to learn:

clinical trials
cross-sectional
case-control
cohort

Measures of disease frequency and effect then need to be chosen based on the study design. A measure of **disease frequency** provides quantitative information about how often a health outcome occurs in subgroups of interest. A **measure of effect** allows for a comparison among subgroups.

Terms to learn:

rate	**risk ratio**
proportion	**odds ratio**
risk	**rate ratio**
odds	**prevalence ratio**
prevalence	
incidence	

We must also consider the potential **biases** of a study. Are there any flaws in the study design, the methods of data collection, or the methods of data analysis that could lead to spurious conclusions about the exposure-disease relationship?

Terms to learn:

selection bias
information bias
confounding bias

Finally, we must perform the appropriate **data analysis**, including stratification and mathematical modeling as appropriate. Analysis of epidemiologic data often includes taking into account other previously known risk factors for the health outcome. Failing to do this can often distort the results and lead to incorrect conclusions.

Terms to learn:

Logistic regression
Risk Factors
Confounding
Effect Modification

Summary: Important Methodological Issues

- What is the study question?
- How should the study variables be measured?
- How should the study be designed?
- What measures of disease frequency should be used?
- What kinds of bias are likely?
- How do we analyze the study data?

96
64
H7

The Study Question

Epidemiology is primarily concerned with identifying the important factors or variables that influence a health outcome of interest. Therefore, an important first step in an epidemiologic research study is to carefully state the key study question of interest.

The study question needs to be stated as clearly and as early as possible, particularly to indicate the variables to be observed or measured. A typical epidemiologic research question describes the relationship between a health outcome variable, **D**, and an exposure variable, **E**, taking into account the effects of other variables already known to predict the outcome (**C,** control variables).

D = health outcome variables
E = exposure variables
C = control variables

A simple situation, which is our primary focus throughout the course, occurs when there is only one **D** and one **E**, and there are several control variables. Then, the typical research question can be expressed as shown below, where the arrow indicates that the variables **E** and the controls (**C**s) on the left are the variables to be evaluated as predictors of the outcome **D**, shown on the right.

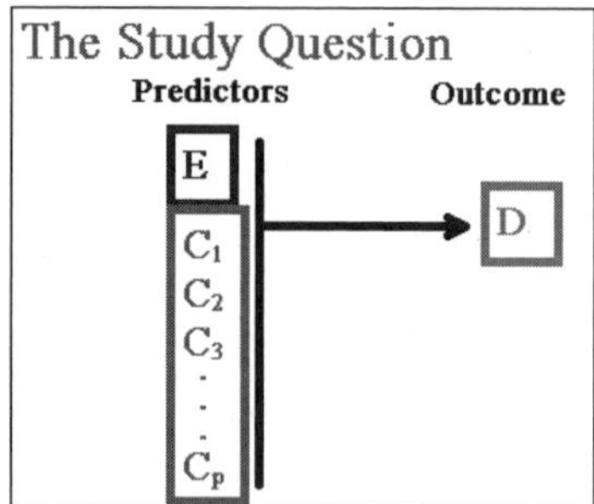

In the Sydney Beach Users Study, the health outcome variable, **D**, of interest is whether or not a person swimming at a beach in Sydney develops an acute infectious illness such as a cough, cold, flu, ear infection, or eye infection, within one week of swimming at the beach.

The study subjects could be classified as either:

D=0 for those did not get ill, or
D=l for those became ill.

A logical choice for the exposure variable is the exposure variable *swimming status*, which is set to:

E=0 for non-swimmers and
E=1 for swimmers during the time period of the study.

(Note that other coding schemes could be used other than 0/1, such as 1/2, Y/N, or +/-, but we will use 0/1).
Control variables might include pollution level at the beach, age of the subject, and duration of swimming.

Generally speaking, a study will not be very useful unless a question or hypothesis of some kind can be formulated to justify the time and expense needed to carry out the study.

Thus, the research question of this study example is to describe the relationship of swimming to the development of an infectious illness, while taking into account the effects of relevant control variables such as pollution level, age of subject and duration of swimming.

Because several variables are involved, we can expect that a complicated set of analyses will be required to deal with all the possible relationships among the variables involved.

Summary: The Study Question

- An important first step in an epidemiologic research study is to carefully state the key study question of interest.
- The general question: To what extent is there an association between one or more exposure variables (**E**s) and a health outcome (**D**), taking into account (i.e., controlling for) the possible influence of other important covariates (**C**s)?
- We can expect a complicated set of analyses to be required to deal with all possible relationships among the variables involved.

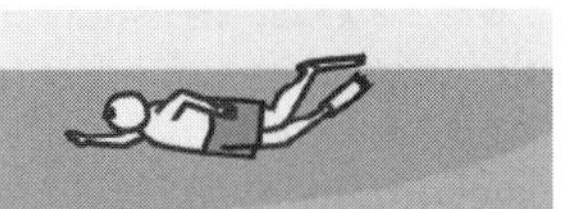

Quiz (Q2.1)

In the Sydney Beach Users study, exposure was alternatively defined by distinguishing those who swam in polluted water from those who swam in non-polluted water and from those who did not swim at all. Based on this scenario, fill in the missing information in the following statement:

1. The exposure variable has **???** categories, one of which is **???**

Choices:
2 **3** **4** **5** **did not swim** **polluted water** **swam** **water not polluted**

2. When considering both swimming and pollution together, which of the following choices is appropriate for defining the exposure variable in the Sydney Beach Users study: **???**

Choices:
a) E=O if did not swim, E=1 if swam in polluted water
b) E=O if did not swim, E=1 if swam in non-polluted water
c) E=O if did not swim, E=1 if swam in polluted water, E=2 if swam in non-polluted water
d) E=O if did not swim, E=1 if swam

In the Sydney Beach Users study, the illness outcome was whether or not an acute infectious illness developed 1 week after swimming at the beach. Also, in addition to age, another control variable was whether or not a study subject swam on days other than the day he or she was interviewed.
Fill in the missing information:

3. The health outcome has **???** categories.
4. There are at least **???** control variables.
5. Which of the following choices is not a control variable: **???**
   a) Age
   b) Swimming status on other days
   c) Swimming status on day of interview

Choices:
2 **3** **4** **5** **a** **b** **c**

2-2 Methodologic Issues (continued)

Measuring the Variables

Another important issue is: How do we measure the variables to be studied? Several measurement issues are now introduced.

Once the study question is determined, the investigators must determine how to measure the variables identified for the study and any other information that is needed. For example, how will the exposure variable be measured? If a subject went into the water but never put his head under the water, does that count as swimming? How much time is required to spend in the water to be counted as swimming? Is it feasible to observe each subject's swimming status on the day of initial interview, and if not, how should swimming status be determined?

After considering these questions, the study team defined swimming as *any immersion of the face and head in the water*. It was decided that subject self-reporting of swimming was the only feasible way to obtain swimming information.

Measuring Exposure Variables

Definition of Swimming:

Any immersion of the face & head in the water

Measuring Swimming Status:

Subject self-reporting

How will the health outcome be measured? Should illness be determined by a subject's self-report, which might be inaccurate, or by a physician's confirmation, which might not be available? The study team decided to use self-reported symptoms of illness obtained by telephone interview of study subjects 7 to 10 days after the initial interview.

Another measurement issue concerned how to determine water quality at the beach. Do water samples need to be collected? What time of day should they be collected? How will such information be linked to study subjects? The study team decided that health department surveyors would collect morning and evening samples at the midpoint of each of three sectors of the beach.

As nearly as could practicably be achieved, study subjects were to be interviewed during the period in which water samples were taken. A standard protocol was determined for how much water was to be sampled and how samples were to be assessed for water quality.

A final measurement issue concerned what information should be obtained from persons interviewed at the beach for possible inclusion into the study? The study team decided to collect basic demographic data including age, sex, and postcode, to ask whether or not each respondent had been swimming anywhere in the previous 5 days, and had any condition that precluded swimming on the day of the interview.

Interview Variables

age
sex
postcode

swimming history
health status

Subjects were excluded from the study if they reported swimming in the previous 5 days or having an illness that prevented them from swimming. Subjects were included if they were at least 15 years old and agreed to both an initial beach interview and a follow-up telephone interview.

All the measurement issues described above must be addressed prior to data collection to ensure standardized information is collected and to provide a study that is both cost and time efficient.

Study Questions (Q2.2)

1. What other variables might you also consider as control variables in the Beach Users Study?
2. How do we decide which variables to measure as control variables?
3. Why should age be considered?
4. How would you deal with subjects who went to the beach on more than one day?

Summary: Measuring the Variables

General measurement issues:

- ❖ How to operationalize the way a measurement is carried out?
- ❖ Should self-reporting of exposure and/or health outcome be used?
- ❖ When should measurements be taken?
- ❖ How many measurements should be taken on each variable and how should several measurements be combined?
- ❖ How to link environmental measures with individual subjects?

The Study Design, including the Sampling Plan

Another important issue is: What ***study design*** *should be used and how should we select study subjects? Several study design issues are now introduced.*

There are a variety of study designs used in epidemiology. The Sydney Beach Users study employed a **cohort** design. A key feature of such a design is that subjects without the health outcome are followed-up over time to determine if they develop the outcome. Subjects were selected from 12 popular Sydney beaches over 41 sampling days. An initial interview with the study subjects took place on the beach to obtain consent to participate in the study and to obtain demographic information.

Persons were excluded from the study if they had an illness that prevented them from swimming on that day or if they had been swimming within the previous 5 days. It was not considered feasible to determine swimming exposure status of each subject on the day of initial interview. Consequently, a follow-up telephone interview was conducted 7 to 10 days later to obtain self-reported swimming exposure as well as illness status of each subject.

Study Questions (Q2.3)

1. How might you criticize the choice of using self-reported exposure and illnesses?
2. How might you criticize the decision to determine swimming status from a telephone interview conducted 7 to 10 days after being interviewed on the beach?

A complex sample survey design was used to obtain the nearly 3000 study participants. Six beaches were selected on any given day and included 2 each from the northern, eastern and southern areas of Sydney. Each beach was divided into three sectors, defined by the position of the swimming area flags erected by the lifeguards. Trained interviewers recruited subjects, starting at the center of each sector and moving in a clockwise fashion until a quota for that sector had been reached. Potential subjects had to be at least 3 meters apart.

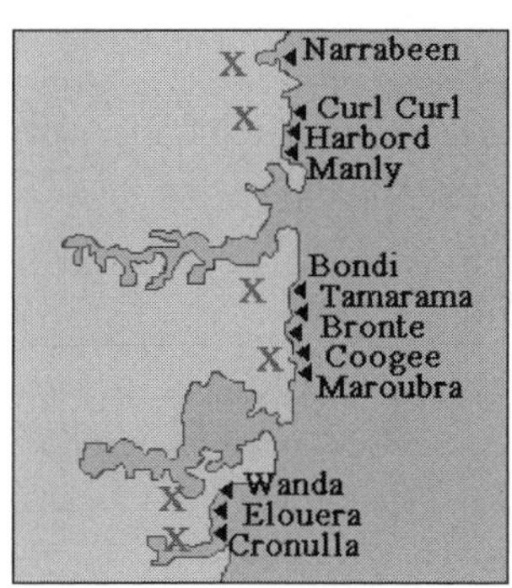

Study Questions (Q2.4)

1. Why do you think potential subjects in a given sector of the beach were specified to be at least 3 meters apart?
2. Why is the Sydney Beach Users Study a cohort study?
3. A fixed cohort is a group of people identified at the onset of a study and then followed over time to determine if they developed the outcome. Was a fixed cohort used in the Sydney Beach Users Study? Explain.
4. A case-control design starts with subjects with and without an illness and looks back in time to determine prior exposure history for both groups. Why is the Sydney Beach Users study *not* a case-control study?
5. In a cross-sectional study, both exposure and disease status are observed at the same time that subjects are selected into the study. Why is the Sydney Beach Users study not a cross-sectional study?

Summary: Study Design

❖ Two general design issues:
  - Which of several alternative forms of **epidemiologic study designs** should be used (e.g., cohort, case-control, cross-sectional)?
  - What is the **sampling plan** for selecting subjects?

Measures of Disease Frequency and Effect

Another important issue is: What ***measure of disease frequency*** *and* ***measure of effect*** *should be used? These terms are now briefly introduced.*

Once the study design has been determined, appropriate measures of disease frequency and effect can be specified. A measure of disease frequency provides quantitative information about how often the health outcome has occurred in a subgroup of interest.

For example, in the Sydney Beach Users Study, if we want to measure the frequency with which those who swam developed the illness of interest, we could determine the number of subjects who got ill and swam and divide by the total number who swam. The denominator represents the total number of study subjects among swimmers that had the opportunity to become ill. The numerator gives the number of study subjects among swimmers who actually became ill. Similarly, if we want to measure the frequency of illness among those who did *not* swim, we could divide the number of subjects who got ill and did not swim by the total number of non-swimming subjects.

Measure of Disease Frequency
Sydney Beach Users Study

Swimmers: $\dfrac{\text{\# ill swimmers}}{\text{total \# swimmers}}$

Non-Swimmers: $\dfrac{\text{\# ill non-swimmers}}{\text{total \# non-swimmers}}$

The information required to carry out the above calculations can be described in the form of a two-way table shown below. A simple summary of the required information can be given in a two-way table. This table shows the number who became ill among swimmers and non-swimmers. We can calculate the proportion ill among the swimmers to be 0.277 or 27.7 percent. We can also calculate the proportion ill among the non-swimmers as 0.165 or 16.5 percent.

		Swim Yes	No	Total
Ill	Yes	532	151	683
	No	1392	764	2156
	Total	1924	915	2839

proportion ill (swimmers): $\frac{532}{1924} = .277$ or 27.7%

proportion ill (non-swimmers): $\frac{151}{915} = .165$ or 16.5%

Each proportion is a measure of disease frequency called a **risk**. **R(E)** denotes the risk among the exposed for developing the health outcome. **R(not E)** [or **R(** $\bar{E}$ **)**] denotes the risk among the ***un***exposed. There are measures of disease frequency other than risk that will be described in this course. The choice of measure (e.g., risk, odds, prevalence, or rate) primarily depends on the type of study design being used and the goal of the research study.

If we want to compare two measures of disease frequency, such as two risks, we can divide one risk by the other, say, the risk for swimmers divided by the risk for non-swimmers. We find that the ratio of these risks in our study is 1.68; this means that swimmers have a risk for the illness that is 1.68 times the risk for non-swimmers.

Risk

proportion ill (swimmers): 27.7%

proportion ill (non-swimmers): 16.5%

$$\frac{R(E)}{R(\text{not } E)} = \frac{27.7\%}{16.5\%} = 1.68$$

$$\text{Risk}_{(\text{swimmers})} = 1.68 \times \text{Risk}_{(\text{non-swimmers})}$$

Such a measure is called a **measure of effect**. In this example, the effect of interest refers to the effect of one's swimming status on becoming or not becoming ill. If we divide one risk by the other, the measure of effect or association is called a **risk ratio**. There are other measures of effect that will be described in this course (e.g., such as the risk ratio, odds ratio, prevalence ratio, rate ratio, risk difference, and rate difference). As with measures of disease frequency, the choice of effect measure depends on the type of study design and the goal of the research study.

Summary: Measures of Disease Frequency and Effect

- A **measure of disease frequency** quantifies how often the health outcome has occurred in a subgroup of interest.
- A **measure of effect** quantifies a comparison of measures of disease frequency for two or more subgroups.
- The choice of measure of disease frequency and measure of effect depends on the type of study design used and the goal of the research study.

Bias

Another important issue is: What are the potential biases of the study? The concept of bias is now briefly introduced.

The next methodologic issue concerns the potential biases of a study. Bias is a flaw in the study design, the methods of data collection, or the methods of data analysis that may lead to spurious conclusions about the exposure-disease relationship. Bias may occur because of: the **selection** of study subjects; incorrect information gathered on study subjects; or failure to adjust for variables other than the exposure variable, commonly called **confounding.**

Bias

A flaw in
1. the study design.
2. the methods of data collection.
3. the methods of data analysis.

that leads to spurious conclusions

Sources of bias:
1. Selection
2. Information
3. Confounding

In the Sydney Beach Users Study, all 3 sources of bias were considered. For example, to avoid **selection bias**, subjects were excluded from the analysis if they were already ill on the day of the interview. This ensured that the sample represented only those healthy enough to go swimming on the day of interview. Sometimes selection bias cannot be avoided. For example, subjects had to be excluded from the study if they did not complete the follow-up interview. This **non-response bias** may affect how representative the sample is.

There was also potential for **information bias** since both swimming status and illness status were based on self-reporting by study subjects. Swimming status was determined by self-report at least seven days after the swimming occurred. Also, the report of illness outcome did not involve any clinical confirmation of reported symptoms.

Confounding in the Beach Users Study concerned whether all relevant variables other than swimming status and pollution level exposures were taken into account. Included among such variables were age, sex, duration of swimming for those who swam, and whether or not a person swam on additional days after being interviewed at the beach. The primary reason for taking into account such variables was to ensure that any observed effect of swimming on illness outcome could not be explained away by these other variables.

Summary

- Bias is a flaw in the study design, the methods of data collection, or the methods of data analysis that may lead to spurious conclusions about the exposure-disease relationship.
- Three general sources of bias occur in:
  - Selection of study subjects
  - Incorrect information gathered on study subjects
  - Failure to adjust for variables other than the exposure variable (confounding)

Analyzing the data

Another important issue is: How do we carry out the data analysis? We now briefly introduce some basic ideas about data analysis.

The final methodologic issue concerns the data analysis. We must carry out an appropriate analysis once collection and processing of the study data are complete. Since the data usually come from a sample of subjects, the data analysis typically requires the use of statistical procedures to account for the inherent variability in the data. In epidemiology, data analysis typically begins with the calculation and statistical assessment of simple measures of disease frequency and effect. The analysis often progresses to more advanced techniques such as stratification and mathematical modeling. These latter methods are typically used to control for one or more potential confounders.

Statistics

Frequency:	Effect:
risk	risk ratio
proportion	odds ratio
rate	prevalence ratio

Stratification

Mathematical modeling

3) Is the risk ratio of 1.33 significantly different from a risk ratio of 1? That is, could the risk ratio estimate of 1.33 have occurred by chance?

Let's consider the data analysis in the Sydney Beach Users Study. We had previously compared swimmers with non-swimmers. Now, we may wish to address the more specific question of whether those who swam in polluted water had a higher risk for illness than those who swam in non-polluted water. We can do this by separating the swimmers into two groups. The non-swimmers represent a baseline comparison group with which the two groups of swimmers can be compared.

Based on the two-way table, we can estimate the risk for illness for each of the three groups by computing the proportion that got ill out of the total for each group. The three risk estimates are 0.357, 0.269 and 0.165, which translates to 35.7 percent, 26.9 percent and 16.5 percent, respectively.

Sydney Beach Users Study

		Swim			
		Yes-P	Yes-NP	No	Total
Ill	Yes	55	477	151	683
	No	99	1293	764	2156
	Total	154	1770	915	2839
risk for illness:		35.7%	26.9%	16.5%	

The risk ratio that compares the Swam-Polluted (Yes-P) group with the Swam-Nonpolluted (Yes-NP) group is 1.33 indicating that persons who swam in polluted water had a 33 percent increased risk than persons who swam in nonpolluted water.

risk ratio: (P vs. NP) $\frac{35.7\%}{26.9\%} = 1.33$

Also, the risk ratio estimates obtained by dividing the risks for each group by risk for non-swimmers are 2.16, 1.63, and 1. This suggests what we call a dose-response effect, which means that as the exposure is increases, the risk increases.

risk ratio:	$\frac{35.7\%}{16.5\%}$	$\frac{26.9\%}{16.5\%}$	$\frac{16.5\%}{16.5\%}$
	2.16	**1.63**	**1.00** (referent)

Dose-response effect

The analysis just described is called a "crude" analysis because it does not take into account the effects of other known factors that may also affect the health outcome being studied. A list of such variables might include age, swimming duration, and whether or not a person swam on additional days. The conclusions found from a crude analysis might be altered drastically after adjusting for these potentially confounding variables.

Several questions arise when considering the control of many variables:

- Which of the variables being considered should actually be controlled?
- What is gained or lost by controlling for too many or too few variables?
- What should we do if we have so many variables to control that we run out of numbers?
- What actually is involved in carrying out a stratified analysis or mathematical modeling to control for several variables?
- How do the different methods for control, such as stratification and mathematical modeling, compare to one another?

These questions will be addressed in later activities.

Study Questions (Q2.5)

1. How do you interpret the risk ratio estimate of 1.33?
2. Does the estimated risk ratio of 1.33 indicate that swimming in polluted water poses a health risk?
3. Given the relatively small number of 154 persons who swam in polluted water, what statistical question would you need to answer about the importance of the estimated risk ratio of 1.33?

1) The risk of illness for persons who swam in polluted water is estimated to be 1.33 times the risk for persons who swam in non-polluted water.

2) Not neccessarily, the importance of any risk ratio estimate depends on the clinical judgement of the investagors & the size of similar risk ratio estimates that have been found in previous studies

Summary: Analyzing the Data

- ❖ The data analysis typically requires the use of statistical procedures to account for the inherent variability in the data.
- ❖ In epidemiology, data analysis often begins with assessment and comparison of simple measures of disease frequency and effect.
- ❖ The analysis often progresses to more advanced techniques such as stratification and mathematical modeling.

Alcohol Consumption and Breast Cancer in the Nurses Health Study

The Harvard School of Public Health followed a cohort of about 100,000 nurses from all over the US throughout the 1980s and into the 1990s. The investigators in this Nurses Health Study, were interested in assessing the possible relationship between diet and cancer. One particular question concerned the extent to which alcohol consumption was associated with the development of breast cancer.

Nurses identified as being 'disease free' at enrollment into the study were asked about the amount of alcohol they currently drank. Other relevant factors, such as age and smoking history, were also determined. Subjects were followed for four years, at which time it was determined who developed breast cancer and who did not. A report of these findings was published in the New England Journal of Medicine in 1987.

Recall that the first methodologic issue is to define the **study question**. Which of the study questions stated here best addresses the question of interest in this study?

A. Is there a relationship between drinking alcohol and developing breast cancer?
B. Are alcohol consumption, age, and smoking associated with developing breast cancer?
C. Are age and smoking associated with developing breast cancer, after controlling for alcohol consumption?
D. Is alcohol consumption associated with developing breast cancer, after accounting for other variables related to the development of breast cancer?

The best answer is "D": Is alcohol consumption associated with developing breast cancer, after accounting for other variables related to the development of breast cancer?" Although "A. Is there a relationship between drinking alcohol and developing breast cancer?" is also correct.

In stating the study question of interest, we must identify the primary variables to be measured.

Study Questions (Q2.6)

Determine whether each of the following is a:

Health outcome variable (D)
Exposure variable (E)
Control variable (C)

1. Smoking history
2. Whether or not a subject develops breast cancer during follow-up
3. Some measure of alcohol consumption
4. Age

Once we have specified the appropriate variables for the study, we must determine how to measure them. The health outcome variable in this example, **D**, is simply *yes* or *no* depending on whether or not a person was clinically diagnosed with breast cancer. The investigators at Harvard interviewed study subjects about their drinking habits, **E**, and came up with a quantitative measurement of the amount of alcohol in units of grams per day that were consumed in an average week around the time of enrollment into the study. How to treat this variable for purposes of the analysis of the study data was an important question considered. One approach was to categorize the alcohol measurement into 'high' versus 'low'. Another approach was to categorize alcohol into 4 groups: non-drinkers; less than 5 grams per day; between 5 and 15 grams per day; and 15 or more grams per day.

Age, denoted $\mathbf{C_1}$, is inherently a quantitative variable, although many of the analyses treated age as a categorical variable in three age groups, shown here:

34 to 44 years
45 to 54 years
55 to 59 years

Smoking history, $\mathbf{C_2}$, was categorized in several ways; one was *never* smoked versus *ever* smoked.

The research question in the nurse's health study can thus be described as determining if there is a relationship between alcohol consumption, **E**, and breast cancer, **D**, controlling for the effects of age, $\mathbf{C_1}$, and smoking history, $\mathbf{C_2}$, and possibly other variables (C_3, C_4, etc.).

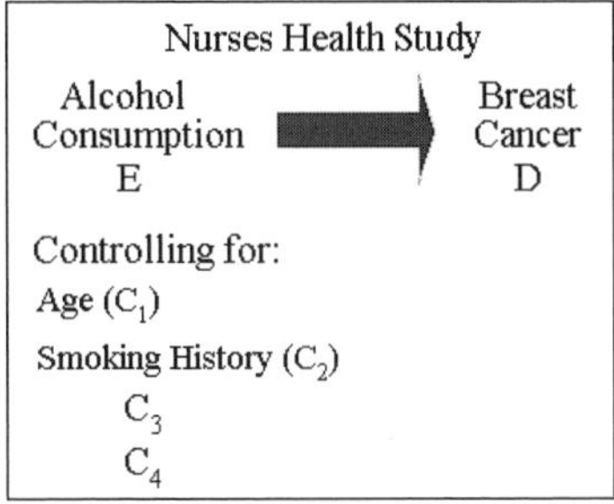

Although a detailed analysis is not described here, the data did provide evidence of a significant association between alcohol use and development of breast cancer. For heavy drinkers, when compared to non-drinkers, there was about an 80% increase in the risk of developing breast cancer. Moderate drinkers were found to have about a 50% increase in risk, and light drinkers had an increased risk of about 20%.

	Compared to Non-drinkers:
Heavy drinkers	80% increased risk
Moderate drinkers	50% increased risk
Light drinkers	20% increased risk

Note: The Nurses Health Study provides an example in which the exposure variable, alcohol consumption, has several categories rather than simply binary. Also, the control variable age and smoking history can be a mixture of different types of variables. In the Nurses Health Study, age is treated in three categories, and smoking history is treated as a binary variable.

The Bogalusa Outbreak

On October 31, 1989, the Louisiana State Health Department was notified by two physicians in Bogalusa, Louisiana, that over 50 cases of acute pneumonia had occurred within a three-week interval in mid to late October, and that six persons had died. Information that the physicians had obtained from several patients suggested that the illness might have been Legionnaires Disease.

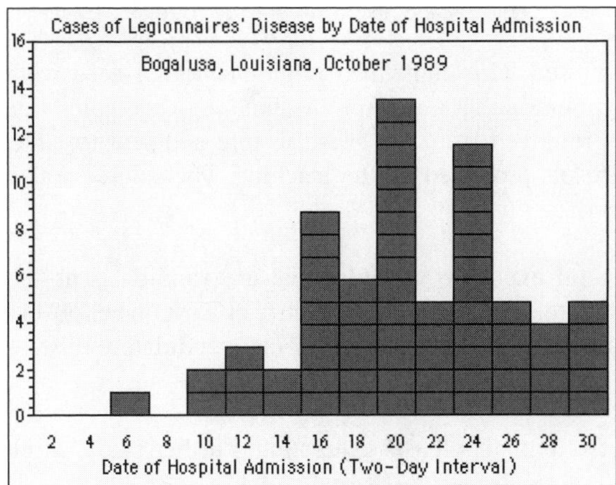

In 1989, Bogalusa was a town of about 16,000 persons. The largest employer was a paper mill located in the center of town adjacent to the main street. The paper mill included five prominent cooling towers. The mill also had three paper machines that emitted large volumes of aerosol along the main street of town. Many people suspected that the cooling towers and or the paper mill were the cause of the outbreak, since they were prominent sources of outdoor aerosols where the legionnaire's bacteria could have been located.

Recall that the first methodologic issue is to define the **study question** of interest. Which of the study questions stated here best addresses the question of interest in this study?

A. Was the paper mill the source of the outbreak of Legionnaires Disease in Bogalusa?
B. What was the source of the outbreak of Legionnaires Disease in Bogalusa?
C. Why did the paper mill cause the outbreak of Legionnaires Disease in Bogalusa?
D. Was there an outbreak of Legionnaires Disease in Bogalusa?

The most appropriate study question is "B. What was the source of the outbreak of Legionnaires Disease in Bogalusa?" Even though the paper mill was the suspected source, the study was not limited to that variable only, otherwise, it might have failed to collect information on the true source of the outbreak.

In stating the study question, we identify the primary variables to be considered in the study.

<u>**Study Questions (Q2.7)**</u>

Determine whether each of these variables is the health outcome variable, **D**, an exposure variable, **E**, or a control variable, **C**:

1. Exposure to the cooling towers of the paper mill?
2. Exposure to emissions of the paper machines?
3. Age of subject?
4. Visited grocery store A?
5. Visited grocery store B?
6. Diagnosed with Legionnaires Disease?
7. Visited drug store A?
8. Visited drug store B?
9. Ate at restaurant A?

The health outcome variable, **D**, indicates whether or not a study subject was clinically diagnosed with Legionnaires Disease during the three week period from mid to late October. The exposure variable is conceptually whatever variable indicates the main source of the outbreak. Since this variable is essentially unknown at the start of the study, there is a large collection of exposure variables, all of which need to be identified as part of the study design and investigated as candidates for being the primary source of the outbreak. We denote these exposure variables of interest $\mathbf{E_1}$ through $\mathbf{E_7}$. One potential control variable of interest was age, which we denoted as $\mathbf{C_1}$.

The general research question of interest in the Bogalusa outbreak can thus be described as evaluating the relationship of one or more of the exposure variables to whether or not a study subject developed Legionnaires Disease, controlling for age.

A **case-control study**, was carried out in which 28 **cases** diagnosed with confirmed Legionnaires Disease were compared with 56 non-cases or **controls**. This investigation led to the hypothesis that a misting machine for vegetables in a grocery store was the source of the outbreak. This misting machine was removed from the grocery store and sent to CDC where laboratory staff was able to isolate Legionella organisms from aerosols produced by the machine. This source was a previously unrecognized vehicle for the transmission of Legionella bacteria.

Note: The Bogalusa study provides an example in which there are several exposure variables that are candidates as the primary source of the health outcome being studied. Hopefully, the investigators will be able to identify at least one exposure variable as being implicated in the occurrence of the outbreak. It is even possible that more than one candidate exposure variable may be identified as a possible source.

The case-control study of this and many other outbreaks can often be viewed as hypothesis generating. Further study, often using laboratory methods, clinical diagnosis, and environmental survey techniques, must often be carried out in order to confirm a suspected exposure as the primary source of the outbreak. The Centers for Disease Control and Prevention has a variety of scientists to provide the different expertise and teamwork that is required, as carried out in the Bogalusa study.

The Rotterdam Study

The Rotterdam study has been investigating the determinants of chronic disabling diseases, including Alzheimer's disease, during the 1990s and beyond.

In the early 1990s, the Department of Epidemiology of the Erasmus University in Rotterdam, the Netherlands, initiated the Rotterdam Study. A cohort of nearly 8000 elderly people was selected. They continue to be followed to this day. The goal of the study is to investigate determinants of chronic disabling diseases, such as Alzheimer's and cardiovascular disease. One particular study question of interest was whether smoking increases the risk of Alzheimer's disease.

Subjects who were free of dementia at a first examination were included in the study. This excluded anyone diagnosed at this exam with Alzheimer's or any other form of dementia due to organic or psychological factors. Approximately two years later, the participants were asked to take a brief cognition test. If they scored positive, they were further examined by a neurologist. The investigators could then determine whether or not a participant had developed Alzheimer's disease, the health outcome variable **D** of interest, since the start of follow-up.

The primary exposure variable, **E**, was smoking history. Three categories of smoking were considered: current smokers at the time of the interview; previous but not current smokers; and, never smokers. Control variables considered in this study included age, gender, education, and alcohol consumption.

Rotterdam Study

Study subjects:

- free of dementia at 1st exam
- cognition test- 2 years later
- neurologist exam (if test +)
- health outcome: Alzheimer's (D)
- exposure variable: smoking history (E)

3 categories:
current smokers, previous smokers, never smokers

We define the study question of interest as: *Is there a relationship between smoking history and Alzheimer's disease, controlling for the effects of age, gender, education and alcohol consumption?*

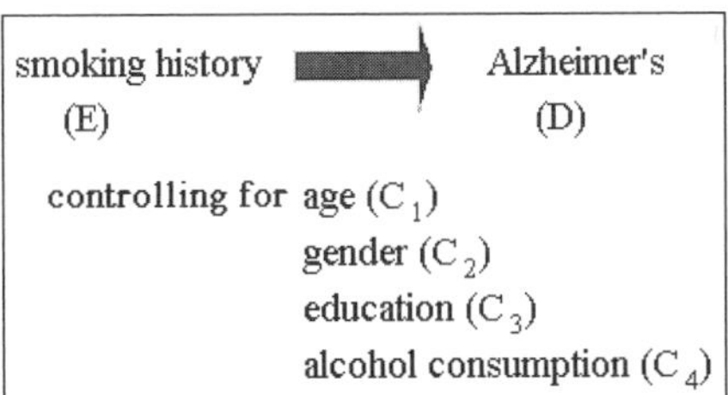

Recall that one of the important methodologic issues is to determine the study design.

How would you define the design of this study?

1. Cohort design
2. Case-control design
3. Cross-sectional design
4. Clinical trial

This is a cohort design because participants without the health outcome of interest, in this case Alzheimer's disease, are followed up over time to determine if they develop the outcome later in life.

Which of the following is influenced by the design of the study?

A. The assessment of confounding
B. The choice of the measures of disease frequency and effect
C. A decision regarding the use of stratified analysis
D. The analysis is not influenced in any way by the study design used

The answer is B. We determine the appropriate measures of disease frequency and effect based on the study design characteristics. Choices A and C are incorrect because they are typically considered regardless of the study design used.

The investigators found that 105 subjects developed Alzheimer's disease. After taking the control variables into account, the risk of Alzheimer's disease for current smokers was 2.3 times the risk for subjects who had never smoked. For subjects who had smoked in the past but who had given up smoking before the study started, the risk of Alzheimer's disease was 1.3 times the risk for subjects who had never smoked.

Results

- **105 subjects developed Alzheimer's**
- risk for current smokers was 2.3 times risk for never smokers
- risk for previous smokers was 1.3 times risk for never smokers

1) The increased risk of 2.3 translates to a 130% increase in the risk for current smokers compared to never smokers.

2) the increased risk of 1.3 translates to a 30% increase in the risk for previous smokers compared to never smokers.

Study Questions (Q2.8)

Based on the above results:

1. What is the *percent increase* in the risk for current smokers when compared to the risk for never smokers?
2. What is the *percent increase* in the risk for previous smokers when compared to the risk for never smokers?

Because these results were statistically significant and controlled for previously established predictors of Alzheimer's, the study gave support to the hypothesis that smoking history was a significant risk factor in the development of Alzheimer's disease.

Analyzing Data in Data Desk

Note that there are two activities in the Lesson that provide information on how to analyze data using the Data Desk statistical program. These activities are not summarized in this ActivEpi Companion Textbook.

Nomenclature

C	Control variable or covariate
D	Disease or outcome variable
E	Exposure variable
R(E)	Risk among the exposed for developing the health outcome
R(not E) or **R($\bar{E}$)**	Risk among the ***non***exposed for developing the health outcome
RR	Risk ratio

References

For the Sydney Beach Users Study:

Corbett SJ, Rubin GL, Curry GK, Kleinbaum DG. The health effects of swimming at Sydney Beaches. The Sydney Beach Users Study Advisory Group. Am J Public Health. 1993;83(12): 1701-6.

For the Nurses Health Study:

Willett WC, Stampfer MJ, Colditz GA, Rosner BA, Hennekens CH, Speizer FE. Moderate alcohol consumption and the risk of breast cancer. N Engl J Med. 1987;316(19):1174-80.

For the Bogalusa Outbreak:

Mahoney FJ, Hoge CW, Farley TA, Barbaree JM, Breiman RF, Benson RF, McFarland LM. Communitywide outbreak of Legionnaires' disease associated with a grocery store mist machine. J Infect Dis. 1992;165(4):736-9.

For The Rotterdam Study:

Hofman A, Grobbee DE, de Jong PT, van den Ouweland FA. Determinants of disease and disability in the elderly: the Rotterdam Elderly Study. Eur J Epidemiol. 1991;7(4); 403-22.

Ott A, Slooter AJ, Hofman A, van Harksamp F, Witteman JC, Van Broeckhoven C, van Duijin CM, Breteler MM. Smoking and risk of dementia and Alzheimer's disease in a population-based cohort study: the Rotterdam Study. Lancet. 1998;351(9119):1840-3.

Homework Exercises

ACE-1. What is Epidemiology? What is the origin of the word "epidemiology" (and why does it have nothing to do with the study of skin)?

ACE-2. Causation. For each of the following excerpts, indicate which of the criteria for causation (proposed by A. B. Hill, circa 1964 or earlier) is/are being addressed (you may choose more than one). Note that these criteria are presented on page 38, Lesson 3:

A. Strength of Association
B. Consistency
C. Temporality
D. Dose response, or biologic gradient
E. Biologic plausibility
F. Specificity
G. Coherence
H. Experiment
I. Analogy

1. [From a study of whether Hispanics are more likely than whites to experience disability]. "Mexican-American participants in the 1978-1980 Health Interview Survey were more likely than non-Hispanic whites to report limitations in their activity. Data from the 1987 National Medical Expenditure Survey suggested the opposite pattern, with Hispanics reporting less functional limitation than non-Hispanic whites. Haan and Weldon presented

data suggesting that Hispanic disability may be more evident among persons with at least two of the chronic illnesses of diabetes, stroke and hypertension."

2. [From the study above] "Among community-dwelling residents, Hispanics were 2-5 times as likely as non-Hispanic whites to need assistance with IADL (Instrumental Activities of Daily Living) tasks. However, a larger proportion of disabled non-Hispanic whites were in nursing homes, and estimates that included nursing home residents suggested a more modest Hispanic excess that was generally less than twofold."

3. [From a study of preconception paternal x-ray exposure and birth outcome] "The exposure variable was generated from an item on the partner's questionnaire asking about specific medical x-ray studies performed any time within 12 months preceding conception."

4. "The pronounced increase in risk of preeclampsia among type I diabetics is consistent with that from previous reports and may be due to microvascular changes impairing the placental perfusion. Our finding that type I diabetes is significantly, albeit less strongly, associated with gestational hypertension may reflect a common metabolic pathway in the pathogenesis of preeclampsia and gestational hypertension."

5. [From a study of predictors of gallbladder disease in men] "Higher levels of BMI (body mass index) were progressively associated with increased risk of disease, and men with BMI > = 24.0 units had a significant, 46 percent increased risk when compared with their counterparts with BMI < 20.0."

6. "An association between cancers of the human nasal cavity and paranasal sinuses and cigarette smoking has been described in recent studies in the United States and China. To date, limited evidence from two studies conducted in Japan suggests that exposure to environmental tobacco smoke is also a risk factor for nasal sinus cancer. The study reported here was designed to test the hypothesis that exposure to environmental tobacco smoke in the home increases the risk for cancer of the nasal cavity and paranasal sinuses in pet dogs ... The risk for nasal cancer was also examined according to histologic type. Dogs with sarcomas had a higher adjusted risk than dogs with carcinomas for the highest tertile of the exposure index."

7. "Studies have often found a lower risk of large bowel cancer associated with higher coffee consumption, although this finding has not been universal. Coffee's composition is quite complex, and varied constituents have potential genotoxic, mutagenic, and anitmutagenic properties. In addition, coffee modulates various physiologic processes, such as large bowel motility, that could alter colonic exposure to potential fecal carcinogens."

8. [From study of coffee and colorectal cancer] "Another possible explanation for the results is that individuals at high risk for developing colorectal cancer, or who have symptoms from undiagnosed cancer of the large bowel, avoid coffee consumption. Rosenberg et al. found similar results whether coffee consumption of the prior year or of 3 years previously was analyzed."

9. "Observational epidemiologic studies of dietary calcium and fractures are inconsistent. There have been at least 14 studies of hip fracture and dietary calcium, and only three of these found a clearly protective effect. On the other hand, two small randomized trials have found a reduced rate of radiographic vertebral fractures among subjects given calcium supplements, and another small study found a nonsignificant reduction in risk of symptomatic vertebral and nonvertebral fractures. A large French trial found that a combination of calcium and vitamin D supplements halved the hip fracture rate among women living in nursing homes."

ACE-3. Causal Exposure/Disease Association. Under what circumstances could an exposure/disease association be causal without being biologically plausible?

ACE-4. A CDC Website. The Centers for Disease control has a website called **EXCITE**, which stands for **Excellence in Curriculum Integration through Teaching Epidemiology.** The website address is

http://www.cdc.gov/excite/

Open up this website on your computer and look over the various features and purposes of the website described on the first page you see. Then click on the item (on menu on left of page) **Disease Detectives at Work** and read the first two articles entitled *Public Health on Front Burner After Sept 11* and *USA's 'Disease Detectives' Track Epidemics Worldwide.*

Then click on the item (on menu on left of page) **Classroom Exercises** and go through the exercise on Legionnaires Disease in Bogalusa, Louisiana. The specific website address for this exercise is:

http://www.cdc.gov/excite/legionnaires.htm

Answers to Study Questions and Quizzes

Q2.1

1. 3, did not swim
2. C
3. 2
4. 2
5. C

Q2.2

1. General health status, smoking status, diet, including what a subject might have eaten at the beach.
2. Choose variables that are already known determinants of the health outcome. This will be discussed later under the topic of **confounding**.
3. Younger subjects might be less likely to get ill than older subjects.
4. In the actual study, the investigators chose to exclude subjects from the analysis if they visited the beach on days other than the day they were interviewed on the beach.

Q2.3

1. Self-reported information may be inaccurate and can therefore lead to spurious study results.
2. As with the previous question, the information obtained about exposure much later than when the actual exposure occurred may be inaccurate and can lead to spurious study results.

Q2.4

1. To minimize the inclusion in the study of a family or social groups.
2. Subjects without the health outcome, that is, healthy subjects selected at the beach, were followed-up over time to determine if they developed the outcome.
3. No, the Sydney Beach User's Study did not use a fixed cohort. Study subjects were progressively added over the summer of 1989-90 to form the cohort.
4. Because the study started with exposed and unexposed subjects, rather than ill and not-ill subjects, and went forward rather than backwards in time to determine disease status.
5. Exposure and disease status were observed at different times for different subjects. Also, each subject was selected one week earlier than the time his or her exposure and disease status were determined.

Q2.5

1. The risk of illness for persons who swam in polluted water is estimated to be 1.33 times the risk of illness for persons who swam in non-polluted water.
2. Not necessarily. The importance of any risk ratio estimate depends on the clinical judgment of the investigators and the size of similar risk ratio estimates that have been found in previous studies.
3. Is the risk ratio of 1.33 significantly different from a risk ratio of 1? That is, could the risk ratio estimate of 1.33 have occurred by chance?

Q2.6

1. C
2. D
3. E
4. C

Q2.7

1. E
2. E
3. C
4. E
5. E
6. D
7. E
8. E
9. E

Q2.8

1. The increased risk of 2.3 translates to a 130% increase in the risk of current smokers compared to never smokers.
2. The increased risk of 1.3 translates to a 30% increase in the risk for previous smokers compared to never smokers.

LESSON 3

EPIDEMIOLOGIC STUDY DESIGNS

A key stage of epidemiologic research is the ***study design****. This is defined to be the process of planning an empirical investigation to assess a* ***conceptual hypothesis*** *about the relationship between one or more* ***exposures*** *and a* ***health outcome****. The purpose of the study design is to transform the conceptual hypothesis into an* ***operational hypothesis*** *that can be empirically tested. Since all study designs are potentially flawed, it is therefore important to understand the specific strengths and limitations of each design. Most serious problems or mistakes at this stage cannot be rectified in subsequent stages of the study.*

3-1 Study Types/Options

Types of Epidemiologic Research

Epidemiologic research can be put into two broad categories depending on whether or not ***randomization*** *is used:* ***experimental*** *studies use randomization;* ***observational*** *studies do not involve randomization.*

There are two broad types of epidemiologic studies, **experimental** and **observational**. An experimental study uses **randomization** to allocate subjects to different categories of the exposure. An observational study does not use randomization. (For additional information on randomization, please refer to the end of this activity.) In experimental studies, the investigator, through randomization, determines the exposure status for each subject, then follows them and documents subsequent disease outcome. In an observational study, the subjects themselves, or perhaps their genetics, determine their exposure, for example, whether to smoke or not. The investigator is relegated to the role of simply observing exposure status and subsequent disease outcome.

Experimental studies in epidemiology usually take the form of **clinical trials** and **community intervention trials**. The objective of most *clinical trials* is to test the possible effect, that is, the efficacy, of a therapeutic or preventive treatment such as a new drug, physical therapy or dietary regimen for either treating or preventing the occurrence of a disease. The objective of most *community intervention trials* is to assess the effectiveness of a prevention program. For example, one might study the effectiveness of fluoridation, of sex education, or of needle exchange.

Most epidemiologic studies are observational. Observational studies are broadly identified as two types: **descriptive** and **analytic**. Descriptive studies are performed to describe the natural history of a disease, to determine the allocation of health care resources, and to suggest hypotheses about disease causation. Analytic studies are performed to test hypotheses about the determinants of a disease or other health condition, with the ideal goal of assessing causation. (See the end of this activity for additional information on disease causation.)

Summary

- There are two broad types of epidemiologic studies: experimental and observational
- Experimental studies use randomization of exposures
- Observational studies do **not** use randomization of exposures
- In experimental studies, the investigator pro-actively determines the exposure status for each subject.
- In observational studies, the subject determines his/her exposure status.
- Experimental studies are usually clinical trials or community intervention trials.
- Observational studies are either descriptive or analytic.

Randomization

Randomization is an allocation procedure that assigns subjects into (one of the) the exposure groups being compared so that each subject has the same probability of being in one group as in any other. Randomization tends to make demographic, behavioral, genetic, and other characteristics of the comparison groups similar except for their exposure status. As a result, if the study finds any difference in health outcome between the comparison groups, that difference can only be attributable to their difference in exposure status.

For example, if subjects are randomly allocated to either a new drug or a standard drug for the treatment of hypertension, then it is hoped that other factors such as age and sex might have approximately the same distribution for subjects receiving the new drug as for subjects receiving the standard drug. Actually, there is no guarantee even with randomization that the distribution of, for example age, will be the same for the two treatment groups. The investigator can always check the data to see what has happened regarding any such characteristic, providing the characteristic is measured or observed in the study. If the age distribution is found to be different between the two treatment groups, the investigator can take this into account in the analysis, for example, by stratifying on age.

The advantage of randomization is what it offers with regard to those characteristics not measured in one's study. Variables that are not measured obviously cannot be taken into account in the analysis. Randomization offers insurance, though no guarantee, that such unmeasured variables are evenly distributed among the different exposure groups. In observational studies, on the other hand, the investigator can account for only those variables that are measured, allowing more possibility for spurious conclusions because of unknown effects of important unmeasured variables.

Causation

In any research field involving the conduct of scientific investigations and the analysis of data derived from such investigations to test etiologic hypotheses, the assessment of **causality** is a complicated issue. In particular, the ability to make **causal inferences** in the health sciences typically depends on synthesizing results from several studies, both epidemiologic and non-epidemiologic (e.g., laboratory or clinical findings).

Instigated by a governmental sponsored effort in the United States to assess the health consequences of smoking, health scientists in the late 1950's and 1960's began to consider defining objective criteria for evaluating causality. The particular focus of this effort was how to address causality based on the results of studies that consider exposures that cannot be randomly assigned, i.e., observational studies.

In 1964, a report was published by the US Department of Health, Education and Welfare that reviewed the research findings dealing with the health effects of smoking, with the objective of assessing whether or not smoking could be identified as a "cause" of lung cancer and perhaps other diseases. The type of synthesis carried out in this report has been referred to in the 1990's as a meta analysis, so that this report was in essence, one of the earliest examples of a **meta analysis** conducted in the health sciences.

The 1964 document based much of its conclusions about smoking causation on a list of general criteria that was formalized by Bradford Hill and later incorporated into a famous 1971 textbook by Hill. The criteria are listed as follows:

1. **Strength of the Association**: The stronger the observed association, the less likely the association is due to bias; weaker associations do not provide much support to a causal interpretation.
2. **Dose-response Effect**: If the disease frequency increases with the dose or level of exposure, this supports a causal interpretation. (Note, however, that the absence of a dose-response effect may not rule out causation from alternative explanations, such as a threshold effect.)
3. **Lack of Temporal Ambiguity**: The hypothesized cause must precede the occurrence of the disease.
4. **Consistency of Findings**: If all studies dealing with a given relationship produce similar results, a causal interpretation is advanced. (Note: Inconsistencies may be due to different study design features, so that perhaps some kind of weighting needs to be given to each study.)
5. **Biological Plausibility of the Hypothesis**: If the hypothesized effect makes sense in the context of current biological knowledge, this supports a causal interpretation. (Note, however, the current state of biological knowledge may be inadequate to determine biological plausibility.)
6. **Coherence of the Evidence**: If the findings do not seriously conflict with our understanding of the natural history of the disease or other accepted facts about disease occurrence, this supports a causal interpretation.
7. **Specificity of the Association**: If the study factor is found to be associated with only one disease, or if the disease is found to be associated with only one factor, a causal interpretation is supported. (However, this criterion cannot rule out a causal hypothesis, since many factors have multiple effects and most diseases have multiple causes.) Examples include vinyl chloride and angiosarcoma of the lever; DES by women and vaginal cancer in offspring.

Continued on next page

8. **Experimentation:** use of experimental evidence, such as clinical trials in humans, animal models, and in vitro laboratory experiments. May support causal theories when available, but its absent does not preclude causality.
9. **Analogy:** when similar relationships have been shown with other exposure-disease relationships. For example, the offspring of women given DES during pregnancy were more likely to develop vaginal cancer. By analogy, it would seem possible that other drugs given to pregnant women could cause cancer in their offspring.

Quiz (Q3.1)

Fill in the blanks with either **Experimental** or **Observational**

1. A strength of the **???** study is the investigator's control in the assignment of individuals to treatment groups.
2. A potential advantage of an **???** study is that they are often carried out in more natural settings, so that the study population is more representative of the target population.
3. The major limitation of **???** studies is that they afford the investigator the least control over the study situation; therefore, results are generally more susceptible to distorting influences.
4. A weakness of an **???** study is that randomization to treatment groups may not be ethical if an arbitrary group of subjects must be denied a treatment that is regarded as beneficial.
5. One community in a state was selected by injury epidemiologists for a media campaign and bicycle helmet discount with any bicycle purchase. A similar community about 50 miles away was identified as a comparison community. The epidemiologists compared the incidence of bicycle-related injuries through emergency room surveillance and telephone survey. This is an example of an **???** study.
6. Researchers administered a questionnaire to all new students at a large state university. The questionnaire included questions about behaviors such as seat belt use, exercise, smoking, and alcohol consumption. The researchers plan to distribute follow-up questionnaires at graduation and every five years thereafter, asking about health events and conditions such as diabetes and heart disease. This is an example of an **???** study.

Directionality

The directionality of a study refers to when the exposure variable is observed relative in time to when the health outcome is observed. In a study with forward directionality, the investigator starts by determining the exposure status for subjects selected from some population of interest and then follows these subjects over time to determine whether or not they develop the health outcome. Cohort studies and clinical trials always have forward directionality.

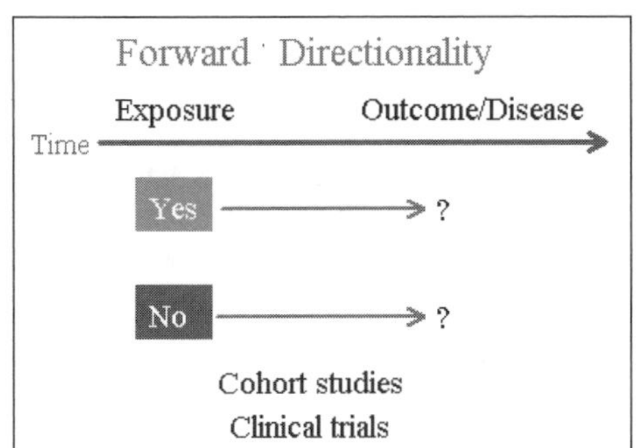

In a backwards design, the investigator selects subjects on the basis of whether or not they have the health outcome of interest, and then obtains information about their previous exposures. Case-control studies always have backwards directionality.

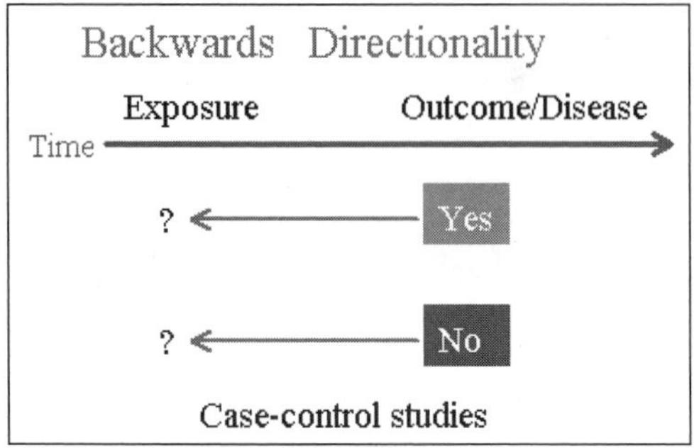

In a non-directional design, the investigator observes both the study factor and the health outcome simultaneously, so that neither variable may be uniquely identified as occurring first. A cross-sectional study is always non-directional.

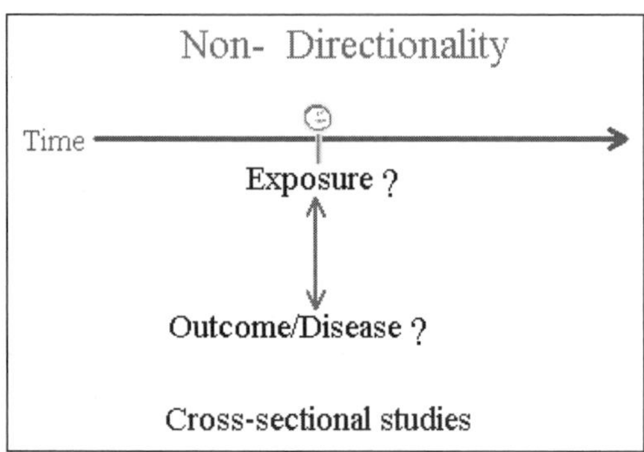

The directionality of a study affects the researcher's ability to distinguish antecedent from consequent. This is important for evaluating **causality**. Also, the directionality chosen affects the way subjects can be selected into the study. Designs that are backwards or non-directional have more potential for **selection bias** than forward designs. Selection bias will be addressed in more detail in a later lesson of this program.

Summary

- ❖ Directionality answers the question: when did you observe the exposure variable relative in time to when you observed health outcome?
- ❖ Directionality can be forward, backward, or non-directional.
- ❖ Directionality affects the researcher's ability to distinguish antecedent from consequent.
- ❖ Directionality also affects whether or not a study will have selection bias.

Timing

Timing concerns the question of whether the health outcome of interest has already occurred before the study actually began. If the health outcome has occurred before the study is initiated, the timing is **retrospective**. For example, let's say a case-control study is initiated to investigate cases of a disease that occurred in the previous year; this would be an example of a **retrospective** case control study.

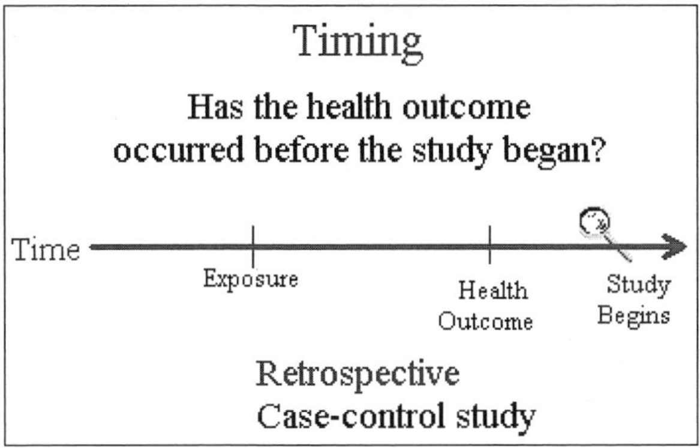

If, on the other hand, the health outcome occurs after the onset of the study, then the timing is **prospective**. Clinical trials are always *prospective*.

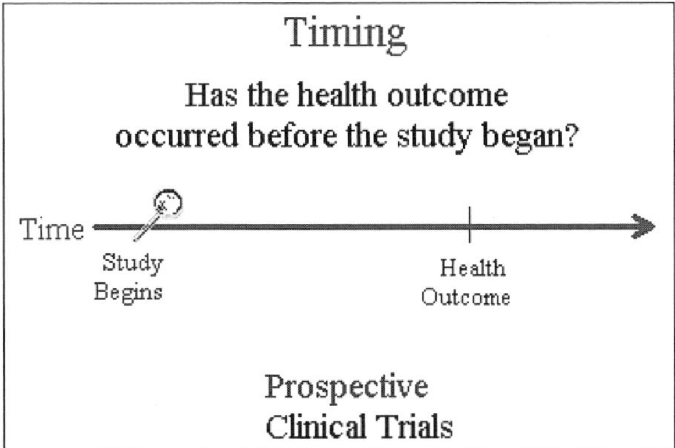

Cohort and case-control studies may *either* be retrospective or prospective since the study may begin either before or after the health outcome has occurred. The timing of a study can have important implications for the quality of the data. Retrospective data are often based on personal recall, or on hospital or employment records, and are therefore more likely than prospective studies to involve measurement errors. Measurement errors frequently lead to information bias, which we discuss in a later lesson.

Some studies may have elements of both **prospective** and **retrospective** timing, sometimes referred to as **mixed** timing.

Summary

- Timing answers the question: has the health outcome of interest already occurred before the study actually began?
- If the health outcome occurs before the study is initiated, the timing is retrospective.
- If the health outcome occurs after the onset of the study, the timing is prospective.
- Timing affects measurement error and information bias.

Clinical Trials

The clinical trial is the epidemiologic design that most closely resembles a laboratory experiment. The major objective is to test the possible effect of a therapeutic preventive intervention.

A clinical trial is an *experimental study* designed to compare the therapeutic or health benefits of two or more treatments. The major objective of a clinical trial is to test the efficacy of a **preventive** or **therapeutic** intervention. The long-range goal of a preventive trial is to prevent disease; the long-range goal of a therapeutic trial is to cure or control a disease. Examples of preventive trials include studies of vaccine efficacy, use of aspirin to prevent coronary heart disease, smoking cessation, diet modification, and exercise. Therapeutic trials are typically performed by pharmaceutical companies to test new drugs for treating disease.

Key features of any clinical trial are **randomization**, **blinding**, **ethical concerns**, and the use of **intention to treat analysis**. **Randomization** is used to allocate subjects to treatment groups so that these groups are comparable on all factors except for exposure status. **Blinding** means that either the patient or the investigator is unaware of the treatment assigned. Single-blinding means either the patient or investigator are unaware of the treatment assignment and double blinding means that both the patient and the investigator are unaware of the treatment assignment. Blinding helps to eliminate bias. The study must be **ethical,** treatments that may be harmful are not used. Stopping rules are planned that would end a trial early if it becomes clear that one of the treatments is superior. An **intention-to-treat analysis** requires that the investigators "analyze what they randomize", that is, analysis should be compared to the originally randomized treatment groups, even if study subjects switch treatments during the study period.

<u>Summary</u>

- The major objective of a clinical trial is to test the efficacy of a preventive or therapeutic intervention.
- Key features of any clinical trial are:
  - Randomization
  - Blinding
  - Ethical concerns
  - Intention to treat analysis

Clinical Trial Example

A clinical trial involving 726 subjects conducted in 1993 compared *standard* insulin therapy with *intensive* insulin therapy involving more frequent insulin injections and blood glucose monitoring for the treatment of diabetes mellitus. The outcome studied was retinopathy resulting in blindness, defined as either present or absent for each patient.

Subjects were randomized to treatment groups using a computerized random number generator. Double blinding could not be used in this clinical trial since both the patient and their physician would know which treatment group the patient was randomized. However, the individuals who graded the fundus photographs to determine the presence or absence of retinopathy were <u>un</u>aware of treatment-group assignments. The randomization resulted in the standard and intensive therapy groups having very similar distributions of baseline characteristics, such as age and sex.

An intention-to-treat analysis compared the originally randomized treatment groups with regard to the occurrence of retinopathy. It was found that 24% of the 378 subjects on standard therapy developed retinopathy, whereas 6.7% of the 348 subjects on intensive therapy developed retinopathy.

Clinical Trial

Treatment:

Standard Therapy	Vs. Intensive Therapy
n=378	n=348
24%	6.7%

These data and more complicated analyses that controlled for several other important predictors indicated that intensive therapy had a much lower risk than standard therapy for retinopathy.

<u>Summary</u>

- A clinical trial involving 726 subjects conducted in 1993 compared standard insulin therapy with intensive insulin therapy.
- The outcome studied was retinopathy resulting in blindness, defined as either present or absent for each patient.
- 24% of subjects on standard therapy developed retinopathy whereas 6.7% of subjects on intensive therapy developed retinopathy.

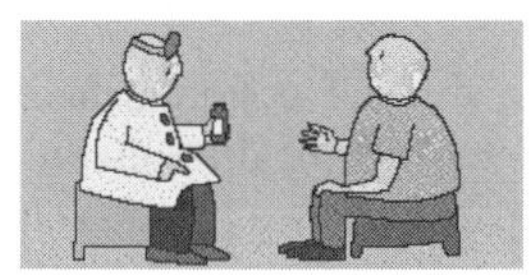

Quiz (Q3.2)

Fill in the Blanks

1. **???** Therapeutic trials are conducted on individuals with a particular disease to assess a possible cure or control for the disease. For example, we may wish to assess to what extent, if at all, a new type of chemotherapy prolongs the life of children with acute lymphatic leukemia.
2. **???** Preventive trials can be conducted on either individuals or entire populations. An example is a study in which one community was assigned (at random) to receive sodium fluoride added to the water supply, while the other continued to receive water without supplementation. This study showed significant reductions in the development of tooth decay in the community receiving fluoride.

Choices

Preventive **Therapeutic**

For each of the following features, choose the option that applies to ***clinical trials***:

1. The investigator's role regarding exposure: . . **???** a
   a. assign b. observe
2. Subject selection into groups: . . . . **???** b
   a. self-selection b. randomization
3. Directionality: . . . . . . **???** b
   a. backwards b. forwards c. non-directional
4. Timing: . . . . **???** a
   a. prospective b. retrospective c. either
5. Blinding: . . . . **???** c
   a. single b. double c. either
6. Topic: . . . . **???** c
   a. medication b. vaccine c. either
7. Analysis by: . . . . . . **???** a
   a. original assignment b. actual experience

3-2 Observational Study Designs

There are three general categories of observational designs:

- Basic Designs: Cohort, Case-Controls, Cross-Sectional
- Hybrid Designs: Nested Case-Control, Case-Cohort
- Incomplete Designs: Ecologic, Proportional

Cohort Studies

In 1948, a long-term observational study began in Framingham Massachusetts. Fifty-one hundred subjects without cardiovascular disease (CVD) were selected and examined, and information about potential risk factors for this disease was recorded. Subjects were then re-examined if possible every 2 years over the next 50 years. This classic study became known as the *Framingham Heart Study* and has been the source of much of our knowledge about risk factors for cardiovascular disease. The Framingham Heart study is an example of a **prospective** cohort study.

The cohort study is a basic observational study design most similar to a clinical trial

Always forward - start with exposure & look to see if health outcome occurs

A cohort design starts with subjects who do not have a health outcome of interest and are followed forward to determine health outcome status. A key feature of a cohort study is that subjects are grouped on the basis of their exposure characteristics prior to observing the health outcome, that is, the directionality of the study is always forward.

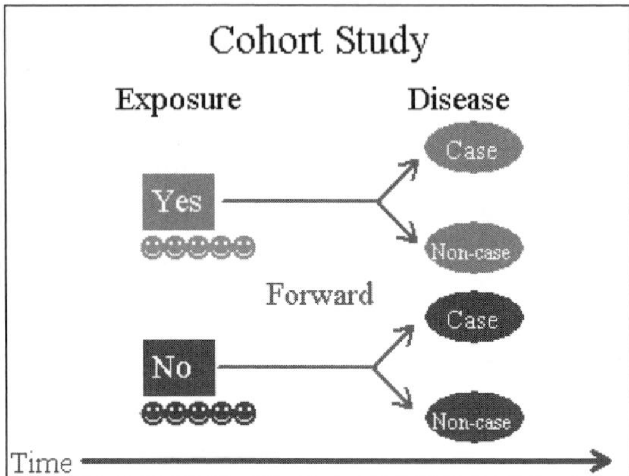

A cohort study may be retrospective or prospective. The Framingham Heart study is an example of a prospective study since the study began before the health outcome occurred.

Summary

- The Framingham Heart Study is a classic example of a cohort study.
- The cohort design is always a follow-up study with forward directionality.
- A cohort study can be prospective or retrospective.
- The Framingham study is a prospective cohort study because the study began before the health outcome occurred.

The Exposure Variable in Cohort Studies

If all exposure variables of interest are fairly common, as were those measured in the original Framingham study, the cohort is typically determined by sampling persons from a large population and, after excluding those already with the health outcome, dividing the remaining sample into exposed and unexposed study subjects.

If the exposure is rare, as when studying a specific occupational exposure, the exposed are usually sampled from a special population, such as a worksite. The unexposed are then determined from an external comparison group, which is as similar as possible to the exposed subjects with respect to factors other than exposure that may be related to the disease.

Also, employed persons are, on average, healthier than unemployed. Consequently if exposed workers are compared with the general population, the effect of an exposure will tend to be under-estimated.

Advantages of a Cohort Study

The primary advantage of a cohort study is its forward directionality. The investigator can be reasonably sure that the hypothesized cause preceded the occurrence of disease. In a cohort study, disease status cannot influence the way subjects are selected, so a cohort study is free of certain selection biases that seriously limit other types of studies.

A prospective cohort design is less prone than other observational study designs to obtaining incorrect information on important variables. Cohort studies can be used to study several diseases, since several health outcomes can be determined from follow-up.

Cohort studies are also useful for examining *rare* exposures. Since the investigator selects subjects on the basis of exposure, he can ensure a sufficient number of exposed subjects. A *retrospective* cohort study can be relatively low-cost and quick. Occupational studies that are based on employment records and death certificates or insurance and worker's comp records are an example.

Disadvantages of a Cohort Study

A prospective cohort study is often quite costly and time-consuming. A potential problem in any cohort study is the loss of subjects because of migration, lack of participation, withdrawal, and death. Such attrition of the cohort over the follow-up period can lead to biased results.

A cohort design is statistically and practically inefficient for studying a *rare* disease with long latency because of the long follow-up time and the number of subjects required to identify a sufficient number of cases. However, a retrospective cohort study may find enough cases since the study events of interest have already occurred.

Another problem in cohort studies is that the exposed may be followed more closely than the unexposed; if this happens, the outcome is more likely to be diagnosed in the exposed. This might create an appearance of exposure-disease relationship where none exists.

Summary: Cohort Study +'s (advantages) and –'s (disadvantages)

- (+) Prospective cohort study: least prone to bias when compared with other observational study designs.
- (+) Can address several diseases in the same study.
- (+) Retrospective cohort study: can be relatively low-cost and quick; frequently used in occupational studies.
- (-) Loss to follow-up is a potential source of bias
- (-) Prospective cohort study: quite costly and time-consuming; may not find enough cases if disease is rare.
- (-) If exposed are followed more closely than unexposed, the outcome is more likely to be diagnosed in exposed.

Example of a Retrospective Cohort Study, VDT's and Spontaneous Abortions

The relationship between adverse pregnancy outcomes and the use of video display terminals (VDT's) became a public health concern in the 1980's when adverse pregnancy outcomes were reported among several clusters of women who used VDT's. A more comprehensive study of the effect of VDT's was reported in the New England Journal of Medicine in 1991. This study, conducted by the National Institute for Occupational Safety and Health (NIOSH) used a retrospective cohort design to examine the hypothesis that electromagnetic energy produced by VDT's might cause spontaneous abortions.

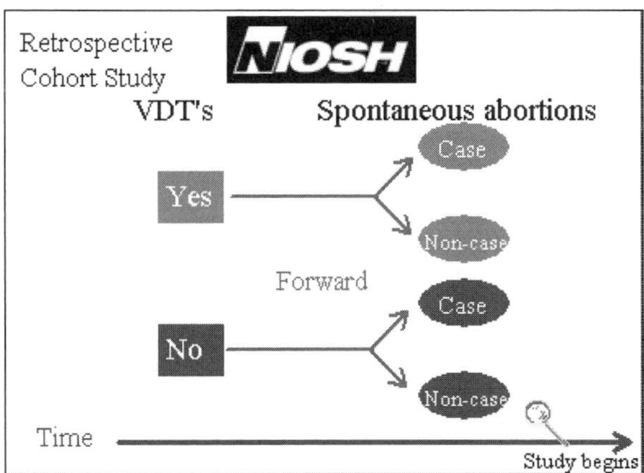

In the NIOSH study, a cohort of female telephone operators who were employed between 1983 and 1986 was selected from employers' personnel records at two telephone companies in eight southeastern states in the US. In this cohort, there were 882 women who had pregnancies that met the inclusion criteria for the study. Of these women, the pregnancy outcomes of 366 directory assistance operators who used VDT's at work were compared with 516 general telephone operators who did not use VDT's.

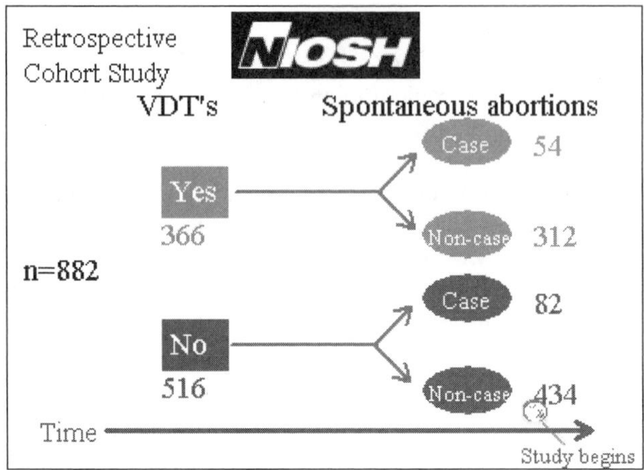

Study Questions (Q3.3)

1. What percentages of women developed spontaneous abortions for VDT uses and VDT non-users separately?

The results of the study showed no excess risk of spontaneous abortion among women who used VDT's during their first trimester of pregnancy. No dose-response relation was found from the analysis of the women's hours of VDT use per week either. Also, no excess risk was associated with VDT use when other relevant characteristics of the study subjects were taken into account. The investigators therefore concluded that the use of VDT's and exposure to electromagnetic fields they produce were not associated with an increased risk of spontaneous abortion.

Summary

- ❖ A 1991 study used a retrospective cohort design to examine the hypothesis that electromagnetic energy produced by video display terminals (VDT's) might cause spontaneous abortions.
- ❖ The pregnancy outcomes of 366 directory assistance operators who used VDT's at work were compared with 516 general telephone operators who did not use VDT's
- ❖ The results of the study showed no excess risk of spontaneous abortion among women who used VDT's.

Example of a Prospective Cohort Study, Rotterdam Study on Alzheimer's Disease

Inflammatory activity in the brain is thought to contribute to the development of Alzheimer's disease. This hypothesis suggests that long-term use of nonsteroidal anti-inflammatory drugs, or NSAIDs, may reduce the risk of this disease.

This hypothesis was investigated within the Rotterdam Study, a cohort study of the elderly that started in the Netherlands in 1990. At that time, 7,000 participants did not have Alzheimer's disease. During eight years of follow-up, 293 of the participants developed the disease.

Study Questions (Q3.4)

1. What is the directionality of this study?
2. Is the timing prospective or retrospective?

To avoid information bias from measuring NSAIDs, the investigators used computerized pharmacy records instead of interview data to determine the total number of months during which participants had used NSAIDs after the study onset. Controlling for age, gender, and smoking status, the investigators found that the risk of Alzheimer's for participants who had used NSAIDs for more than 24 months was significantly less than the risk of Alzheimer's disease for participants who used NSAIDs for less than or equal to 24 months. The investigators concluded that long-term use of NSAIDs has a beneficial effect on the risk of Alzheimer's disease.

Summary

- ❖ The Rotterdam study examined the hypothesis that long-term use of nonsteriodal anti-inflammatory drugs (NSAIDs) may reduce the risk of Alzheimer's disease.
- ❖ The study used a prospective cohort design that followed 7,000 participants without Alzheimer's disease in 1990 over eight years.
- ❖ The risk of Alzheimer's disease for subjects using NSAIDs for more than 24 months was significantly smaller than for subjects using NSAIDs less than or equal to 24 months.

Quiz (Q3.5)

Fill in the Blanks
For each of the following features, choose the option that applies to ***cohort studies***:

1. The investigator's role regarding exposure: . . **???**
   a. assign b. observe
2. Subject selection into groups: . . . . **???**
   a. self-selection b. randomization
3. Directionality: . . . . . . **???**
   a. backwards b. forwards c. non-directional
4. Timing: . . . . **???**
   a. prospective b. retrospective c. either
5. Analysis by: . . . . **???**
   a. original assignment b. actual experience.

For each of the following characteristics (strengths or weaknesses) of a study, choose the type of cohort study with that characteristic:

6. Less expensive: . . . . . . **???**
7. Quicker: . . . . **???**
8. More accurate exposure information: . . . . **???**
9. Appropriate for studying rare exposures: . . . **???**
10. Appropriate for studying rare diseases: . . . **???**
11. Problems with loss to follow-up: . . . . . **???**
12. Better for diseases with long latency: . . . . **???**

Choices
Both **Neither** **Prospective Cohort** **Retrospective Cohort**

3-3 Case-Control and Cross-Sectional

Case-Control Studies

The case-control study is a basic observational study design that is always a retrospective study. It is often quite inexpensive and quick to carry out but is very prone to bias when compared to a cohort design.

In case control studies, subjects are selected based on their disease status. The investigator first selects *cases* of a particular disease and then chooses *controls* from persons without the disease. Ideally, cases are selected from a clearly defined population, often called the **source population**, and controls are selected from the same population that yielded the cases. The prior exposure histories of cases and controls are then determined. Thus, in contrast to a cohort study, a case-control study works backwards from disease status to prior exposure status. While case-control studies are always backward in directionality, they can be either prospective or retrospective in timing.

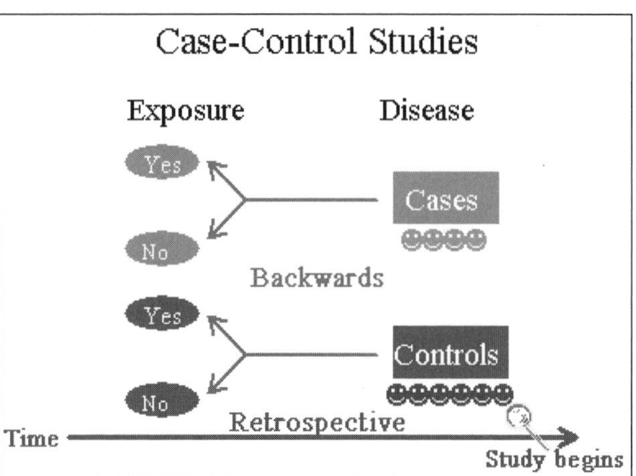

In addition to being both cheaper and quicker than cohort studies, case-control studies have other advantages:

- They are feasible for obtaining sufficient numbers of cases when studying chronic or other rare diseases or diseases with long latency periods.
- They tend to require a smaller sample size than other designs.
- They can evaluate the effect of a variety of different exposures.

There are, nevertheless, several disadvantages of case-control studies:

- They do not allow several diseases to be evaluated, in contrast to cohort studies.
- They do not allow the risk of disease to be estimated directly because they work backwards from disease to exposure status.
- They are more susceptible to selection bias than other designs since the exposure has already occurred before cases and controls are selected.
- They are more susceptible to information bias than cohort studies because they are always retrospective.
- They are not efficient for studying rare exposures

Summary

- ❖ Start with cases and non-cases of a disease or other health outcome and proceed backwards to determine prior exposure history.
- ❖ Popular primarily because cheaper and less time-consuming than cohort studies.

Summary continued on next page

- Other advantages include providing sufficient numbers of cases for rare diseases with long latencies and allowing several exposures to be evaluated.
- Disadvantages include being susceptible to both selection and information bias, not allowing estimation of risk, not considering more than one disease, and not feasible for rare exposures.

Incident versus Prevalent Cases in a Case-Control Study?

Cases can be chosen to be either incident or prevalent. Incident cases are new cases of a disease that develop over the time-period covered by the case-control study. When used in case-control studies, incident cases are typically obtained from an institutional or population-based disease registry, such as a cancer registry, or a health maintenance organization that continuously records new illnesses in a specified population.

Prevalent cases are existing cases of a disease at a point in time. When used in case-control studies, prevalent cases are usually obtained from hospital or clinic records.

An advantage of using of incident cases in case-control studies is that an exposure-disease relationship can be tied only to the development rather than the prognosis or duration of the disease.

In contrast, for prevalent cases, the exposure may affect the prognosis or the duration of the illness. If prevalent cases were used, therefore, an estimate of the effect of exposure on disease development could be biased because of failure to include cases that died before case-selection.

Choosing Controls in a Case-Control Study

One must select a comparison or control group carefully when conducting a case-control study. The ideal control group should be representative of the population from which the cases are derived, typically called the **source population**. This ideal is often hard to achieve when choosing controls.

Two common types of controls are **population-based controls** and **hospital-based controls**. In population-based case-control studies, controls are selected from the community. Methods used to select such controls include random telephone dialing, friend or neighborhood, and department of motor vehicle listings. An advantage of a population-based case-control study is that cases and controls come from the same source population, so they are similar in some way. A disadvantage is that it is difficult to obtain population lists and to identify and enroll subjects. Increasing use of unlisted numbers and answering machines increases non-response by potential controls.

In a hospital-based case-control study, controls are selected from hospital patients with illnesses *other than the disease of interest*. Hospital controls are easily accessible and tend to be more cooperative than population-based controls. Hospital-based studies are much less expensive and time-consuming than population-based studies. But, hospital-based controls are not likely to be representative of the source population that produced the cases. Also, hospital-based controls are ill and the exposure of interest may be a determinant of the control illness as well as the disease of interest. If so, a real association of the exposure with the disease of interest would likely be missed.

Summary

1. The ideal control group should be representative of the *source population* from which the cases are derived.
2. Two common types of controls are *population-based* controls and *hospital-based* controls.
3. In population-based case-control studies, cases and controls come from the same source population.
4. Hospital controls are easily accessible, tend to be cooperative, and are inexpensive.
5. Hospital controls are not usually representative of the source population and may represent an illness caused by the exposure.

Case-Control Studies – an Example of Reye's Syndrome

Several studies in the 1970's and 1980's used a case-control design to assess whether the use of aspirin was associated with the occurrence of Reye's syndrome in children with viral illnesses.

Reye's syndrome is a rare disease affecting the brain and liver that can result in delirium, coma, and death. It usually affects children, and typically occurs following a viral illness. To investigate whether aspirin is a determinant of Reye's

Syndrome, investigators in the nineteen seventies and nineteen eighties decided that using a clinical trial would not be ethical.

Why might a clinical trial on aspirin use and Reye's syndrome be unethical?

A. Children are involved.
B. Harmful consequences of the use of aspirin.
C. Double blinding may be used.
D. Clinical trials are never ethical.

The answer is B, because of the potential harmful consequences of the use of aspirin. A cohort study was also considered inefficient:

Why would a cohort study of aspirin and Reye's syndrome be inefficient?

A. The outcome is rare (would require a lot of subjects).
B. Requires at least 5 years of follow-up.
C. The exposure is rare.
D. Cohort studies are always inefficient

The answer is A, because the outcome is so rare. Consequently, a case-control study was preferred, since such a study could be accomplished over a shorter period, provide a sufficient number of cases, yet require fewer subjects overall than a cohort study.

The original investigation of Reye's Syndrome that identified aspirin as a risk factor was a case-control study conducted in Michigan in 1979 and 1980. This study involved 25 cases and 46 controls. Controls were children who were absent from the same school, in a similar grade, had a similar time of preceding illness, had the same race, the same year of birth, and the same type of preceding illness. A larger 1982 study attempted to confirm or refute the earlier finding. Investigators used a statewide surveillance system to identify all cases with Reye's syndrome in Ohio. This study thus used newly developed, or incident, cases. Population-based controls were selected by identifying and then sampling subjects in the statewide community who had experienced viral illnesses similar to those reported by the cases but had not developed Reye's syndrome. Parents of both cases and controls were asked about their child's use of medication during the illness.

Another study published in 1987 selected cases from children admitted with Reye's syndrome to any of a pre-selected group of tertiary care hospitals. Hospital-based controls were selected from children from these same hospitals who were admitted for a viral illness but did not develop Reye's syndrome. Parents were interviewed to assess previous use of aspirin.

As a result of this case-control research on the relationship between use of aspirin and Reye's syndrome, health professionals recommended that aspirin ***not*** be used to treat symptoms of a viral illness in children. Subsequently, as the use of aspirin among children declined, so did the occurrence of Reye's syndrome.

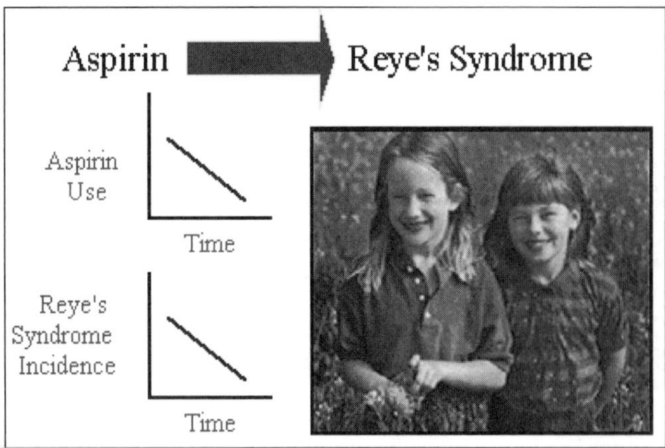

Summary

- In the 1970s and 1980s, case-control studies were used to assess whether the use of aspirin for the treatment of viral illnesses in children was a determinant of Reye's syndrome.
- These studies started with subjects with the disease (i.e., Reye's syndrome) and similar subjects without the disease.
- Parents of both cases and controls were asked about their child's use of medication over a comparable time period preceding the child's first symptoms of Reye's syndrome.
- Health professionals recommended that aspirin not be used to treat symptoms of viral illnesses in children.
- As the use of aspirin among children declined, so did the occurrence of Reye's syndrome.

Case-Control Studies – an Example of Creutzfeldt-Jakob Disease (CJD)

Creutzfeldt-Jakob disease (CJD) is a rare disease characterized by rapidly progressive dementia. In the 1990's, a new variant of CJD in humans was discovered in Europe following an epidemic in cattle of mad cow disease, the animal form of CJD. Subsequently, the European Union organized a study to investigate whether a diet containing animal products is a risk factor for CJD.

Because CJD is a very rare disease with a long latency period, the investigators chose a case-control study design. They collected data on 405 cases of CJD that had occurred in the European Union. An equal number of control participants were recruited from the hospitals where the patients with CJD had been diagnosed. Due to the mental deterioration of patients from the disease, diet information on cases had to be collected by interviewing one of the cases' next of kin.

How do you think the investigators collected diet information on control subjects? Even though the control participants were perfectly capable of giving information about their diets themselves, the investigators interviewed one of the control participants' next of kin instead. This way, they tried to avoid information bias by keeping the quality of the data on diet similar for both cases and controls.

Remember that one of the advantages of a case-control study is the opportunity to evaluate the effect of a variety of different exposures. In this study, the investigators examined separately whether consumption of sausage, raw meat, raw fish, animal blood products, milk, cheese, as well as other specified animal products, increased the risk of CJD. None of these food products significantly increased the risk of CJD, so, the investigators concluded that it is unlikely that CJD is transmitted from animals to man via animal products.

Quiz (Q3.6)

For each of the following features, choose the option that applies to *case-control* studies:

1. The investigator's role regarding exposure: . . **???**.
   a. assign b. observe
2. Subject selection into groups: . . . . **???**.
   a. self-selection b. randomization
3. Directionality: . . . . . . **???**
   a. backwards b. forwards c. non-directional
4. Timing: . . . . **???**
   a. prospective b. retrospective c. either
5. Analysis by: . . . . . **???**
   a. original assignment b. actual experience.

Quiz continued on next page

For each of the following characteristics (strengths or weaknesses) of a study, choose the type of study with that characteristic:

6. Less expensive: . . . . . . **???**
7. Quicker: . . . . . . **???**
8. More accurate exposure information: . . . **???**
9. Appropriate for studying rare exposures: . . **???**
10. Appropriate for studying rare diseases: . . **???**
11. Can study multiple outcomes: . . . **???**
12. Requires a smaller sample size: . . . **???**
13. Can estimate risk: . . . . **???**

Choices
Case-control **Prospective cohort**

Determine whether each of the following statements is **true** or **false**:

14. Ideally, controls should be chosen from the same population that gave rise to the cases. . **???**
15. Ideally, controls should be selected from hospitalized patients . . . . **???**
16. Population-based controls include only neighbors and persons identified by calling random telephone numbers. . . . . **???**

Cross-Sectional Studies

The cross-sectional study is a basic observational design in which all variables are observed or measured at a single point in time. It is usually the least expensive and quickest to carry out among observational study designs, but is also very prone to bias when compared with a cohort design.

In a cross-sectional study, subjects are sampled at a fixed point or within a short period of time. All participating subjects are examined, observed, and questioned about their disease status, their current or past exposures, and other relevant variables. A cross-sectional study provides a snapshot of the health experience of a population at a specified time and is therefore often used to describe patterns of disease occurrence. A cross-sectional sample is usually more representative of the general population being studied than are other study designs. A cross-sectional study is a convenient and inexpensive way to look at the relationships among several exposures and several diseases. If the disease of interest is relatively common and has long duration, a cross-sectional study can provide sufficient numbers of cases to be useful for generating hypotheses about exposure-disease relationships. Other more expensive kinds of studies, particularly cohort and clinical trials, are used to test such hypotheses.

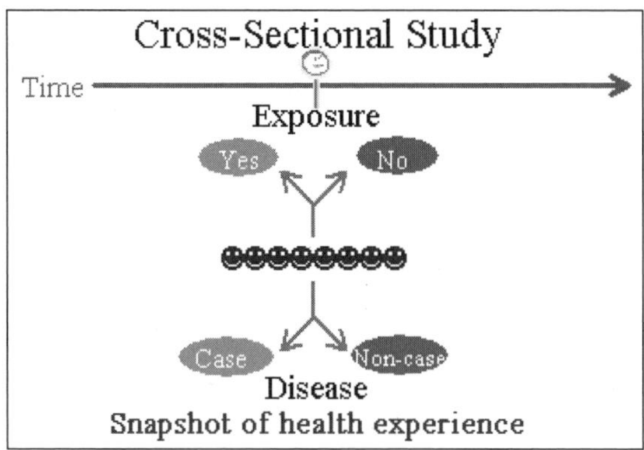

There are some disadvantages to cross-sectional studies. For example, such a study can identify only *existing* or *prevalent* cases at a given time, rather than *new* or *incident* cases over a follow-up time period. Therefore, a cross-sectional study cannot establish whether the exposure preceded the disease or whether the disease influenced the exposure.

Because only existing cases are allowed, a cross-sectional study includes only cases that survive long enough to be available for study. This could lead to a misleading conclusion about an exposure-disease relationship since non-survivors are excluded (see note at the end of this activity on this issue).

Short-duration diseases, such as the common cold or influenza, especially those that occur during a particular season, may be under-represented by a cross-sectional study that looks at the presence of such a disease at a point in time.

Summary: Cross-Sectional Studies

- Subjects are sampled at a fixed point or short period of time: a snapshot.

Advantages

- Convenient and inexpensive.
- Can consider several exposures and several diseases.
- Can generate hypotheses.
- Usually represents the general population.

Disadvantages

- Cannot establish whether the exposure preceded disease or disease influence exposure.
- Possible bias since only survivors are available for study.
- May under-represent diseases with short duration.

How Can Bias Occur from Survivors in a Cross-sectional Study?

In a cross-sectional study, bias can result because only cases that survive long enough are available for such a study. To illustrate this point, suppose that everyone with a certain disease who does not do strenuous physical exercise regularly dies very quickly. Suppose, also, that those who have the disease but do strenuous physical exercise regularly survive for several years.

Now consider a cross-sectional study to assess whether regular strenuous physical activity is associated with the disease. Since this type of study would contain only survivors, we would likely find a low proportion of cases among persons not doing strenuous physical exercise. In contrast, we would likely find a relatively higher proportion of cases among persons who do strenuous physical exercise. This would suggest that doing strenuous physical exercise is harmful for the disease, even if, in fact, it were protective.

Example of a Cross-Sectional Study – Peripheral Vascular Disease, Scotland

A 1991 study examined a sample of 5000 Scottish men for the presence of peripheral vascular disease (PVD). Other characteristics, including whether or not a subject ever smoked, were also determined for each subject during the exam.

This was a cross-sectional study since all study subjects were selected and observed at one point in time. Even though physical exams were performed, the study cost and time was much less than that required if disease-free subjects were followed over time to determine future PVD status. The sample was representative of the Scottish male population.

The study found that 1.3 percent of 1727 ever-smokers had PVD whereas only 0.6 percent of 1299 never-smokers had PVD. Dividing .013 by .006, we see that ever-smokers were 2.2 times more likely to have PVD than never-smokers.

Ever smoked? → PVD

	Smoked: Ever	Never	Total
PVD	23	8	31
No PVD	1704	1291	2995
Total	1727	1299	3026

$$\frac{.013}{.006} = 2.2$$

Ever smokers:

2.2 x more likely to have PVD than never-smokers.

These results suggested that smoking may contribute to developing PVD. Yet, the results are just a snapshot of subjects at a point in time, 1991. Subjects without PVD have not been followed over time. So, how do we know from this snapshot whether PVD leads to smoking or smoking leads to PVD? This illustrates one of the problems with cross-sectional studies - they are always **non-directional**. Also, persons who died from PVD prior to the time that subjects were selected are not allowed in the study. Therefore, the study results may be biased because only PVD survivors are being counted.

Summary

- ❖ An example of a cross-sectional study is a 1991 study of peripheral vascular disease (PVD) in Scotland.
- ❖ Results show that ever-smokers are 2.2 times more likely to have PVD than never-smokers.
- ❖ This study was much cheaper and quicker then a cohort study.
- ❖ Cannot determine whether PVD leads to smoking or smoking leads to PVD.
- ❖ The study results may be biased because only PVD survivors are considered.

Quiz (Q3.7)

For each of the following features, choose the option that applies to *cross-sectional* studies:

1. The investigator's role regarding exposure: . . ??? b
   a. assign b. observe
2. Subject selection into groups: . . . . ??? a
   a. self-selection b. randomization
3. Directionality: . . . . . . ??? c
   a. backwards b. forwards c. non-directional
4. Timing: . . . . ??? b
   a. prospective b. retrospective c. either

Quiz continued on next page

Determine whether each of the following statements is **true** or **false**:

5. Cross-sectional studies are better suited to generating hypotheses about exposure-disease relationships than to testing such relationships. . . . . . . **???**
6. Because exposure and disease are assessed at the same time, cross-sectional studies are not subject to survival bias. . . . . **???**
7. Because exposure and disease are assessed at the same time, cross-sectional studies may not be able to establish that exposure preceded onset of the disease process. . **???**
8. Cross-sectional studies can examine multiple exposures and multiple diseases. . **???**

3-4 Hybrid Designs

Hybrid Designs

Hybrid designs combine the elements of at least two basic designs, or extend the strategy of one basic design through repetition. Two popular hybrid designs are the ***case-cohort study*** *and the* ***nested case-control study****. Both these designs combine elements of a cohort and case-control study. A more recently developed hybrid design, called the* ***case-crossover*** *design, is described.*

Case-Cohort Study

The case-cohort study is a hybrid design that is less prone to bias than the standard case-control design because of the way controls are chosen.

A case-cohort study uses a hybrid design that combines elements of a cohort and a case-control study. A case-cohort population is followed over time to identify new or incident cases of a disease. The control group consists of non-cases sampled from the original cohort. Prior exposure status is then determined for both cases and controls.

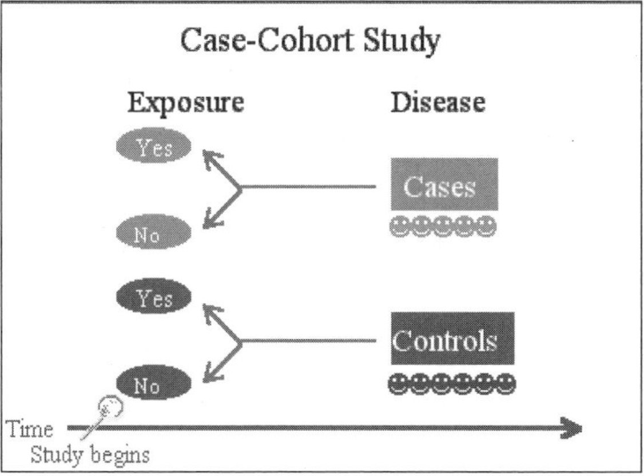

Study Questions (Q3.8)

1. What is the directionality of this study? **Forward** or **Backward**?
2. What is the timing of this study? **Prospective** or **Retrospective**?

As an example, a 1995 study of risk factors for gastric cancer involved a cohort of 9,775 men in Taiwan on whom blood samples were taken and frozen at recruitment into the study. Subsequent follow-up based on cancer registry data identified 29 cases of gastric cancer. A control group of 220 controls that did not develop gastric cancer was sampled from the original cohort. One exposure variable of interest was the presence or absence of Helicobacter pylori infection, which could be assessed by unfreezing and analyzing the blood samples from cases and controls. Thus, the cost and time of the laboratory work in determining exposure status was greatly reduced from having to consider the entire cohort.

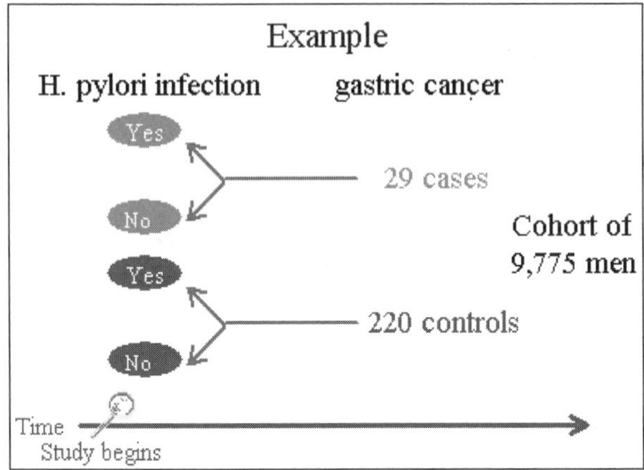

In general, a case-cohort design provides insurance that the controls derive from the source population from which the cases developed. Also, since cases are determined from follow-up, several diseases can be studied, which is not always possible in the typical case-control study. Furthermore, a case-cohort study is more cost and time-efficient than a cohort study, since a much smaller number of non-cases are observed.

Nevertheless, a case-cohort design is more prone to measurement error than a cohort study if exposure status is determined retrospectively after cases and controls are selected. This kind of study can be much more expensive and time-consuming than a case-control study since the latter does not require identifying an original cohort for selecting controls.

Summary

- The case-cohort is a hybrid design that combines features of both case-control and cohort designs.
- In a case-cohort design, controls are sampled from the original cohort.
- Cases are new or incident cases of a disease.
- Controls are chosen from the source population from which the cases derive.
- Several diseases can be studied, in contrast to a case-control study.
- Smaller number of non-cases than in cohort study.
- More prone to measurement error than cohort.
- More expensive than case-control study.

Nested Case-Control Study

The **nested case-control study**, also called a **density-type case-control study**, is a variation of the case-cohort study. This type of study can be used if the time at which subjects become cases is known. In this design, controls are matched to the cases *at the time of case diagnosis*. We select one or more controls for each case from subjects in the original cohort who are still at risk at the time a case is identified. This selection is often referred to as **density sampling of controls**. Of course, controls for a given case may later become cases after the time they are selected as controls. An advantage of using density-type controls over the case-cohort design is that density-sampled controls were at risk for becoming a case for the same amount of time as the case to which they are matched was at risk.

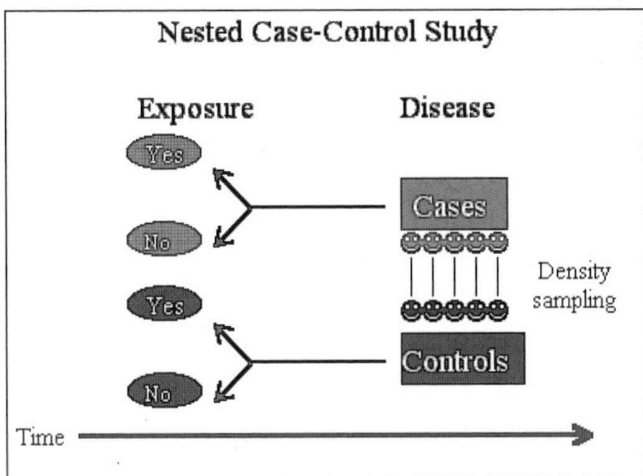

In a 1993 density-type case-control study of cancer risk from serum copper levels, baseline blood specimens and risk factor information were obtained on 5000 telephone employees. A cancer surveillance system identified 133 cancer cases that developed from this cohort. The time of case-diagnosis was determined and used to choose a sample of 241 density-type controls to be compared to the cases with regard to serum copper level and other covariates of interest.

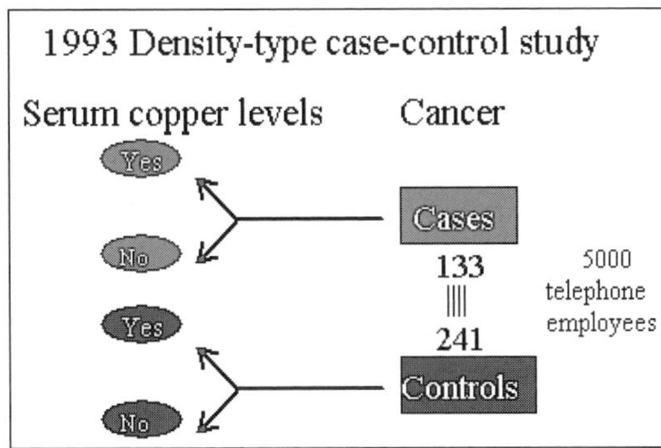

Question: The number of controls is much greater than the number of cases in this example. How is this possible using density sampling? Two or more controls were selected for some cases.

As with a case-cohort study, a nested-case control design provides insurance that the controls derive from the source population from which the cases developed. Also, the nested case-control study is more cost and time-efficient than a cohort study, since a much smaller number of non-cases are observed.

The nested case-control study, like the case-cohort study, is more prone to measurement error than a cohort study if exposure status is determined retrospectively after cases and controls are selected. However, this will not be a problem if, as in the nested case-control study just described, exposure information is obtained upon enrollment in the initial cohort.

Summary

- ❖ The nested case-control design is a variation of the case-cohort design in which controls are chosen using density sampling.
- ❖ Density sampling requires that controls be matched to cases at the time of case diagnosis.
- ❖ Advantage: density-type controls are at risk for same amount of time as its matched case.
- ❖ Disadvantage: more prone to measurement error than cohort study.

Quiz (Q3.9)

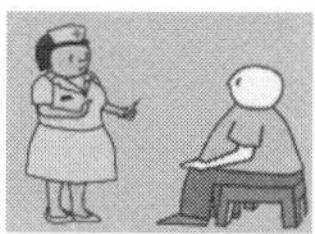

Choose whether each of the following is characteristic of a **case-cohort study**, a **nested case-control study**, **both**, or **neither**.

1. Usually less expensive than a prospective cohort study: . ???
2. Comparison group from same population as cases: . ???
3. Usually less expensive than a case-control study: . ???
4. Respective timing: . . . . ???
5. Backward directionality: . . . . . ???
6. Controls matched to cases at time of diagnosis: . . ???
7. Compares exposure experience of cases versus controls: ???
8. Density sampling: . . . . . ???
9. Must have same number of controls as cases: . ???

---

10. Suppose that, at the time of enlistment in a military service, a sample of blood was drawn from each enlistee and stored. After the Persian Gulf War, some soldiers developed a constellation of symptoms that came to be known as Persian Gulf War Syndrome. If investigators then examined the blood of all soldiers with the syndrome, and blood from twice as many soldiers without the syndrome, this would be an example of: . . . . . ???

Choices
case-cohort study **nested case-control study** **prospective cohort study** **retrospective cohort study**

The Case-Crossover Design

The case-crossover design is a variant of the matched case-control study (described in Lesson 15 on Matching) that is intended to be less prone to bias than the standard case-control design because of the way controls are selected. The design incorporates elements of both a matched case-control study and a nonexperimental retrospective crossover experiment. (Note: In, a crossover design, each subject receives at least two different exposures/treatments at different occasions.) The fundamental aspect of the case-crossover design is that each case serves as its own control. Time-varying exposures are compared between intervals when the outcome occurred (case intervals) and intervals when the outcome did not occur within the same individual.

The case-crossover design was designed to evaluate the effect of brief exposures with transient effects on acute health outcomes when a traditional control group is not readily available. The primary advantage of the case-crossover design lies in its ability to help control confounding. Self-matching subjects against themselves automatically eliminates confounding between subjects and from both measured and unmeasured fixed covariates.

As an example of a case-crossover design, Redlemeier and Tibshirani studied whether the use of a cellular telephone while driving increases the risk of a motor vehicle collision. In their abstract, they say, "We studied 699 drivers who had cellular telephones and who were involved in motor vehicle collisions resulting in substantial property damage but no personal injury. Each person's cellular telephone calls on the day of the collision and during the previous week were analyzed through the use of detailed billing records... The risk of a collision when using a cellular telephone was four times higher than the risk when a cellular telephone was not being used (relative risk, 4.3; 95 percent confidence interval, 3.0 to 6.5... Calls close to the time of the collision were particularly hazardous (relative risk, 4.8 for calls placed within 5 minutes of the collision …"

Incomplete Designs

***Incomplete designs** are studies in which information is missing on one or more relevant factors.*

Ecologic Studies

*An **ecologic study** is an incomplete design for which the unit of analysis is a group, often defined geographically, such as a census tract, a state, or a country.*

In an ecologic study, the unit of analysis is a *group*, often defined geographically, rather than an individual. That is, the basic data are typically percentages or other summary statistics for each group, rather than measurements of characteristics on individuals. The groups might be census tracts, states, or countries.

The advantage of an ecologic study is that it can often be done quickly and inexpensively using existing data, usually mortality data. Ecologic studies are often used to generate hypotheses about exposure-disease relationships. They are also used to evaluate the impact of intervention programs on the health status of target populations.

The primary criticism of an ecologic study is that data are not available on individuals. In particular, an ecologic study has data on the number of exposed persons and the number of cases within each group but does not have the number of exposed cases.

Ecologic Study

Disadvantage:

- Data not available on individuals.

	Exposure Yes	Exposure No	Total
D	×		√
not D			
Total	√		

Consequently, conclusions obtained from ecologic studies about determinants of a health outcome may not carry over to individuals. This problem is called the **ecologic fallacy**.

The graph shown below shows a scatter plot of incidence rates of AIDS in 13 US states during 1989 compared to corresponding incidence rates of tuberculosis for the same year. The graph indicates that the states that had a high incidence of AIDS also have a high incidence of TB. And that states with a low incidence of AIDS tended to have a low incidence of TB. The relationship depicted here therefore suggests that the occurrence of AIDS may influence the development of TB or vice-versa.

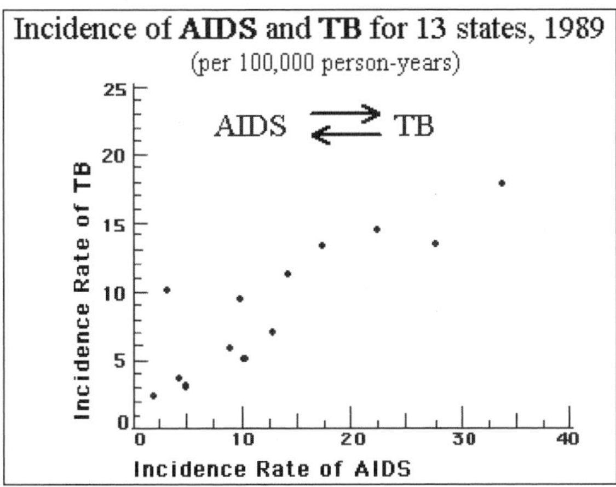

But, these data show a relationship that uses states as the unit of analysis. It is possible that few individuals have both AIDS and TB, even when the incidences of AIDS and TB, separately, are high. That information can only be obtained from data on individuals.

Summary

- ❖ In an ecologic study, the unit of analysis is a group, often defined geographically.
- ❖ Conclusions obtained from ecologic studies may not carry over to individuals (the ecologic fallacy).
- ❖ In an ecologic study, data are available on the number of exposed persons and the number of cases within each group, but not on the number of exposed cases.

Proportional Studies

A **proportional *morbidity*** or **proportional *mortality*** study only includes observations on cases but lacks information about the candidate population at risk for developing the health outcome. If the design involves incident cases, the study is a proportional morbidity study. If deaths are used, the study is a proportional mortality study.

Proportional mortality studies are used to generate new hypotheses or to conduct preliminary tests of etiologic hypotheses without collecting much additional data. Because proportional morbidity or mortality studies do not provide non-cases and the candidate population at risk is not available, it is not possible to compute traditional measures of effect, such as risk ratios or odds ratios used to evaluate exposure-disease relationships.

A 1997 study of nuclear power workers tested the possible relationship between low levels of ionizing radiation and cancer among 3,500 certified deaths that occurred among plant workers between 1944 and 1972.

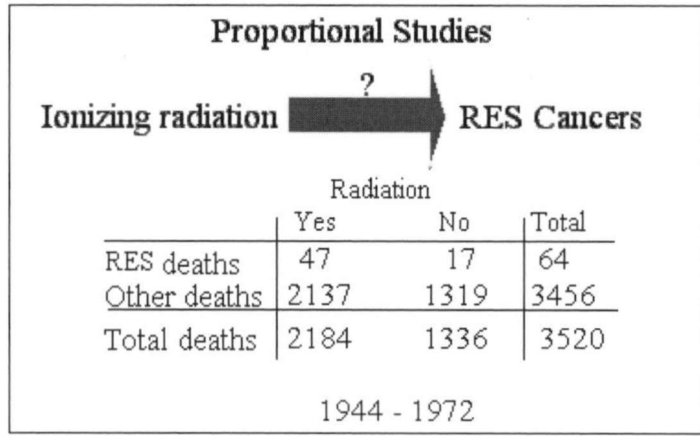

	Radiation Yes	No	Total
RES deaths	47	17	64
Other deaths	2137	1319	3456
Total deaths	2184	1336	3520

1944 - 1972

The data in this study were mortality data exclusively, and did not contain information on the size of the candidate population at risk for dying. Without such information, it was not possible to demonstrate that exposure was positively associated with RES cancer risk. However, these data did show that a significantly greater proportion of workers exposed to radiation at the plant had died of RES compared to unexposed workers. This was certainly a step in the direction of implicating the potential harm of radiation exposure on RES cancer risk.

Summary

- ❖ A proportional morbidity or mortality study only includes observations on cases without information about the candidate population-at-risk.
- ❖ Proportional studies are useful to generate hypotheses.
- ❖ Proportional studies are useful for conducting preliminary tests of etiologic hypotheses.
- ❖ In proportional studies, traditional measures of effect such as risk ratios cannot be computed.

Quiz (Q3.10)

Determine whether each of the following statements is **true** or **false**

1. An ecologic study is one in which populations are compared with individuals: . **???**
2. An examination of the distribution of causes of deaths as listed on the death certificate of all persons who died in Georgia in 1999 is an example of a proportional mortality study: . . **???**
3. Ecologic studies and proportional mortality studies are both better suited to generating causal hypotheses than to testing hypotheses . . . . . . **???**
4. Ecologic studies and proportional mortality studies both have information on populations, but not on individuals: . . . . **???**
5. Ecologic studies and proportional mortality studies can often be done quickly, because both mostly use existing, readily available data: . . . . . . **???**
6. Neither ecologic studies nor proportional morbidity / mortality studies can yield estimates of an individual's risk of disease or death: . . . . **???**

References

Reference for Diabetes Research Group Study (Clinical Trial)

The Diabetes Control and Complications Trial Research Group. The effect of intensive treatment of diabetes on the development and progression of long-term complications in insulin-dependent diabetes mellitus. N Engl J Med 1993; 329(14):977-86.

Reference for Framingham Heart Study (Prospective Cohort Study)

Feinleib M, The Framingham study: sample selection, follow-up, and methods of analysis, in National Cancer Institute Monograph, No. 67, Greenwald P (editor), US Department of Health and Human Services, 1985.

Dorgan JF, Brown C, Barrett M, et al. Physical activity and risk of breast cancer in the Framingham Heart Study, Am J Epidemiol 1994;139(7): 662-9.

Margolis JR, Gillum RF, Feinleib M, Brasch RC, Fabsitz RR. Community surveillance for coronary heart disease: the Framingham Cardiovascular Disease Survey. Methods and preliminary results. Am J Epidemiol 1974;100(6):425-36.

Reference for VDT use and Spontaneous Abortion (Retrospective Cohort Study)

Schnorr TM, Grajewski BA, Hornung RW, Thun MJ, Egeland GM, Murray WE, Conover DL, Halperin WE. Video display terminals and the risk of spontaneous abortion. N Engl J Med 1991;324(11):727-33.

Reference for Nonsteroidal anti-inflammatory drugs and Alzheimer's disease (Prospective Cohort Study)

in t' Veld BA, Ruitenberg A, Hofman A, Launer LJ, van Duijn CM, Stijnen T, Breteler MM, Stricker BH. Nonsteroidal antiinflammatory drugs and the risk of Alzheimer's disease. N Engl J Med 2001;345(21):1515-21.

References for Reye's Syndrome (Case-Control Studies)

Waldman RJ, Hall WN, McGee H, Van Amburg G. Aspirin as a risk factor in Reye's syndrome. JAMA 1982;247(22):3089-94.

Halpin TJ, Holtzhauer FJ, Campbell RJ, Hall LJ, Correa-Villasenor A, Lanese R, Rice J, Hurwitz ES. Reye's syndrome and medication use. JAMA 1982;248(6):687-91.

Daniels SR, Greenberg RS, Ibrahim MA. Scientific uncertainties in the studies of salicylate use and Reye's syndrome. JAMA 1983;249(10):1311-6.

Hurwitz ES, Barrett MJ, Bregman D, Gunn WJ, Pinsky P, Schonberger LB, Drage JS, Kaslow RA, Burlington DB, Quinnan GV, et al. Public Health Service study of Reye's syndrome and medications. Report of the main study. JAMA 1987;257(14):1905-11.

Forsyth BW, Horwitz RI, Acampora D, Shapiro ED, Viscoli CM, Feinstein AR, Henner R, Holabird NB, Jones BA, Karabelas AD, et al. New epidemiologic evidence confirming that bias does not explain the aspirin/Reye's syndrome association. JAMA 1989;261(17):2517-24

References for Creutzfeldt-Jakob Disease (Case-Control Studies)

van Duijn CM, Delasnerie-Laupretre N, Masullo C, Zerr I, de Silva R, Wientjens DP, Brandel JP, Weber T, Bonavita V, Zeidler M, Alperovitch A, Poser S, Granieri E, Hofman A, Will RG. Case-control study of risk factors of Creutzfeldt-Jakob disease in Europe during 1993-95. European Union (EU) Collaborative Study Group of Creutzfeldt-Jakob disease (CJD). Lancet 1998;351(9109):1081-5.

Will RG, Ironside JW, Zeidler M, Cousens SN, Estibeiro K, Alperovitch A, Poser S, Pocchiari M, Hofman A, Smith PG. A new variant of Creutzfeldt-Jakob disease in the UK. Lancet 1996;347(9006):921-5.

General epidemiologic design

Checkoway H, Pearce N, Dement JM. Design and conduct of occupational epidemiology studies: II. Analysis of cohort data. Am J Ind Med 1989;(15(4):375-94.

Greenberg RS, Daniels SR, Flanders WD, Eley JW, Boring JR. Medical Epidemiology (3rd Ed). Lange Medical Books, New York, 2001.

Kleinbaum DG, Kupper LL, Morgenstern H. Epidemiologic Research: Principles and Quantitative Methods. John Wiley and Sons Publishers, New York, 1982.

Steenland K (ed.). Case studies in occupational epidemiology. Oxford University Press, New York, 1993.

Example of Cross-Sectional Studies

Smith WCS, Woodward M, Tunstall-Pedoe H. Intermittent claudication in Scotland, in Epidemiology of Peripheral Vascular Disease. (ed FGR Fowkes.), Springer-Verlag, Berlin, 1991

Hybrid designs

Coates RJ, Weiss NS, Daling JR, Rettmer RL, Warnick GR. Cancer risk in relation to serum copper levels. Cancer Res 1989;49(15): 4353-6.

Linn JT, Wang LY, Wang JT, Wang TH, Yang CS, Chen CJ. A nested case-control study on the association between Helicobacter pylori infection and gastric cancer risk in a cohort of 9775 men in Taiwan. Anticancer Res 1995;15:603-6.

Maclure M, Mittleman MA. Should we use a case-crossover design? Annu Rev Public Health 2000;21:193-221.

Maclure M. The case-crossover design: a method for studying transient effects on the risk of acute events. Am J Epidemiol 1991;133(2):144-53.

Redelmeier DA, Tibshirani RJ. Association between cellular-telephone calls and motor vehicle collisions. N Eng J Med 1997;336(7):453-8.

Ecologic

CDC. Summary of notifiable diseases in the United States. MMWR Morb Mortal Wkly Rep 1990;38(54):1-59.

Morgenstern H. Ecologic studies in epidemiology: concepts, principles, and methods. Annu Rev Public Health 1995;16:61-81.

Proportional Mortality

Mancuso TF, Stewart A, Kneale G. Radiation exposures of Hanford workers dying from cancer and other causes. Health Phys 1977;33:369-85.

Causation and Meta Analysis

Blalock HM, **Causal Inferences in Nonexperimental Research**, Chapter 1, Norton Publishing, 1964.

Chalmers I, Altman DG (eds.), **Systematic Reviews**, BMJ Publishing Group, London, 1995.

Chalmers TC. Problems induced by meta-analyses. Stat Med 1991;10(6):971-80.

Hill AB, Principles of Medical Statistics, 9th Edition, Chapter 24, Oxford University Press, 1971.

Lipsey MW, Wilson DB. Practical meta-analysis. Applied Social Research Methods Series; Vol. 49. Sage Publications, Inc., Thousand Oaks, CA: 2001.

Mosteller F, Colditz GA. Understanding research synthesis (meta-analysis). Annu Rev Public Health 1996;17:1-23.

Petitti DB, Meta-analysis Decision Analysis and Cost-Effectiveness Analysis; Methods for Quantitative Synthesis in Medicine, Oxford University Press, 1994.

Popper KR, The Logic of Scientific Discovery, Harper and Row Publishers, 1968.
Rothman KJ. Causes. Am J Epidemiol 1976;104(6): 587-92.
Susser M, Causal Thinking in the Health Sciences, Oxford University Press, 1973.
U.S. Department of Health, Education, and Welfare, Smoking and Health, PHS Publ. No. 1103, Government Printing, Washington DC, 1964.
Weiss NS. Inferring causal relationships: elaboration of the criterion of 'dose-response'". Am J Epidemiol 1981;113(5):487-90.

Homework Exercises

ACE-1. Study Type. State the type of study described by each of the following paragraphs:

a. To investigate the relationship between egg consumption and heart disease, a group of patients admitted to a hospital with myocardial infarction were questioned about their egg consumption. Another group of patients admitted to a fracture clinic and matched on age and sex with the first group were also questioned about their egg consumption using an identical protocol.
b. To investigate the relationship between certain solvents and cancer, all employees at a factory were questioned about their exposure to an industrial solvent, and the amount and length of exposure measured. These subjects were regularly monitored, and after 10 years a copy of the death certificate for all those who died was obtained.
c. A survey was conducted of all nurses employed at a particular hospital. Among other questions, the questionnaire asked about the grade of the nurse and whether or not she was satisfied with her career prospects.
d. To evaluate a new school (i.e., approach) for treating back pain, patients with lower back pain were randomly allocated to either the new school or to conventional occupational therapy. After 3 months, they were questions about their back pain, and observed lifting a weight by independent monitors.
e. A new triage system has been set up at the local Accident and Emergency Unit. To evaluate this new system, the waiting times of patients were measured for 6 months and compared with the waiting times at a comparable nearby period.
f. The Tumor Registry in a certain US state was used to identify all primary cases of bladder cancer in the state during a given period. These cases were compared to a sample of non-cases from the same state that have been matched on age and time of diagnosis. All subjects or their surviving relatives in both groups were interviewed to collect information on saccharin consumption and other known risk factors for bladder cancer.
g. In the Hanford study of nuclear power workers (1977), 3500 certified deaths occurred among plant workers between 1944 and 1972. Among these deaths, a significantly greater proportion of workers exposed to low levels of ionizing radiation than unexposed workers had died of RES (reticuloendothelial system) cancers.

ACE-2. Case-control vs. Prospective Cohort. Which of the following choices is **not** an advantage of using a case-control study as opposed to a prospective cohort study? (There may be more than one correct answer here.)

a. Less expensive
b. Can be completed more rapidly
c. More appropriate for the study of rare diseases,
d. More appropriate for the study of diseases that develop slowly
e. More appropriate for the study of several exposures.
f. More appropriate for the study of several diseases.
g. Allows more accurate assessment of exposure.

ACE-3. Randomization. A randomized clinical trial was designed to compare two different treatment approaches for irritable–bowel syndrome. The purpose of randomization in this study was to:

a. increase patient compliance with treatment
b. obtain comparison groups that are similar on other variables that may influence the disease.
c. obtain comparison groups that are similar on any other variables measured in the study.
d. Increase the likelihood of finding a significant effect of treatment
e. obtain a representative sample in the study.

ACE-4. Clinical Trial. In a randomized clinical trial designed to compare two treatments for asthma, the clinicians knew which treatment the patients received, but the patients themselves did not know which treatment they received. This is an example of:

a. compliance
b. intention-to-treat
c. double-blinding
d. placebo effect
e. none of the above

ACE-5. Case-Control Study: TB. (Primarily for medical students/clinicians) The following questions apply to the article " Variations in the NRAMPI gene and susceptibility to tuberculosis in West Africans." (Bellamy R, Rowende t al., **New Eng J of Med, 38 (10),** pp. 640-643, March 1998).

a. Who were the patients in this study?
b. Who were the controls?
c. Did the controls and cases differ in any major aspect other than disease status?
d. What was the design of this study?
e. State the "null" hypothesis, either symbolically or in words.
f. What do the investigators conclude about the null hypothesis?

ACE-6. Prospective Cohort Study. In a famous prospective cohort investigation, the population to be studied encompassed all physicians listed in the **British Medical Register** and resident in England and Wales as of October 1951. Information about present and past smoking habits was obtained by questionnaire. Information about lung cancer came from death certificates and other mortality data recorded during ensuing years.

a. What makes this study prospective? List two advantages and two disadvantages of this approach.
b. What advantages and disadvantages come with selecting physicians as a cohort for follow-up?

ACE-7. Intention to Treat. A clinical trial was conducted to compare a new hypertension therapy to the standard therapy. At the end of the follow-up period, the investigators performed two separate analyses. For the first analysis, they followed the "intention to treat" rule. For the second analysis, they included only those patients known to have taken the prescribed therapy throughout the study period. The results of the two analyses differed substantially. The most likely explanation for the discrepancy is: [Choose one best answer]

a. The randomization was unsuccessful
b. The new therapy was not effective
c. There was a significant degree of recall bias
d. There was a problem with patient compliance

ACE-8. Blinding. A clinical trial was conducted to compare the performance of two treatments. Describe a situation in which it would NOT be feasible for the trial to be **blinded**. [Be sure that your answer indicates an understanding of what it means for a trial to be blinded.]

ACE-9. Randomization: Clinical Trials. What is the purpose of randomization in a clinical trial? [Choose one best answer.]

a. To make the diseased and non-diseased as similar as possible with respect to all variables except the exposure of interest
b. To reduce the number of subjects who are lost to follow-up
c. To isolate the effect of the exposure of interest
d. To encourage compliance with the assigned treatment regimen

ACE-10. Study Design: Adiposity and CHD. (Primarily for medical students/clinicians) .The following questions apply to the article: "Abdominal adiposity and coronary heart disease in women." (Rexrode KM, Carey VJ et al., **JAMA, 280 (21),** pp 1643-1646, December 1998).

a. What is the design of this study?
b. What are the two principal null hypotheses for this study?
c. How many women were included in the final analysis? What percentage of the entire cohort does this represent?
d. The study used self-reported weight. How assured are you that this information is accurate?
e. Given the results in Table 2, which of the two variables, waist-hip ratio or waist circumference, seems to be the better predictor of CHD risk? Why?

ACE-11. Density Sampling. Which of the following is NOT true of "density sampling" of controls in a case-control study? [Choose one best answer.]

a. A subject identified as a control may later be identified as a case.
b. A subject identified as a case may later be identified as a control.
c. The odds ratio calculated from such a study is likely to be a good estimate of an incidence measure of association.
d. An individual subject may serve as a control for more than one case.

ACE-12. Control Group. What is the purpose of the control group in a case-control study? [Choose one best answer.]

a. To provide an estimate of the background risk or rate of disease.
b. To provide an estimate of the exposure frequency among the population that produced the cases.
c. To provide an estimate of the magnitude of the placebo effect.
d. To provide an estimate of the expected number of cases among the unexposed.

ACE-13. Study Design: Breast Cancer. A paper entitled "Electric Blanket Use and Breast Cancer Risk among Younger Women" appeared in a recent issue of the *American Journal of Epidemiology*. The methods section included the following information:

Cases were women newly diagnosed with in situ or invasive breast cancer between May 1, 1990, and December 31, 1992, who were residents of three U.S. geographic areas. Controls were women identified by random digit dialing and frequency-matched to cases by 5-year age group and geographic area. All women were asked about whether they had ever regularly used electric blankets, electric mattress pads, or heated waterbeds. A positive response referred to the aggregate use of any or all of the devices at any time in the respondent's life prior to enrollment in the study.

Which one of the following best describes the design of this study?

a. Cross-sectional
b. Cohort
c. Nested Case-Control
d. Population-based Case-Control
e. Descriptive

Answers to Study Questions and Quizzes

Q3.1

1. Experimental
2. Observational
3. Observational
4. Experimental
5. Experimental
6. Observational

Q3.2

1. Therapeutic
2. Preventive

1. a
2. b
3. b
4. a

5. c
6. c
7. a

Q3.3

1. For VDT users the percentage is (54/366) x 100 = 14.8% whereas for VDT non-users the percentage is (82/516) x 100 = 15.9%. The two percentages differ by only 1%.

Q3.4

1. In a cohort study, the directionality is always forward.
2. The timing is prospective, since the health outcome, in this case Alzheimer's disease, occurs after the onset of the study.

Q3.5

1. b
2. a
3. b
4. c
5. b
6. Retrospective
7. Retrospective
8. Prospective
9. Both
10. Neither
11. Both
12. Retrospective

Q3.6

1. b
2. a
3. a
4. c
5. b
6. case-control
7. case-control
8. prospective cohort
9. prospective cohort
10. case-control
11. prospective cohort
12. case-control
13. prospective cohort
14. T – If controls are chosen from a different population from which the cases came, there may be selection bias.
15. F – Hospital controls have an illness; such controls are typically not representative of the community from which the cases came.
16. F – Population-based controls can be obtained from random dialing of telephone numbers in the community from which the cases are derived. There is no guarantee that neighbors of cases will be chosen.

Q3.7

1. b
2. a
3. c
4. b
5. T
6. F
7. T – A cross-sectional study includes only cases that survive long enough to be available for study. This could lead to a misleading conclusion about an exposure-disease relationship since non-survivors are excluded.
8. T

Q3.8

1. Backward
2. Prospective

Q3.9

1. Both
2. Both
3. Neither
4. Neither
5. Both
6. Nested case-control study
7. Both
8. Nested case-control study
9. Neither
10. Case-cohort study

Q3.10

1. F – The unit of analysis in an ecologic study is a group (e.g., census tract, state, country) and data on both exposure and disease is not simultaneously obtained on individuals
2. T – A proportional mortality study includes observations on deaths without information about the candidate population (i.e., denominators).
3. T – Both ecologic and proportional mortality studies use "incomplete" designs.
4. F – Proportional mortality studies use information on deaths about individuals.
5. T
6. T

2 measures of incidence: Cumulative incidence which is a proportion & a risk, but not a rate. Incidence density rate which is not a risk or proportion, but a time rate by definition.

2 measures of prevalence: point prevalence which is a proportion, but not a risk or a rate, & is defined as the number of existing cases at a particular point in time / population size at that same time. Period prevalence is also a proportion, but not a risk or a rate, & is defined as the number of individuals who were observed to have the health outcome anytime during the follow-up period / size of the population for this same time period

LESSON 4

MEASURES OF DISEASE FREQUENCY

In epidemiologic studies, we use a measure of disease frequency to determine how often the disease or other health outcome of interest occurs in various subgroups of interest. We describe two basic types of measures of disease frequency in this chapter, namely, measures of incidence and measures of prevalence. The choice typically depends on the study design being used and the goal of the study.

4-1

Incidence versus Prevalence

There are two general types of measures of disease frequency, **incidence** (**I**) and **prevalence** (**P**). Incidence measures ***new*** cases of a disease that develop over a period of time. Prevalence measures ***existing*** cases of a disease at a particular point in time or over a period of time.

To illustrate how incidence and prevalence differ, we consider our experience with AIDS. The number of annual incident cases of AIDS in gay men decreased in the US from the mid-1980s to the late 1990s. This has resulted primarily both from recent anti-retroviral treatment approaches and from prevention strategies for reducing high-risk sexual behavior. In contrast, the annual prevalent cases of AIDS in gay men has greatly increased in the US during the same period because recent treatment approaches for AIDS have been successful in prolonging life of persons with the HIV virus and/or AIDS.

Prevalence rate is oxymoron

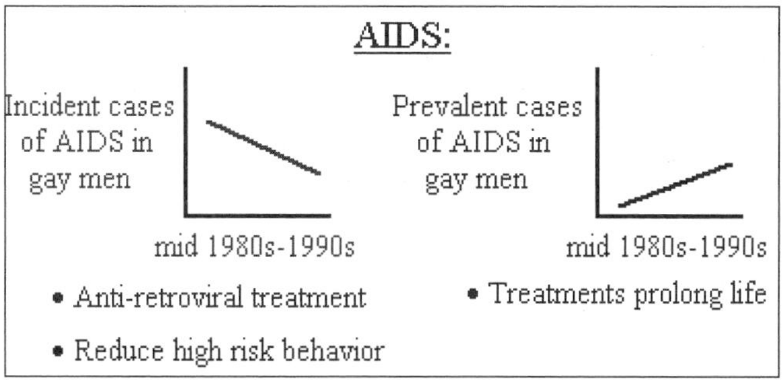

Prevalence can also be viewed as describing a pool of disease in a population, whereas incidence describes the input flow of new cases into the pool, and fatality and recovery reflects the output flow from the pool.

Prevalence is not risk. It is a proportion of individuals in a population that have a particular health outcome of interest.

Person time is the amount of disease free time that an individual is followed. Prevalence looks at existing cases rather than new ones.

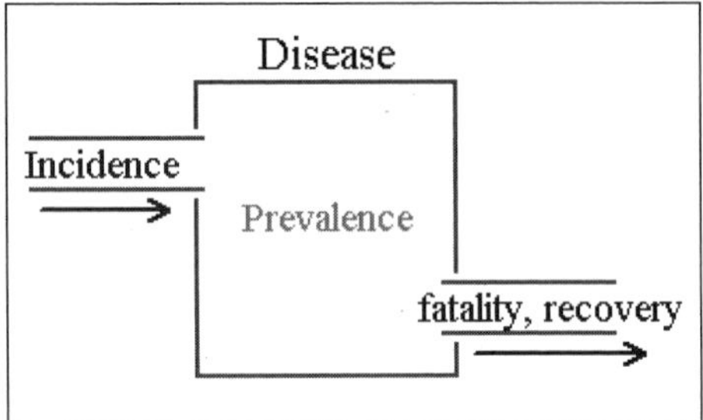

Incidence measures are useful for identifying risk factors and assessing disease etiology. Typically, incidence measures are estimated from clinical trials and from cohort studies, which involve the follow-up of subjects over time.

Prevalence measures are not as useful as incidence measures for assessing etiology because prevalence does not consider persons who die from the disease before the prevalence study begins. Typically, prevalence measures are estimated from cross-sectional studies and from case-control studies that use prevalent, rather than incident, cases. Since the number of prevalent cases indicates demand for health care, prevalence measures are most useful in the planning of health services.

Summary

- Incidence concerns new cases of a disease or other health outcome over a period of follow-up.
- Prevalence concerns existing cases of a disease at a point in time.
- Incidence measures are useful for identifying risk factors and assessing disease etiology
- Prevalence measures are most useful in the planning of health services

Mortality Might Be Used Instead of Disease Incidence

We discuss incidence and prevalence in terms of new or existing cases of a **disease**, whether or not these cases eventually die or not during or after the period of study. There are many situations, however, when the use of strictly **mortality** information is also worthwhile.

Mortality measures are an important tool for **epidemiologic surveillance**. Today such surveillance programs have been applied to monitor the occurrence of a wide variety of health events, including deaths, in large populations. Mortality statistics are also convenient for **evaluating etiologic hypotheses**, especially when incidence data are not available. In particular, for diseases with a low rate of cure or recovery, such as lung cancer, mortality measures give a reasonable approximation to incidence measures.

Use of mortality information for any of the above purposes has several pragmatic advantages:

- Mortality data are widely collected and virtually complete since registration of deaths is compulsory in most industrialized countries and few deaths are not reported.
- Mortality data are defined using standardized nomenclature. In particular, the International Classification of Diseases (ICD) is used to promote uniformity in reporting causes of death.
- Recording of mortality data is relatively inexpensive.

House Guests Example

Suppose guests arrive at your house at the rate of two per day and stay exactly five days. How many people will be in your house after a week?

Let's see what happens day by day. On the first day, two guests arrive and none depart, so there are 2 guests in your house at the end of the first day. On the second day two more guests arrive, and none depart, so there are now 4 guests in your house after 2 days. Similarly, there are 6 guests after 3 days, 8 after 4 days and 10 guests in your house after five days, with no guests departing up to this point. But, on the sixth day, two new guests arrive, but the two guests that came on day 1, having been there for five days, now depart, leaving you again with 10 guests in the house. At the end of the seventh day, there will still be 10 guests in the house, which answers the question raised at the start of all this.

This scenario illustrates the fundamental difference between incidence and prevalence. In the example, after 5 days, a "steady state" is reached at which point there are 10 houseguests as long as the arrival rate is 2 per day. This steady state of 10 houseguests is a prevalence, which describes the existing count of guests, at any point in time after a steady state has been reached. The arrival rate of 2 guests per day is an incidence, which describes how quickly new guests are arriving. The duration of five days that guests stay in your house is the information needed to link the incidence to the prevalence.

Prevalence can be linked to incidence with the following formula:

P = I x D

In our example, **P** is the number of guests in the house on any day after day five, **I** is the arrival rate of 2 guests per day, and **D** is the duration of 5 days for each guest. The formula works in this example since 2 times 5 equals 10.

We can see from this formula that for a given incidence, the prevalence will increase or decrease as the duration increases or decreases. For example, if guests stayed for 8 days rather than 5 days, with the same arrival rate, the number of guests at the house at steady state would be 2 times 8, which equals 16, rather than 10.

For a given duration, the prevalence will increase or decrease as the incidence increases or decreases. Thus, if the guests arrive at the rate of only 1 guest per day rather than 2, and stay 8 days, the prevalence will be 1 times 8, which equals 8, instead of 16.

Summary

- A scenario involving houseguests who arrive at 2 per day and stay five days illustrates the fundamental difference between incidence and prevalence.
- A steady state of 10 houseguests illustrates *prevalence*, which describes the existing count of guests at any point in time after steady state is reached.
- The arrival of 2 guests per day illustrates *incidence*, which describes how quickly new guests are arriving.
- The *duration* of 5 days is the information needed to link how incidence leads to prevalence
- Prevalence is obtained as the product of incidence and duration (P = I x D)

The Relationship between Prevalence and Incidence

In the example involving "house guests", the formula

$$P = (I \times D)$$

was used to demonstrate that the steady state number of guests in the house after 7 days was equal to the product of the number of guests arriving each day times the duration that each guest stayed in the house.

The terms **P**, **I**, and **D** in this formula represent the concepts of prevalence, incidence and duration, respectively, but, as used in the example, they each do not strictly conform to the epidemiologic definitions of these terms. As described in later activities in this lesson (i.e., chapter) on measures of disease frequency, the strict definitions of prevalence and incidence require denominators, whereas the "house guest" scenario described here makes use only of numerator information.

Specifically, ***prevalence*** is estimated using the formula:

$$P = \frac{C}{N}$$

Continued on next page

The Relationship between Prevalence and Incidence (continued)

and ***incidence*** uses one of the following two possible formulas depending on whether risk or rate is the incidence measure chosen:

$$CI = \frac{I}{N} \quad \text{or} \quad IR = \frac{I}{PT}$$

In the above formulae, **P**, **C**, and **N** denote the prevalence, number of existing cases, and steady state population-size, respectively. Also, **CI** denotes cumulative incidence, which estimates risk, **I** denotes the number of new (incident cases), and **N** denotes the size of a disease-free cohort followed over the entire study period. Further, **IR** stands for incidence rate, and **PT** for accumulated person-time information. All these formulae are described and illustrated in later activities.

The important point being made here is that all three of the above formulae have denominators, which were not used in the houseguest example, but are required for computing prevalence and incidence in epidemiology.

The term **D** in the formula at the top of this page was used in the houseguest example to define the duration of stay that was assumed for each houseguest. In the epidemiologic use of this formula, **D** actually denotes the average duration of illness for all subjects in the population under study, rather than being assumed to be the same for each person in the population.

Nevertheless, using the stricter epidemiologic definitions of prevalence and incidence measures and using average duration, the above formula that relates prevalence to incidence and duration still holds, provided the population is in steady state and the disease is rare. By steady state, we mean that even though the population may be dynamic, the number of persons who enter and leave the population for whatever reasons are essentially equal over the study period, so that the population does not change. If the disease is not rare, a modified formula relating prevalence to incidence is required instead, namely:

$$P = \frac{I \times D}{(I \times D) + 1}$$

Quiz (Q4.1)

For each of the following scenarios, determine whether it is more closely related to **incidence** or to **prevalence**.

1. Number of campers who developed gastroenteritis within a few days after eating potato salad at the dining hall? . . . . **???**
2. Number of persons who reported having with diabetes as part of the National Health Interview Survey? . . . . **???**
3. Occurrence of acute myocardial infarction (heart attack) among participants during the first 10 years of follow-up of the Framingham Study? . . . . **???**
4. Number of persons who died and whose deaths were attributed to Hurricane Floyd in North Carolina in 1999? . . . . **???**
5. Number of children who have immunity to measles, either because they had the disease or because they received the vaccine? . . . . **???**

Quiz continued on next page

Suppose a surveillance system was able to accurately and completely capture all new occurrences of disease in a community. Suppose also that a survey was conducted on July 1 that asked every member of that community whether they currently had that disease. For each of the following conditions, determine whether **incidence** (per 1,000 persons per year) or **prevalence** (per 1,000 persons on July 1) is *likely to be higher*.

6. Rabies (occurs rarely and has a short duration, e.g., death within one week)? . **???** Incidence
7. Multiple sclerosis (rare occurrence, long duration [many years])? . . . **???** Prevalence
8. Influenza (common but winter-seasonal occurrence, short duration)? . . . **???** Incidence
9. Poison ivy dermatitis (common spring/summer/fall occurrence, 2-week duration)? . . **???** Incidence
10. High blood pressure (not uncommon occurrence, lifelong duration)? . . **???** Prevalence

4-2

Risk

The term risk is commonly used in everyday life to describe the likelihood, or probability, that some event of interest will occur. We may wonder, for example, what is the risk that the stock market will crash or that we will be involved in a serious auto collision? We may worry about our risk for developing an undesirable health condition, such as a life-threatening illness, even our risk for dying.

In epidemiology, risk is the probability that an individual with certain characteristics, say, age, race, sex, and smoking status, will develop or die from a disease, or even more generally, will experience a health status change of interest over a specified follow-up period. When the health outcome is a disease, this definition assumes that the individual does not have the disease at the start of follow-up and does not die from any other cause during follow-up. Because risk is a probability, it is a number between 0 and l, or, correspondingly, a percentage.

Risk = Cummulative incidence

$$0 \leq \text{RISK} \leq 1$$

$$(\ 0 \leq \text{Percentage} \leq 100\ )$$

When describing risk, it is necessary to specify a period of follow-up, called the **risk period**. For example, to describe the risk that a 45 year-old male will develop prostate cancer, we must state the risk period, say, 10 years of follow-up, over which we want to predict this risk. If the risk period were, for example, 20 years instead of 10 years, we would expect our estimate of risk to be larger than the 10-year risk since more time is being allowed for the disease to develop.

Study Questions (Q4.2)

1. What is the meaning of the following statement? The 10-year risk that a 45-year-old male will develop prostate cancer is 5%? (State your answer in probability terms and be as specific as you can in terms of the assumptions required.)

2. Will the 5-year risk for the same person described in the previous question be larger or smaller than the 10-year risk? Explain briefly.

1) A 45 year old male free of prostate cancer has a probability of 0.05 of developing prostrate cancer over the next 10 years if he does not die from any other cause.

2) Smaller, because the 5 year risk involves a shorter time period for the same person to develop prostate cancer.

Summary

- ❖ Risk is the probability than an individual will develop or die from a given disease or, more generally, will experience a health status change over a specified follow-up period.
- ❖ Risk assumes that the individual does not have the disease at the start of the follow-up and does not die from any other cause during the follow-up.
- ❖ Risk must be some value between 0 and 1, or correspondingly, a percentage
- ❖ When describing risk, it is necessary to give the follow-up period over which the risk is to be predicted.

Confusing Risk with Rate

The term ***rate*** has often been used incorrectly to describe a measure of ***risk***. For example, the term ***attack rate*** is frequently used in studies of outbreaks to describe an estimate of the probability of developing an infectious illness, when in fact, an estimate of ***risk*** is computed. Also, the term ***death rate*** has been confused with ***death risk*** in mortality studies. In particular, the term ***case-fatality rate*** has often been misused to describe the proportion of cases that die, i.e., such a proportion is actually estimating a ***risk***.

The terms risk and rate have very different meanings, as described in other activities in this lesson. Ideally, the correct term should be applied to the actual measure being used. This does not always happen in the publication of epidemiologic findings. Consequently, when reading the epidemiologic literature, one should be careful to determine the actual measure being reported.

Cumulative Incidence

The most common way to estimate risk is to divide the number of newly detected cases that develop during follow-up by the number of disease-free subjects available at the start of follow-up. Such an estimate is often called ***cumulative incidence*** or **CI**. When describing cumulative incidence, it is necessary to give the follow-up period over which the risk is estimated.

$$\mathrm{CI} = \frac{\mathrm{I}}{\mathrm{N}} = \frac{\text{\# of new cases during follow - up}}{\text{\# of disease - free subjects at start of follow - up}}$$

Technically speaking, cumulative incidence is not equivalent to individual risk, but rather is an estimate of individual risk computed from either an entire population or a sample of a population. However, we often use the terms risk and cumulative incidence interchangeably, as we do throughout this course.

We usually put a hat ("^") over the CI when the estimate of cumulative incidence is based on a sample; we leave off the hat if we have data for the entire population.

$$\hat{\mathrm{CI}} = \text{CI “hat”}$$

The cumulative incidence formula, with or without a "hat", is always a proportion, so its values can vary from 0 to 1. If the cumulative incidence is high, as in an outbreak, the CI is sometimes expressed as a percent.

As a simple example, suppose we followed 1000 men age 45 and found that 50 developed prostate cancer within 10 years of follow-up and that no subject was lost to follow-up or withdrew from the study. Then our estimate of simple cumulative incidence is 50 over 1000, or 0.05, or, 5 %.

$$\hat{\mathrm{CI}} = \frac{\mathrm{I}}{\mathrm{N}} = \frac{50}{1000} = .05 = 5\%$$

In other words, the 10-year risk, technically the cumulative incidence for a 45 year-old male is estimated to be 5%. The formula we have given for computing risk is often referred to as **Simple Cumulative Incidence** because it is a simple proportion that assumes a fixed cohort. Nevertheless, the use of simple cumulative incidence is not always appropriate in all kinds of follow-up studies. Problems with simple cumulative incidence and methods for dealing with such problems are discussed in activities to follow.

Summary

- Cumulative incidence (CI) is a population-based estimate of individual risk
- Cumulative incidence is always a proportion
- When describing cumulative incidence, it is necessary to give the follow-up period over which the risk is estimated.
- The formula for simple cumulative incidence is CI=I/N, where I denotes the number of new cases of disease that develop over the follow-up period and N denotes the size of the disease-free population at the start of follow-up.
- The terms cumulative incidence and risk are used interchangeably in this course, even though technically, they are different.

Using Population Data to Calculate Risk

Suppose we follow 1000 men age 45 to estimate the 10-year risk of developing prostate cancer to be .05 or 5 %. To obtain this estimate, we must use information from a group of subjects, all of who happen to be exactly the same age, to predict the risk for a single individual. We might get a much better estimate if we knew specific characteristics of the individual, for example, his diet, whether or not he smokes or drinks alcohol. Nevertheless, even if we knew more characteristics, we would still have to rely on an estimate of risk based on data obtained on a group of subjects on which we measured or observed these additional characteristics. This is how epidemiologists work to estimate individual risk, that is, they must rely on accumulating evidence based on population data.

Shifting the Cohort

The formula for *simple cumulative incidence* implicitly assumes that the cohort is "**fixed**" in the sense that no entries into the cohort are allowed during the follow-up period. What we should do if we do allow new entries into the cohort?

For example, in the Sydney Beach Users study described in Lesson 2, subjects were selected from 12 popular Sydney beaches over 41 sampling days throughout the summer months of 1989-90. Subjects could progressively enter the cohort on different days during the summer, after which self-reported exposure and disease information were obtained one week later.

To illustrate, consider these six subjects. Each subject is followed for the required 7 days. Subjects 1 and 5 (going from the bottom individual to the top individual) are the only subjects who reported becoming ill.

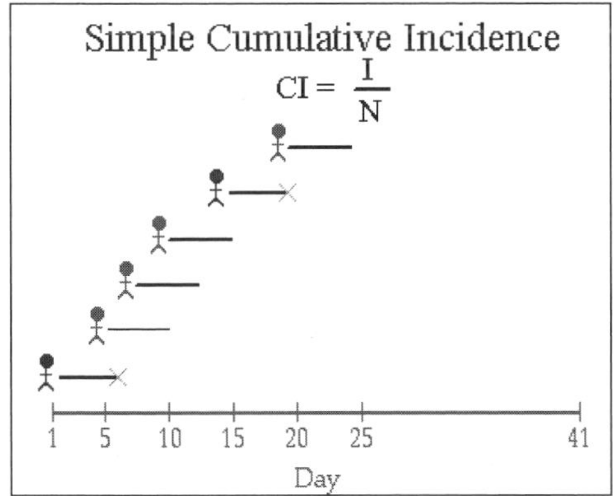

We can restructure these data by shifting the line of follow-up for each person to the left margin so that the horizontal time axis now reflects days of observation from the start of observation for each subject, rather than the actual calendar days at which the observations occurred. This conforms to the follow-up of a fixed cohort, for which the cumulative incidence is estimated to be 2/6 or one-third.

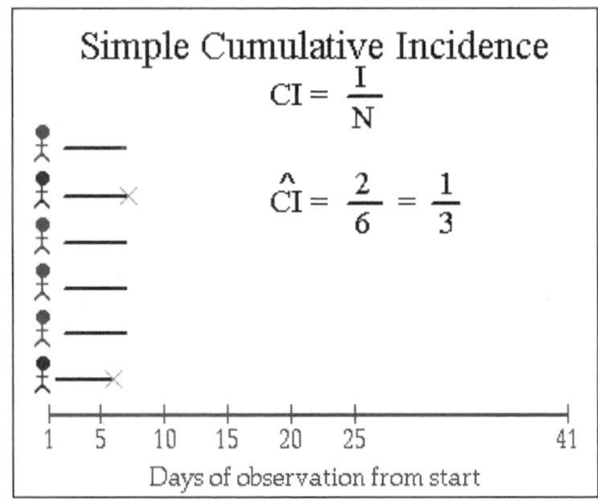

We often have a cohort that allows subjects to progressively enter the study at different calendar times. We can restructure the cohort to be fixed by shifting the data for each subject to reflect the time of observation since initial entry into the study rather than calendar time. We can then use the simple cumulative incidence formula to estimate risk.

Study Questions (Q4.3)

1. After we have shifted the cohort, do we have to assume that subjects who became cases were followed for the same amount of time as subjects who remained disease-free?

Suppose after shifting the cohort, one subject remained disease-free during 4 years of follow-up whereas another subject in the cohort remained disease-free but was only followed for 2 years.

2. Is there a problem with computing the cumulative incidence that includes both these subjects in the denominator of the CI formula?
3. After we have shifted the cohort, do we have to assume that ALL subjects, including those who became cases, were followed for the same amount of time in order to compute cumulative incidence (CI)?

Summary

- ❖ If subjects progressively enter the study at different calendar times, the data can be ***shifted*** to reflect the time of observation since initial entry.
- ❖ Simple cumulative incidence can be used to estimate risk for a ***shifted*** cohort.
- ❖ After shifting the cohort, we can compute cumulative incidence provided all subjects who remained disease-free throughout follow-up are followed for the entire length of follow-up.

Problems with Simple Cumulative Incidence

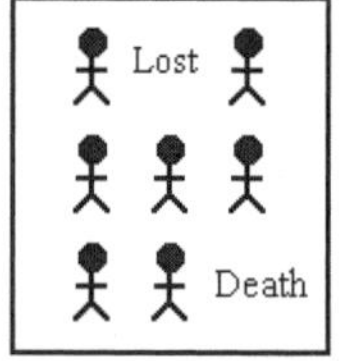

There are several potential problems with assuming a fixed cohort when using the formula for simple cumulative incidence to estimate risk. One problem occurs because the size of a fixed cohort is likely to be reduced during the follow-up period as a result of deaths or other sources of attrition such as loss to follow-up or withdrawal from the study. We don't know whether a subject lost during follow-up developed the disease of interest.

Another problem arises if the population studied is a **dynamic population** rather than a **fixed cohort**. A fixed cohort is a group of subjects identified at some point in time and followed for a given period for detection of new cases. The cohort is "fixed" in the sense that no entries are permitted into the study after the onset of follow-up, although subsequent losses of subjects may occur for various reasons such as withdrawal, migration, and death. But, a dynamic population is continually changing, allowing for both the addition of new members and the loss of previously entered members during the follow-up period.

The denominator in the simple cumulative incidence formula does not reflect the continually changing population size of a dynamic population. And the numerator in the simple cumulative incidence formula does not count new cases that may arise from those persons who entered a dynamic population after the beginning of follow-up.

Another difficulty for either a fixed or dynamic cohort is that subjects may be followed for *different periods of time* so that a cumulative incidence estimate will not make use of differing follow-up periods. This problem can occur when subjects are lost to follow-up or withdraw from the study. It could also occur if subjects enter the study after the study start and are disease-free until the study ends, or if the follow-up time at which a subject develops the disease varies for different subjects.

To illustrate these problems, let's consider a hypothetical example involving 12 initially disease-free subjects who are followed over a 5- year period from 1990 to 1995.

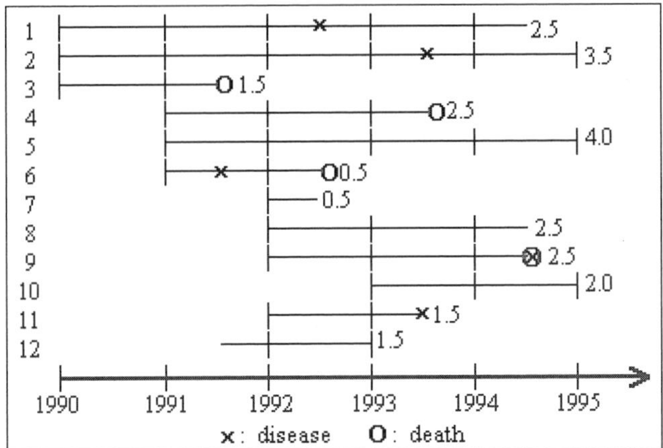

An **X** denotes the time at which a subject was diagnosed with the disease and a **circle** (**O**) denotes the time of death *that could be due to the disease (circle with an* **X** *inside) or due to another cause (circle without an* **X**). Those subjects that have no X or circle on their time line either withdrew from the study, or were lost to follow-up, or were followed until the end of the study without the developing the disease. The value to the right of each subject's time line denotes that subject's follow-up time period until either the disease was diagnosed, the subject withdrew or was lost to follow-up, or until the study ended. Based on this information, answer the following questions:

Study Questions (Q4.4)

The questions below refer to the figure above:

1. What type of cohort is being studied, **fixed** or **dynamic**?

2a. Which of these subjects was diagnosed with the disease?
Subject 2 **Subject 3** **Subject 5** **Subject 7**

2b. Which of these subjects was lost or withdrawn?
Subject 2 **Subject 3** **Subject 5** **Subject 7**

2c. Which of these subjects died with disease?
Subject 3 **Subject 5** **Subject 7** **Subject 9**

2d. Which of these subjects died with*out* the disease?
Subject 3 **Subject 5** **Subject 7** **Subject 9**

2e. Which one was with*out* the disease and alive at the end?
Subject 3 **Subject 5** **Subject 7** **Subject 9**

Study questions continued on next page

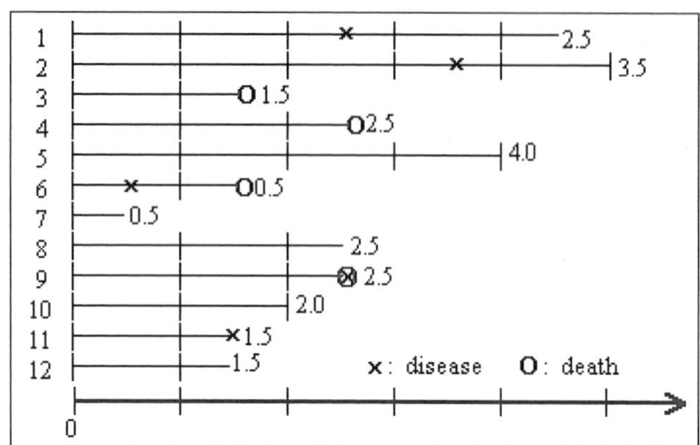

3. If we could shift the cohort, what is your estimate of simple cumulative incidence of disease diagnosis *in percent*?

4. What is your estimate of simple cumulative incidence of death *in percent with no decimal places*?

5. *Using the "unshifted" graph from the previous page*, Subjects 5, 8, 10 and 12 have which of the following in common:

   A. Same amount of observed follow-up time
   B. Entered study at same calendar time
   C. Withdrew from the study
   D. Did not develop disease during follow-up

Computing simple cumulative incidence for the previously shown data is a problem because ...

6a. Not all subjects developed the disease	**Yes**	**No**
6b. Not all subjects died	**Yes**	**No**
6c. The cohort is dynamic	**Yes**	**No**
6d. Some subjects died from another cause	**Yes**	**No**
6e. Some subjects were lost or withdrew	**Yes**	**No**
6f. Some subjects developed disease at different follow-up times	**Yes**	**No**
6g. Subjects not developing the disease had different follow-up times	**Yes**	**No**

Summary

- ❖ There are problems with assuming a fixed cohort when using the formula for simple cumulative incidence to estimate risk.
- ❖ If there is attrition of a fixed cohort, we will not know whether a subject lost during follow-up developed the disease.
- ❖ For a dynamic cohort, the denominator in the simple cumulative incidence formula does not reflect the continually changing population size
- ❖ Simple cumulative incidence does not allow subjects to be followed for different periods of time.

Quiz (Q4.5)

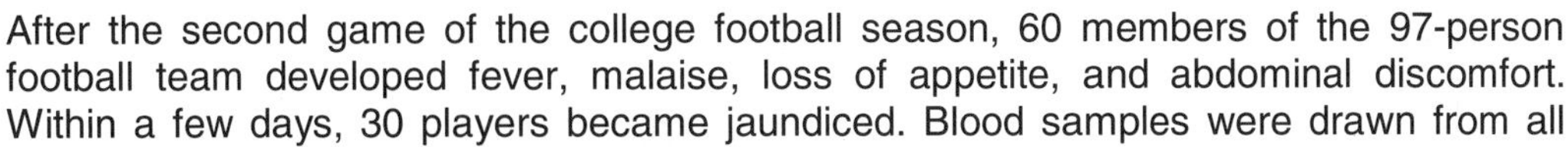

After the second game of the college football season, 60 members of the 97-person football team developed fever, malaise, loss of appetite, and abdominal discomfort. Within a few days, 30 players became jaundiced. Blood samples were drawn from all members of the team to test for antibodies to hepatitis A (the presumptive diagnosis) and to test for elevation of liver enzymes

1. What is the cumulative incidence of jaundice? . . . . . . **???** 30/97
2. If you assume that all persons with symptoms had hepatitis A, even those that did not develop jaundice, what is the presumed cumulative incidence of hepatitis A? . . . . . **???** 60/97
3. Laboratory testing revealed that 91 had elevated liver enzymes of which 90 had IgM antibody indicative of acute hepatitis A infection. Two players with normal liver enzymes had IgG antibody, indicating that they had previously been exposed to hepatitis A and are now immune. What is the cumulative incidence of hepatitis A? **???** 90/95

Choices

30/60	**30/97**	**60/97**	**90/91**	**90/95**	**90/97**	**91/95**	**91/97**

Label each of the following statements as **True** or **False**:

4. Cumulative incidence is always a proportion, even for a cohort with staggered entry ("shifted cohort"). . . . . **???** True
   the numerator is a subset of the denominator
5. Cumulative incidence is a useful measure for diseases with short incubation periods in well-defined populations. . . . . **???** True
   Because the incubation is short, subjects are not likely to be lost to follow up
6. Cumulative incidence is a less-than-ideal measure for diseases with long incubations periods in dynamic populations. . . . . **???** True
   Long incubation means subjects may be lost & not detected. For a dynamic cohort the denominator
7. If a fixed population has substantial loss-to-follow-up, cumulative incidence will overestimate the true risk of disease. . . . . **???** False
   The CI will underestimate

6. does not reflect the continually changing population.

4-3

Rate

***Rate** is a measure of disease frequency that describes how rapidly health events such as new diagnoses of cases or deaths are occurring in a population of interest. Synonyms:* ***hazard, incidence density**.*

Concept of a Rate

The concept of a rate is not as easily understood as risk, and is often confused with risk. Loosely speaking, a rate is a measure of how quickly something of interest happens. When we want to know how fast we are traveling in our car, how quickly the stock market prices are increasing, or how steadily the crime rate is decreasing, we are seeking a rate.

Suppose we are taking a trip in a car. We are driving along an expressway and we look at our speedometer and see we are going 65 miles per hour. Does this mean that we will cover exactly 65 miles in the next hour? Of course not. The speedometer reading tells us how fast we are traveling at the moment of time we looked at the reading. If we were able to drive exactly this way for the next hour without stopping for gas or a rest or slowing down for heavy traffic, we would cover 65 miles in the next hour. The reading of 65 miles per hour on our speedometer is the velocity at which we are traveling, and velocity is an example of a rate.

Actually, velocity is an example of an **instantaneous rate**, since it describes how fast we are traveling at a particular instant of time. There is another kind of rate, called an **average rate**, which we can also illustrate by continuing our car trip. If we actually traveled along the highway for the next hour and covered 55 miles during that time, the average rate, often called the speed that we traveled over the one-hour period, would be 55.

In epidemiology, we use a rate to measure how rapidly new cases of a disease are developing, or alternatively, how rapidly persons with a disease of interest are dying. As with velocity or speed, we might want to know either the instantaneous rate or the average rate. With epidemiologic data, it is typically easier to determine an average rate than an instantaneous rate. We could hardly expect to have a speedometer-like device that measures how fast a disease is occurring at a particular moment of time in a cohort of subjects. Consequently, in epidemiologic studies, we typically measure the average rate at which a disease is occurring over a period of time.

Because a rate is a measure of how quickly something is occurring, it is always measured in units of time, say, days, weeks, months, or years. This clarifies its interpretation. If we describe a rate of 50 new cases per 10,000 person-years, we mean that an average of 50 cases occurs for every 10,000 years of disease free follow-up time observed on a cohort of subjects. The 10,000 figure is obtained by adding together the follow-up times for all subjects in the cohort.

If the unit of time was *months* instead of *years*, the interpretation of the rate can be quite different. A rate of 50 new cases per 10,000 person months indicates a much quicker rate than 50 new cases per 10,000 person years.

Study Questions (Q4.6)

1. Which of the following rates is *not* equivalent to a rate of 50 new cases per 10,000 person years?

   A. 100 new cases per 20,000 person years
   B. 50 new cases per 120,000 person months
   C. 50 new cases per 52,000 person weeks

2. Determine whether or not each of the following statements describes a rate:

A.	5 new cases per 100 person days	**Yes**	**No**
B.	40 miles per hour	**Yes**	**No**
C.	10 new cases out of 100 disease-free persons	**Yes**	**No**
D.	60 new murders per year	**Yes**	**No**
E.	60 deaths out of 200 clung cancer patients	**Yes**	**No**

Summary

- ❖ Generally, a rate is a measure of how quickly something of interest is happening
- ❖ In epidemiology, a rate is a measure of how rapidly are new cases of a disease developing, or alternatively, how rapidly persons with a disease of interest are dying.
- ❖ An **instantaneous rate**, like velocity, describes how rapidly disease or death is occurring at a moment in time
- ❖ An **average rate**, like speed, describes how rapidly disease or death has been occurring as an average over a period of time.
- ❖ In epidemiology, we typically use average rates rather than instantaneous rates.
- ❖ Rates must be measured in units of time.

Incidence Density- The Concept

The term incidence density (ID) has been proposed (Miettinen OS, Am J Epidemiol 1976;103(2):226-35) to provide an intuitive interpretation of the concept of an **average incidence rate**.

The diagram below illustrates incidence density as the concentration (i.e. **density**) of new case occurrences in an accumulation (or *sea*) of person-time. **Person-time** (**PT**) is represented by the area under the curve **N(t)** that describes number of disease-free persons at time t during a period of follow-up from time T_0 to T_1.

Each new case is denoted by a small circle located within the sea of person-time at the time of disease occurrence. The concentration of circles within the sea represents the **density of cases**. The higher the concentration, the higher is the average rate during the period of follow-up.

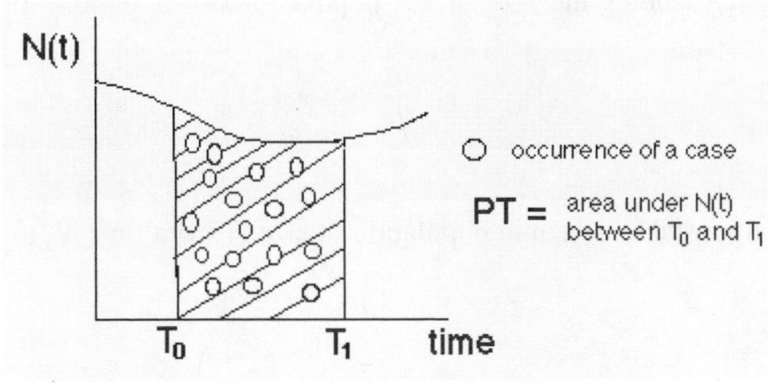

The Instantaneous Rate in Epidemiology

A variety of mathematical definitions have been used to define a **rate**. In epidemiology, where the incidence of a heath condition is of interest, the following definition is commonly used:

$$Rate = \lim_{\Delta t \to 0} \frac{\left[\text{\# of new cases in}(t,\, t + \Delta t)\right] / \Delta t}{N(t)}$$

where **t** denotes time, **Δt** denotes a small time change, and **N(t)** denotes the size of the **population-at-risk** (e.g., the disease-free cohort) at any given time at time **t**. This is a general definition of an **instantaneous rate**, and it applies to both fixed cohort and dynamic populations.

This definition can be interpreted as the **instantaneous potential at time t** (as defined by the limit statement) for the number of new cases that would develop between times t and t + Δt **per unit time relative** to the population-at-risk at time **t**.

A special feature of this definition is the involvement of **N(t)**, which is not required in other more popular uses of the term rate outside of epidemiology, as for example, when describing the velocity observed on a car's speedometer. Velocity denotes how much distance one would potentially cover per time period of travel at a particular instant of time. The distance covered corresponds to the number of new cases developed over a time period of length Δt. But there is no term similar to N(t) involved in the interpretation of velocity.

The Average Rate in Epidemiology

In epidemiology, where the incidence of a heath condition is of interest, the following general definition of average rate is commonly used:

$$\text{Average Rate} = \frac{[\text{\# of new cases in}(\mathbf{T_0}, \mathbf{T_1})]}{\mathbf{PT}}$$

where $\mathbf{T_0}$ and $\mathbf{T_1}$ denote the starting and ending time points of follow-up, and **PT** denotes the amount of disease-free person-time accumulated during the time interval from $\mathbf{T_0}$ to $\mathbf{T_1}$. Mathematically, PT gives the area under the curve that describes how N(t) changes over time between times $\mathbf{T_0}$ and $\mathbf{T_1}$, where N(t) denotes the size of the population-at-risk at time **t**. Technically, the formula for PT is given by:

$$PT = \int_{T_0}^{T_1} N(t)dt$$

As a simple example of the calculation of PT, if we are studying a stable dynamic population of size N from time $\mathbf{T_0}$ to $\mathbf{T_1}$, then

$$PT = N(T_1 - T_0)$$

Alternatively, if we are studying a fixed disease-free cohort of size N and know the individual observed follow-up times of each subject in the cohort, then

$$PT = \sum_{i=1}^{N} \Delta T_1$$

as i goes from 1 to N.

Calculation of a Rate

To calculate a rate, we must follow a cohort of subjects, count the number of new (or incident) cases, **I**, of a disease in that cohort, and compute the total time, called person-time or **PT**, that disease-free individuals in the cohort are observed over the study period. The estimated incidence rate ($\hat{\text{IR}}$) is obtained by dividing **I** by **PT**:

$$\hat{\text{IR}} = \frac{\text{I}}{\text{PT}}$$

This formula gives an average rate, rather than the more difficult to estimate instantaneous rate. The formula is general enough to be used for any outcome of interest, including death. If the outcome is death instead of disease incidence, the formula gives the **mortality incidence rate** rather than the **disease incidence rate**.

For example, consider again the following hypothetical cohort of 12 initially disease-free subjects followed over a 5-year period from 1990 to 1995.

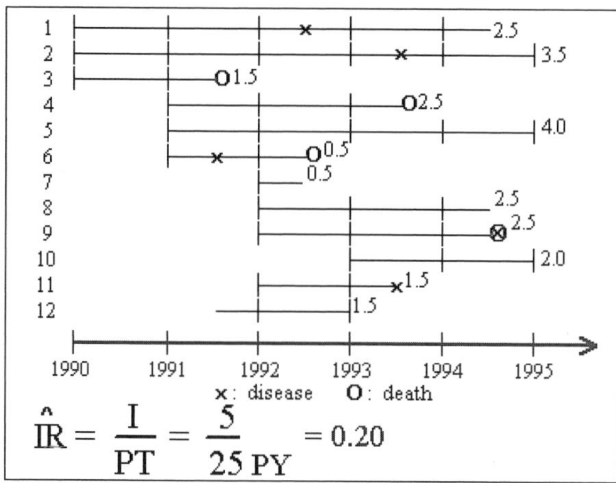

To be complete, the estimated incidence rate is 0.2 per person-year.

From these data, the number of new cases is 5. The total person-time, in this case person-years, is obtained by adding the individual observed disease-free follow-up times this gives a total of 25 person years. The rate is therefore 5 divided by 25 or 0.20, which can be translated as 20 new cases per 100 person years of follow-up.

Study Questions (Q4.7)

1. In this example, is the value of 0.20 a proportion?
2. In this example, does the value of 0.20 represent the risk of developing disease?
3. Which of the following rates is not equivalent to a rate of 20 new cases per 100 person years?
   A. 5 new cases per 25 years
   B. 40 new cases per 200 person years
   C. 480 new cases per 2400 person months
   D. 20 new cases per 1200 person months

Summary

- A rate is calculated using the formula **I/PT**, where **I** denotes the number of incident cases and **PT** denotes the accumulated person-time of observed follow-up over the study period.
- This formula gives an average rate, rather than the more difficult to estimate instantaneous rate.
- A rate is always greater than zero and has no upper bound.
- The rate is always stated in units of person-time.
- A rate of .20 cases per person year is equivalent to 20 cases per 100 person-years as well as 20 cases per 12,000 person-months

The Big-Mac Assumption about Person-Time

We have seen that the general formula for calculating an **average rate** (**R**) is:

$$R = \frac{I}{PT}$$

where **I** is the number of new cases and **PT** is the accumulated person-time over a specified period of follow-up. When individually observed follow-up times are available, PT is determined by summing these individual times together for all N subjects in the disease-free cohort.

For example, if 100 persons are each followed for 10 years, then PT=1000 person-years. Also, if 1000 persons are each followed for 1 year, we get PT=1000.

A key assumption about PT is that both of these situations provide equivalent person-time information. In other words, the rate corresponding to a specified value of PT should not be affected by how the total person-time is obtained. We call this assumption the **Big-Mac assumption** because it is similar to assuming that eating 50 fast-food hamburgers costing $2.00 each is equivalent to eating: $100 gourmet meal at the best-rated restaurant in town.

The Big-Mac assumption for PT will not hold, however, if the average time between first exposure and detection of the disease (i.e., the latency) is longer than the average individually observed follow-up time. In such a case, we would expect the rate to be lower in a large cohort that accumulates the same amount of PT as a smaller cohort with larger individual follow-up times.

For example, if the latency were 2 years, we would expect an extremely low rate for 1000 persons followed for one-year each but a much larger rate for 100 persons followed for two-years each. Individuals in the larger cohort would not be followed long enough to result in many new cases.

Determining Person Time Information

There are a number of ways to determine the person-time denominator in the formula for a rate. As illustrated in the previous activity, when individual follow-up times are available on each person in the cohort, the person-time is calculated by summing (Σ) individual follow-up times over the entire disease-free cohort.

$$\hat{IR} = \frac{I}{PT}$$

When individual follow-up times are not available, one method for computing person-time information uses the formula:

$$PT = N^* \times \Delta t$$

where N* is the average size of the disease-free cohort over the time period of study and Δt is the time length of the study period. This formula is particularly useful if the study cohort is a large population, such as a city, where individual person time information would be very difficult to obtain. For such a large cohort, it would also be difficult to exclude existing cases of the disease at the start of the study period as well as to determine the number of disease-free persons that are not followed for the entire period of study.

Nevertheless, it may be that relatively few persons in the population develop the disease. And, we may be able to assume that the population is a stable dynamic cohort, that is, the population undergoes no major demographic shifts during the time period of interest. If so, the average size of the disease free cohort can be estimated by the size of the entire population based on census data available close to the time period of the study, which is what we have denoted N* in our person-time formula.

As an example, suppose a stable population of 100,000 men is followed for a period of 5 years, during which time 500 new cases of bladder cancer are detected. The accumulated person-years for this cohort can then be estimated as 100,000 times 5, or 500,000 person-years. Consequently, the average incidence rate for the 5-year period is given by 500 divided by 500,000, or 0.001 per year, or equivalently 1 new case per 1000 person years.

Example

$$\hat{IR} = \frac{I}{PT} = \frac{500}{500{,}000} = 0.001 \text{ per year}$$

$$PT = N^* \times \Delta t$$
$$= 100{,}000 \times 5 = 500{,}000 \text{ person-years}$$

$N^* = 100{,}000$ men

$\Delta t = 5$ years

$I = 500$ cases

Summary

- ❖ There are alternative ways to determine person-time information required in the denominator of a rate when individual follow-up times are not available.
- ❖ One method uses the formula PT = N* x Δt, where N* denotes the average size of a stable dynamic cohort based on census data available close to the chronological time of the study, and Δt is the time period of the study.
- ❖ This formula is useful if the study cohort is a large population for which individual person time information would be difficult to obtain.

A Third Method for Determining PT

Here we describe another method for determining the accumulated person-time information (**PT**) when individual follow-up time information is not available. This approach allows for shifting the time of entry of persons who progressively enter the cohort after the start of the study. Assume that you know:

N, the number at risk at the start of follow-up,
W, the number of withdrawals during the study period,
D, the number of deaths from other diseases during the study period, and
I, the number of new cases of the disease during the study period.

The person-time information is then calculated using the formula:

$$PT = \left(N - \frac{W}{2} - \frac{D}{2} - \frac{I}{2}\right) \times \Delta t$$

where **Δt** denotes the time length of the study. This formula gives the size of the initial cohort less half the number of subjects that were not followed for the entire risk period.

This formula essentially gives the **effective number** of subjects at risk that would produce **I** new cases of the disease if all subjects could be followed for the entire period. The values **W/2**, **D/2**, and **I/2** are used to assume that the average follow-up time for those not followed for the entire study occurs at the midpoint of the follow-up period.

To illustrate this approach, consider once again the hypothetical example described in the previous activity involving 12 initially disease-free subjects that are followed over a 5 year period from 1990 to 1995: Suppose that you don't know individual follow-up times, but rather that out of the 12 disease-free subjects, 5 withdrew from the study, 2 died, and 5 were diagnosed with the disease. Here:

N=12 W=5 D=2 I=5 and Δt=5

Substituting these values into the formula, we compute PT to be 30. Since there were 5 new cases, the estimated rate then becomes 5 over 30 or .17 per person-year. This is not that far off from the estimated rate of .20 per person-year rate obtained if we use individual follow-up times to compute PT equal to the correct value of 25.

Incidence Rate of Parkinson's Disease

Parkinson's disease is a seriously disabling disease characterized by a resting tremor, rigidity, slow movements, and disturbed reflexes. A cohort of more than 6,500 Dutch elderly people who did not have Parkinson's disease at the start of the study was followed for six years to determine the incidence rate at which new cases of Parkinson's disease develop. During the follow-up period, 66 participants were diagnosed with Parkinson's disease.

Because Parkinson's disease has a subtle onset, it was difficult to determine exactly when the disease process had begun. Therefore, the investigators calculated the time of onset as the midpoint between the time of diagnosis and the time at which a participant was last known to be free of Parkinson's. They could then calculate the total number of disease-free person-years in this study by adding up the number of person-years that each of the 6,500 participants had contributed to the study until he or she either:

1. Developed Parkinson's disease
2. Died
3. Reached the end of the study period alive without having developed Parkinson's disease.

This resulted in a total of 38,458 disease-free person-years. In this study, the average incidence rate of Parkinson's disease for the 6-year study period is:

66 / 38,458 = 0.0017 cases per person-year

This means that, 1.7 new cases of Parkinson's disease develop per 1,000 person-years.

Study Questions (Q4.8)

1. Using the formula PT = N* x (Δt), how many person-years would have been computed for this study population had no detailed information on each individual's contribution to the total amount of person-years been available?
2. Using the number of person-years from the previous question, what is the incidence rate?

1. N* is the average size of the cohort & Δt is the length of the study time so 6,500 x 6 = 39,000 person-years
2. 66/39,000 = 0.0017 or 1.7 per 1,000 person years

Summary

- ❖ A cohort of more than 6,500 Dutch elderly people who did not have Parkinson's disease at the start of the study was followed for six years to determine the rate at which new cases develop.
- ❖ The results indicate that 1.7 new cases of Parkinson's disease develop for every 1,000 person-years of follow-up.
- ❖ The person-years calculation used the formula PT = N* x Δt since there was no detailed information on each individual's person-years.

Quiz (Q4.9)

Label each of the following statements as **True** or **False**.

1. Rate is not a proportion. . . . . ??? True
2. Rate has units of 1/person-time, and varies from zero to one. . . . ??? False
3. A rate can only be calculated if every person in a cohort is followed individually to count and add up the person-time. . . . . ??? False
4. Rate can be calculated for a dynamic cohort, but not for a fixed, stable cohort. . ??? False

1&2 Rate can range from 0 to infinity whereas a risk can range from 0 to 1
3. There are alternative methods
4 A rate can be calculated from either a dynamic or fixed cohort depending on person-time information available.

Risk versus Rate

Incidence can be measured as either ***risk*** *or* ***rate****. Which of these types to use is an important choice when planning an epidemiologic study.*

We have seen two distinct measures for quantifying disease frequency -risk and rate. **Risk** is a probability, lying between 0 and 1 that gives the likelihood of a change in health status for an individual over a specified period of follow-up.

$0 \leq \text{Risk} \leq 1$

Rate describes how rapidly new events are occurring in a population. An **instantaneous rate**, which is rarely calculated, applies to a fixed point in time whereas an **average rate** applies to a period of time. A rate is *not* a probability, is always non-negative but has no upper bound, and is defined in units of time, such as years, months, or days.

$0 \leq \text{Rate} \leq \infty$

When planning an epidemiologic study, which measure do we want to use, risk or rate? The choice depends on the objective of the study, the type of disease condition being considered, the nature of the population of interest, and the information available.

If the study objective is to predict a change in health status for an individual, then risk is required. In particular, risk is relevant for assessing the prognosis of a patient, for selecting an appropriate treatment strategy, and for making personal decisions about health-related behaviors such as smoking, exercise, and diet. By contrast, a rate has no useful interpretation at the individual level.

If the study objective is to test a specific hypothesis about disease etiology, the choice can be either risk or rate depending on the nature of the disease and the way we observe new cases. If the disease is a chronic disease that requires a long period of follow-up to obtain sufficient case numbers, there will typically be considerable loss to follow-up or withdrawals from the study. Consequently, individual observed follow-up times tend to vary considerably. A rate, rather than a risk, can address this problem.

However, if an acute disease is considered, such as an outbreak due to an infectious agent, there is likely to be minimal loss to follow-up, so that risk can be estimated directly. With an acute illness, we are not so much interested in how rapidly the disease is occurring, since the study period is relatively short. Rather, we are interested in identifying the source factor chiefly responsible for increasing individual risk.

If the population being studied is a large dynamic population, individual follow-up times, whether obtainable or not, will vary considerably for different subjects, so rate must be preferred to risk. However, if individual follow-up times are not available, even a rate cannot be estimated unless it is assumed that the population size is stable, the disease is rare, and a recent census estimate of the population is available.

Risk is often preferred to rate because it is easier to interpret. Nevertheless, rate must often be the measure of choice because of the problems associated with estimating risk.

Summary

- ❖ Risk is the probability that an individual will develop a given disease over a specified follow-up period.
- ❖ Rate describes how rapidly new events are occurring in a population.
- ❖ Risk must be between 0 and 1 whereas rate is always non-negative with no upper bound, and is defined in units of time.
- ❖ Risk is often preferred to rate because it is easier to interpret.
- ❖ Rate must often be the measure of choice because of problems with estimating risk.
- ❖ The choice of risk versus rate depends on the study objective, the type of disease, the type of population, and the information available.

Mortality might be used instead of Disease Incidence

As with incidence measures of disease frequency, incidence measures of mortality frequency can take the form of risk or rate depending on the study design and the study goals. Mortality measures are described in a later activity (on page 4-4).

Quiz (Q4.10)

Determine whether the following statements best define a **rate**, **risk**, or **both**:

1. More useful for individual decision-making. . . . ??? Risk
2. Numerator is number of new cases during a period of follow-up. . ??? Both
3. Lowest possible value is zero. . . . . . ??? Both
4. No upper bound. . . . . . . ??? Rate
5. Can be expressed as a percentage. . . . . ??? Risk
6. Better for studies with variable periods of follow-up. . . ??? Rate
7. Traditionally calculated in the acute outbreak (short follow-up) setting. ??? Risk
8. Measures how quickly illness or death occurs in a population. . ??? Rate
9. Cumulative incidence. . . . . . . ??? Risk
10. Measure of disease occurrence in a population. . . . ??? Both

4-4 Prevalence and Mortality

Prevalence

Prevalence measures existing cases of a health condition and is the primary design feature of a ***cross-sectional study****. There are two types of prevalence,* ***point prevalence****, which is most commonly used, and* ***period prevalence****.*

In epidemiology, prevalence typically concerns the identification of existing cases of a disease in a population and is the primary design feature of cross-sectional studies. Prevalence can also more broadly concern identifying persons with any characteristic of interest, not necessarily a disease. For example, we may wish to consider the prevalence of smoking, immunity status, or high cholesterol in a population.

The most common measure of prevalence is **point prevalence**, which is defined as the probability that an individual in a population is a case at time **t**.

$$\hat{P} = \frac{C}{N} = \frac{(\#\text{ of observed cases at time t})}{(\text{Population size at time t})}$$

Point prevalence is estimated as the proportion of persons in a study population that have a disease at a particular point in time (**C**). For example, if there are 150 individuals in a population and, on a certain day, 15 are ill with the flu, the estimated prevalence for this population is 10%.

$$\hat{P} = \frac{15}{150} = 10\%$$

Study Questions (Q4.11)

1. Is point prevalence a proportion? yes, it is a value from 0 to 1 often expressed as a percentage

2. A study with a large denominator, or one involving rare events, may result in very low prevalence. For example, suppose that 13 people from a population of size 406,245 had a particular disease at time **t**. What is the point prevalence of this disease at time **t**?

   A. 0.0032 C
   B. 32%
   C. 0.000032
   D. 0.0000032

3. Which of the following expressions is equivalent to the point prevalence estimate of 0.000032?

   A. 3.2 per 1,000 B
   B. 3.2 per 100,000
   C. 32 per 100,000

When measuring point prevalence, it is essential to indicate *when* the cases were enumerated by specifying a point calendar time or a fixed point in a time sequence, such as the third post-operative day. Prevalence measures are very useful for assessing the health status of a population and for planning health services. This is because the number of existing cases at any time is a determinant of the demand for healthcare.

However, prevalence measures are not as well suited as incidence measures, such as risk or rate, for identifying risk factors. This is because prevalence concerns only survivors, so that cases that died prior to the time that prevalence is measured are ignored.

PREVALENCE
Useful for: Assessing the health status of a population. Planning health services.
Not useful for: Identifying risk factors.

Summary

- Prevalence concerns existing cases of a disease at a point or period of time.
- Prevalence measures are primarily estimated from cross-sectional surveys.
- Point prevalence is the probability that an individual in a population is a case at time **t**.
- Point prevalence is estimated using the formula **P = C/N**, where **C** is the number of existing cases at time **t**, and **N** is the size of the population at time **t**.
- Prevalence measures are useful for assessing the health status of a population and for planning health services.
- Prevalence measures concern survivors, so they are not well suited for identifying risk factors.

Period Prevalence

An alternative measure to **point prevalence** is **period prevalence (PP)**, which requires the assumption of a stable dynamic population for estimation. **PP** is estimated as the ratio of the number of persons **C*** who were observed to have the health condition (e.g., disease) anytime during a specified follow-up period, say from times T_0 to T_1, to the size **N** of the population for this same period, i.e., the formula for period prevalence is:

Continued on next page

Period Prevalence (continued)

$$PP = \frac{C^*}{N} = \frac{C+I}{N}$$

where **C** denotes the number of prevalent cases at time $\mathbf{T_0}$ and **I** denotes the number of incident cases that develop during the period. For example, if we followed a population of 150 persons for one year, and 25 had a disease of interest at the start of follow-up and another 15 new cases developed during the year, the period prevalence for the year would be:

PP = (25 + 15)/ 150 = .27, or 27%,

whereas the estimated point prevalence at the start of the period is:

P = 25/150 = .17, or 17%

and the estimated cumulative incidence for the one year period is:

CI = 15/125 = .12, or 12%

<u>**Quiz (Q4.12)**</u>

Label each of the following statements as **True** or **False**

1. Prevalence is a more useful measure for health planning than for etiologic research. . **???**
True. Prevalence considers existing cases rather than incident cases
2. Like cumulative incidence, prevalence is a proportion that may range from zero to one. . **???**
True
3. Prevalence measures are most commonly derived from follow-up studies. . . **???**
False, Cross sec. studies are for short time
4. Whereas incidence usually refers to occurrence of illness, injury, or death, prevalence may refer to illness, disability, behaviors, exposures, and genetic risk factors. . . . . **???**
True Prevalence may concern a health outcome or any other characteristic of a subject

Select the correct answer:

5. The formula for point prevalence is: d

    a. # new cases / # persons in population
    b. # new cases / # persons who did not have the disease at the starting point of observation
    c. # new cases / # person-time of follow-up
    d. # current cases / # persons in population
    e. # current cases / # persons who did not have the disease at the starting point of observation
    f. # current cases / # person-time of follow-up

Mortality

As with incidence measures of disease frequency, incidence measures of mortality frequency can take the form of **risk** or **rate** depending on the study design and the study goals. Mortality risk can be measured in a number of ways, including **disease-specific mortality risk**, **all-causes mortality risk**, and **case-fatality risk**. For each measure, the formula for simple cumulative incidence can be used. Here, **I** denotes the number of deaths observed over a specific study period in an initial cohort of size **N.**

Mortality Risk

Disease-Specific Mortality Risk
All-Causes Mortality risk
Case-Fatality Risk

$$CI = \frac{I \text{ (number of deaths)}}{N \text{ (size of cohort)}}$$

Study Questions (Q4.13)

1. For a **disease-specific mortality risk**, what does the **I** in the formula **CI=I/N** represent. **???**

   A. The number of deaths from all causes
   B. The number of deaths due to the specific disease of interest
   C. The number of persons with a specific disease
   D. The size of the initial cohort regardless of disease status

For estimating disease-specific mortality risk, **I** is the number of deaths due to the specific disease of interest, and **N** is the size of the initial cohort regardless of disease status.

Study Questions (Q4.13) continued

2. For **all-causes mortality risk**, what does the **I** in the formula **CI=I/N** represent. **???**

   A. The number of deaths from all causes
   B. The number of deaths due to the specific disease of interest
   C. The number of persons with a specific disease
   D. The size of the initial cohort regardless of disease status

I is the number of deaths from all causes, and **N** is the size of the initial cohort, regardless of disease status.

Case-fatality risk is the proportion of people with a particular disease who die from that disease during the study period.

Study Questions (Q4.13) continued

3. For *case-fatality* risk, what does the **I** in the formula **CI=I/N** represent. **???**

   A. The number of deaths from all causes
   B. The number of deaths due to the specific disease of interest
   C. The number of persons with a specific disease
   D. The size of the initial cohort regardless of disease status

I is the number of persons who die from the given disease, and **N** is the number of persons with this disease in the initial cohort.

Similarly, **mortality rate** can be measured using the general formula for average rate.

$$IR = \frac{I}{PT}$$

Here, **I** denotes the number of deaths observed over a specified study period in an initial cohort that accumulates person-time **PT**. For estimating disease-specific mortality rate, **PT** is the person-time for the initial cohort, regardless of disease condition. For estimating **all-cause mortality rate**, **PT** again is the person-time for initial cohort, regardless of disease condition.

For estimating **case-fatality rate**, **PT** is the person time for an initial cohort of persons *with the specific disease of interest* that is followed to observe mortality status.

As an example of the calculation of mortality risk estimates, suppose you observe an initial cohort of 1000 persons aged 65 or older for three years. One hundred out of the 1000 had lung cancer at the start of follow-up, and 40 out of these 100 died from their lung cancer. In addition, 15 persons developed lung cancer during the follow-up period and 10 died. Of the remaining 885 persons without lung cancer, 150 also died.

Mortality

	Start	Follow-up	Died
1000 persons Age: 65+ 3 years	100 LC		40
	900 no LC	15 LC	10
		885 no LC	150
Total	1000		200

$$\text{Lung-cancer specific mortality risk} = \frac{40+10}{1000} = 5\%$$

$$\text{All-cause mortality risk} = \frac{200}{1000} = 20\%$$

$$\text{Case-fatality risk} = \frac{40}{100} = 40\%$$

- The lung-cancer specific mortality risk for this cohort is 50/1000 or 5%.
- The all-cause mortality risk is 200/1000 or 20%, and
- The case-fatality risk for the 100 lung cancer patients in the initial cohort is 40/100 or 40%.

Study Questions (Q4.14)

For the lung cancer example just presented, answer the following questions:

1. From the data, what is the estimated risk for the incidence of lung cancer over the three-year period?
2. Why is the estimated incidence of lung cancer (LC) different from the estimated LC mortality of 5%?
3. Under what circumstances would you expect the LC incidence and LC mortality risk to be approximately equal?

1. 15/900 = .017 or 1.7%
2. The 5% mortality estimate counts the 40 prevalent LC cases & does not count the 5 new LC cases that didn't die. Futhermore the denominators are different.

Summary

- ❖ Incidence measures of mortality frequency can take the form of risk or rate depending on the study design and the study goals.
- ❖ Mortality risk or rate can be measured in a number of ways, including disease-specific mortality risk or rate, all-causes mortality risk or rate, and case-fatality risk or rate.
- ❖ For measuring mortality risk, the formula used for simple cumulative incidence, namely, CI = I / N, can be used.
- ❖ Similarly, mortality rate can be measured using the general formula for average rate, namely IR = I / PT.

3. If the disease is quickly fatal so there would be few if any prevalent cases & all new cases would have died before the end of followup.

Quiz (Q4.15)

During the past two years, a total of exactly 2,000 residents died in a retirement community with a stable, dynamic population of 10,000 persons.

1. Given these data, the best choice for measure of mortality is the mortality rate. . . **???** T
2. Since mortality is often expressed per 1,000, one could express this mortality measure as 200 per 1,000 per year. . . . . **???** T
3. The disease-specific mortality risk is the number of deaths attributable to a particular disease, divided by the number of persons with that disease. . . . . **???** F
   The denominator is the size of the cohort regardless of disease status.
4. The denominator for the all-cause mortality risk and the cause-specific mortality risk is the same. **???** T
5. The denominator for case-fatality risk is the numerator of the prevalence of the disease. . **???** T

Age-adjusted rate

Most epidemiologic studies involve a comparison of measures of disease frequency among two or more groups. For example, to study the effect of climate conditions on mortality, we might compare mortality risks or rates in two or more locations with different climates. Let's focus on two U.S. states, Arizona and Alaska. This would allow a comparison of mortality in a cold, damp climate with mortality in a hot dry climate.

The crude mortality rates for these two states for the year 1996 were:

Alaska	426.57 deaths per 100,000 population
Arizona	824.21 deaths per 100,000 population

You might be surprised, particularly considering the climates of the two states, that Arizona's death rate is almost twice as high as Alaska's. Does that mean that it's far more hazardous to live in Arizona than Alaska?

Study Questions (Q4.16)

1. What do you think? Is if far more hazardous to live in Arizona than Alaska? This is the crude rate for the entire population. It would be premature to make a conclusion.

A little knowledge of the demographic make-up of these two states might cause you to question such an interpretation. Look at the age distribution of the two states:

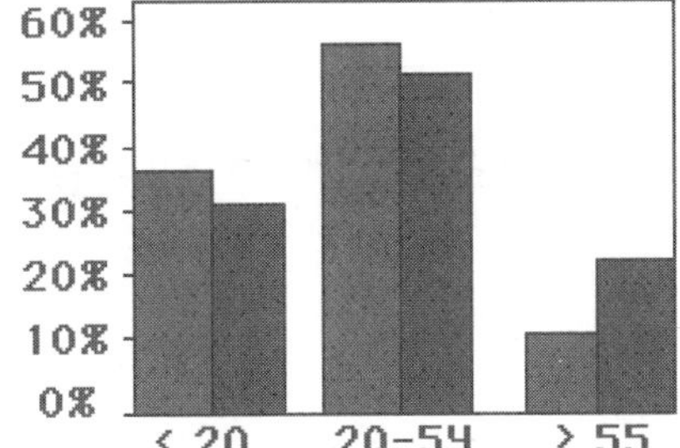

(Note: Alaska is the left bar for each of the clustered bars, Arizona the right bar.)

Study Questions (Q4.16) continued

2. Which population is older?
3. Why should we expect relatively more deaths in Arizona than in Alaska?

The variable age in this situation is called a **confounder** because it distorts the comparison of interest. We should correct for such a potentially misleading effect. One popular method for making such a correction is **rate adjustment**. If the confounding factor is age, this method is generally called **age-adjustment**, and the corrected rates are called **age-adjusted rates**.

The goal of age adjustment is to modify the crude rates so that any difference in mortality rates of Alaska and Arizona cannot be explained by the age differences in the two states. The most popular method of rate adjustment is the **direct method**. This method forces the comparison of the two populations to be made on a **common age distribution**. The confounding factor age is removed by re-computing the rates substituting a common age distribution for the separate age distributions. The two populations are then compared as if they had the same age structure.

The common age distribution is determined by identifying a **standard population**. A logical choice here would be the 1996 total United States population. Other choices for the standard are also possible and usually won't make a meaningful difference in the comparison of adjusted rates.

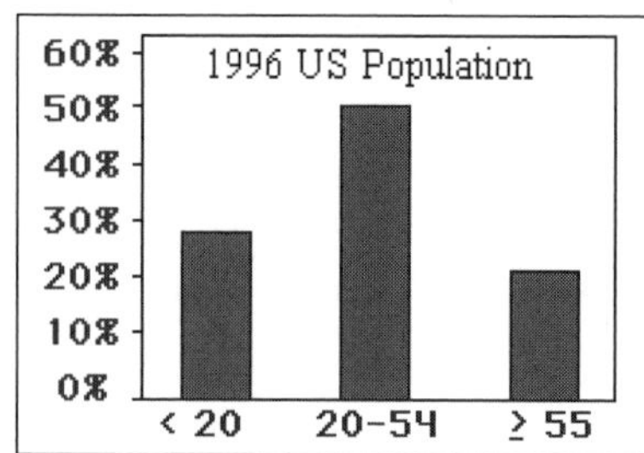

(Note: The actual calculation of the age-adjusted rates is not shown here. For details on the calculation of the age-adjusted rates for this example, on the CD click on the asterisk on the lesson page or see the example at the end of this activity.)

The age-adjusted death rates obtained from using the direct adjustment method with the 1996 US population as the standard are shown here together with the crude rates:

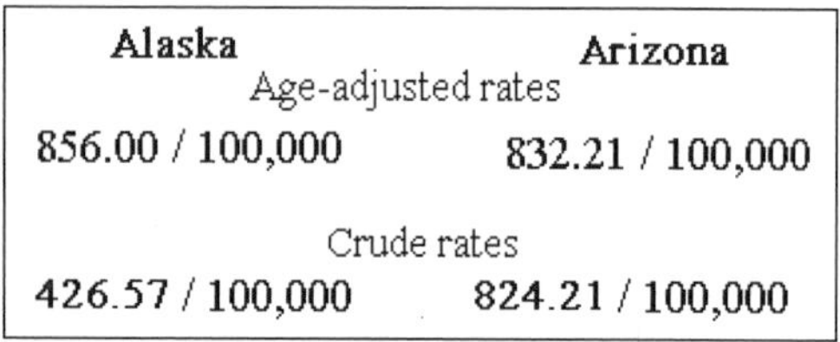

	Alaska	Arizona
Age-adjusted rates	856.00 / 100,000	832.21 / 100,000
Crude rates	426.57 / 100,000	824.21 / 100,000

When we remove age as a factor, the age-adjusted death rate in Arizona is actually lower than in Alaska.

Study Questions (Q4.17)

1. How do we interpret these new age-adjusted results?
2. Based on these results, how do you think the age distribution of Alaska compares to that of the 1996 US population?
3. How do you think the age distribution of Arizona compares to that of the 1996 US population?

Summary

❖ Comparing crude rates for two or more groups may be misleading because such rates do not account for the effects of confounding factors.

Continued on next page

- ❖ If the confounding factor is age, this method is called age-adjustment, and the corrected rates are called age-adjusted rates.
- ❖ The goal of age adjustment is to modify the crude rates so that any difference in rates cannot be explained by age distribution of the comparison groups.
- ❖ The direct method of age-adjustment re-computes the rates by substituting a common age distribution for the separate age distributions of the groups being compared.
- ❖ The common age distribution is determined by identifying a standard population.

Terminology about Adjustment

The rates described for Alaska and Arizona are actually **risks**. We have purposely used the term **rates** in this example to conform to the terminology typically used in published reports/papers that carry out **age adjustment**. In any case, the procedure used for (age) adjustment can be applied to any measure of disease frequency: risk, rate and/or prevalence.

Moreover, potential confounding factors of interest other than age, e.g., race and sex, can also be adjusted, both individually and simultaneously. We generally use the term **rate adjustment** to describe adjustment involving any type or number of confounding factors and any type of measure of disease frequency, whether a risk, rate, or prevalence.

Age-Adjustment – A Worked Example

The method of direct **age-adjustment** involves the following steps:

1. Select a **standard population** whose age structure is known. By convention, the standard distribution used for age-adjustment of mortality rates in the United States is the US age distribution in the year closest to the year of the rates being compared.
2. Multiply the age-specific mortality rates for each group being compared by the corresponding age-specific numbers of persons in the standard population. The result is the expected number of deaths in each group.
3. Sum the expected numbers of deaths within each age group to yield a total number of expected deaths for each group being compared.
4. Divide the total number of expected deaths in each group by the total size of the standard population to yield summary **age-adjusted mortality rates**.

We illustrate direct adjustment by comparing mortality rates for Alaska and Arizona in 1996. The age-specific rates and overall crude rates for these two states are given as follows:

Population size, all-cause mortality, and mortality rates by age, Alaska and Arizona, 1996.

	Alaska 1996			**Arizona 1996**		
Age	**n**	**d**	**r**	**n**	**d**	**r**
<1	10037	72	717.34	75322	575	763.38
1-4	40445	18	44.50	290256	127	43.75
5-9	54359	12	22.07	344886	67	19.42
10-14	52437	14	26.69	328220	95	28.94
15-19	49475	53	107.12	313422	322	102.73
20-24	44690	60	134.25	294762	372	126.20
25-34	84864	137	161.43	657439	1022	155.45
35-44	116015	238	205.14	684967	1700	248.18
45-54	81857	306	373.82	509569	2271	445.67
55-64	40162	359	893.87	347841	3632	1044.15
65-74	20668	518	2506.28	333235	7639	2292.67
75-84	8337	509	6105.31	199416	10494	5262.36
85+	1947	286	14689.26	55929	8240	14731.96
Total/Crude	605285	2582	426.57	4435264	36556	824.21

n = # of persons, **d** = # of deaths, and **r** = (d/n) x 100,000, i.e., deaths per 100,000 persons)

Continued on next page

Age-Adjustment – A Worked Example (continued)

Step 1. The age distribution of the standard population (1996 US) is given next together with age distribution percentages for Alaska and Arizona. Notice that the age distribution for Arizona is quite similar to the age distribution of the US, whereas Alaska's age distribution is somewhat different, with a small percentage at older ages.

US, Alaska, and Arizona age distributions, 1996.

	US 1996		**Alaska 1996**	**Arizona 1996**
Age	n	%	%	%
<1	3891494	1.5	1.6	1.6
1-4	15516482	5.8	6.7	6.5
5-9	19441182	7.3	9.9	7.8
10-14	18981045	7.2	8.7	7.4
15-19	18662151	7.0	8.1	7.1
20-24	17559730	6.6	7.4	6.6
25-34	40368234	15.2	14.0	14.8
35-44	43393341	16.4	19.2	15.4
45-54	32369791	12.2	13.5	11.4
55-64	21361460	8.5	6.6	7.8
65-74	18669337	7.0	3.4	7.5
75-84	11429984	4.3	1.4	4.5
85+	3761561	1.4	0.3	1.3
Crude/Total	265405792	100.0	100.0	100.0

The basic idea in computing a directly adjusted rate for a given state, say Alaska, is to compute what the hypothetical crude rate would be for Alaska if it had the same age structure as the standard population (US in 1996). Since neither Alaska nor Arizona actually have the same age structure as the US, their adjusted rates (using the US as the standard) are actually hypothetical, but they are now at least comparable, because the same standard is used for both states.

Step 2. The expected number of deaths for a given age group in Alaska is obtained by multiplying the size of standard population for that age group by the age-specific death rate for Alaska. For example, for the age group 25-34, we must multiply 40368234 (i.e., US population for ages 25-34) by 161.43/100,000 (i.e., the death rate in Alaska for ages 25-34), which gives 65,166.4 expected deaths in Alaska for this age group. The corresponding expected number of deaths for Arizona is computed by multiplying 40368234 by 155.45/100,000 (i.e., the death rate in Arizona for ages 25-34), yielding 62,752.4 expected deaths in Arizona for this age group.

Step 3. We must then sum the expected numbers of deaths over all age groups separately for Alaska and Arizona to yield a total number of expected deaths for each state. Without showing the calculations, these summed expected values are 2271873.6 for Alaska and 2208441.6 for Arizona.

Step 4. Finally, for each state separately, we divide the total expected numbers of deaths by the total size of the standard population (i.e., 265405792) to get the adjusted rates for each state. We thus obtain (2271873.6/265405792) = 856.0/100,000 for Alaska and (2208441/265405792) = 832.1/100,000 for Arizona.

Summarizing the crude and adjusted rates (per 100,000 persons) for each state, we see the following:

	Crude	**Adjusted**
Alaska	426.6	856.0
Arizona	824.2	832.1

The adjusted rate for Alaska is higher than the adjusted rate for Arizona, whereas the crude rate for Alaska was less than have the crude rate for Arizona!

Quiz (Q4.18)

Label each of the following statements as **True** or **False**.

1. Age-adjustment is a method to eliminate disparities in age between two populations. . **???**
2. Age-adjustment always brings two disparate rates closer together. . . . **???**
3. When age-adjusting, one should use the U.S. population as the standard population. . **???**
4. Age-adjustment can be used for one rate, two rates, or many rates. . . **???**
5. If the age distributions of two populations are very similar, their age-adjusted rates will also be similar. . . . . **???**
6. If the age distributions of two populations are very similar, the comparison of the age-adjusted rates will not be very different from the comparison of the crude rates. . . . . **???**

In the early 1990s, 7,983 elderly Dutch men and women were included in a prospective cohort study. The investigators computed how many person-years each participant had contributed to the study until January 2000. The total was 52,137 person-years. During follow-up, 2,294 of the participants died, and of these, 477 were due to coronary heart disease.

7. What is the all-cause mortality rate in this population? **???** per 1000 person-years.
8. What is the coronary heart disease-specific mortality rate? **???** per 1000 person-years

Choices
2294 **44** **477** **9.1**

The crude all-cause mortality rate for men was 47.4 per 1000 person-years (PY) and for women was 41.9 per 1000 person-years. After making the age distribution in the women comparable to the age distribution in men (by standardizing the rates using the age distribution of the men), the mortality rate for women was only 27.8 per 1000 PY.

9. Based on these figures, the women must be considerably **???** than the men in this population.

Choices
older **younger**

Analyzing Person-Time Data in Data Desk

On the ActivEpi CD ROM, there is an activity that describes how to perform analyses with person-time data using the Data Desk program.

Nomenclature

C	Number of prevalent cases at time T
C*	C + I (number of prevalence cases at time T plus incident cases during study period)
CI	Cumulative incidence ("risk"): CI=I/N
D	Duration of disease
I	Incidence
IR	Incidence rate ("rate"): IR=I/PT
N	Size of population under study
P	Prevalence: P=C/N
PP	Period prevalence: PP=C*/N
PT	Person-time
R	Average rate
T or **t**	Time

References

General References

Greenberg RS, Daniels SR, Flanders WD, Eley JW, Boring JR. Medical Epidemiology (3rd Ed). Lange Medical Books, New York, 2001.

Kleinbaum DG, Kupper LL, Morgenstern H. Epidemiologic Research: Principles and Quantitative Methods. John Wiley and Sons Publishers, New York, 1982.

Ulm K. A simple method to calculate the confidence interval of a standardized mortality ratio (SMR). Am J Epidemiol 1990;131(2):373-5.

References on Rates

Miettinen O. Estimability and estimation in case-referent studies. Am J Epidemiol 1976;103(2):226-35

Giesbergen PCLM, de Rijk MC, van Swieten JC, et al. Incidence of parkinsonism and parkinson's disease in a general population: the Rotterdam Study. Am J Epidemiol (in press).

References on Age-Adjustment

Dawson-Saunders B, Trapp RG. Basic and Clinical Biostatistics, 2nd ed., Appleton and Lange, Stamford, CN, 1994.

Woodward M. Epidemiology: Study Design and Analysis, Chapter 4, pp. 157-167, Chapman and Hall, Boca Raton, FL, 1999.

Reference for Prevalence and Incidence of HIV:

Horsburgh CR, Jarvis JQ, McArthur T, Ignacio T, Stock P. Serconversion to human immunodeficiency virus in prison inmates. Am J Public Hlth 1990;80(2):209-210.

Homework Questions

ACE-1. Measures of Disease Frequency

1. What is the purpose of a measure of disease frequency?
2. What is the difference between incidence and prevalence?
3. How are incidence and prevalence interrelated?
4. What is the difference between cumulative incidence and incidence density?
5. What does it mean to say that a person's 2-year risk is .03?
6. What does it mean to say that the rate in a certain population is .03/year?
7. Under what (design) circumstances would you want to measure risk?
8. Under what (design) circumstances would you want to measure rate?
9. Why do we carry out age-adjustment of risks or rates?
10. How does the direct method of age-adjustment work?

ACE-2. Person-time

What are two ways to calculate person-time in the estimation of a rate (i.e., incidence density)? Under what circumstances would you use each formula? Describe an example of the use of each formula.

ACE-3 Incidence vs. Prevalence

Determine whether each if the following statements requires measurement of INCIDENCE or PREVALENCE.

a. A new oral vaccine, which is purported to prevent cholera, has been introduced into a certain health district. The district health officer wants to monitor an appropriate measure to determine whether the vaccine is working.
b. A school psychologist wants to determine if there is an association between the reading of pornographic materials and teenage sexual violence. She is able to collect interview data on the amount of pornography regularly read and the number of violent sexual encounters experienced by the students.
c. An HMO (Health Maintenance Organization) is considering offering a community-oriented diabetic clinic. It will be necessary to determine how many patients would be interested in utilizing the service.
d. A pharmaceutical company has developed a new drug that is purported to cure asthma. The company wants to monitor the product's effectiveness.
e. A nurse-midwife decides to examine the relationship between home deliveries and post–partum infection. She is able to follow a group of women through the pregnancy and the first week after the birth of their children.
f. Quaker Oats has an ad campaign claiming that a diet high in grains helps prevent colon cancer. An epidemiologist wants to evaluate the validity of this claim.
g. A company is considering a new worksite smoking cessation program. A questionnaire is distributed among employees to determine how many people would be interested in taking part in such a program.
h. School administrators are informed that the school system in a given state is obligated by law to provide Special Education classes for all public school children with learning disabilities. The board wants to estimate how many Special Education teachers will need to be hired in order to meet this obligation.
i. An investigator is interested in assessing whether pregnant women exposed to environmental tobacco smoke are more likely to deliver low birth-weight babies.

ACE-4. Incidence and Prevalence: HIV

A study published in 1990 (**Amer. J. Pub. Health 80**:pp209-10) investigated the occurrence of HIV infection among prisoners in Nevada. Of 1105 prison inmates who were tested for HIV upon admission to the prison system, 36 were found to be infected. All uninfected prisoners were followed for a total of 1207 person-years and retested for HIV upon release from prison. Two of the uninfected inmates demonstrated evidence of new HIV infection. Assuming that the 2 prisoners were infected during their time in prison:

a. Based on the above information, calculate the incidence rate of HIV infection among prisoners in the Nevada prisons.
b. Express the incidence rate calculated in part a in terms of cases per 1000 person-years.

c. Why can't you obtain an estimate of risk based on the information provided?
d. Why would estimating risk likely be inappropriate for these data?
e. Calculate the prevalence of HIV infection among incoming prisoners in the Nevada prisoners under study.
f. Why is the estimate of prevalence calculated in part e not necessarily equal to an appropriate measure of risk that might be calculated for these data?

ACE-5. Interpreting Incidence and Prevalence

The following graph indicates the changing incidence rate and prevalence for disease "X" over time:

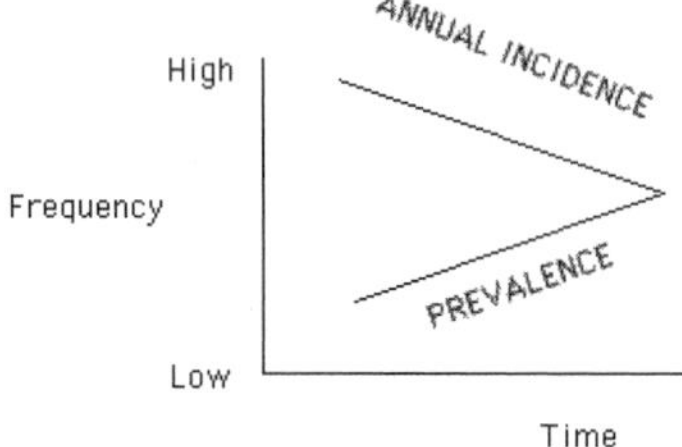

For each statement below, indicate whether the statement is consistent (yes or no) with the information portrayed in the graph.

a. Persons acquiring this disease are being cured quicker.
b. Efforts to prevent this disease appear to be succeeding.
c. The disease is becoming more chronic over time.

ACE-6. Calculate Measures of Disease Frequency

The following data were obtained in a study in which 1000 nurses were followed for 20 years to examine the hypothesis that use of a certain diet pill is a risk factor for heart attack.

	Diet Pill Use		
Heart Attack	Yes	No	Total
Yes	30	11	41
No	470	489	959
Total	500	500	1000

a. Estimate the number of woman-years contributed by the unexposed group.
b. What information would you need in order to obtain a better estimate of the number of women-years?
c. What is the 20-year risk of heart attack among those who used diet pills?
d. What is the prevalence of diet pill use among those who did not have a heart attack?
e. Do the data suggest that Diet Pill Use is a risk factor for heart attack? Explain.

ACE-7. Exercise vs. CHD

A group of epidemiologists was interested in investigating the relationship between exercise and development of coronary heart disease (CHD) among women. A healthy population of women aged 35 to 75 years was polled to assess their exercise habits. They were then followed for a period of 15 years to determine incidence of CHD. Here are the results:

	Frequency of Exercise – Times per week		
	Twice	Once	No exercise
CHD	4	40	23
Person-Years	25,111	117,205	32,843
Rate per 10,000 Person-Years	______	______	______

a. What proportion of women who developed CHD had exercised once per week?
b. Complete the table by calculating the rates per 10,000 and filling in the three empty cells. Express answers to two decimal places.
c. What can you conclude from these data about the relationship between exercise and CHD?

ACE-8. Standardized Rates: Hypertension

An investigator is interested in comparing rates of hypertension in two populations. Which of the following should be taken into account when deciding whether it is necessary to standardize the rates by race? (There may be more than one correct answer here.)

a. Whether the rate of hypertension differs by race.
b. Whether the racial distribution differs in the two populations.
c. Whether the rate of hypertension differs in the two populations.
d. The rate of hypertension in the standard population.

ACE-9. Rates and Rate Adjustment

For each statement below, indicate whether it is true or false.

a. Two populations with the same age-specific rates of death could have different crude (i.e., overall) rates of death.
b. Two populations with the same crude (i.e., overall) rates of death could have different age-specific rates of death.
c. The process of direct adjustment of rates utilizes stratum-specific rates from the standard population.
d. A crude rate is a weighted average of stratum-specific rates.

ACE-10. Rate Adjustment: Standard Populations

Use the data provided below and, carrying all calculations to one decimal, complete the following:

a. Obtain age-adjusted total leukemia incidence rates in Mesa and Weld Counties using their pooled population as the standard.
b. Obtain age-adjusted total leukemia incidence rates in Mesa and Weld counties using the 1970 Colorado population (expressed in percentages) as the standard.
c. Are the age-adjusted rates for each county the same in parts a and b above?
d. Could a standard population be chosen such that the age-adjusted incidence rate for Weld county is higher than the age-adjusted incidence rate for Mesa county?
e. Regardless of the standard population used above, the age-adjusted rate for Weld county is similar to the unadjusted (i.e., crude) rate. What can you conclude from this?
    - o Age adjustment was necessary only for Mesa County, not for Weld county.
    - o Leukemia incidence rates are similar in Weld county and the standard population.
    - o The age structure is similar in Weld county and the standard population.

Age Group	Colorado 1970 Population (%)	Weld County 1970 Pop.	Weld County Leukemia AAIR*	Mesa County 1970 Pop.	Mesa County Leukemia AAIR*
< 5	8.4	7,491	9.5	3,754	7.6
5-19	30.6	28,452	3.0	16,852	4.2
20-34	22.5	20,382	1.4	9,253	4.6
35-49	17.2	13,859	1.0	9,329	3.1
50-64	12.8	11,219	6.4	8,685	11.5
65+	8.5	7,894	12.7	6,501	43.9
Total	100.0	89,297	4.2	54,374	10.2

* Average annual leukemia incidence rates per 100,000 population for the interval 1970-76 based upon 1970 population enumeration.

Answers to Study Questions and Quizzes

Q4.1

1. Incidence – Here we are interested in the number of new cases after eating the potato salad.
2. Prevalence – Here we are interested in the number of existing cases.
3. Incidence – Here we are interested in the number of new cases that occur during the follow-up.
4. Incidence – Here we are interested in the number of new deaths attributed to the hurricane.
5. Prevalence – Here we are interested in the existing number of children who have immunity to measles.
6. Incidence – Since rabies has a short duration, we would expect the prevalence on a particular day to be low relative to the incidence.
7. Prevalence – The incidence of multiple sclerosis would be low, but since it has a long duration, we would expect the prevalence to be higher.
8. Incidence – The incidence of influenza would be high, but since it is of short duration the prevalence would be low.
9. Incidence – Since the duration of poison ivy is relatively short the prevalence would be low, and since it is a common occurrence, the incidence would be high.
10. Prevalence – Since high blood pressure is common and of long duration, both incidence and prevalence would be high, however the prevalence would be higher.

Q4.2

1. The statement means that a 45-year-old male free of prostate cancer has a probability of .05 of developing prostate cancer over the next 10 years if he does not die from any other cause during the follow-up period.
2. Smaller, because the 5-year risk involves a shorter time period for the same person to develop prostate cancer.

Q4.3

1. No, subjects should be counted as new cases if they were disease-free at the start of follow-up and became a case at any time during the follow-up period specified.
2. Yes, there is a problem, since a subject followed for 2 years does not have the same opportunity for developing the disease as a subject followed for 4 years.
3. No, but we have to assume that those subjects that do not develop the disease have the same amount of follow-up time. Otherwise, we can get a misleading estimate of CI because not all subjects will have the same opportunity to develop the disease over the follow-up period.

Q4.4

1. Dynamic
2. Subject 2
3. Subject 7
4. Subject 9
5. Subject 3
6. Subject 5
7. 5/12=42%
8. 4/12=33%
9. D
10. No
11. No
12. Yes
13. Yes
14. Yes
15. No
16. Yes

Q4.5

1. 30/97
2. 60/97
3. 90/95
4. True – The numerator of the CI formula is a subset of the denominator.
5. True – Because the incubation period is short, subjects are not likely to be lost to follow-up.
6. True – The long incubation period means subjects are likely to be lost to follow-up, and hence cases may not be detected. For a dynamic cohort, the denominator in the CI formula does not reflect the continually changing population size.
7. False – the estimated CI will underestimate the true risk of disease.

Q4.6

1. C
2.
    a. Yes
    b. Yes
    c. No
    d. Yes
    e. No

Q4.7

1. No, the denominator of 25 does not describe 25 persons, but rather the accumulated follow-up time for 12 persons.
2. No, the risk in this example would be calculated as 5/12 or 0.42. However, using risk would be questionable here because different subjects have different follow-up times.
3. C

Q4.8

1. N* is the average size of the disease-free cohort and Δt is the time length of the study period. Therefore, a rough estimate of the total amount of person-years contributed by the study is 6,500 *6 = 39,000 person-years.
2. The incidence rate is 66/39,000 = 0.0017, or 1.7 per 1,000 person-years.

Q4.9

1. True – For questions 1 & 2: a rate can range from 0 to infinity, whereas a risk (which is a proportion) ranges from 0 to 1 (or 0% to 100%).
2. False
3. False – There are alternative ways to calculate person-time information when individual follow-up time is unavailable.
4. False – A rate can be calculated for either a dynamic cohort or fixed cohort, depending on the person-time information available.

Q4.10

1. Risk
2. Both
3. Both
4. Rate
5. Risk
6. Rate
7. Risk
8. Rate
9. Risk
10. Both

Q4.11

1. Yes, its value can range from 0 to 1 and it is often expressed as a percentage
2. C. The prevalence of disease is 13/406,245 = 0.000032
3. B. 3.2 per 100,000 is an equivalent expression and is easier to interpret

Q4.12

1. True – Prevalence considers existing cases rather than incident cases.
2. True – Since the numerator is contained in the denominator, prevalence is a proportion and must range from 0 to 1 (or 0% to 100%).
3. False – Cross-sectional studies are carried out at essentially a single (or short) point in time.
4. True – Prevalence may concern a health outcome or any other characteristic of a subject.
5. d

Q4.13

1. B
2. A
3. B

Q4.14

1. The estimate of LC incidence is calculated as CI = 15/900 = .017 or 1.7%
2. The 5% mortality estimate counts the 40 prevalent LC cases and does not count the 5 new LC cases that did not die. Furthermore, the denominators are different.
3. The LC incidence and mortality risks would be about equal if the disease was quickly fatal, so that there would be few if any prevalent cases in the initial cohort and all new cases would have died before the end of follow-up.

Q4.15

1. True
2. True
3. False – The denominator of a disease-specific mortality risk is the size of the initial cohort regardless of disease status.
4. True
5. True

Q4.16

1. The two rates are crude rates because they represent the overall mortality experience in 1996 for the entire population of each state. Crude rates do not account for any differences in these populations on factors such as age, race, or sex that might have some influence on mortality. Without consideration of such factors, it would be premature to make such a conclusion.
2. Arizona. The dry, warm climate of Arizona attracts many older persons than does Alaska.
3. There are relatively older persons living in Arizona, and older persons are at high risk of dying.

Q4.17

1. Controlling for any age differences in the two populations, the overall mortality rate is higher in Alaska with a cold, damp climate, then in Arizona where the climate is warm and dry.
2. The population of Alaska must be much younger than the US population since the age-adjusted rate was so much higher than the crude rate.
3. The rate for Arizona did not change much from crude to adjusted because Arizona's age distribution was only slightly older than that of the entire US in 1996.

Q4.18

1. True – If age-adjustment is not used, then a difference in risk or rates between two populations may be primarily due to age differences in the two populations.
2. False – There is no guarantee that two adjusted measures will be either closer or further from each other than were corresponding crude measures.
3. False – The choice of standard population depends on the characteristics of the populations being considered.
4. True – There is no limitation on the number populations that could be age-adjusted.
5. False – For questions 5 & 6: If the crude rates are quite different whereas the age distributions are similar, then the adjusted rates are likely to be quite different.
6. True
7. 44
8. 9.1
9. older – Women must be older than men in this case. The mortality rate drops substantially in women when we standardize the rate using the age distribution of men. In other words, if we take age out of the picture, the rates for women drop. If the women were younger we would expect to see the adjusted rate increase once we remove age as a factor.

LESSON 5

MEASURES OF EFFECT

5-1 Risk Ratio versus Odds Ratio

*In epidemiologic studies, we compare disease frequencies of two or more groups using a **measure of effect**. We will describe several types of measures of effect in this chapter. The choice of measure typically depends on the study design being used.*

Ratio Versus Difference Measures of Effect

Our focus in Lesson 5 is on ratio measures of effect, which are of the form M_I/M_0, where M_I and M_0 are two measures of disease frequency, e.g., risks, rates, or prevalences that are being compared.

We consider difference measures of effect, which are of the form M_I-M_0, in Lesson 6 on "Measures of Potential Impact". Difference measures are also called measures of attributable risk.

Ratio measures are typically used in epidemiologic studies that address the etiology of a disease/health outcome, whereas difference measures are used to quantify the public health importance of factors that are determinants of a disease/health outcome.

Smoking and Lung Cancer

Cigarette smoking became increasingly popular in America after World War I when cigarettes were handed out to soldiers as a way to boost morale. But along with the rise in smoking, came a disturbing rise in the lung cancer rate and some early warnings from a handful of doctors about possible dangers of smoking. Early studies in the 1930s and 1940s of the possible relationship between smoking and lung cancer were **case-control studies**. It became quite apparent that lung cancer patients smoked much more than controls. In one study in particular, lung cancer patients were 17 times more likely than controls to be two-pack-a-day smokers.

In the early 1950s, doctors Horn and Hammond of the American Cancer Society conducted one of the first **cohort studies** on the harmful effects of smoking. About 200,000 people were given a smoking questionnaire and then followed for four years. Death rates and cause of death for smokers and for non-smokers were compared. The preliminary study published in 1958 caused quite a sensation. It was the largest study on smoking that had been done, and it showed that smokers were ten times more likely than nonsmokers to get lung cancer.

Both the cohort and case-control studies attempted to assess the proposed relationship between smoking and lung cancer by deriving a measure of effect that quantified the extent of this relationship. The measure described in the **case-control study** is called an **odds ratio.** The measure described in the **cohort study** is called a **risk ratio**. The activities that follow discuss these two fundamental measures of effect.

Summary

- ❖ The odds ratio and the risk ratio are two fundamental measures of effect.
- ❖ These measures were used in epidemiologic studies of the relationship between smoking and lung cancer.
- ❖ The odds ratio is typically the measure of effect used in case-control studies.
- ❖ The risk ratio is typically the measure of effect used in cohort studies.

The Risk Ratio

The table below summarizes the results of a five-year follow-up study to determine whether or not smokers who have had a heart attack will reduce their risk for dying by quitting smoking. A cohort of 156 heart attack patients was studied, all of whom were regular smokers up to the time of their heart attack. Seventy-five of these patients continued to smoke after their attack. The other 81 patients quit smoking during their recovery period. Of the 75 patients that continued smoking, 27 died, so the proportion of these patients that died is 0.36. Of the 81 patients who quit smoking, 14 died, so the corresponding proportion is 0.17. These proportions estimate the five-year risks of dying for these two groups of patients. We may wonder whether those heart attack patients who continue smoking are more likely to die within 5 years after their first heart attack than those who quit.

Heart Attack Patients

	Smoke	Quit	Total
Death	27	14	41
Survival	48	67	115
Total	75	81	156

5-year risks of dying

continuing smokers:	27/75 = 0.36
smokers who quit:	14/81 = 0.17

A measure of effect gives a numerical answer to this question. Such a measure allows us to make a comparison of two or more groups, in this case, continuing smokers and smokers who quit. For follow-up studies such as described here, the typical measure of effect is a **risk ratio**. To calculate a risk ratio, we take the ratio of the two risks being compared, that is, we simply divide one risk by the other. Actually, we are getting an "estimate" of the risk ratio, which we indicate by putting a "hat" symbol over the RR notation. $\hat{RR}$ is an estimate because we are using two estimates of risk based on samples from the two groups being compared. In our example, therefore, we divide 0.36 by 0.17 to get 2.1.

$$\text{Estimated } \hat{RR} = \frac{\text{Estimated Risk for continuing smokers}}{\text{Estimated Risk for smokers who quit}} = \frac{0.36}{0.17} = 2.1$$

The estimated risk ratio of 2.1 tells us that continuing smokers are about twice as likely to die as smokers who quit. In other words, for heart attack patients the five-year risk for continuing smokers is about twice the corresponding risk for smokers who quit.

Study Questions (Q5.1):

Using the five-year follow-up study comparing mortality between smokers and quitters example:

1. How would you interpret a Risk Ratio of 4.5?
2. What if the Risk Ratio was 1.1?
3. How about if the Risk Ratio was less than 1, say 0.5?
4. How would you interpret a value of 0.25?

If our estimated risk ratio had been 1.1, we would have evidence that the risk for continuing smokers was essentially equal to the risk for smokers who quit. We call a risk ratio of 1 the **null value** of the risk ratio. This is the value that we get for the risk ratio when there is no effect, that is, the effect is null.

Summary

- ❖ The risk ratio (RR) is the ratio of the risk for one group, say group 1, to the risk for another group, say group 0.
- ❖ The value of RR can be greater than one, equal to one, or less than one.
- ❖ If the RR is greater than one, the risk for group 1 is larger than the risk for group 0.
- ❖ If the RR is below one, the risk for group 1 is less than the risk for group 0.
- ❖ And, if the RR is equal to 1, the risks for group 1 and 0 are equal, so that there is no effect of being in one group when compared to the other.

Risk Ratio Numerator and Denominator

In general, the risk ratio that compares two groups is defined to be the risk for one group divided by the risk for the other group. It is important to clearly specify which group is in the numerator and which group is in the denominator.

If, for example, the two groups are labeled group 1 and group 0, and the risk for group 1 is in the numerator, then we say that the risk ratio compares group 1 to group 0. On the other hand, if the risk for group 0 is in the numerator, then we say that the risk ratio compares group 0 to group 1.

Quiz (Q5.2)

For heart attack patients, the risk ratio is defined to be the risk for continuing smokers divided by the risk for smokers who quit. For the following scenarios what would be the risk ratio?

1. Continuing smokers are twice as likely to die than smokers who quit. . . . **???**
2. Continuing smokers are just as likely to die as smokers who quit. . . . **???**
3. Smokers who quit are twice as likely to die than continuing smokers. . . . **???**

Choices
0 **0.1** **0.2** **0.5** **1** **2**

Let's consider the data from a randomized clinical trial to assess whether or not taking aspirin reduces the risk for heart disease. The exposed group received aspirin every other day whereas the comparison group received a placebo. A table of the results is shown below.

		Aspirin		**Placebo**		**Total**
		n	Column %	n	Column %	
Developed Heart Disease	**Yes**	104	(1.04)	189	(2.36)	293
	No	9,896	(98.96)	7,811	(97.64)	17,707
	Total	10,000	(100.00)	8,000	(100.00)	18,000

Quiz continued on next page

4. The estimated risk for the aspirin group is . . . . 0.0104 ???

5. The estimated risk for the placebo group is . . . . 0.0236 ???

6. The estimated risk ratio that compares the aspirin group to the placebo group is given by 0.44 ???

Choices
0.0104 0.0236 0.44 104/189 2.269 98.96/97.64

The Odds Ratio

Epidemiologists in the Division of Bacterial Diseases at CDC, the Centers for Disease Control and Prevention in Atlanta, investigate the sources of outbreaks caused by eating contaminated foods. For example, a case-control study was carried out to determine the source of an outbreak of diarrheal disease at a Haitian Resort Club from November 30 to December 8, 1984.

The investigators wondered whether eating raw hamburger was a primary source of the outbreak. Because this is a **case-control study** rather than a follow-up study, the study design starts with **cases**, here, persons at the resort who had diarrhea during the time period of interest. The **controls** were a random sample of 33 persons who stayed at the resort but did not get diarrhea during the same time period. There were a total of 37 cases during the study period. All 37 cases and the 33 controls were interviewed by a team of investigators as to what foods they ate during their stay at the resort.

Of the 37 cases, 17 persons ate raw hamburger, so that the proportion of the cases that ate raw hamburger is 0.46. Of the 33 controls, 7 ate raw hamburger, so the corresponding proportion is 0.21. We may wonder, then, whether these data suggest that eating raw hamburger was the source of the outbreak.

	Raw Hamburger Ate	Did not eat	Total
Cases	17	20	37
Controls	7	26	33
Total	24	46	70

Proportion of cases: 17/37 = 0.46
Proportions of controls: 7/33 = 0.21

Because this is a case-control study rather than a follow-up study, these proportions do not estimate risks for cases and controls. Therefore, we can***not*** compute a risk ratio. So, then, what measure of effect should be used in case-control studies? The answer is the **odds ratio (OR),** which is described in the next activity.

Summary

- A case-control study was used to investigate a foodborne outbreak at a Caribbean resort.
- In a case-control study, we cannot estimate risks for cases and controls.
- Consequently, we cannot use the risk ratio (RR) as a measure of effect, but must use the odds ratio (OR) instead.

Why can't we use a risk ratio in case-control studies?

In a case-control study, we cannot estimate risk, but rather, we estimate **exposure probabilities** for cases and controls. The exposure probability for a case is the probability that a subject is exposed given that he/she is a case; this is not equivalent to the probability that a subject is a case given that he/she is exposed, which is the risk for exposed.

In other words, using conditional probability notation:

Continued on next page

Why can't we use a risk ratio in case-control studies? (Continued)

Pr(exposed | case) ≠ Pr(case | **exposed**), where "|" denotes "given".

Similarly the exposure probability for a control is not equivalent to 1 minus the risk for exposed. That is,

Pr(**exposed** | control) ≠ 1 - Pr(case | **exposed**).

The ratio of two exposure probabilities is, unfortunately, not a risk ratio. Therefore, in case-control studies we must use a different measure of effect, namely the odds ratio.

The Odds Ratio (continued)

To understand odds ratios, we must start with the concept of an **odds**. The term **odds** is commonly used in sporting events. We may read that the odds are 3 to 1 against a particular horse winning a race, or that the odds are 20 to 1 against Spain winning the next World Cup, or that the odds are 1 to 2 that the New York Yankees will reach the World Series this year. When we say that the odds against a given horse are 3 to 1, what we mean is that the horse is 3 times more likely to lose than to win.

The odds of an event are easily calculated from its probability of occurrence. The odds can be expressed as **P**, the probability that the event will occur, divided by 1 - P, the probability that the event will not occur.

$$\text{Odds} = \frac{P}{1 - P} = \frac{\text{P(Event will occur)}}{\text{P(Event will not occur)}}$$

In our horse race example, if P denotes the probability that the horse will lose, then 1 - P denotes the opposite probability that the horse will win. So, if the probability that the horse will lose is 0.75, then the probability that the horse will win is 0.25, and the odds are 3, or 3 to 1.

$$\text{Odds} = \frac{P}{1 - P} = \frac{\text{P(horse will lose)}}{\text{P(horse will win)}} = \frac{0.75}{0.25} = 3 \text{ or } \frac{3}{1}$$

In the Haitian resort case-control study, recall that the event of interest occurs if a study subject ate raw hamburger, and, if so, we say this subject is **exposed**. The estimated probability of exposure for the cases was 0.46, so the estimated **odds of being exposed for cases** is 0.46 divided by 1 - 0.46:

$$\hat{\text{Odds}}_{\text{Cases}} = \frac{0.46}{1 - 0.46} = .85$$

Similarly, the estimated probability of exposure for controls was 0.21, so the estimated odds for controls is 0.21 divided by 1 - 0.21:

$$\hat{\text{Odds}}_{\text{Controls}} = \frac{0.21}{1 - 0.21} = .27$$

The **estimated odds ratio** for these data is the ratio of the odds for cases divided by the odds for controls, which equals 3.2.

$$\text{Odds Ratio (OR)} = \frac{\hat{\text{Odds}}_{\text{Cases}}}{\hat{\text{Odds}}_{\text{Controls}}} = \frac{.85}{.27} = 3.2$$

How do we interpret this odds ratio estimate? One interpretation is that the **exposure odds for cases** is about 3.2 times the **exposure odds for controls**. Since those who ate raw hamburger are the exposed subjects, the odds that a case ate raw hamburger appear to be about 3.2 times the odds that a control subject ate raw hamburger.

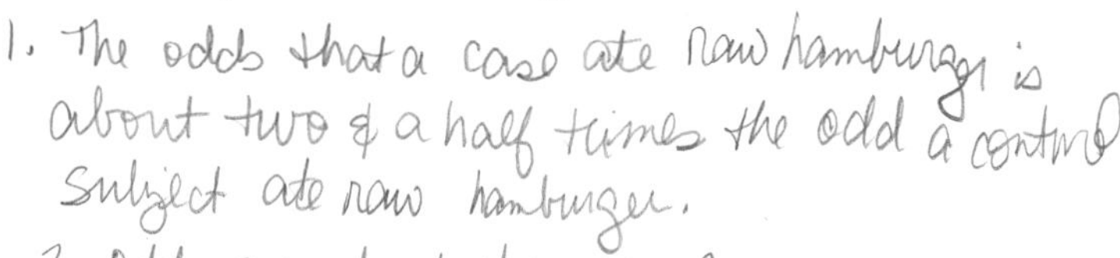

Study Questions (Q5.3)
Using the Haiti case-control study example:

1. How would you interpret an odds ratio of 2.5?
2. What if the odds ratio was 1.1?
3. How about if the odds ratio less than 1, say 0.5?
4. How would you interpret a value of 0.25?

Odds ratios, like risk ratios, can be greater than one, equal to one, or less than one. An odds ratio greater than one says that the exposure odds for cases is **larger** than the exposure odds for controls. An odds ratio below one says that the exposure odds for cases is **less** than the exposure odds for controls. An odds ratio equal to 1 says that the exposure odds for cases and controls are equal.

$$\hat{OR} = \hat{Odds}_{Cases} / \hat{Odds}_{Controls}$$

$\hat{OR}$ greater than 1 ?

$$\hat{Odds}_{Cases} > \hat{Odds}_{Controls}$$

$\hat{OR}$ equal to 1 ?

$$\hat{Odds}_{Cases} = \hat{Odds}_{Controls} \quad \text{No effect}$$

$\hat{OR}$ less than 1 ?

$$\hat{Odds}_{Cases} < \hat{Odds}_{Controls}$$

Summary

- The odds of an event can be calculated as P/(1-P) where P is the probability of the event.
- The odds ratio (OR) is the ratio of two odds.
- In case-control studies, the OR is given by the exposure odds for the cases divided by the exposure odds for controls.
- Odds ratios, like risk ratios, can be greater than 1, equal to 1, or less than 1, where 1 is the null value.

Quiz (Q5.4)

A causal relationship between cigarette smoking and lung cancer was first suspected in the 1920s on the basis of clinical observations. To test this apparent association, numerous studies were conducted between 1930 and 1960. A classic **case-control study** was done in 1947 to compare the smoking habits of lung cancer patients with the smoking habits of other patients.

1. In this case-control study, it is **???** to calculate the risk of lung cancer among smokers, and thus, the appropriate measure of association is the **???**.

Choices
Not possible **odds ratio** **possible** **risk ratio**

Let's consider the data below from this classic case-control study to assess the relationship between smoking and lung cancer. Cases were hospitalized patients newly diagnosed with lung cancer. Controls were patients with other disorders. This 2 x 2 table compares smoking habits for the male cases and controls.

2. The probability of being a smoker among cases is . . **???**

3. The probability of being a smoker among controls is . **???**

4. The odds of smoking among cases is . . . **???**

5. The odds of smoking among controls is . . . **???**

6. The odds ratio is . . . . . **???**

Choices
0.11 **1.04** **10.50** **1296/1357** **1350/1357** **1350/2646** **192.86** **21.25** **7/68** **9.08**

	Cigarette Smoker	Non-Smoker	Total
Cases	1350	7	1357
Controls	1296	61	1357
Total	2646	68	2714

In a case-control study to find the source of an outbreak, the odds ratio for eating coleslaw is defined to be the odds for cases divided by the odds for controls. For the following scenarios what would be the odds ratio?

7. Cases have an odds for eating coleslaw three times higher than controls . . . **???**

8. Cases have the same odds for eating coleslaw as controls . . . . **???**

9. Controls have three times the odds for eating coleslaw as cases . . . **???**

Choices
0 **0.25** **0.333** **1** **3** **4**

5-2 Odds Ratio Calculation/Different Study Designs

Calculating the Odds Ratio

This layout for a two by two table provides a more convenient way to calculate the odds ratio. The formula is ***a*** times ***d*** over ***b*** times ***c***. It is called the **cross product ratio** formula because it is the ratio of one product that crosses the table divided by the other product that crosses the table.

General Form of the 2 by 2 Table

	Exposure Status Yes	No	Total
Cases	a	b	m_1
Controls	c	d	m_0
Total	n_1	n_0	m

Cross Product Ratio

$$\hat{OR} = \frac{a \times d}{b \times c}$$

To illustrate this formula consider the data from the Haitian resort outbreak. The cross product formula gives us the same result, 3.2, as we obtained originally from the ratio of exposure odds for cases and controls.

Case-control Study: Outbreak of Diarrheal Disease at a Haitian Resort Club

	Raw Hamburger Yes	No	Total
Cases	a=17	b=20	m_1=37
Controls	c=7	d=26	m_0=33
Total	n_1=24	n_0=46	m=70

Cross Product Ratio

$$\hat{OR} = \frac{a \times d}{b \times c} = \frac{(17)(26)}{(20)(7)} = 3.2$$

$$= \frac{\hat{Odds}_{Cases}}{\hat{Odds}_{Controls}}$$

Study Question (Q5.5)

1. Should we calculate the OR for other foods eaten during the outbreak before we blame raw hamburger as the source? Yes, it is possible that something else caused the outbreak

Although the odds ratio must be computed in case-control studies for which the risk ratio cannot be estimated, the odds ratio can also be computed in follow-up studies.

	OR	RR
Case-Control Studies	✓	☒
Follow-up (Cohort)	✓	✓

(Note that the OR and RR can also be calculated in randomized clinical trials that have cumulative incidence measures.)

For example, let us consider the "quit smoking" study for heart attack patients. The study design here is a follow-up study. We previously estimated that the risk for patients who continued to smoke was 2.1 times greater than the risk for those who quit.

Follow-up (Cohort) Study

	Smoke	Quit	Total
Death	27	14	41
Survival	48	67	115
Total	75	81	156

$$\hat{RR} = \frac{\hat{\text{Risk}} \text{ for continuing smokers}}{\hat{\text{Risk}} \text{ for smokers who quit}} = \frac{27/75}{14/81} = \frac{0.36}{0.17} = 2.1$$

$$\hat{OR} = \frac{a \times d}{b \times c} = \frac{(27)(67)}{(14)(48)} = 2.7$$

Using the cross product formula on these follow-up data yields 2.7. The fact that these two numbers (the risk ratio and odds ratio) are not equal should not be surprising, since the risk ratio and odds ratio are two different measures. But the values in this example are not very different. In fact, these two estimates have similar interpretations since they both suggest that there is a moderate relationship between quit smoking status and survival status.

Summary

- A convenient formula for the OR is the cross product ratio: (ad)/(bc)
- The OR can be estimated in both case-control and follow-up studies using the cross-product formula.

(See the activities on page 5-4 of this Lesson for discussion of how the risk ratio can be approximated by the odds ratio.)

Quiz (Q5.6)

To study the relationship between oral contraceptive use and ovarian cancer, CDC initiated the Cancer and Steroid Hormone Study in 1980. It was a case-control study.

1. Using the cross product ratio formula, the OR comparing the exposure status of cases versus controls is (93) * (**???**) / (**???**) * (959) which equals **???**.
2. This means that the **???** of **???** among the cases was **???** the **???** of exposure among the **???**.

Choices
0.23 **0.77** **1.3683** **86** **cases** **controls** **disease** **exposed** **exposure** **greater than** **less than** **non-exposed** **odds** **risk**

	Ever Used OCs	Never Used OCs	Total
Cases	93	86	179
Controls	959	683	1642
Total	1052	769	1821

The Odds Ratio in Different Study Designs

The odds ratio can be computed for both case-control and follow-up (cohort) studies. Because a case-control study requires us to estimate exposure probabilities rather than risks, we often call the odds ratio computed in case-control studies the **exposure odds ratio** (**EOR**). In contrast, because a follow-up study allows us to estimate risks, we often call the odds ratio computed from follow-up studies the **risk odds ratio** (**ROR**).

The odds ratio can also be computed for cross-sectional studies. Since a cross-sectional study measures prevalence or existing conditions at a point in time, we usually call an odds ratio computed from a cross-sectional study a **prevalence odds ratio** (**POR**).

Odds Ratio

Case-control studies (exposure probabilities):
Exposure odds ratio (EOR)

Follow-up (Cohort) studies (risks):
Risk odds ratio (ROR)

Cross-sectional studies (prevalences):
Prevalence odds ratio (POR)

As an example of the computation of a prevalence odds ratio for cross-sectional data, consider these data that were collected from a cross-sectional survey designed to assess the relationship between coronary heart disease and various risk factors, one of which was personality type. For these cross-sectional data, we can use the general cross product ratio formula to compute a prevalence odds ratio. The odds of having a type A personality among those with coronary heart disease is 5 times the odds of those without the disease.

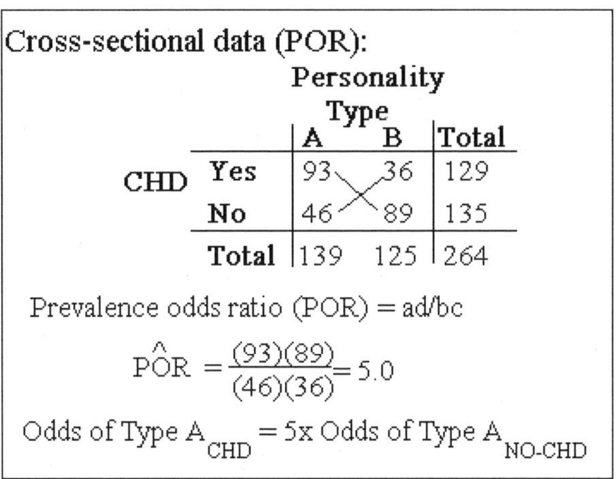

Cross-sectional data (POR):

		Personality Type A	B	Total
CHD	Yes	93	36	129
	No	46	89	135
	Total	139	125	264

Prevalence odds ratio (POR) = ad/bc

$$\hat{POR} = \frac{(93)(89)}{(46)(36)} = 5.0$$

Odds of Type A_{CHD} = 5x Odds of Type $A_{NO\text{-}CHD}$

In general we can use the cross product ratio formula to compute an exposure odds ratio, a risk odds ratio, or a prevalence odds ratio depending on the study design used.

Summary

- The OR computed from a case-control study is called the exposure odds ratio (EOR).
- The OR computed from a follow-up study is called the risk odds ratio (ROR)
- The OR computed from a cross-sectional study is called the prevalence odds ratio (POR)
- We can use the general cross-product ratio formula to calculate the EOR, ROR, or POR depending on the study design used.

Does ROR = EOR = POR?

Not necessarily. Although the calculation formula (i.e., ad/bc) is the same regardless of the study design, different values of the estimated odds ratio from a 2 x 2 table might be obtained for different study designs. This is because of the possibility of selection bias (described in Lesson 8). For example, a case-control study that uses prevalent cases could yield a different odds ratio estimate than a follow-up study involving only incident cases.

Quiz (Q5.7)

Data is shown below for a cross-sectional study to assess whether maternal cigarette smoking is a risk factor for low birth weight.

1. Calculate the odds ratio that measures whether smokers are more likely than non-smokers to deliver low birth weight babies. OR=**???** 2.18
2. This odds ratio estimate suggests that smokers are **???** than non-smokers to have low birth weight babies. more likely
3. This odds ratio is an example of a(n) **???** odds ratio. prevalance

Choices
0.48 **2.04** **2.18** **exposure** **less likely** **more likely** **prevalence** **risk**

	Smokers	Non-Smokers	Total
Low Birth weight	1,556	14,974	16,530
High Birth weight	694	14,532	15,226
Total	2,250	29,506	31,756

Compute Measures of Association with Data Desk

This activity teaches users how to compute measures of association using the Data Desk program.

5-3 Comparing Odds Ratio with Risk Ratio Approximations

Comparing the Risk Ratio and the Odds Ratio

We have described two widely used measures of effect, the risk ratio and the odds ratio. Risk ratios are often preferred because they are easier to interpret. But, as we have seen, in case-control studies, we cannot estimate risks and must work instead with an exposure odds ratio (EOR). In follow-up studies, however, we have the option of computing both a risk ratio and a risk odds ratio (ROR). Which should we prefer?

It can be shown mathematically that if a risk ratio estimate is equal to or greater than one, then the corresponding risk odds ratio is at least as large as the risk ratio. For example, using the follow-up data for the quit smoking study of heart attack patients, we saw that the estimated risk ratio was 2.1, which is greater than one; the corresponding odds ratio was 2.7, which is larger than 2.1.

Follow-up (Cohort) Studies: RR vs. ROR

If $\hat{RR} \geq 1$, then $\hat{ROR} \geq \hat{RR}$

Quit Smoking Data for Heart Attack Patients

	Smoke	Quit	Total
Death	27	14	41
Survival	48	67	115
Total	75	81	156

$\hat{RR} = 2.1$

$\hat{ROR} = 2.7$

Similarly if the risk ratio is less than one, the corresponding odds ratio is as small or smaller than the risk ratio. For example, if we switch the columns of the quit smoking table, then the risk ratio is 0.48, which is less than one, and the corresponding odds ratio is 0.37, which is less than 0.48.

$\text{If } \hat{RR} \leq 1, \text{ then } \hat{ROR} \leq \hat{RR}$

Quit Smoking Data for Heart Attack Patients

	Quit	Smoke	Total
Death	14	27	41
Survival	67	48	115
Total	81	75	156

$\hat{RR} = 0.48$

$\hat{ROR} = 0.37$

$$\text{If } \hat{RR} \geq 1, \text{ then } \hat{ROR} \geq \hat{RR}$$
$$\text{If } \hat{RR} \leq 1, \text{ then } \hat{ROR} \leq \hat{RR}$$

It can also be shown that if a disease is "rare", then the risk odds ratio will closely approximate the risk ratio. For follow-up studies, this *rare disease assumption* means that the risk that any study subject will develop the disease is small enough so that the corresponding odds ratio and risk ratio estimates give essentially the same interpretation of the effect of exposure on the disease.

Typically a rare "disease", is considered to be a disease that occurs so infrequently in the population of interest that the risk for any study subject is approximately zero. For example, if one out of every 100,000 persons develops the disease, the risk for this population is zero to 4 decimal places. Now that's really rare!

1. Depends on the disease being considered & on the time-period of follow-up over which the risk is computed. However for most chronic diseases & short time periods, a risk of 0.01 is not rare.

Study Questions (Q5.8)

1. Is a risk of .01 rare?
2. Suppose that for a given follow-up study, the true risk is not considered to be rare. Is it possible for the ROR and RR to be approximately the same? 2. Yes, because even though the risk may not be rare, it may be small enough so that the ROR & RR, are approximately the same

We can write a formula that expresses the risk odds ratio in terms of the risk ratio:

$$ROR = RR \text{ x } f$$

where

$$f = \frac{(1 - R_0)}{(1 - R_1)}$$

R_0 is the risk for the unexposed
R_1 is the risk for the exposed
$RR = R_1/R_0$

This formula says that the risk odds ratio is equal to the risk ratio multiplied by the factor f, where f is defined as 1 minus the risk for the unexposed group (R_0) divided by 1 minus the risk for the exposed group (R_1). You can see from this equation that if both R_1 and R_0 are approximately 0, then f is approximately equal to one, and the risk odds ratio is approximately equal to the risk ratio.

Study Questions (Q5.9)

1. In the quit smoking example, where R_0 is 0.17 and R_1 equals 0.36, what is f? $1 - .17/1 - .36 = \frac{.83}{.64} = 1.3$
2. For this value of f, is the ROR close to the RR? No, since for these data the estimated RR = 2.1 & ROR = 2.7
3. What happens to f if the risks are halved, i.e., $R_0 = 0.17/2 = 0.085$ and $R_1 = 0.36/2 = 0.180$?

$f = 1 - 0.085 / 1 - 0.180 = \frac{.915}{.82} = 1.12$

Study questions continued on next page

4. Are the ROR and RR estimates close for this *f*? Yes, since the estimated RR is again 2.1, but ROR is 2.4
5. What happens to f if we again halve the risks, so that R_0=0.0425 and R_1=0.09? f= 1-.0425/1-.09= 1.05
6. Is the approximation better? Yes, since the estimated ROR is now 2.2
7. Based on your answers to the above questions, how "rare" do the risks have to be for the odds and risk ratios to be approximately equal? For this example, risks below 0.10 for both groups indicate a 'rare' disease

Summary

- ❖ If an estimate of RR ≥ 1, then the corresponding estimate of ROR is at least as large as the estimate of the RR.
- ❖ If an estimate of RR ≤ 1, then the corresponding estimate of ROR is as small or smaller than the estimate of RR.
- ❖ In follow-up (cohort) studies, the "rare disease assumption" says that the risk for any study subject is approximately zero.
- ❖ Under the rare disease assumption, the risk odds ratio (ROR) computed in a follow-up study approximates the risk ratio (RR) computed from the same study.

Comparing the RR and the OR in the Rotterdam Study

Osteoporosis is a common disease in the elderly, and leads to an increased risk of bone fractures. To study this disease, a cohort consisting of nearly 1800 postmenopausal women living in Rotterdam, the Netherlands, was followed for four years. The Rotterdam Study investigators wanted to know which genetic factors determine the risk of fractures from osteoporosis. They focused on a gene coding for one of the collagens that are involved in bone formation. Each person's genetic make-up consists of two alleles of this gene, and each allele can have one of two alternative forms, called allele A or allele B. The investigators showed that women with two A alleles had a higher bone mass than women with at least one B allele. They therefore hypothesized that the risk of fractures would be higher in women with allele B.

Of the 1194 women with two A alleles, 64, or 5.36%, had a fracture during follow-up. Of the 584 women with at least one B allele, 47, or 8.05%, had a fracture.

Study Questions (Q5.10)

1. Calculate the risk ratio for the occurrence of fractures in women with at least one B allele compared to women with two A alleles. The risk ratio in this study is 0.0805/0.0536=1.5

Because the risk ratio estimate is greater than one, we expect the risk odds ratio to be at least as large as the risk ratio.

Study Questions (Q5.10) continued

2. Calculate the risk odds ratio for the occurrence of fractures in women with at least one B allele compared to women with two A alleles. Risk ratio is 47/537 /64/1130 = 1.54

Note that the risk of fractures is relatively rare in this population, therefore the risk odds ratio is approximately equal to the risk ratio. Recall the formula ROR = RR * f. Here, f is defined as 1 minus the risk in women with two A alleles divided by 1 minus the risk in women with at least one B allele.

$$\mathrm{ROR} = \mathrm{RR} \times \frac{1 - \mathrm{R(2\ A\ alleles)}}{1 - \mathrm{R(1\ B\ allele)}}$$

Study Questions (Q5.10) continued

3. Using the formula ROR = RR x f, can you show that we computed the correct risk odds ratio?
f=(1-0.053)/1 - 0.0805 = 1.03 The ROR= 1.03*RR= 1.03*1.50=1.54

In this study, both the risk ratio and the risk odds ratio lead to the same conclusion: women with at least one B allele have a 50% higher chance of fractures than women with two A alleles. The Rotterdam Study investigators concluded that genetic make-up can predispose women to osteoporotic fractures.

Quiz (Q5.11): RR versus OR in follow-up studies

A questionnaire was administered to those persons who attended a social event in which 39 of the 87 participants became ill with a condition diagnosed as salmonellosis. The 2 x 2 table below summarizes the relationship between consumption of potato salad and illness.

1. The risk ratio comparing the exposed to the non-exposed is . **???**
2. The odds ratio is . . . . . **???**
3. Does the odds ratio closely approximate the risk ratio? . **???**
4. Do you consider this illness to be "rare"? . . **???**

Choices
0.25 **1.7** **3.7** **36.0** **9.8** **no** **yes**

	Exposed	Non-Exposed	Total
Ill	36	3	39
Well	12	36	48
Total	48	39	87

Let's consider data from a classic study of pellagra. Pellagra is a disease caused by dietary deficiency of niacin and characterized by dermatitis, diarrhea, and dementia. Data comparing cases by gender are shown below.

5. The risk ratio of pellagra for females versus males is (to one decimal place) . **???**
6. The odds ratio is (to one decimal place) . . . . **???**
7. Does the odds ratio closely approximate the risk ratio? . . . **???**
8. Do you consider this illness to be "rare"? . . . . **???**

Choices
1.4 **2.4** **2.5** **24.2** **no** **yes**

	Females	Males	Total
Ill	46	18	64
Well	1438	1401	2839
Total	1484	1419	2903

Odds Ratio Approximation in Case-Control Studies

Comparing the RR and OR in Case-Control Studies

We have already seen that, for follow-up studies, if the disease is "rare", then the risk odds ratio will be a close approximation to the risk ratio computed from the same follow-up data. However, in case-control studies, a risk ratio estimate cannot be computed, and an exposure odds ratio must be used instead. So, for case-control data, if the disease is "rare", does the exposure odds ratio approximate the risk ratio that would have resulted from a comparable follow-up study? The answer is yes, depending on certain conditions that must be satisfied, as we will now describe.

This two-way table categorizes lung cancer and smoking status for a cohort of physicians in a large metropolitan city that are followed for 7 years. Forty smokers and twenty non-smokers developed lung cancer. The risk ratio is 2. Also, for this population, the risk odds ratio is equal to 2.02, essentially the same as the risk ratio. Since these are measures of effect for a population, we have not put the hat symbol over the risk ratio and risk odds ratio terms.

Cohort of Physicians (7-year follow-up)

	Smoker ? Yes	No	Total
LC	40	20	60
No LC	1960	1980	3940
Total	2000	2000	4000

population measures

$$RR = \frac{40/2000}{20/2000} = 2$$

$$ROR = \frac{40 \times 1980}{1960 \times 20} = 2.02$$

We now consider the results that we would expect to obtain if we carried out a case-control study using this cohort as our source population. We will assume that the 7-year follow-up has occurred. We also assume that there exists a comprehensive cancer registry, so that we were able to find al1 60 incident cases that developed over the 7year period. These would be our cases in our case-control study. Now suppose we randomly select 60 controls from the source population as our comparison group. Since half of the entire cohort of 4000 physicians was exposed and half was unexposed, we would expect 30 exposed and 30 unexposed out of the 60 controls.

Case-Control Study

	E	not E	Total
(incident) Cases	40	20	60
Controls	30	30	60
Total	70	50	120

We can use the cross product ratio formula to compute the expected exposure odds ratio, which turns out to be 2. This value for the exposure odds ratio obtained from case-control data is the same that we would have obtained from the risk ratio and the risk odds ratio if we had carried out the follow-up study on this population cohort. In other words, the expected EOR from this case-control study would closely approximate the RR from a corresponding follow-up study, even if the follow-up study was never done!

$$\hat{EOR} = 2.0 \approx \hat{RR} = 2$$

We may wonder whether the EOR would approximate the RR even if the 60 controls did not split equally into exposed and unexposed groups as expected. This can occur by chance from random selection or if we do a poor job of picking controls. For example, suppose there were 40 exposed and 20 unexposed among the controls. Then the estimated exposure odds ratio would equal 1 instead of 2, so in this situation, the EOR would be quite different from the RR obtained from a comparable follow-up study.

$\hat{EOR} = 1.0 \not\approx \hat{RR} = 2$

Case-Control Study

		E	not E	Total
(incident)	Cases	40	20	60
	Controls	40	20	60
	Total	80	40	120

$\hat{EOR} = 1.0$

What we have shown by example actually reflects an important caveat when applying the rare disease assumption to case-control data. The choice of controls in a case-control study must be representative of the source population from which the cases developed. If not, either by chance or a poor choice of controls, then the exposure odds ratio will not necessarily approximate the risk ratio even if the disease is rare. There is another important caveat for applying the rare disease assumption in a case-control study. The cases must be incident cases, that is, the cases need to include all new cases that developed over the time-period considered for determining exposure status. If the cases consisted only of prevalent cases at the time of case-ascertainment, then a biased estimate may result because the measure of effect would be estimating prevalence rather than incidence.

Summary

❖ In case-control studies, the EOR approximates an RR when the following 3 conditions are satisfied:

1) The rare disease assumption holds
2) The choice of controls in the case-control study must be representative of the source population from which the cases developed.
3) The cases must be incident cases.

Quiz (Q5.12): Understanding Risk Ratio

In a case-control study, if the rare disease assumption is satisfied, then:

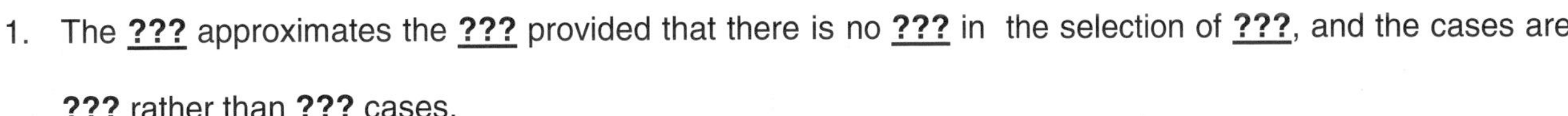

1. The **???** approximates the **???** provided that there is no **???** in the selection of **???**, and the cases are **???** rather than **???** cases.

Choices
EOR **ROR** **RR** **bias** **cases** **controls** **incidence** **prevalent** **randomness**

In a community of 1 million persons, 100 cases of a disease were reported, distributed by exposure according to the table below.

2. Calculate the RR. . . . . . **???**

3. Calculate the ROR. . . . . . **???**

4. Is this a rare disease? . . . . **???**

	Exposed	Non-Exposed	Total
Ill	90	10	100
Well	499,910	499,990	999,900
Total	500,000	500,000	1,000,000

If the exposure status of all one million persons in the study population had not been available, the investigator may have conducted a case-control study. Suppose a random sample of 100 controls were selected.

5. Approximately what percentage of these controls would you expect to be exposed? . . . ???
6. What is the expected EOR in the case-control study? . . . . ???

Choices
0.11 10 50 9.00 90 no yes

The Math Behind the Rare Disease Approximation

The mathematical argument that explains why the exposure odds ratio (EOR) approximates a risk ratio (RR) when the disease of interest is rare can be outlined as follows.

We first demonstrate that the risk odds ratio (ROR) computed from follow-up data approximates the risk ratio (RR) for these same data if the disease is rare. We then show using conditional probabilities that the risk odds ratio and the exposure odds ratio are equal. From this it follows that the exposure odds ratio must also approximate the risk ratio for rare diseases.

Step 1: $\hat{ROR} \approx \hat{RR}$ if rare disease

Step 2: $\hat{ROR} = \hat{EOR}$

Step 3: $\hat{EOR} \approx \hat{RR}$

Let's first describe the risk ratio and risk odds ratio in terms of conditional probabilities. The risk ratio can be expressed as the ratio of the conditional probability of developing the disease, given exposed, divided by the conditional probability of developing disease, given unexposed. To describe the risk odds ratio, we start with the odds of developing the disease for exposed persons and the odds of developing the disease for unexposed persons.

The risk odds ratio is then given by the ratio of these two odds, as shown here. With a little algebra, we can rewrite the risk odds ratio as follows.

Step 1: $\hat{ROR} \approx \hat{RR}$ if rare disease

$$RR = \frac{P(D|E)}{P(D|\text{not } E)}$$

$$ROR = \frac{\text{odds for D given E}}{\text{odds for D given not E}}$$

$$= \frac{\dfrac{P(D|E)}{1 - P(D|E)}}{\dfrac{P(D|\text{not } E)}{1 - P(D|\text{not } E)}} = \frac{P(D|E)}{P(D|\text{not } E)} * \frac{1 - P(D|\text{not } E)}{1 - P(D|E)}$$

We can thus express the risk odds ratio in terms of the risk ratio by replacing the first part of the product term in the expression on the right, by the risk ratio, so that we obtain the following formula.

$$ROR = RR * \frac{1 - P(D|\text{not } E)}{1 - P(D|E)}$$

Now if the disease is rare, the probability of disease is small, regardless of exposure. So, both P(D | not E) and P(D | E) are approximately 0. If we substitute 0 for these terms in the formula for the risk odds ratio, we obtain an expression that says that the risk odds ratio approximates the risk ratio when the disease is rare.

$$\text{ROR} = \text{RR} * \frac{1-0}{1-0} \approx \text{RR} * 1$$

Rare Disease $\Rightarrow$ $P(D|\text{not } E) \approx 0$ and $P(D|E) \approx 0$

We now use a famous theorem about conditional probabilities, called **Bayes Theorem**, to show that the exposure odds ratio equals the risk odds ratio, from which it logically follows that the exposure odds ratio approximates the risk ratio for rare diseases. Bayes theorem expresses conditional probabilities of the form P(D | E) and P(D | not E) in terms of conditional probabilities of the form P(E | D) and P(E | not D). We need to consider P(E | D) and P(E | not D) because in case-control studies the disease status is given first and the conditional probability of prior exposure status is then determined. Using Bayes Theorem, we can express P(D | E) and P(D | not E) as follows:

Step 2: $\hat{\text{ROR}} = \hat{\text{EOR}}$ Bayes Theorem

$$P(D|E) = \frac{P(D)\,P(E|D)}{P(D)\,P(E|D) + [1 - P(D)]\,P(E|\text{not } D)}$$

$$P(D|\text{ not } E) = \frac{P(D)\,[1-P(E|D)]}{P(D)\,[1-P(E|D)] + [1 - P(D)]\,[1-P(E|\text{not } D)]}$$

If we then substitute these expressions for P(D | E) and P(D | not E) into the formula for the risk odds ratio and then do a considerable amount of algebra using the substituted terms, we obtain the expression below. The expression on the right is the exposure odds ratio.

$$\text{ROR} = \frac{\dfrac{P(D|E)}{1-P(D|E)}}{\dfrac{P(D|\text{ not } E)}{1-P(D|\text{ not } E)}} = \frac{\dfrac{P(E|D)}{1-P(E|D)}}{\dfrac{P(E|\text{ not } D)}{1-P(E|\text{ not } D)}} = \text{EOR}$$

We have thus shown that the risk odds ratio equals the exposure odds ratio. Combining the result that risk odds ratio approximates risk ratio when the disease is rare with the result that the risk odds ratio equals exposure odds ratio, it follows that exposure odds ratio approximates the risk ratio when the disease is rare.

√ Step 1: $\hat{\text{ROR}} \approx \hat{\text{RR}}$ if rare disease

√ Step 2: $\hat{\text{ROR}} = \hat{\text{EOR}}$

√ Step 3: $\hat{\text{EOR}} \approx \hat{\text{RR}}$ if rare disease

The mathematical argument we have just completed for the equivalence of the exposure odds ratio and risk odds ratio requires two additional assumptions. These assumptions are needed to carryover from theoretical probabilities to their estimates derived from case-control data. First, the choice of controls in a case-control study must be representative of the source population from which the cases developed. Second, the cases must be incident, rather than prevalent cases.

Summary: The Math Behind the Rare Disease Approximation

- Use algebra involving conditional probabilities and Bayes Theorem.
- Bayes Theorem: conditional probabilities of the form P(D | E) and P(D | notE) in terms of P(E | D) and P(E | not D).
- Two assumptions also required: representative controls and incident cases
- Step 1: ROR computed from follow-up data approximates the RR for these same data if the disease is rare
- Step 2: using Bayes Theorem, the ROR and EOR are equal
- Step 3: Combining Step 1 with Step 2, the EOR approximates the RR for rare diseases

5-4 The Rate Ratio and its Characteristics

The Rate Ratio

A **rate ratio** is a ratio of two average rates. It is sometimes called an **incidence density ratio** or a **hazard ratio**. Recall the general formula for an average rate: **I** denotes the number of new cases of the health outcome, and **PT** denotes the accumulation of person-time over the follow-up.

The general data layout for computing a rate ratio is shown below. $\mathbf{I_1}$ and $\mathbf{I_0}$ denote the number of new cases in the exposed and unexposed groups, and $\mathbf{PT_1}$ and $\mathbf{PT_0}$ denote the corresponding person time accumulation for these two groups. The formula for the **rate ratio** or the **incidence density ratio** (**IDR**) is also provided. We have used the notation IDR instead of RR to denote the rate ratio in order to avoid confusion with our previous use of RR to denote the risk ratio.

Average Rate: $\frac{I}{PT}$

Layout for computing a Rate Ratio (i.e., IDR)

	Exposed	Unexposed	Total
New Cases	I_1	I_0	I
Person Time	PT_1	PT_0	PT

Rate Ratio:

$$IDR = \frac{\frac{I_1}{PT_1}}{\frac{I_0}{PT_0}}$$

As with both the risk ratio and odds ratio measures, the rate ratio can be >1, <1, or =1. If the rate ratio is equal to 1, it means that there is no relationship between the exposure and disease using this measure of effect.

To illustrate the calculation of a rate ratio, we consider data on the relationship between serum cholesterol level and mortality from a 1992 study of almost 40,000 persons from the Chicago area. The data shown compares white males with borderline-high cholesterol levels and white males with normal cholesterol levels. Subjects, including persons from other race and sex categories, were enrolled into the study between 1967 and 1973, screened for cardiovascular disease (CVD) risk factors, and then followed for an average of 14 to 15 years. There were a total of 26 CHD-related deaths based on 36,581 person-years of follow-up among white males aged 25 to 39 with borderline-high cholesterol at entry into the study. This yields a rate of 71.1 deaths per 100,000 person-years. Among the comparison group there were 14 CHD-related deaths based on 68,239 person-years of follow-up, this yields a rate of 20.5 deaths per 100,000 person-years. Thus, white males aged 25-39 with borderline high cholesterol have 3.5 times the mortality rate as those with normal cholesterol, indicating that persons with even moderately high cholesterol carry an increased risk for CHD mortality.

Serum cholesterol → Mortality

1992, 40,000 persons, Chicago (white males, ages 25-39)

	Cholesterol Level	
	Borderline-high	Normal
Deaths	26	14
Person-years	36,581	68,239
$\frac{I}{PT}$	$\frac{71.1}{100{,}000}$ person-yrs	$\frac{20.5}{100{,}000}$ person-yrs

$$\hat{IDR} = \frac{\frac{I_1}{PT_1}}{\frac{I_0}{PT_0}} = \frac{71.1}{20.5} = \mathbf{3.5}$$

Summary: Rate Ratio

- A ratio of two average rates is called a rate ratio (i.e., an incidence density ratio, hazard ratio)
- The formula for the rate ratio (IDR) is given by:

$$IDR = \frac{\frac{I_1}{PT_1}}{\frac{I_0}{PT_0}}$$

where I_1 and I_0 are the number of new cases and PT_1 and PT_0 are the accumulated person-time for groups 1 and 0, respectively.

- As with the RR and the OR, the IDR can be >1, <1, or =1.

Odds Ratio Approximation to Rate Ratio

We have already seen that, in case-control studies, the exposure odds ratio estimates the risk ratio for a corresponding follow-up study provided the health outcome is rare. In this activity, we demonstrate that under certain conditions, the odds ratio from a case-control study estimates the rate ratio from a comparable cohort study that uses person-time information. It is not necessary to assume a rare disease for this approximation to hold. The key condition required is that the source population from which the cases and controls are derived be in **steady state**, a term we will define shortly.

Case-Control		Follow-up (Cohort)
$\hat{EOR}$	≈	$\hat{RR}$ if disease is rare
Case-Control		Follow-up (Cohort)
$\hat{EOR}$	≈	$\hat{IDR}$ no rare disease source population must be in **Steady State**

Suppose the source population at time t_0 contains N_1 disease-free exposed persons and N_0 disease-free unexposed persons. Suppose, further, that after Δt years of follow-up, I_1 new cases of disease develop from those exposed at time T_0 and I_0 new cases develop from those unexposed at time T_0. Suppose further that the population undergoes no major demographic shifts, so that the size of the source population is essentially constant over the Δt years of follow-up. The source population here is then considered to be stable or in steady state.

Follow-up of the Source Population for Δt years

	Exposed	Not Exposed
New Cases in $(t_0, t_0+\Delta t)$	I_1	I_0
Non-Cases	$N_1 - I_1$	$N_0 - I_0$
Total Disease-Free at t_0	N_1	N_0

Source Population is constant $\Rightarrow$ Steady State

Under such steady state conditions, the person-years of observation for each exposure group can be approximated using the formula PT = N* x Δt where N* denotes the size of this stable source population for each exposure group. Thus, the person-years for exposed persons is N_1 x Δt and the person-years for unexposed persons is N_0 x Δt.

Follow-up of the Source Population for Δt years
Steady State conditions

	Exposed	Not Exposed
New Cases in $(t_0, t_0+\Delta t)$	I_1	I_0
Total Disease-Free at t_0	N_1	N_0

$PT = N * \Delta t$

$$\text{Rate Ratio} = \widehat{IDR} = \frac{\frac{I_1}{PT_1}}{\frac{I_0}{PT_0}} = \frac{\frac{I_1}{N_1 \Delta t}}{\frac{I_0}{N_0 \Delta t}}$$

The rate ratio for this cohort, which we have denoted IDR, is then given by substituting the expressions for PT_1 and PT_0 into the IDR formula to obtain the expression shown here.

We can cancel out Δt from this expression to simplify it as follows:

$$\widehat{IDR} = \frac{\frac{I_1}{I_0}}{\frac{N_1}{N_0}}$$

Now, suppose we conduct a case-control study using this source population. We select a random sample of incident cases that developed between time t_0 and $t_0 + \Delta t$ and a random sample of controls from the source population. We then determine prior exposure status for cases and controls. This gives us the following two-way table for the case-control data.

General Form of the 2 by 2 Table

	Exposure Status Yes	No
Cases	a	b
Controls	c	d

Exposure Odds

$a/b = I_1 / I_0$

$c/d = N_1 / N_0$ source pop

$$\hat{IDR} = \frac{I_1/I_0}{N_1/N_0} = \frac{a/b}{c/d} = \frac{ad}{bc} = \hat{EOR}$$

From this table, exposure odds for the cases is given by a/b and the exposure odds for the controls is given by c/d. Assuming that the sampling is "blind" to exposure status, we would expect the proportion of cases in the study who were exposed to be equal, on average, to the proportion of cases in the full cohort who were exposed. With a little algebra it follows that a/b, the exposure odds among cases, equals I_1/I_0, the corresponding exposure odds in the source population. Similarly, it can be shown that the exposure odds for the controls, c/d should, on average, equal the exposure odds among the disease-free persons in the source population. If we now substitute a/b for I_1/I_0 and c/d for N_1/N_0 in the formula for the rate ratio we obtain ad/bc, the odds ratio from the case control study

We have thus shown that under the steady state conditions and using incident cases, the exposure odds ratio will approximate the rate ratio, without requiring the disease to be rare.

Summary: When EOR approximates the IDR

- Under steady state conditions, the odds ratio from a case-control study will approximate the rate ratio from a comparable cohort study that uses person-time information.
- This approximation does not require the rare disease assumption.
- Steady-state means that there is not a major shift in the demographic make-up of the source population.

Quiz (Q5.13)

Data is shown below for a follow-up study to compare mortality rates among diabetics and non-diabetics.

1. The mortality rate for diabetics is . . . **???**
2. The mortality rate for non-diabetics is . . . **???**
3. The rate ratio is . . . . . **???**

Choices

13.9 **13.9 per 1000 person-years** **2.8** **2.8 per 1000 person-years**
38.7 **38.7 per 1000 person-years**

	Diabetic	Non-diabetic	Total
Dead	72	511	583
Alive	146	3,312	3,458
Person-Years	1,862.4	36,532.2	38,394.6

Quiz continued on next page

4. The rate ratio comparing the mortality rates of diabetics with non-diabetics is 2.8. Which of the following is the correct interpretation of this measure?

   A. Those with diabetes are 2.8 times more likely to die than those without.
   B. People are 2.8 times more likely to die of diabetes than any other illness
   C. Death among diabetics is occurring at a rate of 2.8 times that of non-diabetics

Compute Measures of Association for Person-Time Data in Data Desk

In this activity an introduction to computing measures of association for person-time data using the Data Desk program is provided.

Nomenclature

Table setup for cohort, case-control, and prevalence studies:

	Exposed	Not Exposed	Total
Disease/cases	a	b	m_1
No Disease/controls	c	d	m_0
Total	n_1	n_0	n

Table setup for cohort data with person-time:

	Exposed	Not Exposed	Total
Disease (New cases)	I_1	I_0	I
No Disease	-	-	-
Total disease-free person-time	PT_1	PT_0	PT

Δt Change in time
EOR Exposure odds ratio; odds of exposure in diseased divided by the odds of exposure in nondiseased
I Average incidence or total number of new cases
I_0 Number of new cases in nonexposed
I_1 Number of new cases in exposed
IDR Incidence density ratio ("rate ratio"): IDR=rate in exposed/rate in nonexposed
N Size of population under study
N_0 Size of population under study in nonexposed at time zero
N_1 Size of population under study in exposed at time zero
OR Odds ratio: ad/bc
P Probability of an event
$P(D \mid E)$ Probability of disease given exposed
$P(D \mid \text{not } E)$ Probability of disease given not exposed
$P(E \mid D)$ Probability of exposure given diseased
$P(E \mid \text{not } D)$ Probability of exposure given not diseased
POR Prevalence odds ratio; an odds ratio calculated with prevalence data
PT Disease-free person-time
PT_0 Disease-free person-time in nonexposed
PT_1 Disease-free person-time in exposed
R_0 Risk in unexposed
R_1 Risk in exposed
ROR Risk odds ratio; an odds ratio calculated from cohort risk data
RR Risk ratio: risk in exposed divided risk in unexposed
T or **t** Time

Formulae

$IDR = (I_1/PT_1) / (I_0/PT_0)$

$Odds = P / (1-P)$

Odds ratio = ad/bc

$ROR = RR * f$ where $f=(1-R_0)/(1-R_1)$

$RR = R_1 / R_0$

References

Doll R, and Hill AB. Smoking and carcinoma of lungs: preliminary report. Br Med J 1950;2:739-48.

Dyer AR, Stamler J, Shekelle RB. Serum cholesterol and mortality from coronary heart disease in young, middle-aged, and older men and women in three Chicago epidemiologic studies. Ann of Epidemiol 1992;2(1-2): 51-7.

Greenberg RS, Daniels SR, Flanders WD, Eley JW, Boring JR. Medical Epidemiology (3rd Ed). Lange Medical Books, New York, 2001.

Hammond EC, Horn D. The relationship between human smoking habits and death rates. JAMA 1958;155:1316-28.

Johansson S, Bergstrand R, Pennert K, Ulvenstam G, Vedin A, Wedel H, Wilhelmsson C, Wilhemsen L, Aberg A. Cessation of smoking after myocardial infarction in women. Effects on mortality and reinfarctions. Am J Epidemiol 1985;121(6): 823-31.

Kleinbaum DG, Kupper LL, Morgenstern H. Epidemiologic Research: Principles and Quantitative Methods. John Wiley and Sons Publishers, New York, 1982.

Spika JS, Dabis F, Hargett-Bean N, Salcedo J, Veillard S, Blake PA. Shigellosis at a Caribbean Resort. Hamburger and North American origin as risk factors. Am J Epidemiol 1987;126 (6): 1173-80.

Steenland K (ed.). Case studies in occupational epidemiology. Oxford University Press, New York, NY: 1993.

Uitterlinden AG, Burger H, Huang Q, Yue F, McGuigan FE, Grant SF, Hofman A, van Leeuwen JP, Pols HA, Ralson SH. Relation of alleles of the collagen type I alpha1 gene to bone density and the risk of osteoporotic fractures in postmenopausal women. N Engl J Med 1998;338(15):1016-21.

Wynder EL, Graham EA. Tobacco smoking as a possible etiologic factor in bronchogenic carcinoma: a study of six hundred and eighty-four proved cases. JAMA 1950;143:329-36.

Homework

ACE-1. Measures of Effect: Chewing Tobacco vs. Oral Leukoplakia

A study is conducted to investigate the association between chewing tobacco and oral leukoplakia (a precancerous lesion) among currently active professional baseball players in the southeastern United States. A roster of all active players is obtained (n=500). All potential study subjects agree to participate. Each subject has an interview regarding current use of chewing tobacco and has his mouth examined by a dentist. Of the 500 subjects, 125 subjects chew tobacco and 375 do not chew tobacco. Of the chewers, 25 have evidence of oral leukoplakia. Of the non-chewers, 15 have evidence of oral leukoplakia. All 500 players were followed for a period of 5 years. Of those who had evidence of oral leukoplakia, 18 died of some type of oral cancer.

a. Draw a 2 x 2 table demonstrating the relationship between chewing tobacco and oral leukoplakia. In drawing this table, put the exposure variable on the columns and the health outcome variable on the rows.
b. Draw a second 2 x 2 table demonstrating the relationship between chewing tobacco and oral leukoplakia, but this time, put the exposure variable on the rows and the health outcome variable on the columns.
c. Using the table drawn in part a, compute the prevalence ratio and the prevalence odds ratio of oral leukoplakia for chewers compared to non-chewers. Are these two estimates close to one-another? Why are these prevalence measures and not incidence measures?
d. Using the table drawn in part b, compute the prevalence ratio and the prevalence odds ratio of oral leukoplakia for chewers compared to non-chewers. Are these estimates equal to their corresponding estimates computed using the table drawn in part a? Explain.
e. Ignoring the issue of statistical inference and the control of other variables, what do these results say about the relationship between chewing tobacco and the presence of oral leukoplakia?
f. Calculate the case-fatality rate (actually, a risk) in this study. Why is this a measure of risk?
g. Based on the information provided, why can't you evaluate whether tobacco chewers have a higher case-fatality risk than non-chewers?

ACE-2. Rate Ratios: Colon Cancer Deaths

The following table shows the number of colon cancer deaths and person-years of risk by the frequency of aspirin for males and females.

Table 1. Rates of death from colon cancer, according to frequency of aspirin use in the cohort ***before*** patients with illness at enrollment were excluded.

		Aspirin Use (times per month)			
		0	<1	1-15	≥16
Men	Number of Deaths	378	184	127	85
	Person-years at risk	646,346	486,620	389,083	201,636
	Death rate per 100,000	_______	_______	_______	_______
	Rate ratio	1.00	_______	_______	_______
Women	Number of Deaths	284	157	100	73
	Person-years at risk	705,064	671,927	505,854	265,424
	Death rate per 100,000	_______	_______	_______	_______
	Rate ratio	1.00	_______	_______	_______

a. Calculate the death rates and the rate ratios for each of the aspirin-use categories in the above table.
b. Given the results in this table, what is your conclusion about the association between the use of aspirin and fatal colon cancer?

ACE-3. Rate Ratios: NSAIDS's

NSAID's (i.e., non-steroidal anti-inflammatory drugs) are prescribed or taken over-the-counter for acute and chronic, perceived and diagnosed illnesses. For this reason, the investigators in the study described in question 2 also analyzed the data excluding those individuals with selected illnesses at the start of follow-up. Table 2 shows the number of colon cancer deaths by the frequency of aspirin-use after exclusion of those subjects with selected illnesses.

Table 2. Rates of death from colon cancer, according to frequency of aspirin use in the cohort ***after*** excluding patients with illness at time of enrollment.

		Aspirin Use (times per month)			
		0	<1	1-15	≥16
Men	Number of Deaths	171	101	63	28
	Person-years at risk	487,932	385,321	302,106	116,947
	Death rate per 100,000	_______	_______	_______	_______
	Rate ratio	1.00	_______	_______	_______
Women	Number of Deaths	126	98	54	32
	Person-years at risk	521,467	531,469	396,956	175,409
	Death rate per 100,000	_______	_______	_______	_______
	Rate ratio	1.00	_______	_______	_______

a. Calculate the rates and rate ratios for each of the aspirin-use categories in the above table.
b. Given the results in both Table 1 (from question 2) and Table 2, what do you conclude about the association between the use of aspirin and fatal colon cancer?

ACE-4. Incidence Measures of Effect: Quitting Smoking

The following data come from a study of self-help approaches to quitting smoking. Smokers wanting to quit were randomized into one of four groups (C = control, M = quitting manual, MS = manual + social support brochure, MST = manual + social support brochure + telephone counseling). Smoking status was measured by mailed survey at 8, 16, and 24 months following randomization. These are the 16-month results:

Randomization Group

Status	C	M	MS	MST	Total
Quit	84	71	67	109	331
Smoking	381	396	404	365	1546
Total	465	467	471	474	1877

a. The "quit rate" is calculated as the proportion abstinent (quit) at the time of follow-up. What was the overall 16-month quit rate for all subjects? Based upon quit rates, which of the intervention groups was the least successful as of the 16-month follow-up? Justify your answer.
b. Is this "quit rate" a cumulative incidence-type measure or an incidence density-type measure? Justify your answer.
c. Compare the quit rate for the MST group with that of the control group by calculating both a CIR and an OR. Show your work. Provide an interpretation of the CIR.

ACE-5. Incidence Density Ratio: Radiotherapy Among Children

In a study of adverse effects of radiotherapy among children in Israel, 10,834 irradiated children were identified from original treatment records and matched to 10,834 non-irradiated comparison subjects selected from the general population. Subjects were followed for a mean of 26 years. Person-years of observation were: irradiated subjects, 279,901 person-years; comparison subjects, 280,561 person-years. During the follow-up period there were 49 deaths from cancer in irradiated subjects, and 44 in the non-irradiated population comparison subjects.

a. What are the rates of cancer death (per 105 person-years) in each of the two groups?
b. Calculate and interpret the IDR for cancer death comparing irradiated and non-irradiated subjects.

ACE-6. Odds Ratio: Alcohol Consumption vs. Myocardial Infarction

A case-control study was conducted to assess the relationship of alcohol consumption and myocardial infarction (MI). Cases were men aged 40 to 65 years who had suffered their first MI during the six months prior to recruitment into the study. A group of age-matched men who had never experienced an MI were selected as controls. Data from this study are summarized below:

	Exposed	Unexposed	Totals
Diseased	158	201	359
Nondiseased	252	170	422
Totals	410	371	781

a. What is the odds of exposure among the controls?
b. Calculate and interpret the exposure odds ratio for these data.
c. Do you think that the OR calculated in part b above is a good estimate of the corresponding risk ratio for the relationship between alcohol and MI? Why or why not?

Answers to Study Questions and Quizzes

Q5.1

1. The five-year risk for continuing smokers is 4½ times greater than the risk for smokers who quit.
2. The risk ratio is very close to 1.0, which indicates no meaningful difference between the risks for the two groups.
3. Think of an inverse situation.
4. You should have the hang of this by now.

Q5.2

1. 2
2. 1
3. 0.5
4. 0.0104
5. 0.0236
6. 0.44 – In general, the risk ratio that compares two groups is defined to be the risk for one group divided by the risk for the other group. It is important to clearly specify which group is in the numerator and which group is in the denominator. If, for example, the two groups are labeled *group 1* and *group 0*, and the risk for group 1 is in the numerator, then we say the risk ratio compares group 1 to group 0.

Q5.3

1. The odds that a case ate raw hamburger is about two ½ times the odds that a control subject ate raw hamburger.
2. Because the odds ratio is so close to being equal to one, this would be considered a null case, meaning that the odds that a case ate raw hamburger is about the same as the odds that a control subject age raw hamburger.
3. An odds ratio less than one means that the odds that a case subject ate raw hamburger is less than the odds that a control subject ate raw hamburger.
4. You should have the hang of this by now.

Q5.4

1. Not possible, odds ratio – The risk of disease is defined as the proportion of initially disease-free population who develop the disease during a specified period of time. In a case-control study, the risk cannot be determined.
2. 1350/1357
3. 1296/1357
4. 192.86
5. 21.25
6. 9.08 – In general, the odds ratio that compares two groups is defined to be the odds for the cases divided by the odds for the controls. The odds for each group can be calculated by the formula P/(1-P), where P is the probability of exposure.
7. 3
8. 1
9. 0.333

Q5.5

1. Of course! It is possible, for example, that mayonnaise actually contained the outbreak-causing bacteria and maybe most of the cases that ate raw hamburger used mayonnaise.

Q5.6

1. 683, 86, 0.77
2. odds, exposure, less than, odds, controls – If the estimated odds ratio is less than 1, then the odds of exposure for cases is less than the odds of exposure for controls. If the estimated odds ratio is greater than 1, then the odds of exposure for cases is greater than the odds of exposure for controls.

Q5.7

1. 2.18
2. more likely
3. prevalence

Q5.8

1. That depends on the disease being considered and on the time-period of follow-up over which the risk is computed. However, for most chronic diseases and short time periods, a risk of .01 is not rare.
2. Yes, because even though the risk may not be rare, it may be small enough so that the ROR and the RR are approximately the same.

Q5.9

1. f = (1 – 0.17) / (1 – 0.36) = 1.30
2. No, since for these data, the estimated RR equals 2.1 whereas the estimate ROR equals 2.7.
3. f = (1 – 0.085) / (1 – 0.180) = 1.12
4. Yes, since the estimated RR is again 2.1, (0.180/0.085), but the estimated ROR is 2.4.

5. f=1.05
6. Yes, since the estimated ROR is now 2.2.
7. In the context of the quit smoking example, risks below 0.10 for both groups indicate a "rare" disease.

Q5.10

1. The risk ratio in this study is 0.0805 divided by 0.0536, which equals 1.50.
2. The risk odds ratio is 47/537 divided by 64/1130 equals 1.54.
3. f=(1-0.0536) / (1-0.0805) = 1.03. The ROR = 1.03*RR = 1.03*1.50=1.54.

Q5.11

1. 9.8
2. 36.0
3. No
4. No – The risk ratio that compares two groups is defined to be the risk for one group divided by the risk for the other group. The odds ratio can be calculated by the cross product formula ad/bc. In general, a disease is considered "rare" when the OR closely approximates the RR.
5. 2.44
6. 2.49
7. Yes
8. Yes

Q5.12

1. EOR, RR, bias, controls, incident, prevalent
2. 9
3. 9
4. Yes – A disease is considered rare when the ROR closely approximates the RR.
5. 50
6. 9.00

Q5.13

1. 38.7 per 1000 person-years – The mortality rate for diabetics equals 72/1,862.4 person-years = 38.7 per 1000 person-years.
2. 13.9 per 1000 person-years – The mortality rate for non-diabetics equals 511/36,653.2 person-years = 13.9 per 1000 person-years.
3. 2.8 – The rate ratio is 38.7 per 1000 person-years/13.9 per 1000 person-years = 2.8.
4. C

LESSON 6

MEASURES OF POTENTIAL IMPACT

6-1

*In the previous lesson on Measures of Effect, we focused exclusively on **ratio measures of effect**. In this lesson, we consider **difference measures of effect** and other related measures that allow the investigator to consider the potential public health impact of the results obtained from an epidemiologic study.*

The Risk Difference – An Example

*The **risk difference** is the difference between two estimates of risk, whereas the **risk ratio** is the ratio of two risk estimates. We illustrate a risk difference using a cohort study of heart attack patients who either continue or quit smoking after their heart attack.*

Consider again the results of a five-year follow-up study to determine whether or not smokers who have had a heart attack will reduce their risk for dying by quitting smoking. The estimated risk ratio is 2.1, which means that the risk for continuing smokers was 2.1 times the risk for smokers who quit.

The Risk Difference

Heart Attack Patients	Smoke	Quit	Total
Death	27	14	41
Survival	48	67	115
Total	75	81	156

Five-year risks of dying
continuing smokers: 27/75 = 0.36
smokers who quit: 14/81 = 0.17

Risk Ratio = 0.36/ 0.17 = 2.1

Risk Difference = 0.36 - 0.17 = .19

We now focus on the difference between the two estimates of risk, rather than their ratio. What kind of interpretation can we give to this difference estimate? The **risk difference** (**RD**) of 0.19 gives the **excess risk** associated with continuing to smoke after a heart attack. The estimated risk, 0.17, of dying in the quit smoking group is the background or "expected" level to which the risk of 0.36 in the continuing smokers group, is compared.

Study Questions (Q6.1)

1. How many deaths would have occurred among the 75 patients who continued to smoke after their heart attack if these 75 patients had quit smoking instead?
2. How many excess deaths were there among the 75 patients who continued to smoke after their heart attack?
3. What is the proportion of excess deaths among continuing smokers?

The null value that describes "no excess risk" is 0. There would be no excess risk if the two estimated risks were equal. Because the risk difference describes excess risk, it is also called the **attributable risk**. It estimates the additional risk “attributable” to the exposure.

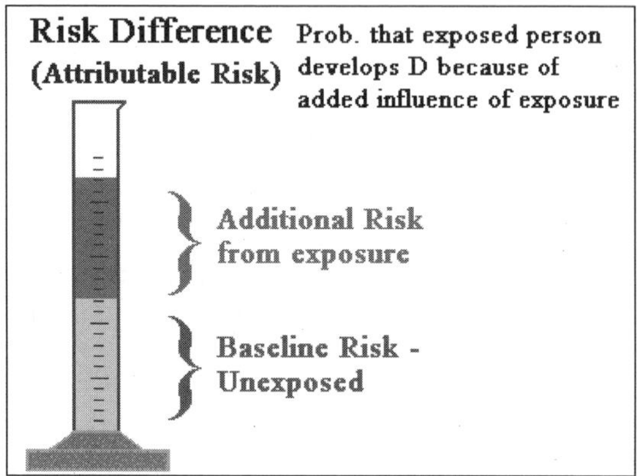

The risk difference, therefore, can be interpreted as the probability that an exposed person will develop the disease because of the additional influence of exposure over the baseline risk. In this example, the five-year attributable risk of 0.19 estimates the probability that continuing smokers will die because they have continued to smoke.

Study Questions (Q6.1) continued

4. If the study involved 1,000 heart attack patients who continued to smoke after their heart attack, how many deaths could be avoided (i.e., attributable to exposure) for a risk difference of 0.19 if all patients quit smoking?
5. How might you evaluate whether the excess risk of 0.19 is clinically (not statistically) excessive
6. Can you think of a reference value to compare with the excess risk? If so, how would you interpret this relative comparison?

Summary

- The risk difference is the difference between two estimates of risk.
- The null value of the risk difference is 0, whereas the null value of the risk ratio is 1.
- The risk difference reflects an **excess risk** attributable to exposure.
- Excess risk describes the proportion of cases that could be avoided among exposed subjects if exposed subjects had the same risk as unexposed subjects.
- The risk difference is also called the **attributable risk**.

The Mathematics Behind Excess Risk

The concept of **excess risk** can be explained mathematically as follows: Consider the following 2 x 2 table that describes data from a cohort study that allows you to estimate individual risk using cumulative incidence:

	Exposed	Not Exposed	Total
Cases	a	b	m_1
Non-cases	c	d	m_0
Total	n_1	n_0	n

From this table, the estimated risks for exposed, unexposed and total groups are given by the following estimated cumulative incidence formulae:

Continued on next page

The Mathematics Behind Excess Risk (continued)

$$\hat{CI}_E = \frac{a}{n_1}, \hat{CI}_{\text{not E}} = \frac{b}{n_0}, \hat{CI}_{\text{Total}} = \frac{m_1}{n} = \frac{(a+b)}{n}$$

From the above definitions, we can compute the excess number of cases attributable to exposure, which we denote **N(AE),** as follows:

$$\begin{aligned} N(AE) &= \text{\# of new cases among exposed - \# of new cases expected if exposure absent} \\ &= n_1 \text{ x } CI_E - n_1 \text{ x } CI_{\text{not E}} \\ &= n_1 \text{ x } [CI_E - CI_{\text{not E}}] \\ &= n_1 \text{ x Risk Difference} \end{aligned}$$

Dividing both sides of the final equation by n_1, it follows that:

$$\text{Risk Difference} = \frac{N(AE)}{n_1} = \text{excess risk among the exposed subjects}$$

An alternative way to determine N(AE) that considers all cases instead of exposed cases is as follows:

$$\begin{aligned} N(AE) &= \text{\# of total cases - \# of total cases expected if exposure is absent} \\ &= n \text{ x } CI_{\text{Total}} - n \text{ x } CI_{\text{not E}} \\ &= n\left(\frac{[n_1 \times CI_E \times CI_{\text{not E}}]}{n}\right) - n \times CI_{\text{not E}} \\ &= [n_1 \text{ x } CI_E + n_0 \text{ x } CI_{\text{not E}}] - (n_1 + n_0) \text{ x } CI_{\text{not E}} \\ &= [n_1 \text{ x } CI_E + n_0 \text{ x } CI_{\text{not E}}] - [n_1 \text{ x } CI_{\text{not E}} + n_0 \text{ x } CI_{\text{not E}}] \\ &= n_1 \text{ x } [CI_E - CI_{\text{not E}}] \\ &= n_1 \text{ x Risk Difference} \end{aligned}$$

Difference Measures of Effect

Difference measures of effect can be computed in randomized clinical trial, cohort, and cross-sectional studies, but *not* in case-control studies. In cohort studies that estimate individual risk using cumulative incidence measures, the difference measure of interest is called the **risk difference**. It is estimated as the difference between $\hat{CI}_1$, the estimated cumulative incidence for the exposed group, and $\hat{CI}_0$, the estimated cumulative incidence for the unexposed group

In cohort studies that estimate average rate using person-time information, the difference measure is the **rate difference**. It can be estimated as the difference between two estimated rates, or incidence densities, $\hat{ID}_1$ and $\hat{ID}_0$.

In cross-sectional studies, the difference measure is called the **prevalence difference**, and is estimated as the difference between two prevalence estimates.

Cohort studies

Individual risk ($\hat{CI}$):

Risk Difference ($\hat{CID}$) = $\hat{CI}_1 - \hat{CI}_0$

Average rate ($\hat{ID}$):

Rate Difference ($\hat{IDD}$) = $\hat{ID}_1 - \hat{ID}_0$

Cross-sectional studies

Prevalence ($\hat{P}$):

Prevalence Difference ($\hat{PD}$) = $\hat{P}_1 - \hat{P}_0$ (exposed, unexposed)

Difference measures of effect cannot be estimated in case-control studies because in such studies neither risk, rate, nor prevalence can be appropriately estimated.

We'll illustrate the calculation of the rate difference. We again consider data on the relationship between serum cholesterol level and mortality from a 1992 study of almost 40,000 persons in Chicago, Illinois. Among white males ages 25-39 with borderline-high cholesterol, there were 71.1 deaths per 100,000 person-years. Among the comparison group, there were 20.5 deaths per 100,000 person-years.

1992, 40,000 persons, Chicago (white males, ages 25-39)

	Cholesterol Level	
	Borderline-high	Normal
Deaths	26	14
Person-years	36,581	68,239
	$\hat{ID}_1 = \frac{71.1}{100{,}000\text{ PY}}$	$\hat{ID}_0 = \frac{20.5}{100{,}000\text{ PY}}$

The estimated rate ratio that compares these two groups is 3.5. The estimated rate difference, or IDD, is 50.6 deaths per 100,000 person years. What kind of interpretation can we give to this rate difference?

$$\hat{IDR} = \frac{\frac{71.1}{100{,}000\text{ PY}}}{\frac{20.5}{100{,}000\text{ PY}}} = 3.5 \qquad \hat{IDD} = \frac{71.1}{100{,}000\text{ PY}} - \frac{20.5}{100{,}000\text{ PY}} = \frac{50.6}{100{,}000\text{ PY}}$$

The rate difference indicates an excess rate of 50.6 deaths per 100,000 person years associated with having a borderline-high cholesterol when compared to normal cholesterol. Here, we are using the estimated rate of CHD-related deaths in the unexposed group as the *background* or *expected* level to which the rate in the exposed group is compared. The rate difference is also called the **attributable rate** since it gives the additional rate attributable to the exposure.

Study Questions (Q6.2)

1. How many CHD-related deaths per 100,000 person years (i.e., py) could be avoided (i.e., attributable to exposure) among persons with borderline-high cholesterol if these persons could lower their cholesterol level to normal values?
2. What is the excess rate of CHD-related deaths per 100,000 py among persons with borderline-high cholesterol?
3. How might you evaluate whether the excess rate of 50.6 is clinically (not statistically) excessive?
4. Can you think of a reference value to compare with the excess rate? If so, how would you interpret this relative comparison?

Summary

- ❖ Difference measures can be computed in cohort and cross-sectional studies, but not in case-control studies.
- ❖ If **risk** is estimated, the difference measure is the **risk difference**.
- ❖ If **rate** is estimated, the difference measure is the **rate difference**.
- ❖ If **prevalence** is estimated, the difference measure is the **prevalence difference**.
- ❖ Difference measures of effect allow you to estimate the (excess) risk attributable to exposure over the background risk provided by the unexposed.

The Number Needed to Treat

The risk difference describes the excess risk of disease that is attributable to exposure. The risk difference can also be used to compute the "number needed to treat" or NNT. The NNT represents the number of patients that must be treated to prevent one outcome from occurring.

As an example, we use data from the British Medical Research Council in a study of patients with mild hypertension. Treatment of hypertension with either a b-antagonist or diuretic was compared to use of a placebo. The 10-year cumulative incidence of stroke in patients getting the placebo was 2.6%; in patients who were treated, the 10-year cumulative incidence was 1.4%.

Treatment

Placebo VS. b-antagonist diuretic

10-year incidence of stroke:

$\hat{CI}_1 = 2.6\%$ $\hat{CI}_0 = 1.4\%$

Study Questions (Q6.3)

1. What is the risk difference?

We now want to know how many patients with hypertension we need to treat in order to prevent one stroke. The formula used to compute the NNT is 1 divided by the risk difference.

$$NNT = \frac{1}{\hat{CI}_1 - \hat{CI}_0}$$

Study Questions (Q6.3) continued

2. How many patients with hypertension do we need to treat to prevent one stroke?
3. For how long do we need to treat these patients?

The results from this study showed that if 83 subjects with hypertension were treated with either b-antagonists or diuretics during 10 years, the incidence of one stroke could be prevented.

Summary

- ❖ The risk difference can be used to compute the "number needed to treat" (NNT).
- ❖ The NNT represents the number of patients that must be treated to prevent one outcome from occurring.
- ❖ The formula used to compute the NNT is 1 divided by the risk difference.

The Number Needed to Treat – Definition and Rationale

In the example used in this activity, the 10-year cumulative incidence of stroke was determined for patients with mild hypertension. For those treated with either a b-antagonist or a diuretic, the estimated CI was 1.4%. For those given a placebo (i.e., not treated), the estimated CI was 2.6%. The risk difference is then calculated as 2.6% - 1.4% = 1.2%. In other words, the risk of stroke attributable to use of a placebo (or non-treatment) was 1.2%

The **number needed to treat** (**NNT**) is defined as the expected number of patients who must be treated with an experimental therapy in order to prevent one additional adverse outcome event (or, depending on the context, to expect one additional beneficial outcome). We can calculate the NNT by inverting the value of the risk difference, i.e.,

NNT = 1/(Risk Difference).

For this example, therefore, NNT = 1/.012, which gives 83.3. In other words, 83 patients must be treated with b-antagonist or diuretic to prevent one stroke.

Using the above example, the rationale for the NNT formula is given as follows: If 1000 patients were treated with either a b-antagonist or a diuretic, we would expect 14 strokes (1.4%) over 10 years of follow-up. However, if these 1000 patients were given a placebo instead of treatment, we would expect 26 strokes (2.6%) over 10 years.

Thus, for 1000 patients followed for 10 years, we would expect to prevent 26 - 14 = 12 strokes (2.6% - 1.4% = 1.2%) if all 1000 patients received the treatment instead of all patients not receiving the treatment (i.e., getting a placebo). Since 1000/12 = 83.3/1, this means that we could expect to prevent one stroke over 10 years for every 83 patients who are treated, i.e., 12 is to 1000 as 1 is to 83.3.

Quiz (Q6.4)

Which of the following terms are synonymous with risk difference?

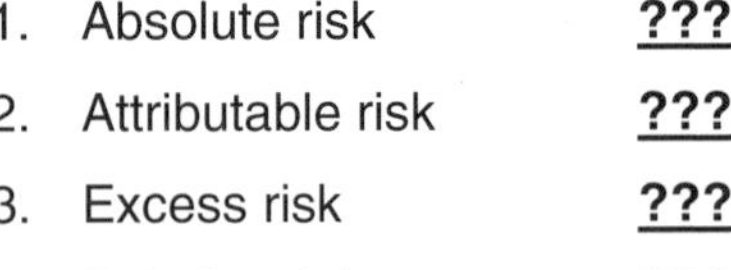

1. Absolute risk ???
2. Attributable risk ???
3. Excess risk ???
4. Relative risk ???

Choices
No **Yes**

During the 1999 outbreak of West Nile encephalitis in New York, incidence varied by location. The reported rates were:

Queens	16.4 per million	Bronx	7.5 per million
Brooklyn	1.3 per million	Manhattan	0.7 per million
Staten Island	0.0 per million	Total NYC	6.1 per million

Quiz continued on next page

To calculate the rate difference for residents of Queens, which location(s) could be used for the baseline or expected rate?

5. Queens **???**
6. Bronx **???**
7. Brooklyn **???**
8. Manhattan **???**
9. Staten Island **???**
10. Total NYC **???**
11. Bronx+Brooklyn+Manhattan+Staten Island **???**

Choices
No **Yes**

12. Calculate the rate difference between Queens and Manhattan. **???**

Choices
15.7 **15.7 per million** **23.4** **23.4 per million**

Investigators interviewed all persons who had attended the Smith-Jones wedding two days earlier, comparing the proportion who developed gastroenteritis among those who did and those who did not eat certain foods. They now want to determine the impact of eating potato salad on gastroenteritis.

13. The appropriate measure of potential impact is **???**.

Investigators conducted a cross-sectional survey, identified respondents who had been diagnosed with diabetes, and calculated an index of obesity using reported heights and weights. They now want to determine the impact of obesity on diabetes.

14. The appropriate measure of potential impact is **???**.

Investigators enrolled matriculating college freshmen into a follow-up study. The investigators administered questionnaires and drew blood each year to identify risk factors for and seroconversion to Epstein-Barr virus (the etiologic agent of mononucleosis). Using person-years of observation, the investigators now want to determine the impact of residing in a co-ed dormitory on EBV seroconversion.

15. The appropriate measure of potential impact is **???**.

Choices

not calculable **odds difference** **prevalence difference** **rate difference** **risk difference**

Difference versus Ratio Measures of Effect

Consider this table of hypothetical information describing the separate relationships of four different exposures to the same disease.

LOCATION	CHEWT	COFFEE	ALCOHOL
CI_1 = 0.010	CI_1 = 0.005	CI_1 = 0.050	CI_1 = 0.050
CI_0 = 0.010	CI_0 = 0.001	CI_0 = 0.046	CI_0 = 0.010
RR = 1.000	RR = 5.000	RR = 1.087	RR = 5.000
RD = 0.000	RD = 0.004	RD = 0.004	RD = 0.040

First focus on location, rural versus urban, for which the risk ratio is one and the risk difference is 0. There is no evidence of an effect of location on the disease, whether we consider the risk ratio or the risk difference. In fact, if the risk ratio is exactly 1, then the risk difference must be exactly 0, and vice versa.

Location: Rural vs. Urban
RR=1.000 RD = 0.000
No Effect

Now, let's look at the effect of chewing tobacco on disease. The risk ratio for chewing tobacco is 5; this indicates a very strong relationship between chewing tobacco and the disease. But, the risk difference of .004 seems quite close to zero, which suggests no effect of chewing tobacco.

Chewing Tobacco
RR = 5.000 RD = 0.004
Strong Effect Small Effect

Thus, it is possible to arrive at a different conclusion depending on whether we use the risk ratio or the risk difference. Does only one of these two measures of effect give the correct conclusion, or are they both correct in some way? Actually, both measures give meaningful information about two different aspects of the relationship between exposure and disease.

Let's now compare the effect of chewing tobacco with the effect of coffee drinking.

Coffee Drinking
RR = 1.087 RD = 0.004

The risk ratios for these two exposures are very different, yet the risk differences are exactly the same and close to zero. There appears to be little, if any, effect of coffee drinking. So, is there or is there not an effect of tobacco chewing?

If we ask whether or not we would consider chewing tobacco to be a strong risk factor for the disease, our answer would be yes, since a chewer has 5 times the risk of a non-chewer for getting the disease. That is, chewing tobacco appears to be associated with the etiology of the disease, since it is such a strong risk factor.

However, if we ask whether chewing tobacco poses a public health burden in providing a large case-load of patients to be treated, our answer would be no. To see the public health implications, recall that the risk difference of .004 for chewing tobacco gives the excess risk that would result if chewing tobacco were completely eliminated in the study population. Thus, out of, say, 1000 chewers, an excess of 1000 times 0.004, or only 4 chewers would develop the disease from their tobacco chewing habit. This is not a lot of patients to have to treat relative to the 1000 chewers at risk for the disease.

Study Questions (Q6.5)

1. Compare the effect of chewing tobacco with the effect of alcohol consumption on the disease. Do they both have the same effect in terms of the etiology of the disease?
2. Do chewing tobacco and alcohol use have the same public health implications on the treatment of disease?
3. Explain your answer to the previous question in terms of the idea of excess risk.

Summary

- ❖ If the risk ratio is exactly 1, then the risk difference is exactly 0, and vice versa, and there is no effect of exposure on the health outcome.
- ❖ If the risk ratio is very different from 1, it is still possible that the risk difference will be close to zero.
- ❖ If the risk difference is close but not exactly equal to 0, it is possible that the risk difference will be large enough to indicate a public health problem for treating the disease.
- ❖ Ratio measures are primarily used to learn about the etiology of a disease or other health outcome.
- ❖ Difference measures are used to determine the public health importance of a disease or other health outcome.

Quiz (Q6.6)

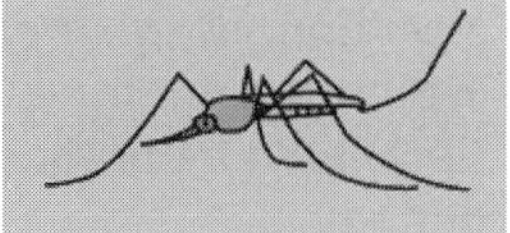

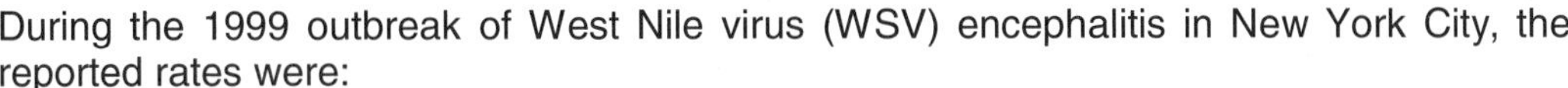

During the 1999 outbreak of West Nile virus (WSV) encephalitis in New York City, the reported rates were:

Queens	16.4 per million population
Rest of NYC	2.4 per million
Total NYC	6.1 per million

Label each of the following statements as **True** or **False**.

1. If Queens had experienced the same WNV rate as the rest of NYC, 10.3 fewer cases per million would have occurred there, i.e., the rate difference is 10.3 per million. . . . ???
2. The excess rate in Queens was 14.0 cases per million (compared to the rest of NYC) ???
3. The attributable rate (i.e., rate difference) in Queens was 16.4 cases per million. . ???
4. The most common measure of effect for comparing Queens to the rest of NYC is 6.8. ???

Determine whether each of the following statements is more consistent with **risk difference**, **risk ratio**, **both**, or **neither**.

5. More of a measure of public health burden . . . . . ???
6. More of a measure of etiology . . . . ???
7. Null value is 0.0 . . . . ???
8. Can be a negative number . . . . ???
9. Can be a number between 0.0 and 1.0 . . . . . . ???
10. Can be calculated from most follow-up studies . . . . . ???
11. Can be calculated from most case-control studies . . . . ???
12. Has no units . . . . ???
13. A value very close to 0.0 indicates a strong effect . . . . ???
14. Synonymous with attributable risk . . . . . . ???

Quiz continued on next page

Consider the data in the table below and the following estimates of risk on smoking and incidence of lung cancer and coronary heart disease (CHD).

	Lung Cancer	CHD
Rate Ratio	12.9	2.1
Rate Difference	79.0/100k/yr	190.4/100k/yr

15. Which disease is most strongly associated with smoking? . ???

16. Elimination of smoking would reduce the most cases of which disease? ???

Incidence of lung cancer

	Smokers	Nonsmokers	Total
New Lung Cancer cases	60,000	10,000	70,000
Estimated person-years	70,000,000	150,000,000	220,000,000
Estimated incidence density	$\hat{ID}_1$ =85.7 per 100,000 person-years	$\hat{ID}_0$ =6.7 per 100,000 person-years	$\hat{ID}$ =31.8 per 100,000 person-years

Incidence of coronary heart disease (CHD)

	Smokers	Nonsmokers	Total
New CHD cases	250,000	250,000	500,000
Estimated person-years	70,000,000	150,000,000	220,000,000
Estimated incidence density	$\hat{ID}_1$ =357.1 per 100,000 person-years	$\hat{ID}_0$ =166.7 per 100,000 person-years	$\hat{ID}$ =227.3 per 100,000 person-years

Analyzing Data in Data Desk

This activity shows how to calculate a risk difference using the Data Desk software program.

6-2 Measures of Potential Impact (continued)

Potential Impact – The Concept

A measure of potential impact provides a public health perspective on an exposure-disease relationship being studied. More specifically, a measure of potential impact attempts to answer the question, by how much would the disease load of a particular population be reduced if the distribution of an exposure variable were changed? By disease load, we mean the number of persons with a disease of interest that would require health care at a particular point in time.

The typical measure of potential impact is a proportion, often expressed as a percentage, of the number of cases that would not have become cases if all persons being studied had the same exposure status. For example, when determining the potential impact of smoking on the development of lung cancer, the potential impact of smoking gives the proportion of new lung cancer cases that would not have developed lung cancer if no one in the population smoked.

Or, one might determine the potential impact of a vaccine on the prevention of a disease, say, HIV, in high-risk persons. The potential impact of the vaccine gives the proportion of all the potential cases of HIV prevented by the vaccine if there had been no vaccine, all of these cases would have occurred.

These examples illustrate two kinds of potential impact measures. A measure of the impact of smoking on lung cancer considers an exposure that is associated with an increased risk of the disease and is called an **etiologic fraction.** A measure of the impact of a vaccine to prevent HIV considers an exposure that is associated with a decreased risk of disease and is called a **prevented fraction**.

Summary

- A measure of potential impact gives a public health perspective about the effect of an exposure-disease relationship.
- In general, measures of potential impact ascertain what proportion of cases developed the disease as a result of the purported influence of the exposure.
- The etiologic fraction is a measure of potential impact that considers an exposure that is a potential cause of disease.
- The prevented fraction is a measure of potential impact that considers an exposure that is preventive of the disease.

Generalized Measures of Potential Impact

This Lesson on measures of potential impact focuses on quantifying the proportional reduction in disease incidence from a particular change in the prevalence of a single binary exposure variable. The **etiologic fraction** (**EF**) considers the potential for future benefits (i.e., the potential impact) resulting from completely eliminating the presence of a harmful exposure, e.g., smoking. The **prevented fraction** (**PF**) considers past benefits from introducing (by completely eliminating the absence of) a protective exposure, e.g., a vaccine or exercise program. These two measures, nevertheless, are limited in a number of ways that have led to more generalized measures of potential impact. Detailed discussion of such generalized measures is beyond the scope of this presentation. However, we briefly describe below several important generalizations.

1. **Multilevel exposures.** The formula for EF can be extended as follows if estimates of risk are available for k+ 1 categories of exposure:

$$EF = 1 - \left(\frac{1}{\sum_{i=0}^{k} (p_i \times RR_i)} \right)$$

where p_i denotes the proportion of the population in exposure group i, where i goes from 0 to k, and RR_i is the risk ratio that compared i-th exposure category to the referent group 0. If the data involves rates instead of risks, substitute IDR_i for RR_i in the above formula.

2. **Adjustment for other factors**. Three approaches:
   a) separate measures for subgroups;
   b) stratified analysis:
      i) use weighted average of EF's for each subgroup, or
      ii) use adjusted RR estimate;
   c) logistic regression.

3. Impact of partial modification (rather than complete elimination) of exposure. Examples:
   a) To measure the reduction in lung cancer mortality if smoking is reduced but not eliminated;
   b) To measure the reduction in coronary heart disease from increased levels of physical activity.

A formula for a generalized impact fraction (**IF**) is given as follows:

$$IF = \frac{\sum_{i=1}^{k} [(p_i * - p_i **) \times RR_i]}{\sum_{i=0}^{k} [p_i * \times RR_i]} \quad \text{where}$$

IF is the reduction in disease risk as a result of a change in the distribution of a multilevel exposure variable,
$\mathbf{p_i}$***** is the proportion of the candidate population in the ith exposure category before the planned intervention or change, and
$\mathbf{p_i}$****** is the proportion of the candidate population in the ith exposure category after the change.

Etiologic Fraction

The **etiologic fraction** answers the question: what proportion of new cases that occur during a certain time period of follow-up are attributable to the exposure of interest? Other names for this measure are the **etiologic fraction *in the population***, **attributable fraction *in the population***, the **population attributable risk**, and the **population attributable risk percent**.

In mathematical terms, the etiologic fraction is given by the formula **I*** divided by **I**, where **I*** denotes the number of new cases attributable to the exposure and **I** denotes the number of new cases that actually occur. The numerator, **I*** can be found as the difference between the actual number of new cases and the number of new cases that would have occurred in the absence of exposure.

$$EF = \frac{I^*}{I}$$

To illustrate the calculation of the etiologic fraction, consider once again the results of a five-year follow-up study to determine whether or not smokers who have had a heart attack will reduce their risk for dying by quilting smoking The estimated risk ratio here is 2.1 and the estimated risk difference is 0.19.

Heart Attack Patients	Smoke	Quit	Total
Death	27	14	41
Survival	48	67	115
Total	75	81	156

Risks			
continuing smokers:	0.36	Risk Ratio	2.1
smokers who quit:	0.17	Risk Difference	.19

$$\hat{EF} = \frac{\hat{CI} - \hat{CI}_0}{\hat{CI}} = \frac{0.263 - 0.17}{0.263} \approx 0.35$$

A computational formula for the etiologic fraction is given here, where $\hat{CI}$ denotes the estimated cumulative incidence or risk for all subjects, exposed and unexposed combined, in the study. And $\hat{CI}_0$ denotes the estimated cumulative incidence for unexposed subjects. Notice that the numerator in this formula is ***not*** the risk difference, which would involve $\hat{CI}_1$, the estimated risk for exposed persons, rather than $\hat{CI}$, the overall estimated risk.

To calculate the etiologic fraction using our data then, we first must compute $\hat{CI}$, which equals .263, or roughly 26%. We already know that $\hat{CI}_0$ is .173 or roughly 17%. Substituting these values into the formula, we find that the etiologic fraction is .35, or 35%.

How do we interpret this result? The etiologic fraction of .35 tells us that 35% of all cases that actually occurred are due to continuing smoking. In other words, if we could have gotten all patients to quit smoking after their heart attack, there would have been a 35% reduction in the total number of deaths. This is why the etiologic fraction is often referred to as the population attributable risk percent. It gives the percent of all cases in the population that are attributable, in the sense of contributing excess risk, to the exposure.

Study Questions (Q6.7)

Based on the smoking example from the previous page:

1. How many cases would have been expected if all subjects had been unexposed?
2. What is the excess number of total cases expected in the absence of exposure?
3. What is I*/I for these data?

Summary

- The etiologic fraction is given by the formula I*/I, where I denotes the number of new cases that actually occur and I* denotes the number of new cases attributable to the exposure.
- The numerator, I*, can be quantified as the difference between the actual number of new cases and the number of new cases that would have occurred in the absence of exposure.
- A computational formula for the etiologic fraction is EF = $(CI - CI_0) / CI$, where CI denotes cumulative incidence.
- EF if often referred to as the population attributable risk percent, because it gives the percent of all cases in the population that are attributable to exposure.

Alternative Formula for Etiologic Fraction

In a cohort study that estimates risk, the etiologic fraction can be calculated from estimates of cumulative incidence for the overall cohort and for unexposed persons. An equivalent formula can be written in terms of the risk ratio and the proportion (p) of exposed persons in the cohort.

Alternative Formula/ Etiologic Fraction

$$\hat{EF} = \frac{\hat{CI} - \hat{CI}_0}{\hat{CI}} = \frac{p(\hat{RR} - 1)}{p(\hat{RR} - 1) + 1}$$

p = proportion exposed

To illustrate this alternative formula, consider once again the results of a five-year follow-up study to determine whether or not smokers who have had a heart attack will reduce their risk for dying by quitting smoking. Using the first formula, we previously computed the etiologic fraction to be .35, or 35%. Thus, 35% of all cases that actually occurred are due to those who continued to smoke.

5 Year Follow-up Study

	Smoke	Quit	Total
Death	27	14	41
Survival	48	67	115
Total	75	81	156

$\hat{RR} = 2.1$

$\hat{RD} = .19$

To use the second formula, we first calculate the proportion of the cohort exposed which is .481.

$$p = 75/156 = .481$$

We now substitute this value and the estimated risk ratio of 2.1 into the second formula. The result is .35, which is exactly the same as previously obtained because both formulas are equivalent.

$$\hat{EF} = \frac{.481(2.1 - 1)}{.481(2.1 - 1) + 1} = 0.35$$

The second formula gives us some additional insight into the meaning of the etiologic fraction. This formula tells us that the size of the etiologic fraction depends on the size of the risk ratio and the proportion exposed. In particular, the potential impact for a strong determinant of the disease, that is, when the risk ratio is high, may be small if relatively few persons in the population are exposed.

Suppose in our example, that only 10% instead of 48% of the cohort were exposed so that p equals .10. Then the etiologic fraction would be reduced to 0.10 or 10%, which indicates a much smaller impact of exposure than 35%. Furthermore, if the entire cohort were unexposed, then the etiologic fraction would be zero.

$$\hat{EF} = \frac{0.10(2.1 - 1)}{0.10(2.1 - 1) + 1} = 0.10 < 0.35$$

Now suppose that 90%, instead of 48%, of the cohort were exposed, so that p equals .90. Then the etiologic fraction increases to 0.50 or 50%. If the entire cohort were exposed the etiologic fraction would increase to its maximum possible value of .52 or 52% for a risk ratio estimate of 2.1.

$$\hat{EF} = \frac{0.90(2.1 - 1)}{0.90(2.1 - 1) + 1} = 0.50$$

In general, for a fixed value of the risk ratio, the etiologic fraction can range between zero, if the entire cohort were unexposed, to a maximum value of RR minus one over RR if the entire cohort were exposed.

For fixed $\hat{RR}$:

$$0 \leq \hat{EF} \leq \frac{\hat{RR} - 1}{\hat{RR}}$$

$p = 0$ (all unexposed) $\quad$ $p = 1$ (all exposed)

Study Questions (Q6.8)

1. Use the formula (RR – 1)/RR to compute the maximum value possible EF when the RR is 2.1
2. What is the maximum value possible for EF when RR is 10?
3. As RR increases towards infinity, what does the maximum possible value of the EF approach?
4. If the RR is very large, say 100, can the EF still be relatively small? Explain.

Summary

- ❖ An alternative formula for the etiologic fraction is EF = p(RR-1) / [p(RR-1)+1], where RR is the risk ratio and p is the proportion in the entire cohort that is exposed.
- ❖ The size of the etiologic fraction depends on the size of the risk ratio and the proportion exposed.
- ❖ For a fixed value of the risk ratio, the etiologic fraction can range between zero to a maximum value of (RR – 1)/RR.
- ❖ The potential impact for a strong determinant of the disease (i.e., high risk ratio) may be small if relatively few persons in the population are exposed.

Etiologic Fraction for Person-Time Cohort Studies

In cohort studies that estimate a rate using person-time information, the **etiologic fraction** can be calculated using estimates of **incidence density** rather than cumulative incidence.

$$\hat{EF} = \frac{\hat{ID} - \hat{ID}_0}{\hat{ID}}$$

In the above formula, $\hat{ID}$ denotes the estimated incidence density or rate for all subjects, exposed and unexposed combined. $\hat{ID}_0$ denotes the estimated incidence density for unexposed subjects.

An equivalent version of this formula can be written in terms of the estimated incidence density ratio, $\hat{IDR}$, and the proportion p*, of total person-time for exposed persons.

$$\hat{EF} = \frac{(p^*)(\hat{IDR} - 1)}{(p^*)(\hat{IDR} - 1) + 1}$$

In this formula, $\hat{IDR}$ is the ratio of the rates for exposed and unexposed groups; and p* is calculated as L_1 divided by $(L_1 + L_0)$, where L_1 and L_0 are the person-time information for exposed and unexposed groups, respectively.

$$\hat{IDR} = \frac{\frac{I_1}{PT_1}}{\frac{I_0}{PT_0}}, \quad p^* = \frac{L_1}{L_1 + L_0}$$

To illustrate the calculation of each formula, we consider again the data on serum cholesterol level and mortality. To compute the etiologic fraction using the formula involving incidence densities, we must first calculate the estimated overall incidence density, which is 38.2 per 100,000 person years ([26+14]/[36,581+68,239]). Substituting 38.2 and 20.5 into the formula for $\hat{ID}$ and $\hat{ID}_0$ respectively, we obtain an etiologic fraction of .463, or 46.3%.

Cholesterol Level

	Borderline-high	Normal
Deaths	26	14
Person-years	36,581	68,239

$$\hat{ID}_1 = \frac{26}{36,581} = \frac{71.1}{100,000} \qquad \hat{IDR} = 71.1 / 20.5 = 3.47$$

$$\hat{ID}_0 = \frac{14}{68,239} = \frac{20.5}{100,000} \qquad \hat{IDD} = \frac{71.1 - 20.5}{100,000} = \frac{50.6}{100,000}$$

$$\hat{EF} = \frac{\hat{ID} - \hat{ID}_0}{\hat{ID}} = \frac{38.2 - 20.5}{38.2} = 46.3\%$$

Using the second formula, we find that p* is .349 and, since the estimated rate ratio is 3.47 (36,581/[36,581+68,239]), the etiologic fraction from this formula is also computed to be 46.3%:

$$\hat{EF} = \frac{(p^*)(\hat{IDR} - 1)}{(p^*)(\hat{IDR} - 1) + 1} = \frac{0.349(3.47 - 1)}{0.349(3.47 - 1) + 1} = 46.3\%$$

How do we interpret this result?

Study Questions (Q6.9)

1. Which of the following statements is **not** correct about the EF (46.3%) computed in the previous example?
   A. Almost half of all deaths are due to persons with borderline-high cholesterol.
   B. The proportion of excess deaths due to exposure (having borderline-high cholesterol) out of total deaths is .463.
   C. The percentage reduction in total deaths if persons with borderline-high cholesterol could lower their cholesterol to normal levels is 46.3.
   D. 46% of all deaths among persons with borderline-high cholesterol are due to their borderline-high cholesterol.

2. In the table below, what is the number of excess deaths out of total deaths if all persons in the cohort had normal cholesterol levels?
   A. 18.5
   B. 40.0
   C. 21.5
   D. 14.0

	Cholesterol Level	
	Borderline-high	Normal
Deaths	26	14
Person-years	36,581	68,239

$\hat{ID}_1 = 26 / 36{,}581 = 71.1/100{,}000$ $\hat{IDR} = 71.1 / 20.5 = 3.47$

$\hat{ID}_0 = 14 / 68{,}239 = 20.5/100{,}000$ $\hat{IDD} = \frac{71.1}{100{,}000} - 20.5 = \frac{50.6}{100{,}000}$

Summary

- In cohort studies that use person-time information, the etiologic fraction can be defined in terms of incidence densities or as a combination of the incidence density ratio and the proportion of total person-time for exposed persons.
- The formula involving incidence densities is given by:

$$\hat{EF} = \frac{\hat{ID} - \hat{ID}_0}{\hat{ID}}$$ where $\hat{ID}$ =overall incidence density and $\hat{ID}_0$ =incidence density among unexposed

- An equivalent formula is given by:

$$\hat{EF} = \frac{(p^*)(\hat{IDR} - 1)}{(p^*)(\hat{IDR} - 1) + 1}$$ where $\hat{IDR}$ is the incidence density ratio.

$$p^* = \frac{L_1}{L_1 + L_0}$$, where L_1 and L_0 are the person-time information for exposed and unexposed groups, respectively.

Etiologic Fraction among the Exposed

There are two conceptual formulations of the etiologic fraction. One, which we have previously described, focuses on the potential impact of exposure on the total number of cases, shown below as **I**. A second focuses on the potential impact of the exposure on the number of exposed cases, which we denote as I_1. This measure is called the **etiologic fraction *among the exposed***, **attributable fraction *among the exposed***, or the **attributable risk percent *among the exposed***. In mathematical terms, the etiologic fraction among the exposed, is given by the formula **I*** divided by I_1, where **I*** denotes the number of exposed cases attributable to the exposure and I_1 denotes the number of exposed cases that actually occur.

1. Potential Impact on total # of cases (I)

$$EF = \frac{I^*}{I} = \frac{\hat{CI} - \hat{CI}_0}{\hat{CI}} = \frac{p(\hat{RR} - 1)}{[p(\hat{RR} - 1) + 1]}$$

2. Potential Impact on the # of exposed cases (I_1)

$$EF_e = \frac{I^*}{I_1}$$

I^* → # exposed cases attributable to exposure

I_1 → # exposed cases

The denominator (in the **EFe** formula) is the number of exposed cases. This is different from the denominator in **EF**. That's because the referent group for **EFe** is the number of exposed cases that occur in the cohort rather than the total number of cases in **EF**. The numerator in both formulas is the same, namely **I***. In particular, the **I*** in both **EF** and **EFe** can be quantified as the difference between the actual number of cases and the number of cases that would have occurred in the absence of exposure

To illustrate the calculation of the etiologic fraction among the exposed, consider once again the results of a five-year follow-up study to determine whether or not smokers who have had a heart attack will reduce their risk of dying by quitting smoking. The previously computed etiologic fraction, or equivalently, the population attributable risk percent computed for these data was 35%.

Heart Attack Patients	Smoke	Quit	Total
Death	27	14	41
Survival	48	67	115
Total	75	81	156

Risks { continuing smokers: 0.36; smokers who quit: 0.17

Risk Ratio 2.1
Risk Difference .19

$$\hat{EF} = \frac{\hat{CI} - \hat{CI}_0}{\hat{CI}} = 35\%$$

$$\hat{EF}_e = \frac{\hat{CI}_1 - \hat{CI}_0}{\hat{CI}_1} = \frac{.19}{.36} = \mathbf{53\%}$$

The etiologic fraction among the exposed (**EFe**) can be calculated for these same data using the formula shown above. The term $\hat{CI}_1$ denotes the estimated cumulative incidence or risk for exposed subjects in the study and $\hat{CI}_0$ denotes the estimated cumulative incidence for unexposed subjects. The numerator in this formula is the estimated risk difference ($\hat{RD}$). Since the estimated risk difference is .19 and the risk for exposed persons is .36, we can substitute these values into the formula for EFe to obtain .53, or 53%. How do we interpret this result?

The etiologic fraction of .53 tells us that 53% of all deaths among continuing smokers are due to continuing smoking. In other words, if we could have gotten the continuing smokers who died to quit smoking after their heart attack, there would have been a 53% reduction in deaths among these persons.

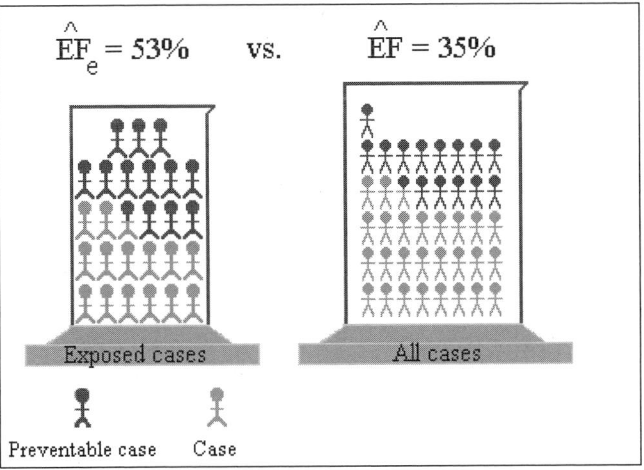

Study Questions (Q6.10)

Heart Attack Patients	Smoke	Quit	Total
Death	27	14	41
Survival	48	67	115
Total	75	81	156

Risks		Measure	
continuing smokers:	0.36	Risk Ratio	2.1
smokers who quit:	0.17	Risk Difference	.19

1. What is the excess number of exposed cases (i.e., deaths among continuing smokers) expected in the absence of exposure?
2. Fill in the blanks: In this example, ________ of the ________ deaths among continuing smokers could have been avoided.
3. Use the formula I^*/I_1 to compute EFe for these data.
4. An alternative formula for the etiologic fraction among exposed is EFe=(RR-1)/RR, where RR is the risk ratio. Use this formula to compute EFe for the heart attack study data.
5. The population attributable risk percent (EF) computed for these data is 35% whereas the attributable risk percent among the exposed (EFe) is 53%. How do you explain these differences?
6. For cohort studies that use person-time information, state a formula for the etiologic fraction among the exposed that involves incidence densities in exposed and unexposed groups.
7. As in the previous question, state an alternative formula for EFe that involves the incidence density ratio.
8. For case-control studies, which cannot estimate risk or rate, can you suggest formulae for EF and EFe?

Summary

- ❖ The etiologic fraction among the exposed, EFe, focuses on the potential impact of the exposure on the number of exposed cases, rather than the total number of cases.
- ❖ EFe is defined as I^*/I_1, where I^* is the excess number of exposed cases due to exposure and I_1 is the actual number of exposed cases.
- ❖ For cohort studies that estimate risk:

$$\hat{EFe} = \frac{(\hat{CI}_1 - \hat{CI}_0)}{\hat{CI}_1} = \frac{\hat{RR} - 1}{\hat{RR}}$$

- ❖ For cohort studies that estimate rate:

$$\hat{EFe} = \frac{(\hat{ID}_1 - \hat{ID}_0)}{\hat{ID}_1} = \frac{\hat{IDR} - 1}{\hat{IDR}}$$

Etiologic Fraction – An Example

Hypothyroidism, a disease state in which the production of thyroid hormone is decreased, is known to increase the risk of cardiovascular disease. In elderly women, the subclinical form of hypothyroidism is highly prevalent. The Rotterdam Study investigators therefore examined the potential impact of subclinical hypothyroidism on the incidence of myocardial infarction.

In this study of nearly 1,000 women aged 55 and over, the prevalence of subclinical hypothyroidism was 10.8%. Consider the two-by-two table depicted here. The cumulative incidence of myocardial infarction is 2.9% (3/103) in women with subclinical hypothyroidism, 1.2% (10/854) in women without hypothyroidism, and 1.4% (13/957) overall.

	Subclinical Hypothyroidism	No Subclinical Hypothyroidism	Total
MI	3	10	13
No MI	100	844	944
Total	103	854	957
	$\hat{CI}_1$ =2.9%	$\hat{CI}_0$ =1.2%	$\hat{CI}$ =1.4%

Study Questions (Q6.11)

1. Using these data, can you calculate the etiologic fraction?

The etiologic fraction is: EF = (1.4 - 1.2) / 1.4 = 14%. This indicates that of all myocardial infarctions that occur in elderly women, 14% are due to the presence of subclinical hypothyroidism. In other words, if subclinical hypothyroidism could be prevented, there would be 14% less myocardial infarctions in this population.

Study Questions (Q6.11) continued

2. Can you calculate the etiologic fraction using the alternative formula:
   EF = [p(RR-1)]/[p(RR-1) + 1]?
3. Can you calculate the etiologic fraction in the exposed?

The etiologic fraction in the exposed (EFe) is (2.9-1.2), which is equal to the risk difference, divided by 2.9, which is 60%. Thus, among the women that are affected, 60% of the myocardial infarctions can be attributed to the presence of subclinical hypothyroidism.

Summary

- The Rotterdam Study investigators examined the potential impact of subclinical hypothyroidism on the incidence of myocardial infarction.
- Of all myocardial infarctions that occur in elderly women, 14% is due to the presence of subclinical hypothyroidism.
- Among women that are affected, 60% of the myocardial infarctions can be attributed to the presence of subclinical hypothyroidism.

Quiz (Q6.12)

Consider data in the table below on smoking and incidence of lung cancer and cardiovascular disease (CHD).

Incidence of lung cancer

	Smokers	**Nonsmokers**	**Total**
New Lung Cancer cases	60,000	10,000	70,000
Estimated person-years	70,000,000	150,000,000	220,000,000
Estimated incidence density	$\hat{ID}_1$ =85.7 per 100,000 person-years	$\hat{ID}_0$ =6.7 per 100,000 person-years	$\hat{ID}$ =31.8 per 100,000 person-years

Incidence of coronary heart disease (CHD)

	Smokers	**Nonsmokers**	**Total**
New CHD cases	250,000	250,000	500,000
Estimated person-years	70,000,000	150,000,000	220,000,000
Estimated incidence density	$\hat{ID}_1$ =357.1 per 100,000 person-years	$\hat{ID}_0$ =166.7 per 100,000 person-years	$\hat{ID}$ =227.3 per 100,000 person-years

Quiz continued on next page

1. The prevalence of smoking in this population is . . . . . **???**.
2. The etiologic fraction for lung cancer is . . . . . **???**.
3. The etiologic fraction for coronary heart disease is . . . . **???**.
4. The etiologic fraction among the exposed for lung cancer is . . **???**.
5. The etiologic fraction among the exposed for CHD is . . . **???**.
6. The proportion of lung cancer among smokers attributable to their smoking is **???**.

Choices
0.0% **26.7%** **31.8%** **53.3%** **79.0%** **92.2%**

Label each of the following as either a **Risk/Rate Difference**, an **Etiologic Fraction** or an **Etiologic Fraction Among the Exposed**.

7. Attributable risk percent among the exposed . . **???**
8. Population attributable risk . . . . **???**
9. Excess risk . . . . . . **???**
10. Influenced by prevalence of the exposure in the population **???**
11. Has same units as measure of occurrence . . **???**
12. Can be a negative number . . . . **???**

---

13. **???** can never be larger than **???**

Choices (for Question 13)
Etiologic Fraction **Etiologic Fraction Among the Exposed**

6-3 Measures of Potential Impact (continued)

Prevented Fraction

The **prevented fraction** gives the proportion of potential new cases that were prevented by the presence of a **protective** exposure. Other names for the prevented fraction are the **prevented fraction *in the population***, the **population prevented risk**, and **population prevented risk percent**.

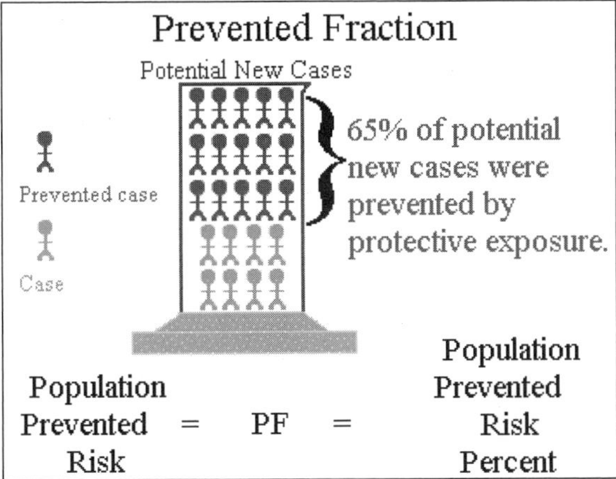

The prevented fraction may be expressed mathematically like this:

$$\hat{PF} = \frac{I^{**}}{I^{**} + I}$$

The numerator **I**** denotes the number of new cases that would be expected in the absence of exposure *minus* the number of new cases that actually did occur. In other words, the numerator gives the number of cases prevented by the presence of a protective exposure.

The **I** in the denominator denotes the total number of new cases actually occurring during follow-up of the cohort. The sum of **I**** and **I** gives the number of new cases that would be expected in the *absence* of exposure. That is, the denominator gives the potential number of new cases.

To illustrate the calculation of a prevented fraction, we consider a study of an epidemic of measles in Texarkana, a city bisected by the state line between Texas and Arkansas. This city never had a measles vaccination campaign although a large proportion of children had been previously vaccinated. Results from the study are shown below:

Texarkana Study	Immunized	Not Immunized	Total
Ill	27	512	539
Well	6323	4323	10646
Total	6350	4835	11185

Cumulative Incidence per 1,000

Immunized: 4.2

Not immunized: 105.9

Overall: 48.2

$\hat{RR}$ = 0.04

$\hat{RD}$ = - 101.7 per 1,000

A computational formula for the prevented fraction is:

$$\hat{PF} = \frac{\hat{CI}_0 - \hat{CI}}{\hat{CI}_0}$$

As in the etiologic fraction formula, $\hat{CI}_0$ denotes the cumulative incidence in the unexposed and $\hat{CI}$ denotes the overall cumulative incidence. This formula differs from the formula for etiologic fraction in *both* the numerator and denominator; it is necessary to quantify the exposure as protective rather than harmful.

In our example, $\hat{CI}_0$ is 105.9 per 1,000 persons and $\hat{CI}$ is 48.2 per 1000 persons. Substituting these values into the formula, we find that the prevented fraction here is .55, or 55 percent.

$$\hat{PF} = \frac{\hat{CI}_0 - \hat{CI}}{\hat{CI}_0} = \frac{105.9 - 48.2}{105.9} = 0.55 = 55\%$$

Study Questions (Q6.13)

Based on the Texarkana example:

1. How many total ill children would be expected none of the subjects had been immunized?
2. How many children actually became ill in the entire cohort?
3. How many cases were prevented by the immunization program?
4. Using the values I**=645.5 and I=539, calculate the prevented fraction?

How do we interpret this result? The prevented fraction .55 tells us that out of a total of 1,184.5 illnesses that would be expected if all subjects were not immunized, 645.4 illnesses were prevented by immunization. That is, we've observed 645.4 fewer illnesses than expected because of immunization. Thus, 55% of the total expected cases were prevented by immunization.

Summary

- The prevented fraction (PF) gives the proportion of potential new cases that were prevented by exposure.
- Other names for the PF are population prevented risk and population prevented risk percent.
- PF=I** / (I** + I) where I** denotes the total cases prevented and I denotes the total cases that occurred.
- Using cumulative incidence data,

$$\hat{PF} = \frac{\hat{CI}_0 - \hat{CI}}{\hat{CI}_0} = p(1 - \hat{RR})$$ where $\hat{CI}$, $\hat{RR}$, and p are as in the EF formula.

- Using person-time data,

$$\hat{PF} = \frac{(\hat{ID}_0 - \hat{ID})}{\hat{ID}_0} = (p^*)(1 - \hat{IDR})$$ where $\hat{ID}_0$, $\hat{ID}$, $\hat{IDR}$,

$$p^* = \frac{L_1}{L_1 + L_0}$$, and L_1 and L_0 are as in the corresponding EF formula.

The Relationship between PF and EF

We have seen the formulae for the **etiologic fraction** (**EF**):

$$EF = \frac{\hat{CI} - \hat{CI}_0}{\hat{CI}} = \frac{\hat{p}(\hat{RR} - 1)}{\hat{p}(\hat{RR} - 1) + 1}$$

The corresponding formulae for the **prevented fraction** (**PF**) are shown here:

$$PF = \frac{\hat{CI}_0 - \hat{CI}}{\hat{CI}_0} = \hat{p}(1 - \hat{RR})$$

With a little algebra, we can write PF in terms of EF as follows:

$$PF = \frac{EF}{EF - 1}$$

We can also write PFe in terms of EFe as follows:

$$PFe = \frac{EFe}{EFe - 1}$$

We illustrate the relationship between PF and EF using the results from the 5-year follow-up study of heart attack patients described in several activities in this Lesson. The EF = .35 and EFe = .53 when measuring the potential impact of continuing to smoke. We computed EF rather than PF because the continuing smokers had a higher estimated risk (.36) than did smokers who quit (.17). Now suppose we turn our potential impact question around by asking what is the prevented fraction (**PF**) for quitting smoking? To answer this, we must switch cumulative incidences so that $\mathbf{CI_0}$ denotes the estimated cumulative incidence in the continuing smokers group (.36) and $\mathbf{CI_1}$ denotes the estimated cumulative incidence in the group that quit smoking (.173). The overall cumulative incidence **CI** remains at .263.

The PF in this case is given by (.36-.263)/.36 = .269 = 26.9%. Thus 26.9% of expected deaths can be prevented by quitting smoking. This is, in contrast to the etiologic fraction of .35, which says that 35% of all deaths are attributable to continuing smoking. Note that the EF for quitting smoking is the negative value given by (.263-.36)/.263=-.368. Substituting -.368 for EF in the formula PF=EF/(EF-1), we obtain PF=-.368/(-.368-1) = .269, which is the prevented fraction for quitting smoking. Thus, the etiologic fraction of 35% for continuing to smoke is not equal to the prevented fraction of 26.9% from quitting smoking. We also will not obtain the value of .269 if we substitute the EF value of .35 into the formula relating PF to EF. This formula only works if we switch the cumulative incidences from CI_1 to CI_0 and vice versa.

Prevented Fraction among the Exposed

As with the etiologic fraction, there are two conceptual formations of the prevented fraction. One, which we have previously described, focuses on the potential impact of a protective exposure on the **total number of cases** that would be expected if exposure was absent.

$$\hat{PF} = \frac{I^{**}}{I^{**} + I} = \frac{\hat{CI}_0 - \hat{CI}}{\hat{CI}_0} = p(1 - RR)$$

A second formulation of the prevented fraction focuses on the potential impact of a protective exposure on the expected number of **exposed cases**. This measure is called the **prevented fraction among the exposed**, or alternatively, the

prevented risk percent among the exposed.

$$\hat{PF}_e = \frac{I^{**}}{I^{**} + I_1}$$

Although the denominator in the prevented fraction among the exposed (PFe) is different from the denominator in the prevented fraction (PF), the numerators are the same.

$$\hat{PF} = \frac{I^{**}}{I^{**} + I} = \frac{\text{Prevented cases}}{\text{Prevented cases} + \text{Actual cases among all subjects}}$$

$$\hat{PF}_e = \frac{I^{**}}{I^{**} + I_1} = \frac{\text{Prevented cases}}{\text{Prevented cases} + \text{Actual cases among the exposed}}$$

To illustrate the calculation of the prevented fraction among the exposed, consider once again the results of an epidemic of measles in Texarkana in 1970 and 1971. The previously computed prevented fraction, or equivalently, the population prevented risk percent these data is 55%.

Prevented Fraction

Texarkana Study	Immunized	Not Immunized	Total
Ill	27	512	539
Well	6323	4323	10646
Total	6350	4835	11185

Cumulative Incidence per 1,000

Immunized: 4.2

Not immunized: 105.9

Overall: 48.2

$\hat{RR} = 0.04$

$\hat{RD} = \frac{-101.7}{1000}$

$\hat{PF} = .55$ or 55%

$$\hat{PF}_e = \frac{\hat{CI}_0 - \hat{CI}_1}{\hat{CI}_0} = \frac{101.7}{105.9} = 0.96 = 96\%$$

The prevented fraction among the exposed can be calculated for these same data using the formula shown above. $\hat{CI}_0$ and $\hat{CI}_1$ denote the estimated cumulative incidences for unexposed and exposed subjects, respectively. The numerator in this formula is the negative of the estimated risk difference. The estimated risk difference is minus 101.7 per 1000 and the cumulative incidence for unexposed ($\hat{CI}_0$) persons is 105.9 per 1000. Thus the prevented fraction among the exposed is 101.7 divided by 105.9, which is .96, or 96%.

Study Questions (Q6.14)

Using the Texarkana data:

1. How many ill children would be among those immunized if immunization was not effective?
2. How many children actually became ill among those who were immunized?
3. How many exposed cases were prevented by the immunization program?
4. Using the values I** = 645.5 and $I_1 = 27$, calculate the prevented fraction among the exposed.

How do we interpret this result? The prevented fraction of .96 among the exposed tells us that 96% of total expected cases among those immunized were prevented by immunization. Out of a total or 672.5 illnesses expected among those immunized, 96%, or 645.4 illnesses were prevented by immunization. That is, there were 645.4 fewer illnesses children than expected because or the immunization.

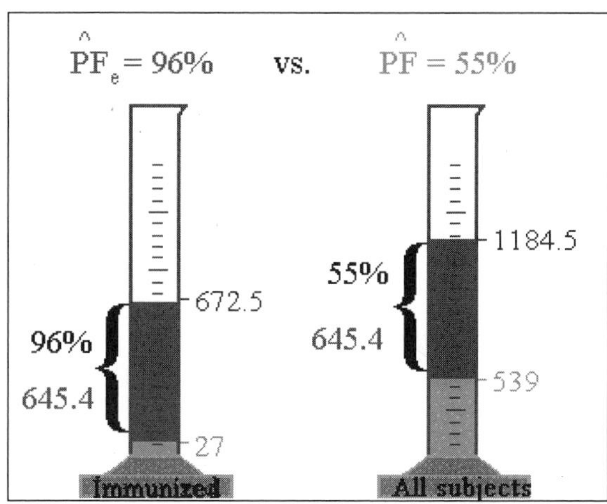

<u>**Study Questions (Q6.15)**</u>

1. An alternative formula for the prevented fraction among exposed is PFe = (1 – RR), where RR is the risk ratio. Use this formula to compute PFe for the immunization data.
2. The population prevented risk percent (PF) computed for these data is 55% whereas the prevented risk percent among the exposed (PFe) is 96%. How do you explain the difference in these two values?
3. For cohort studies that use person-time information, state a formula for the prevented fraction among the exposed that involves incidence densities in the exposed and unexposed groups.
4. As in the previous question, state an alternative formula for PFe that involves the incidence density ratio.
5. For case-control studies, which cannot estimate risk or rate, can you suggest a formulae for PF and PFe?

<u>**Summary**</u>

- The prevented fraction among the exposed (PFe) focuses on the potential impact of the exposure *on the exposed cases,* rather than the total number of cases.
- PFe is defined as I** / (I** + I_1), where I** is the number of exposed cases prevented by exposure and I_1 is the actual number of exposed cases.
- For cohort studies that estimate risk:

$$\hat{PFe} = \frac{\hat{CI}_0 - \hat{CI}_1}{\hat{CI}_0} = (1 - \hat{RR})$$

- For cohort studies that estimate rate:

$$\hat{PFe} = \frac{(\hat{ID}_0 - \hat{ID}_1)}{\hat{ID}_0} = (1 - \hat{IDR})$$

Analyzing Data in Data Desk

This activity shows how to compute measures of potential impact using the Data Desk program.

Quiz (Q6.16)

Label each of the following statements as **True** or **False.**

1. Use prevented fraction rather than etiologic fraction when risk difference < 0. . . ???
2. Prevented fraction may be calculated simply as 1 -EF when risk difference is < 0. . ???
3. Prevented fraction among the exposed may be calculated simply as 1 -RR. . . ???
4. Prevented fraction is based on theoretical or nonexistent cases, i.e., cases that did not occur but would have occurred in the absence of the intervention or exposure. . . . . ???
5. The denominator for both PF and PFe is risk (or rate) in the unexposed group. . . ???
6. Between PF and PFe, only PF is influenced by prevalence of exposure. . . . ???

Using the data in the table below, label each of the following as:
Prevented Fraction, **Prevented Fraction Among the Exposed**, or **Neither**.

7. Could be calculated as ([44/160] -[48/888]) / [44/160] = 80.3%. . . . . ???
8. Could be calculated as ([44/160] -[92/1048]) / [44/160] = 68.1 %... . . . ???
9. If fewer children had been vaccinated, this measure would **increase**. . . . ???
10. If fewer children had been vaccinated, this measure would **decrease**. . . . ???
11. The proportion of potential cases in the community prevented by vaccination. . . ???

	Vaccinated	Unvaccinated	Total
Measles	48	44	92
No Measles	840	116	956
Total	888	160	1048

CI in unvaccinated = 27.5%; CI in vaccinated = 5.4%; Overall CI = 8.8%

Nomenclature:

Table setup for cohort, case-control, and prevalence studies:

	Exposed	Not Exposed	Total
Disease/cases	a	b	n_1
No Disease/controls	c	d	n_0
Total	m_1	m_0	N

Table setup for cohort data with person-time:

	Exposed	Not Exposed	Total
Disease (New cases)	I_1	I_0	I
No Disease	-	-	-
Total disease-free person-time	PT_1 or L_1	PT_0 or L_0	PT

$\hat{CI}$	Cumulative incidence in the population (n_1/n)
$\hat{CI}_0$	Cumulative incidence in the nonexposed (b/m_0)
$\hat{CI}_1$	Cumulative incidence in the exposed (a/m_1)
CID	Cumulative incidence difference or risk difference, $CI_1 - CI_0$
EF	Etiologic fraction
EFe	Etiologic fraction in the exposed
I	Number of new cases that occur
I*	Number of new cases attributable to the exposure
I**	Number of cases that would be expected in the absence of exposure
I_0	Number of new cases in nonexposed
I_1	Number of new cases in exposed
$\hat{ID}$	Incidence density (or "rate") in the population (I/PT)
$\hat{ID}_0$	Incidence density (or "rate") in the not exposed (I_0/PT_0)
$\hat{ID}_1$	Incidence density (or "rate") in the exposed (I_1/PT_1)
IDD	Incidence density difference or rate difference, $ID_1 - ID_0$
IDR	Incidence density ratio or rate ratio: ID_1 / ID_0
L_0	Disease-free person-time in nonexposed
L_1	Disease-free person-time in exposed
n	Size of population under study
n_0	Size of population under study in nonexposed at time zero
n_1	Size of population under study in exposed at time zero
NNT	Number Needed to Treat, $1/(CI_1\text{-}CI_0)$
OR	Odds ratio: ad/bc
P_1	Prevalence in exposed
P_0	Prevalence in unexposed
PD	Prevalence difference
PF	Prevented fraction
PFe	Prevented fraction in the exposed
PT	Disease-free person-time
PT_0	Disease-free person-time in nonexposed
PT_1	Disease-free person-time in exposed
R_0	Risk in unexposed
R_1	Risk in exposed
RD	Risk difference: risk in exposed minus risk in unexposed
RR	Risk ratio: risk in exposed divided by risk in unexposed

Formulae for difference measures of effect by risk data, rate data, and prevalence data.

Risk Data	Rate Data	Prevalence Data
$\hat{CID} = \hat{CI}_1 - \hat{CI}_0$	$\hat{IDD} = \hat{ID}_1 - \hat{ID}_0$	$\hat{PD} = \hat{P}_1 - \hat{P}_0$

Formulae for comparing PF with EF and PFe with EFe and formula for NNT:

$$PF = \frac{EF}{EF-1} \qquad PFe = \frac{EFe}{EFe-1} \qquad NNT = \frac{1}{\hat{CI}_1 - \hat{CI}_0}$$

Formulae for etiologic fraction, etiologic fraction in the exposed, prevented fraction, and prevented fraction in the exposed for risk data, rate data, and case-control studies.

	Risk Data	Rate Data	Case-Control*
EF	$\hat{EF} = \frac{I^*}{I}$	$\hat{EF} = \frac{I^*}{I}$	
	$\hat{EF} = \frac{\hat{CI} - \hat{CI}_0}{\hat{CI}}$	$\hat{EF} = \frac{\hat{ID} - \hat{ID}_0}{\hat{ID}}$	$\hat{EF} = \frac{(p')(\hat{OR} - 1)}{(p')(\hat{OR} - 1) + 1}$
	$\hat{EF} = \frac{(p)(\hat{RR} - 1)}{(p)(\hat{RR} - 1) + 1}$	$\hat{EF} = \frac{(p^*)(\hat{IDR} - 1)}{(p^*)(\hat{IDR} - 1) + 1}$	
EFe	$\hat{EF}_e = \frac{I^*}{I_1}$	$\hat{EF}_e = \frac{I^*}{I_1}$	
	$\hat{EF}_e = \frac{\hat{CI}_1 - \hat{CI}_0}{\hat{CI}_1}$	$\hat{EFe} = \frac{\hat{ID}_1 - \hat{ID}_0}{\hat{ID}_1}$	
	$\hat{EF}_e = \frac{\hat{RD}}{\hat{CI}_1}$	$\hat{EFe} = \frac{\hat{IDD}}{\hat{ID}_1}$	$\hat{EFe} = \frac{\hat{OR} - 1}{\hat{OR}}$
	$\hat{EFe} = \frac{\hat{RR} - 1}{\hat{RR}}$	$\hat{EFe} = \frac{\hat{IDR} - 1}{\hat{IDR}}$	
PF	$\hat{PF} = \frac{I^{**}}{I^{**} + I}$	$\hat{PF} = \frac{I^{**}}{I^{**} + I}$	
	$\hat{PF} = \frac{\hat{CI}_0 - \hat{CI}}{\hat{CI}_0}$	$\hat{PF} = \frac{\hat{ID}_0 - \hat{ID}}{\hat{ID}_0}$	$\hat{PF} = p'(1 - \hat{OR})$
	$\hat{PF} = p(1 - \hat{RR})$	$\hat{PF} = (p^*)(1 - \hat{IDR})$	
PFe	$\hat{PF}_e = \frac{I^{**}}{I^{**} + I_1}$	$\hat{PF}_e = \frac{I^{**}}{I^{**} + I_1}$	
	$\hat{PFe} = \frac{\hat{CI}_0 - \hat{CI}_1}{\hat{CI}_0}$	$\hat{PFe} = \frac{\hat{ID}_0 - \hat{ID}_1}{\hat{ID}_0}$	$\hat{PFe} = 1 - \hat{OR}$
	$\hat{PFe} = 1 - \hat{RR}$	$\hat{PFe} = 1 - \hat{IDR}$	
where	$p = \frac{m_1}{n}$	$p^* = \frac{L_1}{L_1 + L_0}$	$p' = \frac{c}{n_0}$

*In case-control studies, the EF, EFe, PF, PFe based on the odds ratio will be a good estimates when the OR is a good estimate of the RR (e.g., rare disease assumption)

References

Benichou J, Chow WH, McLaughlin JK, Mandel JS, and Fraumeni JF Jr. Population attributable risk of renal cell cancer in Minnesota. Am J Epidemiol 1988;148(5):424-30.

Cook RJ, Sackett DL. The number needed to treat: a clinically useful measure of treatment effect. BMJ 1995;310(6977):452-4.

Greenberg RS, Daniels SR, Flanders WD, Eley JW, Boring JR. Medical Epidemiology (3rd Ed). Lange Medical Books, New York, 2001.

Hak AE, Pols HA, Visser TJ, Drexhage HA, Hofman A, Witteman JC. Subclinical hypothyroidism is an independent risk factor for atherosclerosis and myocardial infarction in elderly women: the Rotterdam Study. Ann Intern Med 2000;132(4):270-8.

Kleinbaum DG, Kupper LL, Morgenstern H. Epidemiologic Research: Principles and Quantitative Methods. John Wiley and Sons Publishers, New York, 1982.

Landrigan PJ, Epidemic measles in a divided city. JAMA 1972;221(6):567-70.

Medical Research Council trial of treatment of mild hypertension: principal results. MRC Working Party. BMJ 1985:291;97-104.

Morgenstern H and Bursic ES,. A method for using epidemiologic data to estimate the potential impact of an intervention on the health status of a target population. J Community Health 1982;7(4):292-309.

Spirtas R, Heineman EF, Bernstein L, Beebe GW, Keehn RJ, Stark A, Harlow BL, Benichou J. Malignant mesothelioma: attributable risk of asbestos exposure. Occup Environ Med 1994;51(12):804-11.

Wacholder S, Benichou J, Heineman EF, Hartge P, Hoover RN. Attributable risk: advantages of a broad definition of exposure. Am J Epidemiol 1994;140(4):303-9..

Walter SD. Calculation of attributable risks from epidemiological data. Int J Epidemiol 1978;7(2):175-82.

Walter SD. Prevention for multifactorial diseases. Am J Epidemiol 1980;112(3):409-16.

Walter SD. Attributable risk in practice. Am J Epidemiol 1998;148(5):411-3.

Walter SD. Number needed to treat (NNT): estimation of a measure of clinical benefit. Stat Med 2001:20(24);3947-62.

Wilson PD, Loffredo CA, Correa-Villasenor A, Ferencz C. Attributable fraction for cardiac malformations. Am J Epidemiol 1998;148(5):414-23.

Homework

ACE-1. Measure of Potential Impact: Quitting Smoking II

The following data come from a study of self-help approaches to quitting smoking. (These data are the same as described in question 4 in the homework exercises for Lesson 5.) Smokers wanting to quit were randomized into one of four groups (C = control, M = quitting manual, MS = manual + social support brochure, MST = manual + social support brochure + telephone counseling). Smoking status was measured by mailed survey at 8, 16, and 24 months following randomization. These are the 16-month results:

	Randomization Group				
Status	C	M	MS	MST	**Total**
Quit	84	71	67	109	331
Smoking	381	396	404	365	1546
Total	465	467	471	474	1877

Compute an appropriate measure of impact, comparing the MST and MS groups. Interpret the result.

ACE-2. Potential Impact: Radiotherapy Among Children II

This is a continuation of question 5 in the homework exercises for Lesson 5.) In a study of adverse effects of radiotherapy among children in Israel, 10,834 irradiated children were identified from original treatment records and matched to 10,834 non-irradiated comparison subjects selected from the general population. Subjects were followed for a mean of 26 years. Person-years of observation were: irradiated subjects, 279,901 person-years; comparison subjects, 280,561 person-years. During the follow-up period there were 49 deaths from cancer in irradiated subjects, and 44 in the non-irradiated population comparison subjects.

a. Assuming causality, how many cancer deaths per 100,000 irradiated subjects per year were due to the effect of radiotherapy?
b. Again assuming causality, what proportion of cancer deaths in irradiated subjects can be attributed to the effect of radiotherapy?

ACE-3. Etiologic Fraction and Etiologic Fraction in the exposed: CVD and Lung Cancer

The following table shows cardiovascular (CVD) mortality and lung cancer (LC) mortality among smoking and non-smoking physicians obtained from a prospective study involving physicians listed in the British Medical Register and living in England and Wales as of October 1951. Information about cause of death was obtained from the death certificate and mortality records over the subsequent 10 years.

a. Complete the following table by calculating the rate ratio, the etiologic fraction (EF), and the etiologic fraction in the exposed (EFe) for each smoking category. Round all calculations to two decimal places.

CVD	Non-smokers	All	Heavy Smokers (≥25 cig/day)	General Population
Mortality rater per 1000 py	7.32	9.51	9.53	9.40
Rate ratio		________	________	
Etiologic fraction (EF)		________	________	
Etiologic fraction in the exposed (EFe)		________	________	
Lung Cancer				
Mortality rater per 1000 py	0.07	1.30	2.27	0.93
Rate ratio		________	________	
Etiologic fraction (EF)		________	________	
Etiologic fraction in the exposed (EFe)		________	________	

b. If you only consider "all smokers" (versus non-smokers), for which of the two diseases is smoking of greater etiologic importance? Why?
c. If you only consider "all smokers" (versus non-smokers), for which of the two diseases is smoking of greater public health importance? Why?
d. How would you answer the previous two parts (i.e., b and c) if you consider "heavy smokers" versus "non-smokers"?

ACE-4. Etiologic Fraction: Asbestos Exposure vs. Mesothelioma

The table below represents the results from a study investigating whether there is an association between exposure to asbestos and the rare cancer mesothelioma:

		Exposure to Asbestos	
		Yes	**No**
Mesothelioma	**Yes**	20	80
	No	3	93

a. Why does it make sense to think that the study design used was a case-control study instead of a cohort study?
b. Assuming that the measure of effect of interest is the odds ratio, calculate measures of the population etiologic fraction and the etiologic fraction among the exposed using the following formulae:

EF = p'(OR - 1) / [p'(OR - 1) +1] and EFe = (OR - 1) / OR, where OR is the odds ratio and p' is the proportion of all controls that are exposed.

c. How do you interpret these measures?

ACE-5. Potential Impact: Cholera

In 1963, the Cholera Research Laboratory in Bangladesh assessed the nutritional status of a probability sample of children ages 12-23 months. These children were followed for two years, and all deaths were identified. Results are presented in the table below:

Number of Deaths During Two Years of Follow-up of Children Ages 12-23 Months at Entry to the Study: Bangladesh, 1963-5.

	Nutritional Status		
	Normal	**Moderate Malnutrition**	**Severe Malnutrition**
Deaths	20	44	47
Survivors	526	1002	380
Total	546	1046	427

a. Calculate the risk for the group with moderate malnutrition that was due to being moderately malnourished (i.e., the risk difference for moderate malnutrition compared to normal nutrition).
b. What is the etiologic fraction for moderately malnourished children (compared to normal children)?
c. Calculate the risk for the group with severe malnutrition that was due to being severely malnourished (i.e., the risk difference for severe malnutrition compared to normal nutrition).
d. What is the etiologic fraction for severely malnourished children (compared to normal children)?
e. In a single sentence and using the figures you have calculated for severely malnourished children as an example, explain what is meant by risk difference.
f. In a single sentence and using the figures you have calculated for severely malnourished children as an example, explain what is meant by etiologic fraction.

ACE-6. Potential Impact: Neural Tube Defects

In 1988, Mulinare et al reported their findings concerning the association of neural tube defects (NTD's) and periconceptional use of multivitamins (**JAMA 260**:3141-3145). They selected several groups of infants. One group consisted of "all live-born or stillborn infants with the diagnosis of anencephaly or spina bifida during the years 1968 through 1980 who were registered in the Metropolitan Atlanta Congenital Defects Program (MACDP). The second group consisted of "live-born" babies without birth defects who were randomly chosen from all live births that occurred in the MACDP surveillance area. "We obtained data on multivitamin use and defined multivitamin use as 'multivitamin or prenatal multivitamin consumption during every month of the entire six month period… from three months before conception through the third month of pregnancy." The following table presents partial results from that report (with a few simplifications for the purpose of this exercise.

Distribution of Periconceptional Vitamin Use Among Mothers Giving Birth to Infants With and Without Neural Tube Defects.

	Multivitamin Use		
	Periconceptional use	**No vitamin use**	**Total**
Cases	24	159	183
Controls	411	1052	1463

a. Using the above data, can you calculate the prevalence of neural tube defects in the MACDP area? If so, do so, if not state why not.
b. Using the above data, can you calculate the risk of neural tube defects in the MACDP area? If so, do so, if not state why not.
c. Using the above data, can you calculate the etiologic fraction for NTDs due to failure of the mother to make periconceptual use of multivitamins? If so, do so, if not state why not.

Answers to Study Questions and Quizzes

Q6.1

1. 75 * .17 = 12.75 deaths would have occurred if the 75 patients had quit smoking.
2. 27 – 12.75 = 14.25 excess deaths among those who continued to smoke
3. p(excess deaths among continuing smokers) = 14.25 / 75 = .19 = risk difference
4. 1,000 x 0.19 = 190 excess deaths could be avoided.
5. The largest possible risk difference is either plus or minus one. Nevertheless, this doesn't mean that 0.19 is small relative to a clinically meaningful reference value, which would be desirable.
6. One choice for a reference value is the risk for the exposed, i.e., 0.36. The ratio 0.19/0.36 = 0.53 indicates that 53% of the risk for the group of continuing smokers would be reduced if this group had quit smoking.

Q6.2

1. 36, 851 x 14 / 68,239 = 7.5 is the expected number of CHD-related deaths per 100,000 py if persons with borderline-high cholesterol had their cholesterol lowered to normal values. Thus 26 – 7.5 = 18.5 CHD-related deaths per 100,000 py could be avoided could have been avoided.
2. 18.5 / 36,581 = 50.6 (71.1-20.5) excess CHD-related deaths per 100,000 person years. This value of 50.6 per 100,000 is the rate difference or attributable rate.
3. The largest possible rate difference is infinite. Nevertheless, this does not mean that 50.6 is small relative to a clinically meaningful reference value.
4. One choice for a reference is 71.1, the rate for the exposed. The ratio 50.6 / 71.1 = .72 indicates that the rate in borderline-high cholesterol group would be reduces by 72% if this group could lower their cholesterol to normal levels.

Q6.3

1. The risk difference is 2.6% – 1.4% = 1.2%
2. 1/0.012 = 83
3. Since we computed the number need to treat using 10-year cumulative incidences, the answer to the previous question indicates that 83 patients must be treated for 10 years to prevent one stroke.

Q6.4

1. No – Absolute risk describes the risk in a particular group rather than the difference in risk from two groups.
2. Yes
3. Yes
4. No – Relative risk is the ***ratio*** of (rather than the ***difference*** between) risk among two groups.
5. No – for questions 5-11: Any location that does not include Queens itself could be used for a baseline or expected rate. So, Queens and New York City would not be good choices since they both include Queens.
6. Yes
7. Yes
8. Yes
9. Yes
10. No
11. Yes
12. 15.7 per million – Since the individual rates are in units of *per million*, the difference in the rates will have the same unit of measurement.
13. Risk difference – In an outbreak such as this, the investigators are comparing two risks. The appropriate measure of impact here is the risk difference.
14. Prevalence difference – In a prevalence study, the appropriate measure of disease frequency is prevalence. A corresponding measure of impact is the prevalence difference.
15. Rate difference – In a follow-up study we can use person-years of observation to calculate a rate. The appropriate measure of potential impact here is the rate difference.

Q6.5

1. Yes, because both chewing tobacco and alcohol use have the same value (5) for the risk ratio.
2. No. Chewing tobacco has little public health effect, whereas alcohol consumption has a much stronger public health effect.
3. Out of 1000 heavy drinkers (i.e., high alcohol consumption), 40 persons would develop the disease because of their drinking. In contract, only 4 tobacco chewers out of 1000 tobacco chewers would develop the disease from chewing tobacco.

Q6.6

1. False – The rate difference between Queens and the rest of NYC is 16.4 per million – 2.4 per million = 14.0 per million. The excess rate (i.e., rate difference) in Queens is therefore 14.0 cases per million population.
2. True – see above for answer
3. False – 16.4 cases per million in not the attributable rate (i.e., rate difference), but the absolute rake of West Nile encephalitis in Queens. The rate

difference is 16.4 – 2.4 = 14.0 per million population.

4. True – The most common measure of effect for comparing Queens to the rest of NYC is 6.8 and this is the rate ratio calculated as 16.4/2.4.
5. Risk difference – a measure of public health burden.
6. Risk ratio – a measure of disease etiology.
7. Risk difference – the null value for the risk difference is 0.0; the null value for the risk ratio is 1.0.
8. Risk difference – The risk difference can be negative if the baseline risk is higher than the risk in the group of interest. The risk ratio can never be negative because it is a ratio of two positive numbers.
9. Both – it is *possible* to have a risk difference and a risk ratio in the range of 0.0 to 1.0. Note that the CD states the correct answer is Neither with the rationale that neither the risk ratio or risk difference is restricted to values between 0.0 to 1.0.
10. Both – Since risk can be calculated from most follow-up studies, then both a risk ratio and risk difference can be calculated.
11. Neither – Since risk cannot be calculated from case-control studies, neither a risk difference nor a risk ratio can be calculated.
12. Risk ratio – The risk ratio has no units since it is a ratio of risks that has the same units.
13. Risk ratio – A risk ratio close to zero would indicate a strong protective effect. A risk difference close to zero would indicate no effect
14. Risk difference
15. Lung cancer – The rate ratio is much higher for lung cancer than for CHD
16. CHD – elimination of smoking would reduce the number of CHD by 190.4 cases per 100,000 per year.

Q6.7

1. 156 x .173 = 27, where .173 is the risk for the unexposed subjects.
2. 41 – 27 = 14 = I*
3. I*/I = 14/41 = .35 = EF

Q6.8

1. (2.1 – 1) / 2.1 = .52
2. (RR – 1) / RR = (10 – 1) / 10 = .90
3. The maximum possible value for the EF approaches 1 as RR approaches infinity.
4. Yes, even if RR is very large, the EF can be small, even close to zero, if the proportion exposed in the population is very small.

Q6.9

1. The **in**correct statement is choice D. This is incorrect because it considers only deaths among exposed persons, that is, those with borderline-high cholesterol. A corrected version of this statement, which would be essentially equivalent to choice A, is 46% of all deaths are due to persons having borderline-high cholesterol.
2. The correct answer is choice A. The calculation of this value is shown here.

Excess deaths = (all deaths) - (expected deaths if all persons have normal cholesterol)

= 4 0 - (104820 * 14 / 68239)

= 4 0 - 21.5

= 18.5

Q6.10

1. 75 x .19 = 14.25 = I*, where 75 is the number of exposed subjects and .19 is the risk difference.
2. In this example, <u>14</u> of the <u>27</u> deaths among continuing smokers could have been avoided.
3. I^*/I_1 = 14.25 / 27 = .53 = EFe.
4. EFe = (2.1 –1) / 2.1 = .52. This is the same as the .53 previously obtained, other than round-off error.
5. The EF considers the potential impact of exposure on 'all cases' in the cohort whereas the EFe focuses on the potential impact of exposure on only 'exposed cases' in the cohort. Both measures are meaningful, but have a different focus.
6. EFe = $(ID_1 - ID_0) / ID_1$, where ID_1 and ID_0 are the incidence densities (i.e., rates) for exposed and unexposed persons in the cohort.
7. EFe = (IDR – 1) / IDR
8. EF = p'(OR - 1)/[p'(OR – 1) + 1] and EFe = (OR – 1) / OR, where OR is the odds ratio and p' is the proportional of all controls that are exposed.

Q6.11

1. The etiologic fraction is (1.4-1.2) / 1.4 = 14%
2. p=0.108
   RR=(3/103)/(10/854)=2.5
   $$EF = \frac{0.108(2.5-1)}{0.108(2.5-2)+1} = 14\%$$
3. EFe=(2.9-1.2)/2.9=60%

Q6.12

1. 31.8% - prevalence of smoking = 70 million/220 million = 31.8%
2. 79.0% - The EF for lung cancer is (31.8-6.7)/21.8=79.0%.
3. 26.7% -

4. 92.2%
5. 53.3%
6. 92.2%
7. Etiologic fraction among the exposed
8. Etiologic fraction
9. Risk/rate difference
10. Etiologic fraction
11. Risk/rate difference
12. Risk/rate difference
13. Etiologic Fraction, Etiologic Fraction Among the Exposed

Q6.13

1. Multiply the total number of study subjects, 11,185, by the cumulative incidence for the unexposed (105.9 per 1,000 persons) to get 1184.5 potential cases.
2. 539, which can be seen in the table or computed by multiplying 11,185 by the overall cumulative incidence of 48.2 per 1,000 persons. This number is **I** in the initial formula for PF.
3. 1184.5 – 539 = 645.5 = I** total cases were prevented.
4. PF = I** / (I** + I) = 645.5 / 1,184.5 = .55 or 55%.

Q6.14

1. Multiply the total number of immunized children, 6,350, by the cumulative incidence for the unexposed (105.9 per 1,000 persons) to get 672.5 expected number of ill children.
2. 27, which can be seen in the table or computed by multiplying 6,350 by the cumulative incidence for immunized children of 4.2 per 1,000 persons. This number is I_1 in the initial formula for PFe.
3. 672.5 – 27 = 645.5 = I** total cases were prevented.
4. PFe = I** / (I** + I_1) = 645.5 / 672.5 = .96, or in percents, 96%.

Q6.15

1. EFe = (1 – 0.4) = .96. This is the same as we obtained using the formula involving CI_0 and CI.
2. The PF considers the potential impact of exposure on 'all cases' in the cohort whereas the PFe focuses on the potential impact of exposure on only 'exposed cases' in the cohort. Both measures are meaningful, but have a different focus.
3. PFe = (ID_0 – ID_1) / ID_1, where ID_0 and ID_1 are the incidence densities (i.e., rates) for unexposed and exposed persons, respectively, in the cohort.
4. PFe = (1 – IDR)
5. PF = p'(1 – OR) and PFe = (1 – OR), where OR is the odds ratio an d p' is the proportion of all controls that are exposed.

Q6.16

1. True – when the risk difference is <0, the exposure appears to be protective, and the prevented fraction is preferred.
2. False – 1-EF is not the same as the prevented fraction. The difference in the two measures occurs both in the numerator and denominator.
3. True – An alternative formula for the PF among the exposed in a cohort or clinical trial is (1-RR).
4. True – Prevented fraction is the proportion of cases that did not occur that would have if the exposure had not been present.
5. True – The denominator for both PF and PFe is CI_0 for risk data or ID_0 for rate data.
6. True – An alternative formula for PF is PF=p(1-RR) where p is the prevalence of exposure.
7. Prevented fraction among the exposed – $$\hat{P}Fe = \frac{\hat{C}I_0 - \hat{C}I_1}{\hat{C}I_0}$$
8. Prevented fraction - $\hat{P}F = \dfrac{\hat{C}I_0 - \hat{C}I}{\hat{C}I_0}$
9. Neither - An alternative formula for PF is PF=p(1-RR) where p is the prevalence of exposure. If fewer children had been vaccinated, the PF would decrease.
10. Prevented fraction - An alternative formula for PF is PF=p(1-RR) where p is the prevalence of exposure. If fewer children had been vaccinated, the PF would decrease.
11. Prevented fraction – The PF is the proportion of potential cases in the community prevented by vaccination.

LESSON 7

VALIDITY

7-1 Validity

*The primary objective of most epidemiologic research is to obtain a **valid estimate** of an **effect measure** of interest. In this activity, we illustrate three general types of **validity** problems, distinguish validity from **precision**, introduce the term **bias**, and discuss how to adjust for bias.*

Examples of Validity Problems

Validity in epidemiologic studies concerns methodologic flaws that might distort the conclusions made about an exposure-disease relationship. Several examples of validity issues are briefly described.

The validity of an epidemiologic study concerns whether or not there are imperfections in the study design, the methods of data collection, or the methods of data analysis that might distort the conclusions made about an exposure-disease relationship. If there are no such imperfections, we say that the study is **valid**. If there are imperfections, then the extent of the distortion of the results from the correct conclusions is called **bias**. Validity of a study is what we strive for; bias is what prevents us from obtaining valid results

In 1946, Berkson demonstrated that case-control studies carried out exclusively in hospital settings are subject to a type of **"selection" bias**, aptly called **Berkson's bias**. Berkson's bias arises because patients with two disease conditions or high-risk behaviors are more likely to be hospitalized than those with a single condition. Such patients will tend to be over-represented in the study population when compared to the community population. In particular, respiratory and bone diseases have been shown to be associated in hospitalized patients but not in the general population. Moreover, since cigarette smoking is strongly associated with respiratory disease, we would expect a hospital study of the relationship between cigarette smoking and bone disease to demonstrate such a relationship even if none existed in the general population.

In the 1980's and 1990's, US Air Force researchers assessed the health effects among Vietnam War veterans associated with exposure to the herbicide Agent Orange. Agent Orange contained a highly toxic trace contaminant known as TCDD. Initially, exposure to TCDD was classified according to job descriptions of the veterans selected for study. It was later determined that this produced substantial misclassification of TCDD. The validity problem here is called **information bias**. Bias could be avoided using laboratory techniques that were developed to measure TCDD from blood serum. The use of such **biologic markers** in epidemiologic research is rapidly increasing as a way to reduce misclassification and, more generally, to improve accuracy of study measurements.

As a final example, we return to the Sydney Beach Users Study described previously. A validity issue in this study concerned whether all relevant variables, other than swimming status and pollution level, were taken into account. Such variables included age, sex, duration of swimming, and additional days of swimming. The primary reason for considering these additional variables is to ensure that any observed effect of swimming on illness outcome could not be explained away by these other variables. A distortion in the results caused by failure to take into account such additional variables is called **confounding bias**.

Summary

- ❖ Validity: The general issue of whether or not there are imperfections in the study design, the methods of data collection, or methods of data analysis that might distort the conclusions made about an exposure-disease relationship.
- ❖ Bias: A measure of the extent of distortion of conclusions about an exposure-disease relationship.
- ❖ Validity issues are illustrated by:
    - Hospital-based case-control studies (Berkson's **selection bias**).
    - Job misclassification to assess TCDD exposure (**information bias**).
    - Control of relevant variables in the Sydney Beach Users Study (**confounding**).

Validity versus Precision

*Validity and precision concern two different sources of inaccuracy that can occur when estimating an exposure-disease relationship: **systematic error** (a validity problem) and **random error** (a precision problem). Systematic and random error can be distinguished in terms of shots at a target.*

Validity and precision are influenced by two different types of error that can occur when estimating an exposure-disease relationship. **Systematic error** affects the **validity**, and **random error**, the **precision**.

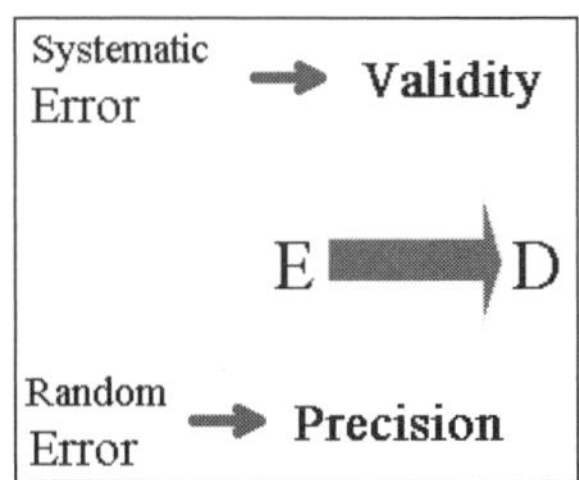

These two types of error can be distinguished by viewing an epidemiologic study as a shot at a target. The blue dot in the middle of the target symbolizes the **true measure of effect** being estimated in a population of interest. (Note: to be consistent with the CD, the use of the term "blue dot" in this text refers to the center of the target or the "bull's eye".) Each shot represents an **estimate** of the true effect obtained from one of possibly many studies in each of three populations.

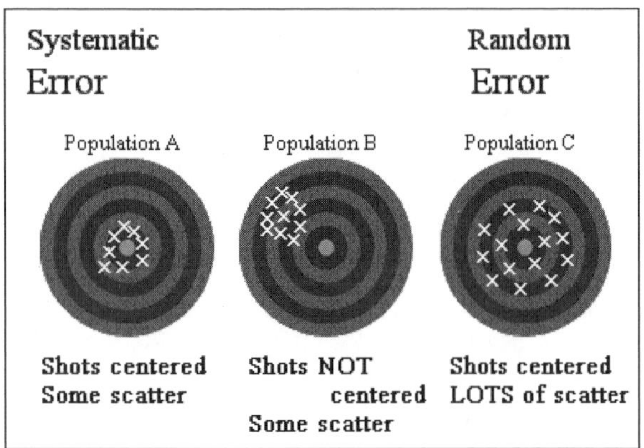

For Target A, the shots are centered around the blue dot, although none of the shots actually hit it and all shots hit a different part of the target. For Target B, the shots are all far off center, but have about the same amount of scatter as the shots at target A. For target C, the shots are centered around the blue dot, but unlike Target A, are more spread out from one another.

Systematic error is illustrated by comparing Target A with Target B. The shots at Target A are aimed at the blue dot, whereas the shots at Target B are not aimed at the blue dot, but rather centered around the red dot. (Note: to be consistent with the CD, the term "red dot" will refer to the dot above and to the left of the bull's eye in Population B). The distance between the blue dot and the red dot measures the systematic error associated with Target B. In contrast, there is no systematic error associated with Target A.

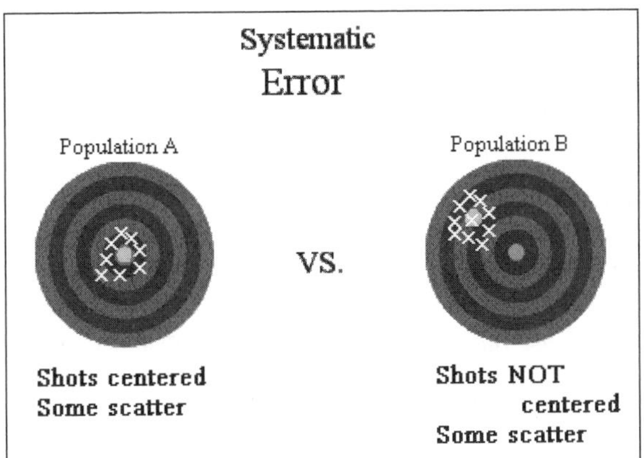

Systematic error occurs when there is a difference between the true effect measure and what is actually being estimated. We say that the study is valid if there is no systematic error. Thus, validity is concerned with whether or not a study is aiming at the correct effect measure, as represented by the bull's eye. Unfortunately in epidemiologic and other research, the bull's eye is usually not known. Consequently, the amount of bias is difficult to determine and the evaluation of bias is to some extent always subjective.

All the targets illustrate random error, which occurs when there is a difference between any estimate computed from the study data and the effect measure actually being estimated. Targets A and B exhibit the same amount of random error because there is essentially the same amount of scatter of shots around the blue dot of Target A as there is around the red dot of Target B. In contrast Target C, in which shots are much more spread out, exhibits more random error than targets A or B.

Thus, the more spread out the shots, the more random error, and the less precision from any one shot. Precision therefore concerns how much individual variation there is from shot to shot, given the actual spot being aimed at. In other words, precision reflects **sampling variability**.

Problems of precision generally concern statistical inference about the parameters of the population actually being aimed at. In contrast, problems of validity concern methodologic imperfections of the study design or the analysis that may influence whether or not the correct population parameter, as represented by the blue dot in each target, is being aimed at by the study

Study Questions (Q7.1)

Consider a cross-sectional study to assess the relationship between calcium intake (high versus low) in one's diet and the prevalence of arthritis of the hip in women residents of the city of Atlanta between the ages of 45 and 69. A sample of female hospital patients is selected from hospital records in 1989, and the presence or absence of arthritis as well as a measure of average calcium intake in the diet prior to enter the hospital are determined on each patient.

1. What is the target population in this study?
2. What does the center of the target (i.e., the bulls-eye) represent in epidemiologic terms?
3. What do we mean by random error associated with this study?
4. What do we mean by systematic error associated with this study?

Summary

- Validity concerns systematic error whereas precision concerns random error.
- Systematic and random error can be distinguished in terms of shots at a target.
- Systematic error: a difference between what an estimator is actually estimating and the effect measure of interest.
- Random error: a difference between any estimate computed from the study data and the effect measure actually being estimated.
- Validity does not consider statistic inference, but rather methodologic imperfections of the study design or analysis.
- Precision concerns statistical inferences about the parameter of the population actually being aimed at.

Use Data Desk to Explore Error

This activity explores the concepts of systematic and random error using Data Desk's dynamic graphics.

A Hierarchy of Populations

To further clarify the difference between validity and precision, we now describe a hierarchy of populations that are considered in any epidemiologic study.

We typically identify different populations when we think about the validity of an epidemiologic study. These populations may be contained within each other or they may simply overlap.

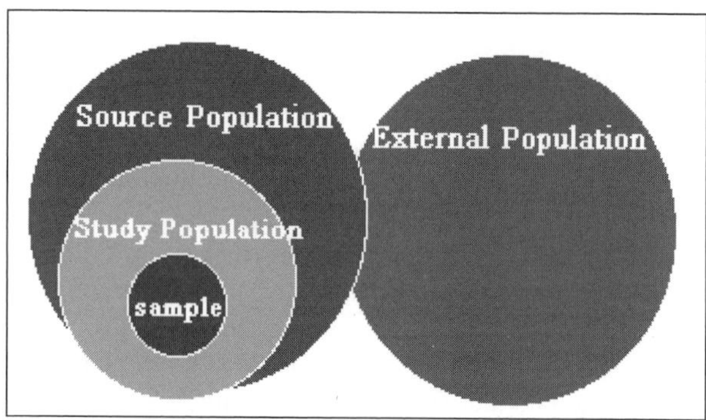

We refer to the collection of individuals from which the study data have been obtained as the **sample**. We use results from the sample to make inferences about larger populations. But what populations can we make these inferences about? What population does the sample represent?

The **study population** is the collection of individuals that our sample actually represents and is typically those individuals we can feasibly study. We may be limited to sampling from hospitals or to sampling at particular places and times. The study population is defined by what is practical, which may not be what we ideally would like.

The **source population** is the collection of individuals of restricted interest; say in a specific city, community, or occupation, who are at risk for being a case. Clearly all cases must come from the source population (if they were not at risk, they would not have become cases). The source population also is likely to include individuals who, although at risk, may not become cases. The source population has been called the **study base** or the **target population**.

We can make statistical inferences from the sample to the study population, but we would like to be able to make inferences from the sample to the source population. Unfortunately, the **study population**, the population actually represented by our sample, may not be representative of the **source population**.

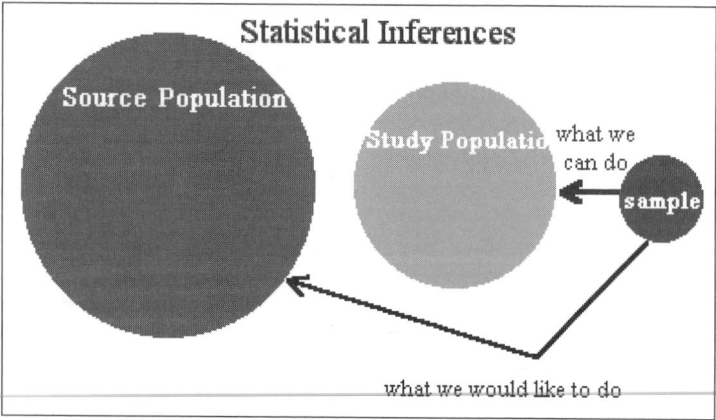

Study Questions (Q7.2)

Consider an epi study carried out in New York City (NYC) to assess whether obesity is associated with hypertension in young adults. The investigators decided that it was not feasible to consider taking a sample from among all young adults in the city. It was decided that fitness centers would provide a large source of young NYC adults. A sample of subjects is taken from several randomly selected fitness centers throughout the city and their blood pressure is measured to determine hypertension status.

1. What is the source population for this study? young adults in NYC
2. What is the study population in this study? young adults who attend fitness centers in NYC
3. Does the sample represent the study population? yes the sample is randomly selected & is therefore representative
4. Does the study population represent the source population? no may not rep all adults in NYC

In a simple case every member of the study population is also in the source population -that is, we are only studying individuals who are in fact at risk. If the study population is representative of the source population and the sample is representative of the study population then there is no bias in inferring from the sample to the source population.

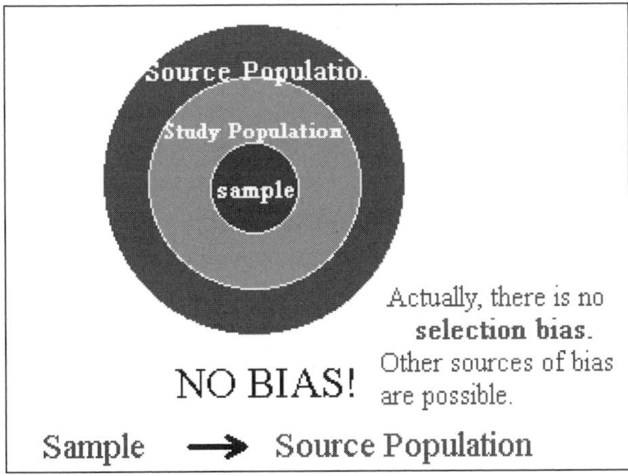
Source Population
Study Population
sample
Actually, there is no selection bias.
Other sources of bias are possible.
NO BIAS!
Sample → Source Population

Study Questions (Q7.2) continued

Recall the epi study carried out in New York City to assess whether obesity is associated with hypertension in young adults. Suppose the investigators decided that it was important to obtain a sample from all young adults within NYC. They used the 2000 census information to get a listing of all young NYC adults. A sample of subjects is taken from several randomly selected city blocks throughout the city and their blood pressure is measures to determine hypertension status.

5. What is the source population for this study? young adults in NYC
6. What is the study population in this study? young adults who provided information
7. Does the sample represent the study population? yes, it is randomly selected
8. Does the study population represent the source population? yes if they participate

Sometimes, however, even a sample from a study population that represents the source population can become biased. For example, in a typical cohort study, even though every member of the initial study population is also in the source population the initial sample may change in the course of the study. The initial sample may suffer from exclusions, withdrawals, non-response, or loss-to-follow-up. The final study population is then only those individuals who are willing and able to stay in such a study, a population that may not represent the initial study population or the source population as well as we might wish.

Study Questions (Q7.2) continued

9. True or False. In a cohort study, the ***study*** population includes persons eligible to be selected but who would have been lost to follow-up if actually followed. False
10. True or False. In a cohort study, the ***source*** population includes persons eligible to be selected but who would have been lost to follow-up if actually followed. True
11. True or False. In a case-control study, the study population may contain persons eligible to be controls who were not at risk for being a case. True
12. True or False. In a cross-sectional study, the study population may contain persons who developed the health outcome but had died prior to the time at which the sample was selected. False, survivors only

In general, it is possible for members of the study population not even to be at risk and we may not be able to tell. We may, for example, draw our sample from a study population of persons who attend a clinic for sexually transmitted diseases. These people may or may not have an STD and may or may not be exposing themselves to STD's. For example, some of these subjects may have partners who are not infected, and thus may not be at risk themselves. If they are not at risk, they are not part of the source population, but we may not know that.

It can also happen that the study population fails to include individuals who are at risk (and thus part of the source population) either because we do not know they are at risk or because it is not practical to reach them. For example when AIDS was poorly understood, a study of gay men at risk for AIDS may have failed to include IV drug users who were also at risk.

Finally, we would often like to generalize our conclusions to a different **external population**. An external population is a population that differs from the study population but to which we nevertheless would like to generalize the results, for example, a different city, community, or occupation. In a public health setting, we are always concerned with the health of the general public even though we must study smaller subpopulations for practical reasons. For statistical conclusions that are based on a sample to generalize to an external population, the study population must itself be representative of the external population, but that is often difficult to achieve.

Study Questions (Q7.2) continued

Consider an epi study carried out in New York City to assess whether obesity is associated with hypertension in young adults. Suppose it was of interest to determine whether the study results carry over to the population of the entire state of New York.

13. Considering the variety of populations described, what type of population is being considered? Explain briefly.

external population

Summary

- There are a variety of populations to consider in any epi study.
- The **sample** is the collection of individuals from which the study data have been obtained.
- The **study population** is the collection of individuals that our sample actually represents and is typically those individuals we can feasibly study.
- The **source population** is the group of restricted interest about which the investigator wishes to assess an exposure-disease relationship.
- The **external population** is a group to which the study has not been restricted but to which the investigator still wishes to generalize the study results.

Internal versus External Validity

Target shooting provides an example that illustrates the difference between **internal** and **external validity**. Internal validity considers whether or not we are aiming at the center of the target. If, we are aiming at this red dot (to the left and above the bulls-eye) rather than at the bulls-eye, then our study is **not** internally valid.

Internal validity is about drawing conclusions about the source population based on information from the study population. Such inferences do not extend beyond the source population of restricted interest.

External validity concerns a different target; in particular, one at which we are not intending to shoot; whose bulls-eye we can't really see. We might imagine this external target being screened from our vision.

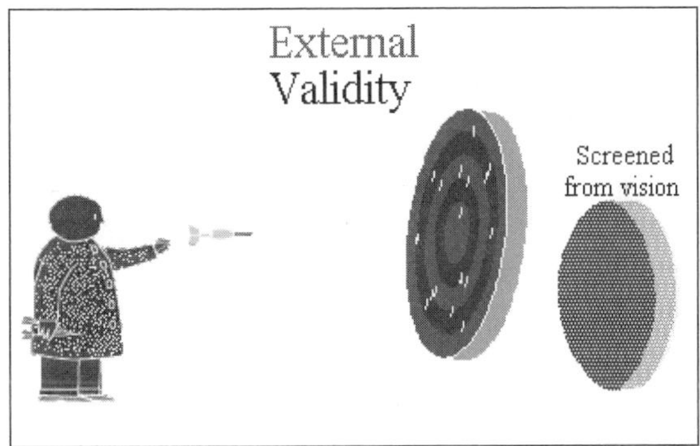

Suppose that this screened target is in line with the target at which we are shooting. Then, by aiming at the bulls-eye of the target we can see, we are also aiming at the bulls-eye of the external target. In this case, the results from our study population can be generalized to this external population, and thus, we have external validity. If the external target is not lined up with our target, our study does not have external validity, and the study results should not extend to this external population.

External validity is about applying our conclusions to an external population beyond the study's restricted interest. Such inferences require judgments about other findings and their connection to the study's findings, conceptualization of the disease process and related biological processes, and comparative features of the source population and the external population. External validity is therefore more subjective and less quantifiable than internal validity.

Study Questions (Q7.3)

Consider an epi study carried out in New York City to assess whether obesity is associated with hypertension in young adults. Subjects are sampled from several fitness centers throughout the city and their blood pressure is measured to determine hypertension status.

1. What is required for this study to be internally valid?

that the study population is representative of the source population

Suppose it was of interest to determine whether the study results carry over to the entire State of New York.

2. Does this concern internal validity or external validity? Explain briefly.

Results from the Lipid Research Clinics Primary Prevention Trial published in 1984 (JAMA, vol. 251) demonstrated a significant reduction in cardiovascular mortality for white men ages 35 to 59 who were placed on a cholesterol-reducing diet and medication.

3. What question might be asked about the results of this study that concerns external validity?
4. What question(s) might be asked about the study results that concern(s) internal validity?

Summary

- ❖ Internal validity concerns whether or not we are aiming at the center of the target we know we are shooting at.
- ❖ External validity concerns a target that we are not intending to shoot at, whose bulls-eye we can't really see.
- ❖ Internal validity concerns the drawing of conclusions about the target population based on information from the study population.
- ❖ External validity concerns drawing conclusions to an external population beyond the study's restricted interest.

Quiz (Q7.4)

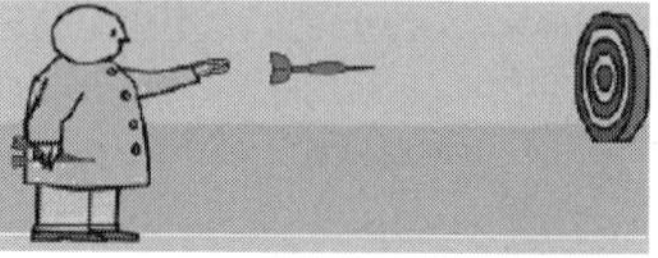

Label each of the following statements as **True** or **False**; for questions 8-11, an additional response option is **It depends**.

1. Random error occurs whenever there is any (non-zero) difference between the value of the odds ratio in the study population and the estimated odds ratio obtained from the sample that is analyzed. **???**
2. Systematic error occurs whenever there is any (non-zero) difference between the value of the effect measure in the source population and the estimate from the sample. . . . **???**
3. In a valid study, there is neither systematic error nor random error. **???**
4. The study population is always a subset of the source population. . . . . **???**
5. The sample is always a subset of the study population. .. . . . . **???**
6. The sample is always a subset of the source population. . . . . . . **???**
7. The estimated effect measure in the sample is always equal to the corresponding effect measure in the study population. . . . . **???**
8. Suppose the risk ratio in the source population is 3.0, whereas the risk ratio estimate in the (study) sample is 1.2. Then the study is not internally valid. . . . . . **???**
9. Suppose the risk ratio in the source population is 3.0, whereas the risk ratio in the study population is 1.2. Then the study is not internally valid. . . . . . . **???**
10. Suppose the risk ratio in the source population is 3.0, whereas the risk ratio estimate in the study population is 3.1. Then the study is not internally valid. . . . . . . **???**
11. Suppose the risk ratio in the source population is 3.0, whereas the risk ratio estimate in the study population is 3.1. Then the study is not externally valid. . . . . . . **???**

7-2 Validity (continued)

Quantitative Definition of Bias

*A **bias** in an epidemiologic study can be defined quantitatively in terms of the **target parameter** of interest and measure of effect actually being estimated in the **study population**.*

A study that is not internally valid is said to have bias. Let's quantify what we mean by bias. The measure of effect in the source population is our target parameter.

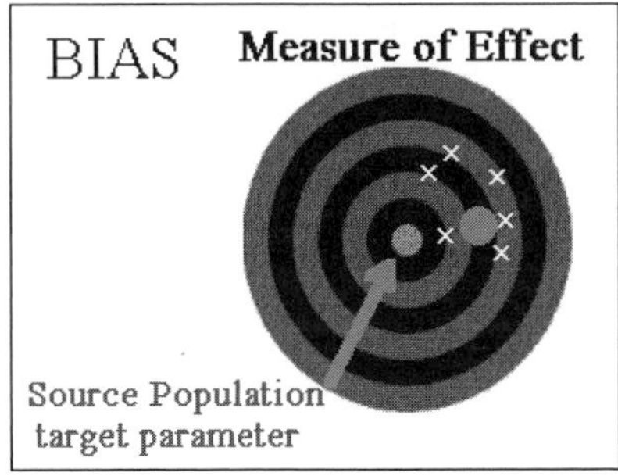

The choice of this parameter depends on the study design features, objectives of the study, and the type of bias being considered. We denote the target parameter with the Greek letter $\boldsymbol{\theta}$ ("theta"). We want to estimate the value of $\boldsymbol{\theta}$ in the source population.

Recalling the hierarchy of populations associated with a given study, we denote as $\boldsymbol{\theta}^0$ the measure of effect in the study population. $\hat{\theta}$ ("theta-hat") denotes the estimate of our measure of effect obtained from the sample actually analyzed. Of course, $\hat{\theta}$, $\boldsymbol{\theta}^0$ and $\boldsymbol{\theta}$ may all have different values.

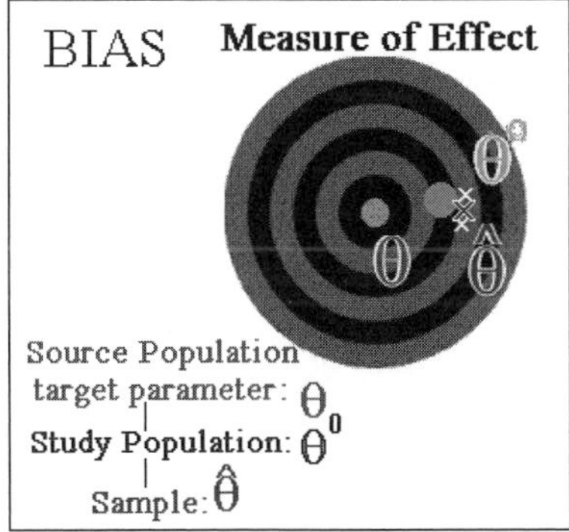

Any difference between $\hat{\theta}$ and $\boldsymbol{\theta}^0$ is the result of random error. Any difference between $\boldsymbol{\theta}^0$ and $\boldsymbol{\theta}$ is due to systematic error.

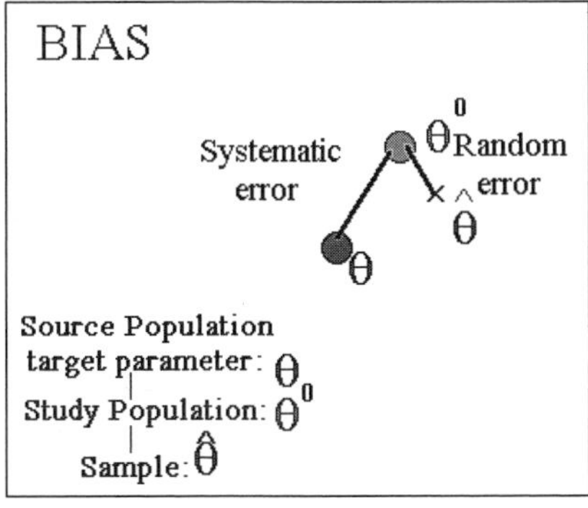

We use $\hat{\theta}$ to estimate $\boldsymbol{\theta}$. We say that $\hat{\theta}$ is a biased estimate of $\boldsymbol{\theta}$ if $\boldsymbol{\theta}^0$ is not equal to $\boldsymbol{\theta}$, and we define the bias to be the difference between these two parameters: Bias $(\hat{\theta},\boldsymbol{\theta}) = \boldsymbol{\theta}^0 - \boldsymbol{\theta}$

Thus, a bias occurs if an estimated measure of effect for the study population differs systematically from the value of the target parameter. The *not equal* sign shown below should not be strictly interpreted as any difference from zero, but rather as a meaningful difference from zero. Such a flexible interpretation of the definition is necessary because bias can rarely be quantified precisely because the target parameter is always unknown.

> $\hat{\theta}$ is a biased estimate of θ if
>
> $\theta^0 - \theta \neq 0$
>
> meaningful difference

Study Questions (Q7.5)

Consider a cohort study to evaluate whether heavy drinking during pregnancy leads to low birth weight babies. Although it is usually unknown, suppose the risk ratio in the source population (i.e., entire cohort) is 3.5. Suppose further that the study sample is representative of the source population at the start of follow-up, but that there is considerable migration out of the target population location. As a result, the risk ratio found in the final sample is 1.5.

1. For this scenario, what are the values of the target (i.e., source) population parameter and the study population parameter?
2. Is there bias in this study? Explain briefly.
3. What is the value of the bias in this study?
4. If the true effect being estimated is *high* (e.g., RR>3.5), but the study data show essentially no effect, does this indicate a bias? Explain briefly.
5. If the true effect estimated indicates *no* association, but the study data show a high association, does this indicate a bias? Explain briefly.

Summary

- Bias measures the extent that study results are distorted from the correct results that would have been found from a valid study.
- Bias occurs if an estimated measure of effect for the study population differs systematically from the value of the target parameter.
- Bias can rarely be quantified precisely primarily because the target parameter is always unknown.
- $\boldsymbol{\theta}$ ("theta") is the target parameter, $\boldsymbol{\theta}^0$ is the study population parameter, and $\hat{\theta}$ is the sample estimate.
- $\hat{\theta}$ is a biased estimate of $\boldsymbol{\theta}$ provided $\boldsymbol{\theta}^0$ is not equal to $\boldsymbol{\theta}$.
- Bias $(\hat{\theta},\boldsymbol{\theta}) = \boldsymbol{\theta}^0 - \boldsymbol{\theta}$

Relating the Target Parameter to Type of Bias

The target parameter in most epidemiologic studies is typically a measure of effect, e.g., some kind of risk ratio, odds ratio, or rate ratio, appropriate for the study design being considered. The choice of parameter depends on whether the type of bias of interest is selection bias, information bias, or confounding bias, or whether more than one of these three types of bias is of concern.

When **selection bias** is the only type of bias being considered, the target parameter is typically the value of the measure of effect of interest in the source population from which the cases are derived.

If, however, **information bias** is the only type of bias considered, then the target parameter is the measure of effect that corrects for possible misclassification or that would result from the absence of misclassification.

If there is only bias due to **confounding**, then the target parameter is the measure of effect estimated when confounding is controlled.

If more than one bias is possible, then the target parameter is the value of the measure of effect after all contributing sources of bias are corrected.

Epidemiologic Bias versus Statistical Bias

The definition of bias we have given using the expression bias = $\boldsymbol{\theta}^0 - \boldsymbol{\theta}$ implicitly identifies $\hat{\theta}$ to be a **statistically consistent estimator**, i.e., $\hat{\theta}$ converges to $\boldsymbol{\theta}$ in probability as the size of the study is hypothetically increased without changing any of the essential features of the study design itself.

This definition is recommended for describing **epidemiologic bias** because statistical consistency essentially reflects only methodological aspects of the study design that might be associated with systematic error rather than random error.

In contrast, a definition of **statistical bias** would be based on the concept of the **statistical expected value**, i.e., bias = Expected value of $\hat{\theta}$ - θ where the "expected value" is the long run average of $\hat{\theta}$ as the sample size increases without limit.

Using **statistical bias**, we would allow bias to be present solely because of the mathematical properties of the estimator. For instance, the expected value of an estimate of the risk ratio is not equal to the population risk ratio it is actually estimating even if methodological flaws are absent. This is not what **epidemiologic bias** is meant to describe, and so the use of expected value theory to describe epidemiologic bias is not desirable.

Direction of the Bias

Although the precise magnitude of bias can never really be quantified, the ***direction of bias*** *can often be determined. The direction of the bias concerns whether or not the target parameters is either overestimated or underestimated without specifying the magnitude of the bias.*

We have defined bias as the difference between the value of the effect measure in our target (i.e., source) population and the value of the effect measure actually being estimated in the study population. Since the target parameter is always unknown and the effect being estimated in the study population has random error, it is virtually impossible to quantify the magnitude of a bias precisely in a given epidemiologic study. Nevertheless, the investigator can often determine the direction of the bias. Such assessment usually requires subjective judgment based on the investigator's knowledge of the variables being studied and the features of the study design that are the sources of possible bias. By direction, we mean a determination of whether the target parameter is **overestimated** or **underestimated** without specifying the magnitude of the bias. If the target parameter is **overestimated**, we say that the direction of the bias is **away from the null**. If the target parameter is **underestimated**, we say that the direction of the bias is **towards the null**.

$$\text{BIAS}(\hat{\theta},\theta) = \theta^0 - \theta$$

Direction of Bias

$\theta^0 > \underset{\text{Target}}{\theta}$ Overestimated → Away from the null

$\theta^0 < \theta$ Underestimated → Towards the null

For example, suppose the target parameter is a risk ratio whose value is 1.5, but the risk ratio actually being estimated from our study is 4. Then the true effect has been overestimated, since the effect from the study appears to be stronger than it really is. Both the target parameter and the study population parameter in this example are greater than the null value of 1 for a risk ratio. Thus, the bias is away from the null, since the incorrect value of 4 is further away from 1 than the correct value of 1.5.

Overestimated → Away from the null

$$\mathbf{RR^0 = 4 > RR = 1.5 > 1}$$

null value

Similarly, if the target risk ratio is less than the null value of 1, say .70 and the risk ratio being estimated is .25, the true effect is also overestimated. In this case, the true effect is protective, since it is less than 1, and the estimated effect of .25

is even more protective than the true effect. Again, the incorrect value of .25 is further away from 1 than the correct value of .70, so the bias is away from the null.

Overestimated ➔ Away from the null

$RR^0 = 0.25 < RR = 0.70 < 1$

More Protective *Protective*

To describe **underestimation**, or **bias towards the null**, suppose the target risk ratio is 4, but the estimated risk ratio is 1.5. Then the true effect is underestimated, since the effect from the study appears to be weaker than it really is. Moreover, the incorrect value of 1.5 in this case is closer to the null value of 1 than is the correct value of 4, so the bias is **towards the null**.

Underestimated ➔ Towards the null

$1 < RR^0 = 1.5 < RR = 4$

Study effect (1.5) appears weaker than it really is (4)

If the target risk ratio is 0.25 but the risk ratio being estimated is 0.70, then once again the correct value is underestimated and the bias is towards the null. In this case, the incorrect value of 0.70 is closer to the null value of 1 than is the correct value of .25.

Underestimated ➔ Towards the null

$1 > RR^0 = 0.70 > RR = 0.25$

0.70 closer to the null than 0.25

Suppose, however, that the target risk ratio is .50 but the risk ratio actually being estimated is 2. These two values are on opposite sides of the null value, so we cannot argue that the bias is either towards or away from the null. In this case, we call this kind of bias a **switchover bias**. In other words, a switchover bias may occur if the exposure appears in the data to have a harmful effect on the disease when it is truly protective. Alternatively, a switchover bias can occur if the exposure appears to be protective when it is truly harmful.

Switchover Bias

$RR^0 = 2 > 1 > RR = 0.50$

Exposure appears to have a harmful effect when truly protective.

OR

$RR^0 = 0.7 < 1 < RR = 1.5$

Exposure appears to be protective when truly harmful.

Study Questions (Q7.6)

Knowing the direction of the bias can be of practical importance to the investigator.

1. Suppose an investigator finds a very strong effect, say the estimated RR (in the study population) is 6, and she can also persuasively argue that any possible bias must be towards the null. Then what can be concluded about the correct (i.e., target) value of RR?
2. Suppose an investigator finds a very weak estimated RR (in the study population) of, say, 1.3, and can argue that any bias must be away from the null. Then what can be concluded about the correct value of the RR?

Summary

- ❖ The direction of the bias concerns whether or not the target parameter is either overestimated or underestimated without specifying the magnitude of the bias.
- ❖ If the target parameter is overestimated, the direction of the bias is away from the null.
- ❖ If the target parameter is underestimated, the direction of the bias is towards the null.
- ❖ A switchover bias occurs if the target parameter is on the opposite side of the null value from the parameter actually being estimated in one's study.

Positive versus Negative Bias

We can characterize the direction of the **bias** as **positive** or **negative** based on our quantitative definition of bias. The definition of bias we have given using the expression **bias** = $\theta^0 - \theta$ where θ^0 is the effect measure actually being estimated by the study sample and θ is the effect measure value in the target population of interest.

To illustrate, suppose the target parameter is the risk ratio, whose value is, say 1.5, and the actual value of the parameter being estimated is 4, i.e.:

RR=1.5 and **RR^o=4**.

Then, using the above formula the bias would be positive. Similarly, if:

RR=0.70 and **RR^o=0.25**

the bias would also be positive. Thus, using our definition, if the target parameter is **overestimated**, as it is in both these examples, then the bias would be **positive**.

In contrast, suppose the target parameter is again the risk ratio, whose value is, say 4, and the actual value of the parameter being estimated is 1.5, i.e.:

RR=4 and **RR^o=1.5**

Then, using our formula for bias, the bias would be **negative**. Similarly, if:

RR=0.25 and **RR^o=0.70**

the bias would also be **negative**. Thus, using our definition, if the target parameter is **underestimated**, as it is in these last two examples, then the bias would be **negative**.

Unfortunately, if we wish to characterize the bias as positive or negative, we must remember that the formula that we are using subtracts the correct value of the parameter from the incorrect value. If we switch these two parameters around, however, the conclusion as to positive versus negative will be switched also.

Consequently, to avoid such possible confusion, the direction of the bias is more commonly described as away from the null (i.e., overestimation) versus towards the null (i.e., underestimation), as discussed in the exposition corresponding to this asterisk.

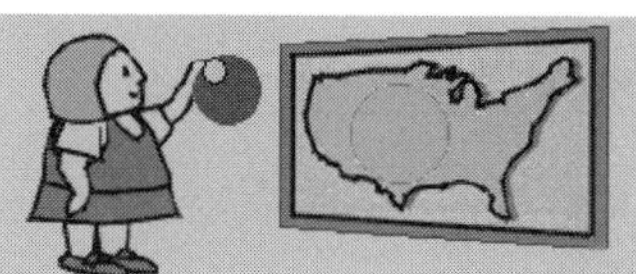

Quiz (Q7.7)

Label each of the following statements as **True** or **False**.

1. If the estimated RR equals 2.7 in the sample and it is determined that there is a bias away from the null, then the RR in the target (i.e., source) population is greater than 2.7. . . . **???**
2. If the estimated RR in the sample is 1.1 and it is determined that there is bias towards the null, then there is essentially no association in the target population (as measured by RR). . . **???**
3. If the estimated RR in the sample is 1.1 and it is determined that there is bias away from the null, then there is essentially no association in the target population. . . . . **???**
4. If the estimated RR equals 0.4 in the sample and it is determined that there is a bias away from the null, then the RR in the target (i.e., source) population is less than 0.4. . . . **???**
5. If the estimated RR in the sample is 0.4 and it is determined that there is bias towards the null, then there is essentially no association in the target population (as measured by RR). . . **???**
6. If the estimated RR in the sample is 0.98 and it is determined that there is bias away from the null, then there is essentially no association in the target population. . . . . **???**

Fill in the Blanks

7. If OR equals 3.6 in the target population and 1.3 in the study population, then the bias is . **???**
8. If IDR is 0.25 in the target population and 0.95 in the study population, then the bias is . **???**
9. If the RR is 1 in the target population and 4.1 in the study population, then the bias is . **???**
10. If the RR is 0.6 in the target population and 2.1 in the study population, then the bias is . **???**
11. If the RR is 1 in the target population and 0.77 in the study population, then the bias is . **???**
12. If the RR is 4.0 in the target population and 0.9 in the study population, then the bias is . **???**

Choices

Away from the null **Switchover** **Towards the null**

What Can be Done About Bias?

The evaluation of bias is typically subjective and involves a judgment about either the presence of the bias, the direction of the bias, or, much more rarely, the magnitude of the bias. Nevertheless, there are ways to address the problem of bias, including adjusting the sample estimate to "correct" for bias. Three general approaches are now described.

Here are three general approaches for addressing bias: 1) a priori study **design** decisions; 2) decisions during the **analysis** stage; and 3) **discussion** during the publication stage.

3 Approaches for Addressing Bias
1) a priori study design decisions
2) decisions during analysis stage
3) discussion during publication stage

When you **design** a study, you can make decisions to minimize or even avoid bias in the study's results. You can avoid **selection bias** by including or excluding eligible subjects, by choice of the source population, or by the choice of the comparison group, say the control group in a case-control study.

Study Questions (Q7.8)

1. What type of bias may be avoided by taking special care to accurately measure the exposure, disease, and control variables being studied, including using pilot studies to identify measurement problems that can be corrected in the main study?
   A. Selection bias
   B. Information bias 
   C. Confounding bias
2. What type of bias may be avoided by making sure to measure or observe variables at the design stage that may be accounted for at the analysis stage?
   A. Selection bias
   B. Information bias 
   C. Confounding bias

At the **analysis** stage, the investigator may be able to determine either the presence or direction of possible bias by logical reasoning about methodologic features of the study design actually used.

Study Questions (Q7.8) continued

3. In the Sydney Beach User's study, both swimming status and illness outcome were determined by subject self-report and recall. This indicates the need to assess the presence or direction of which type of bias at the analysis stage?
   A. Selection bias
   B. Information bias 
   C. Confounding bias

4. Also, in the Sydney Beach Users study, subjects had to be excluded from the analysis if they did not complete the follow-up interview. This non-response may affect how representative the sample is. This is an example of which type of bias?
   A. Selection bias 
   B. Information bias
   C. Confounding bias

At the **analysis** stage, bias can also be reduced or eliminated by *adjusting* a sample estimate by a guestimate of the amount of bias. Such adjustment is typically done for confounding by quantitatively accounting for the effects of confounding variables using **stratified analysis** or **mathematical modeling** methods.

Adjustment for selection bias and information bias is limited by the availability of information necessary to measure the extent of the bias. A simple formula for a "corrected" estimate involves manipulating the equation for bias by moving the target parameter to the left side of the equation. This formula is not as easy to apply as it appears. Most investigators will have to be satisfied with making a case for the direction of the bias instead. The estimated bias depends on the availability of more fundamental parameters, which are often difficult to determine. We discuss these parameters further in the lessons that follow.

Analysis Stage

Adjustment

Selection/Information Bias:

Limited by available information to measure **extent** of bias

$$\theta^{c} = \hat{\theta} - \hat{BIAS}$$

not easy to apply

θ and θ^{0} always unknown

θ^{c} - corrected estimate

The final approach to addressing bias is how you report your study. A description of the potential biases of the study is typically provided in the "Discussion" section of a publication. This discussion, particularly when it concerns possible **selection** or **information** bias, is quite subjective, but judgment is expected because of the inherent difficulty in quantifying biases. Rarely if ever does the investigator admit in the write-up that bias casts severe doubt on the study's conclusions. So, the reader must review this section with great care!

Summary – What Can be Done about Bias?

- The answer depends on the type of bias being considered: selection, information, or confounding.
- Approached for addressing bias are: decisions in the study design stage, the analysis stage, and the publication stage.
- At the study design stage, decisions may be made to avoid bias in the study's results.
- At the analysis stage, one may use logical reasoning about methodologic features of the study design actually used.
- Also at the analysis stage, **confounding** bias can be reduced or eliminated by quantitatively adjusting the sample estimate.
- A simple formula for a corrected estimate: $\theta^{c} = \hat{\theta} - \hat{Bias}$. This formula is not as easy to apply as it looks.
- Potential biases are described in the **Discussion** section of a publication. Beware!

Nomenclature

θ	"theta", the parameter from the target population
$\hat{\theta}$	"theta-hat", the parameter estimate from the sample actually analyzed
θ^{0}	The parameter from the study population
RR	Risk ratio of the target population
RR^{0}	Risk ratio of the study population

References

Berkson J. Limitations of the application of fourfold table analysis to hospital data. Biometrics Bulletin 1946;2: 47-53.

Corbett SJ, Rubin GL, Curry GK, Kleinbaum DG. The health effects of swimming at Sydney Beaches. The Sydney Beach Users Study Advisory Group. Am J Public Health. 1993;83(12): 1701-6.

Greenberg RS, Daniels SR, Flanders WD, Eley JW, Boring JR. Medical Epidemiology (3rd Ed). Lange Medical Books, New York, 2001.

Hill H, Kleinbaum DG. Bias in Observational Studies. In Encyclopedia of Biostatistics, pp 323-329, Oxford University Press, 1999.

Horwitz RI, Feinstein AR. Alternative analytic methods for case-control studies of estrogens and endometrial cancer. N Engl J Med 1978;299(20):1089-94.

Kleinbaum DG, Kupper LL, Morgenstern H. Epidemiologic Research: Principles and Quantitative Methods. John Wiley and Sons Publishers, New York, 1982.

Perera FP, et al. Biologic markers in risk assessment for environmental carcinogens. Environ Health Persp 1991;90:247.

The Lipid Research Clinics Coronary Primary Prevention Trial results: II. The relationship of coronary heart disease to cholesterol lowering. JAMA 1984;251(2):365-74.

Warner L, Clay-Warner J, Boles J, Williamson J. Assessing condom use practices. Implications for evaluating method and user effectiveness. Sex Transm Dis 1998; 25(6):273-7.

Homework

ACE-1. General Validity Issues

a. Describe the difference between systematic error and random error in terms of shooting at a target.
b. Why should consideration of validity take precedence over consideration of precision?
c. Describe an example in which the study population is not a subset of the source population. (Hint: STD clinic sample.)
d. Describe an example in which the study population is a subset of the source population. (Hint: Loss-to follow-up in a cohort study.)
e. If the source population differs from the study population, is it still possible that the study sample can provide an estimate of effect close to the target parameter in the source population? Explain.
f. Why can't you assess external validity based on the design/results of a single study?
g. In statistics terminology, the expected value of a sample statistic is the long-run average of values of the statistic obtained from repeated sampling. Explain why the definition of "epidemiologic bias" in terms of the difference between the study and source population parameter values is not equivalent to saying that the "expected value of the study population parameter equals the source population parameter"?
h. Suppose you are an investigator interested in demonstrating that a certain exposure variable is associated with a health outcome of interest. You conduct case-control study and find that your estimated odds ratio is 3.5. If you can justifiably reason that whatever possible bias that exists is towards the null, why should you feel good about the results of your study? Explain
i. Suppose you are a government scientist reviewing a study your agency has funded to evaluate whether a certain pharmaceutical drug already on the public market causes harmful health consequences. The study finds a statistically significant estimated effect of 3.5 that indicates that the drug is harmful. However, based on your review, you can convincingly argue that any bias that may exist is away from the null. Should you recommend that the drug be taken off the market? Explain.

ACE-2. Bias: Genes and Bladder Cancer

A group of investigators was interested in studying the potential relationship between a newly discovered gene and development of bladder cancer. Suppose it was discovered that people who were positive for the gene and developed bladder cancer tended to die quickly and thus were less likely to be available for inclusion in a case-control study of this relationship. What effect would this tend to have on the observed estimate of the odds ratio? (One of the following choices is correct.)

a. Bias towards the null.
b. Bias away from the null.
c. Bias, but direction cannot be determined.
d. No bias, but a potential problem with external validity.

ACE-3. Type of Bias: Genes and Bladder Cancer II

What kind of bias is being considered in question 2 above?

ACE-4. Direction of Bias: Multivitamins vs. Birth Defects

A case-control study was conducted to evaluate the relationship between a woman's self-reported use of multivitamin supplements (exposure) and subsequent delivery of a child with birth defects (outcome). The following table summarizes the findings from this study:

	Multivitamin Use		
	Periconceptional use	No vitamin use	Total
Birth Defect	164	166	330
No Birth Defect	228	122	350

When the study was published, its authors were criticized for having chosen a control group with an inappropriately high prevalence of multivitamin use. What impact would the use of such a control group have on the observed odds ratio? (Assume that the exposure is truly protective.) Choose the one best answer.

a. It would create bias towards the null.
b. It would create bias away from the null.
c. It would create bias, but the direction cannot be predicted.
d. It would increase the probability of a type II error.

Answers to Study Questions and Quizzes

Q7.1

1. Women residents from the city of Atlanta between 45 and 69 years of age.
2. A measure of effect, either a prevalence ratio or a prevalence odds ratio (i.e., the blue dot or bulls eye), for the association between calcium intake and prevalence of arthritis of the hip in the target population.
3. Random error concerns whether or not the estimated odds ratio in the hospital sample (i.e., the shot at the target) differs from the odds ratio in the population of hospital patients (i.e., the red dot or the center of the actual shots) from which the sample is selected.
4. Systematic error concerns whether or not the odds ratio in the population of hospital patients is being sample (i.e., the red dot) is different from the odds ratio in the target population (i.e., the blue dot).

Q7.2

1. All young adults in New York City.
2. All young adults who attend fitness centers in New York City (NYC) and would eventually remain in the study for analysis.
3. Yes, the sample is randomly selected from the study population and is therefore representative.
4. Probably not. The group of young adults in NYC is different from the group of all young adults in NYC. Since fitness is so strongly related to health, the use of those attending fitness centers for all young adults is probably not the best choice for this study.
5. All young adults in New York City.
6. All young adults in NYC that would eventually remain in the study for analysis.
7. Yes, the sample is randomly selected from the study population (by definition of the study population) and is therefore representative of it. Nevertheless, neither the study population nor the sample may be representative of all young adults in NYC if not everyone selected into the sample participates in the study.
8. Yes, assuming that everyone selected participates (i.e., provides the required data) in the study, the study population is the same as the source population. However, if many of those sampled (e.g., a particular subgroup) do not provide the necessary study data, the final sample and its corresponding study population might be unrepresentative of all young adults in NYC.
9. False. Persons lost-to-follow-up are not found in the study sample, so they can't be included in the study population that is represented by the sample.
10. True. Persons lost-to-follow-up are not found in the sample, but they are still included in the source population of interest.
11. True. In the study population, controls may be specified as persons without the disease, regardless of whether they are at risk for being a case. However, the source population may only contain persons at risk for being a case.
12. False. The study population in a cross-sectional study is restricted to survivors only.
13. The population of young adults in New York State is an external population, because the study was restricted to young adults in New York City. Extrapolating the study results to New York State goes beyond considering the methodological aspects of the actual study.

Q7.3

1. The study will be internally valid provided the study population corresponding to the sample actually analyzed is not substantially distorted from the source population of young adults from fitness

centers in the city. For example, if the sample eventually analyzed is a much healthier population than the source population, internal validity may be questioned.

2. External validity. The study was restricted to persons in New York City. Extrapolating the study results to New York State goes beyond considering the methodological aspects of the New York City study
3. Do the results of the study also apply to women or to men of different ages?
4. Was the study sample representative of the source population? Were the comparison groups selected properly? Did subjects consistently stick to their diet and medication regimen? Were relevant variables taken into account?

Q7.4

1. T
2. F – Systematic error occurs when there is a difference between the true effect measure and that which is actually estimated, i.e., a difference between the source and study populations.
3. F – A valid study means there is no systematic error, but there may still be random error
4. F – Not always. Ideally the study population would be equivalent to the source population. However, it may be that the study population and source population simply overlap.
5. T – The sample is always selected from the study population.
6. F – If the study population is not a subset of or equivalent to the source population, then the sample may not be a subset of (or completely contained in) the source population.
7. F – They may be different due to random error.
8. It depends – The difference may be a result of random error. If the risk ratio in the study population is meaningfully different from 3.0, then the study is not internally valid.
9. T – Any meaningful difference between the study and source population means that the study is not internally valid.
10. F – The difference between 3.0 and 3.1 would not be considered a meaningful difference.
11. It depends – We do not know about the risk ratio in the external population.

Q7.5

1. The target population parameter is 3.5. We don't know the value of the study population parameter, but it is likely to be closer to 1.5 than to 3.5 because the sample is assumed to be representative of the study population.
2. There appears to be bias in the study because the sample estimate of 1.5 is meaningfully different from the population estimate and it is reasonable to think that the final sample no longer represents the source population.
3. The bias can't be determined exactly because 1.5 is a sample estimate; however, the bias in the risk ratio is approximately 1.5 – 3.2 = -2.
4. Yes, provided the reason for the difference is due to systematic error.
5. Yes, provided the reason for the difference is due to systematic error.

Q7.6

1. The correct RR must be even larger than 6.
2. The correct RR must indicate an even weaker, or no effect.

Q7.7

1. F – If the bias is away from the null, the RR in the source population must lie between 1 and 2.7.
2. F – If the bias is towards the null, then the RR in the source population is greater than 1.1. Since we cannot determine how much greater, we cannot conclude that these is essentially no association.
3. T – If the bias is away from the null, then the RR in the source population is between 1 and 1.1. We can thus conclude that there is essentially no association.
4. F – If the bias is away from the null, then the RR in the source population must lies between 0.4 and 1.0.
5. F – If the bias is towards the null, then the RR in the source population must be less than 0.4 and hence there is an association.
6. T – If the bias is away from the null, then the RR in the source population is between 0.98 and 1.0, which means there is essentially no association.
7. Towards the null
8. Towards the null
9. Away from the null
10. Switchover
11. Away from the null
12. Switchover

Q7.8

1. B
2. C
3. B
4. A

LESSON 8

SELECTION BIAS

8-1 Selection Bias

Selection bias *concerns systematic error that may arise from the manner in which subjects are selected into one's study. In his lesson, we describe examples of selection bias, provide a quantitative framework for assessing selection bias, show how selection bias can occur in different types of epidemiologic study designs, and discuss how to adjust for or otherwise deal with selection bias.*

Selection Bias in Different Study Designs

Selection bias is systematic error that results from the way subjects are selected into the study or because there are selective losses of subjects prior to data analysis. Selection bias can occur in any kind of epidemiologic study. In case-control studies, the primary source of selection bias is the manner in which cases, controls, or both are selected and the extent to which exposure history influences such selection. For example, selection bias was of concern in case-control studies that found an association between use of the supplement L-tryptophan and EMS (eosinophilia myalgia syndrome), an illness characterized primarily by incapacitating muscle pains, malaise, and elevated eosinophil counts. The odds ratios obtained from these studies might have overestimated the true effect.

Study Questions (Q8.1)

1. Assuming that the odds ratio relating L-tryptophan to EMS is overestimated, which of the following choices is correct?
    1. The bias is towards the null.
    2. L-tryptophan has a weaker association with EMS than actually observed.
    3. The correct odds ratio is larger than the observed odds ratio.

Consequently, the bias would be away from the null.

A primary criticism of these studies was that initial publicity about a suspected association may have resulted in preferential diagnosis of EMS among known users of L-tryptophan when compared with nonusers.

Study Questions (Q8.1) continued

2. Assuming preferential diagnosis of EMS from publicity about L-tryptophan, which of the following is correct? The proportion exposed among diagnosed cases selected for study is likely to be ??? the proportion exposed among all cases in the source population.
    1. larger than
    2. smaller than
    3. equal to

In cohort studies and clinical trials, the primary sources of selection bias are **loss-to-follow-up**, **withdrawal from the study**, or **non-response**. For example, consider a clinical trial that compares the effects of a new treatment regimen with a standard regimen for a certain cancer. Suppose patients assigned to the new treatment are more likely than those on the standard to develop side effects and consequently withdraw from the study.

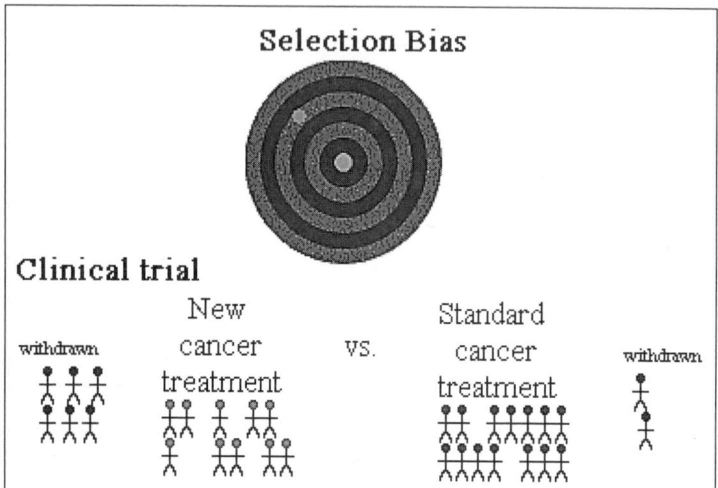

Study Questions (Q8.2)

Clinical trial: new cancer regimen versus standard regimen. Suppose patients on new regimen are more likely to withdraw from study than those on standard.

1. Why might the withdrawal information above suggest the possibility of selection bias in this study?
2. Why won't and intention-to-treat analysis solve this problem?
3. What is the source population in this study?
4. What is the study population in this study?

In cross-sectional studies, the primary source of selection bias is what is called **selective survival.** Only survivors can be included in cross-sectional studies. If exposed cases are more likely to survive longer than unexposed cases, or vice versa, the conclusions obtained from a cross-sectional study might be different than from an appropriate cohort study.

Study Questions (Q8.2) continued

Suppose we wish to assess whether there is selective survival in a cross-sectional study.

5. What is the source population?
6. What is the study population?

Summary

- Selection bias can occur from systematic error that results from the way subjects are selected into the study and remain for analysis.
- The primary reason for such bias usually differs with the type of study used.
- In case-control studies, the primary source of selection bias is the manner in which cases, controls, or both are selected.
- In cohort studies and clinical trials, the primary source of selection bias is loss to follow-up, withdrawal from the study, or non-response.
- In cross-sectional studies, the primary source of selection bias is what is called selective survival.

Example of Selection Bias in Case-Control Studies

In case-control studies, because the health outcome has already occurred, the selection of cases, controls, or both might be influenced by prior exposure status. In the 1970's, there was a lively published debate about selection bias in studies of whether use of estrogen as a hormone replacement leads to endometrial cancer. Early studies were case-control studies and they indicated a strong harmful effect of estrogen. The controls typically used were women with gynecological cancers other than endometrial cancer. Critics claimed that because estrogen often causes vaginal bleeding irrespective of cancer, estrogen users with endometrial cancer would be selectively screened for such cancer when compared to nonusers with endometrial cancer.

Study Questions (Q8.3)

1. If the critics reasoning were correct, why would there be a selection bias problem with choosing controls to be women with gynecological cancers other than endometrial cancer?
2. If the critics reasoning were correct, would you expect the estimated odds ratio obtained from the study to be biased towards or away from the null? Explain briefly.

An alternative choice of controls was proposed; women with benign endometrial tumors, since it was postulated that such a control group would be just as likely to be selectively screened as would the cases.

Study Questions (Q8.3) continued

3. Why would estrogen users with benign endometrial tumors be more likely to be selectively screened for their tumors when compared to nonusers?
4. Assuming that estrogen users with both cancerous and benign endometrial tumors are likely to be selectively screened for their tumors when compared to non-users, what problem may still exist if the latter group is chosen as controls?

Continued research and debate, however, have indicated that selective screening of cases is not likely to contribute much bias. In fact, the proposed alternative choice of controls might actually lead to bias.

Study Questions (Q8.3) continued

Researchers concluded that because nearly all women with invasive endometrial cancer will ultimately have the disease diagnosed, estrogen users will be slightly over-represented, if at all, among a series of women with endometrial cancer. Assume for the questions below that selective screening of cases does not influence the detection of endometrial cancer cases.

5. If the control group consisted of women with benign tumors in the endometrium, why would you expect to have selection bias?
6. Would the direction of the bias be towards or away from the null? Briefly explain.
7. If the control group consisted of women with gynecologic cancers other than in the endometrium, why would you not expect selection bias in the estimation of the odds ratio?

In current medical practice, the prevailing viewpoint is that taking estrogen alone is potentially harmful for endometrial cancer. Consequently, women who are recommended for hormone replacement therapy are typically given a combination of progesterone and estrogen rather than estrogen alone.

Summary

- Selection bias concerns a distortion of study results that occurs because of the way subjects are selected into the study.
- In case-control studies, the primary concern is that selection of cases, controls, or both might be influenced by prior exposure status.
- In the 1970's, there was a lively published debate about possible selection bias among researchers studying whether use of estrogen, the exposure, as a hormone replacement leads to endometrial cancer.
- The argument supporting selection bias has not held up over time; current medical practice for hormone replacement therapy typically involves a combination of progesterone and estrogen rather than estrogen alone.

Example of Selection Bias in Cohort Studies

Selection bias can occur in cohort studies as well as in case-control studies. In prospective cohort studies, the health outcome, which has not yet occurred when exposure status is determined, cannot influence how subjects are selected into the study. However, if the health outcome is not determined for everyone initially selected for study, the study results may be biased. The primary sources of such selection bias are **loss-to-follow-up**, **withdrawal** or **non-response**. The collection of subjects that remain to be analyzed may no longer represent the source population from which the original sample was selected.

Consider this two-way table that describes the five-year follow-up for disease "D" in a certain source population.

Follow-up study

(Source Pop.)	E	not E	Total
D	150	50	200
not D	9850	9950	19800
Total	10000	10000	20000

RR = (150/10000) / (50/10000) = 3.0

Suppose that a cohort study is carried out using a 10% random sample from this population. What would be the expected cell frequencies for this cohort assuming **no** selection bias?

10% random sample: Fill in the blanks!

(Initial Sample)	E	Not E	
D	15	5	
Not D	985	995	
Total	1000	1000	

Study Questions (Q8.4)

1. Assuming this is the sample that is analyzed, is there selection bias?

Assume that the initial cohort was obtained from the 10% sampling. However, now suppose that 20% of exposed persons are lost to follow-up but 10% of unexposed persons are lost. Also, assume that exposed persons have the same risk for disease in the final cohort as in the initial cohort and that the same is true for unexposed persons.

Now assume: 20% lost to follow-up 10% lost to follow-up

10% random sample plus loss-to-follow-up

	E	Not E
D	12	4.5
Not D	788	895.5
Total	800	900

Study Questions (Q8.4) continued

2. Does the sample just described represent the source population or the study population?
3. What is the source population for which this sample is derived?
4. For the above assumptions, is there selection bias?

Suppose that a different pattern of loss-to follow-up results in the two-way table shown here.

10% random sample, different loss-to-follow-up:

	E	Not E
D	14	4
Not D	786	896
Total	800	900

$\hat{RR} = 3.9$

Study Questions (Q8.4) continued

5. Do these results indicate selection bias?
6. Do the exposed persons in the study population have the same risk for disease as in the source population?
7. Do the unexposed persons in the study population have the same risk for disease as in the source population?
8. Do the previous examples demonstrate that there will be selection bias in cohort studies whenever the percent lost to follow-up in the exposed group differs from the percent lost-to-follow-up in the unexposed group?

Summary

- The primary sources of selection bias in cohort studies are loss-to-follow-up, withdrawal, and non-response.
- In cohort studies, the collection of subjects that remain in the final sample that is analyzed may no longer represent the source population from which the original cohort was selected.
- Selection bias will occur if loss to follow-up results in risk for disease in the exposed and/or unexposed groups that are different in the final sample than in the original cohort.

Some Fine Points about Selection Bias in Cohort Studies

Reference: Hill, HA and Kleinbaum, DG, "Bias in Observational Studies", in the Encyclopedia of Biostatistics, P.A. Armitage and T Colton, eds., June 1998.

Selection bias in cohort studies may occur even with a fairly high overall response rate or with very little loss to follow-up. Consider a cohort study in which 95% of all subjects originally assembled into the cohort remain for analysis at the end of the study. That is, only 5% of subjects are lost to follow-up. If losses to follow-up are primarily found in exposed subjects who develop the disease, then despite the small amount of follow-up loss, the correct (i.e., target) risk ratio could be underestimated substantially. This is because, in the sample that is analyzed, the estimated risk for developing the disease in exposed subjects will be less than what it is in the source population, whereas the corresponding risk for unexposed subjects will accurately reflect the source population.

Continued on next page

Some Fine Points about Selection Bias in Cohort Studies (continued)

There may be no selection bias despite small response rates or high loss to follow-up. Suppose only 10% of all initially selected subjects agree to participate in a study, but this 10% represents a true random sample of the source population. Then the resulting risk ratio estimate will be unbiased. The key issue here is whether risks for exposed and unexposed in the sample that is analyzed are disproportionately modified because of non-response or follow-up loss from the corresponding risks in the source population from which the initial sample was selected.

We are essentially comparing two 2x2 tables here, one representing the source population and the other representing the sample:

Source Population

	E	Not E
D	A	B
Not D	C	D
Total	N_1	N_0

Sample for Analysis

	E	Not E
D	a	b
Not D	c	d
Total	n_1	n_0

Selection bias will occur only if, when considering these tables, the risk ratio in the source population, i.e.,

$$\frac{A/N_1}{B/N_0}$$

is meaningfully different from the risk ratio in the sample, i.e.,

$$\frac{a/n_1}{b/n_0}$$

In the first example above (95% loss to follow-up), the argument for selection bias is essentially that the numerator $\mathbf{a/n_1}$ of the risk ratio in the analyzed sample would be less than the corresponding numerator $\mathbf{A/N_1}$ in the source population, whereas the corresponding denominators in these two risk ratios would be equal.

In the second example (10% non-response), the argument for no selection bias is essentially that despite the high non-response, corresponding numerators and denominators in the source population and sample are equal.

Other Examples

Here are a few more examples of studies that are likely to raise questions about selection bias.

Study Questions (Q8.5)

Consider a retrospective cohort study that compares workers in a certain chemical industry to a population-based comparison group for the development of coronary heart disease (CHD).

1. In such a study, selection bias may occur because of the so-called "health worker effect". How might such a bias come about?

Selection bias may result from using volunteers for a study.

2. Explain the above statement in terms of study and source populations.
3. What is an alternative way to view the validity problem that arises when a study is restricted to volunteers?

In assessing long term neurologic disorders among children with febrile seizures, clinic-based studies tend to report a much higher frequency of such disorders than found in population-based studies.

4. Does the above statement indicate that clinic-based studies can result in selection bias? Explain briefly.

Summary

- ❖ Selection bias may occur because of the so-called "healthy worker effect". Workers tend to be healthier than those in the general population and may therefore have a more favorable outcome regardless of exposure status.
- ❖ Selection bias may result from using volunteers, who may have different characteristics from persons who do not volunteer.
- ❖ Clinic-based studies may lead to selection bias because patients from clinics tend to have more severe illness than persons in a population-based sample.

8-2 Selection Bias (continued)

Selection Ratios and Selection Probabilities

A selection probability gives the likelihood that a person from one of the four cells in the source population will be a member of the study population.

To quantify how selection bias can occur, we need to consider underlying parameters called **selection ratios**. There are four such selection ratios, **alpha** (**α**), **beta** (**β**), **gamma** (γ), and **delta** (**δ**), which correspond to the four cells of the two-way table that relates exposure to disease.

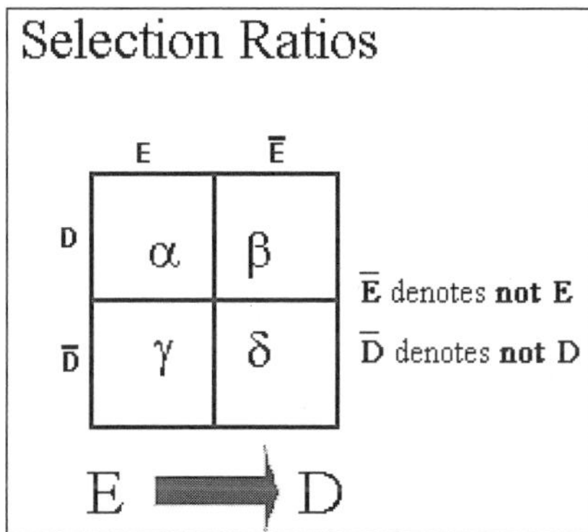

A selection ratio is defined by dividing the cell frequency from one of the four cells in the study population by the corresponding cell frequency in the source population. Using this framework, $\alpha = A^{o}/A$, $\beta = B^{o}/B$, and so on. In many studies, the study population will be a subset of the source population. If so, **α**, **β**, **γ**, and **δ** are typically referred to as selection probabilities.

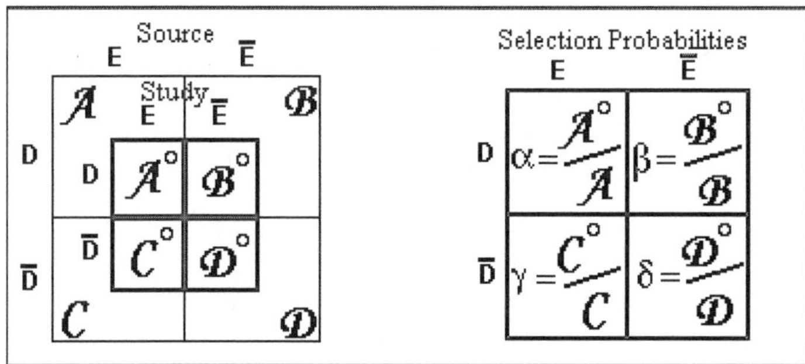

Each selection probability gives the probability that an individual from one of the four cells in the source population will be eligible to be selected into the study population. Thus, **α** denotes the probability that an exposed case in the source population will be selected into the study and similar probability estimates apply to **β**, **γ** and **δ**.

Study Questions (Q8.6)

The tables shown below give the source and study populations for a cohort study in which 20% of the exposed group and 10% of the unexposed group are lost to follow-up. These tables assume that exposed persons have the same risk for disease in the final cohort as in the original cohort and that the same is true for unexposed persons.

1. What are **α**, **β**, **γ**, and **δ** for these data?
2. Is there selection bias in either the odds ratio or the risk ratio?
3. What is the value of the cross-product of selection probabilities, i.e., (αδ)/(βγ)?

Source pop.	E	Not E	Study pop.	E	Not E
D	150	50	D	12	4.5
not D	9850	9950	not D	788	895.5
Total	10000	10000	Total	800	900
	RR = 3.0			$RR^0 = 3.0$	

The tables shown below give source and study populations for another cohort study in which 20% of the exposed group and 10% of the unexposed group are lost to follow-up. In these tables, exposed persons have a higher risk for disease in the study population than in the source population and the opposite is true for unexposed persons.

4. What are **α**, **β**, **γ**, and **δ** for these data?
5. Is there selection bias in either the odds ratio or the risk ratio?
6. What is the value of the cross-product of selection probabilities, i.e., (αδ)/(βγ)?

Source pop.	E	Not E	Study pop.	E	Not E
D	150	50	D	14	4
not D	9850	9950	not D	786	896
Total	10000	10000	Total	800	900
	RR = 3.0			$RR^0 = 3.9$	

Summary

- To quantify how selection bias can occur, we need to consider underlying parameters called selection ratios.
- There are four selection ratios to consider, one for each cell of the 2x2 table relating exposure status to disease status.
- A selection ratio gives the number of subjects from one of the four cells in the study population divided by the corresponding number of subjects in the source population.
- If the study population is a subset of the source population, a selection ratio is typically called a selection probability.
- A selection probability gives the likelihood that a person from one of the four cells in the source population will be a member of the study population.

Quantitative Assessment of Selection Bias

Selection bias can be assessed using a mathematical expression involving the four selection ratios or selection probabilities that relate the source population to the study population. This expression considers bias in estimating an odds ratio.

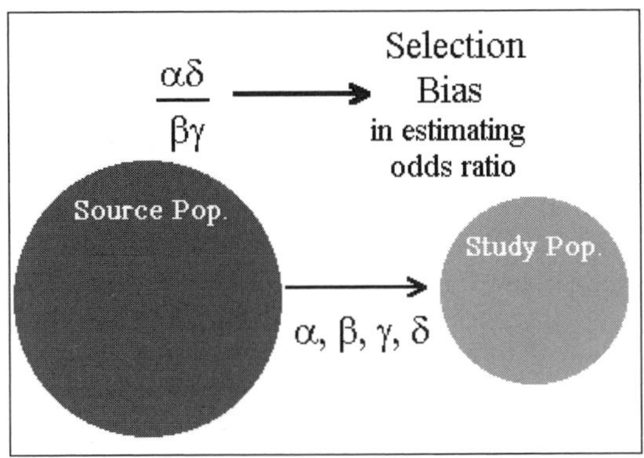

Recall that an estimate of an odds ratio ($\hat{OR}$) is biased if the odds ratio in the study population (OR^o) is meaningfully different from the odds ratio in the source population (OR). The bias is the difference between these two parameters, as shown here:

$$\text{Bias}(\hat{OR}, OR) = OR^0 - OR$$

Now using the framework previously described for the fourfold tables in the source and study populations, we can express the bias by substituting into the bias formula the cell frequencies in each table, as shown here:

$$\text{Bias}(\hat{OR}, OR) = OR^0 - OR = \frac{A^0 D^0}{B^0 C^0} - \frac{AD}{BC}$$

Source

	E	$\bar{E}$
D	A (Study: A^o)	B (Study: B^o)
$\bar{D}$	C (Study: C^o)	D (Study: D^o)

Since the selection ratios α, β, γ, and δ are defined as the relative frequencies for the study population divided by the source population, we can equivalently rewrite $A^o = \alpha \times A$, $B^o = \beta \times B$, $C^o = \gamma \times C$, $D^o = \delta \times D$.

$$\text{Bias}(\hat{OR}, OR) = \frac{A^0 D^0}{B^0 C^0} - \frac{AD}{BC}$$

Source

	E	$\bar{E}$
D	A (Study: A^o)	B (Study: B^o)
$\bar{D}$	C (Study: C^o)	D (Study: D^o)

Selection Probabilities

	E	$\bar{E}$
D	$\alpha = \frac{A^o}{A}$	$\beta = \frac{B^o}{B}$
$\bar{D}$	$\gamma = \frac{C^o}{C}$	$\delta = \frac{D^o}{D}$

$A^o = \alpha A$

$B^o = \beta B$

$C^o = \gamma C$

$D^o = \delta D$

We then substitute these expressions for the cell frequencies in the expression for bias to obtain the result shown here.

$$\text{Bias}(\hat{OR}, OR) = \frac{\alpha A \delta D}{\beta B \gamma C} - \frac{AD}{BC}$$

If we set this expression equal to 0, it follows that the **bias = 0** if and only if the cross product ratio $(\alpha\delta)/(\beta\gamma) = 1$. Thus, we obtain the following rule for assessing selection bias in the odds ratio.

$$\frac{\alpha\delta}{\beta\gamma} = 1 \Longleftrightarrow \text{Bias}_{OR} = 0$$

The bias is either > 0, = 0, or < 0 depending on whether the cross product ratio of selection probabilities is > 1, = 1, or < 1.

$$\text{Bias}_{OR} \begin{cases} > 0 \text{ if } \frac{\alpha\delta}{\beta\gamma} > 1 \\ = 0 \text{ if } \frac{\alpha\delta}{\beta\gamma} = 1 \\ < 0 \text{ if } \frac{\alpha\delta}{\beta\gamma} < 1 \end{cases}$$

The "= 0" part of this rule states conditions for the **absence of bias**, whereas the > 0 and < 0 parts state conditions for the **direction of the bias**. Nevertheless, in interpreting this rule, we must remember that the equal sign means "meaningfully equal" rather than "exactly equal", and similar interpretations should be applied to the greater than and less than signs.

Study Questions (Q8.7)

The tables shown below give source and study populations for a cohort study in which 20% of the exposed group and 10% of the unexposed group are lost to follow-up. The selection ratios for study population 1 are: $\alpha = 12/150 = .08$; $\beta = 4.5/50 = .09$; $\gamma = 788/9850 = .08$; and $\delta = 895.5/9950 = .09$.

1. Is there selection bias in the odds ratio for study population 1? Explain briefly in terms of the cross product ratio of selection ratios.

Target Pop.	E	not E	Study Pop.	E	not E
D	150	50	D	12	4.5
not D	9850	9950	not D	788	895.5
Total	10000	10000	Total	800	900

The tables below consider the same source population as in the previous question but a different study population (#2). The selection ratios for study population 2 are: $\alpha = 14/150 = .093$; $\beta = 4/50 = .08$; $\gamma = 786/9850 = .08$; and $\delta = 896/9950 = .09$.

2. Is there selection bias in the odds ratio for population 2? Explain briefly in terms of the cross product ratio of selection ratios.
3. If your answer to the previous question was yes, what is the direction of the bias?

Target Pop.	E	not E	Study Pop.	E	not E
D	150	50	D	14	4
not D	9850	9950	not D	786	896
Total	10000	10000	Total	800	900

Summary

- Selection bias can be assessed using a mathematical expression involving the four selection probabilities that relate the target to the study populations.
- Bias in estimating the odds ratio = 0 if and only if the cross product ratio $(\alpha\delta)/(\beta\gamma) = 1$.
- The bias is either > 0, = 0, or < 0 depending on whether the cross product ratio of selection probabilities is > 1, = 1, or < 1.

Selection Bias for Risk Ratio

We now discuss how to assess selection bias for the risk ratio. We have shown that the risk ratio from a follow-up study can be approximated by the odds ratio if the disease is rare. Moreover, the odds ratio is often not that much different from the risk ratio even if the disease is not rare. Consequently, this rare disease approximation suggests that assessing selection bias in the risk ratio can more often than not involve the same rule about the cross-product of selection ratios $[(\alpha\delta)/(\beta\gamma)]$ that applies to assessing selection bias in the odds ratio.

Study Questions (Q8.8)

The tables shown below give source and study populations for a cohort study. Here no subjects who developed the disease are lost to follow-up. However, 20% of disease-free subjects are lost to follow-up, irrespective of exposure status.

1. Calculate the 4 selection ratios for this study.
2. Are the selection ratios also selection probabilities?
3. Is the disease in this study a rare disease in both the source population and the study population?
4. Based on selection ratios, is there selection bias in the odds ratio?
5. Is there selection bias in the risk ratio? Explain briefly.

Source pop.	E	Not E
D	100	50
not D	9900	9950
Total	10000	10000

RR = 2.0, OR = 2.01

Study pop.	E	Not E
D	100	50
not D	7920	7960
Total	8020	8010

$RR^0 = 2.0$, $OR^0 = 2.01$

In the example, the rare disease approximation applied to both the source population and the study population. If, however, the rare-disease approximation does not hold in both these populations, then the presence or absence of bias in the odds ratio might not correspond to bias in the risk ratio calculated for the same data.

Study Questions (Q8.8) continued

The tables below give the same source population but a different study population from the previous example. Once again, no subjects who developed the disease are lost to follow-up. However, there is extremely high loss to follow-up (99%) for subjects who do **not** develop the disease, irrespective of exposure status.

6. Calculate the four selection ratios for this study.
7. Based on the selection ratios, is there selection bias in the odds ratio?
8. Is there selection bias in the risk ratio? Explain briefly.
9. Is the disease in this study a rare disease in both the source population and the study population?

Source pop.	E	Not E	Study pop.	E	Not E
D	100	50	D	100	50
not D	9900	9950	not D	99	99.5
Total	10000	10000	Total	199	149.5
RR = 2.0, OR = 2.01			RR^o= 1.503, OR^o= 2.01		

The previous example illustrates that there is no guarantee that the presence or absence of bias in the odds ratio will always correspond to the same degree of bias in the risk ratio. Consequently, one must be careful when attempting to use the cross-product rule for selection ratios to assess whether there is bias in the risk ratio.

Study Questions (Q8.8) continued

The tables below give different source and study populations than previously considered. The study is a cohort study in which no subjects who developed the disease are lost to follow-up. However, 50% of exposed non-cases are lost to follow-up whereas none of the unexposed non-cases are lost to follow-up.

10. Calculate the four selection ratios for this study.
11. Based on the selection ratios, is there selection bias in the odds ratio?
12. Is there selection bias in the risk ratio? Explain briefly.
13. Is the disease in this study a rare disease in both the source population or the study population?
14. Is the extent of bias the same for both risk and odds ratios?

Source pop.	E	Not E	Study pop.	E	Not E
D	40	10	D	40	10
not D	60	90	not D	30	90
Total	100	100	Total	70	100
RR = 4.0, OR = 6.0			RR^o= 5.71, OR^o= 12		

Summary

- ❖ The rare disease approximation suggests that assessing selection bias in the risk ratio can involve the same rule about the cross-product of selection ratios that applies to assessing selection bias in the odds ratio.
- ❖ If, however, the rare disease approximation does not hold in both source and study populations, then the presence or absence of bias in the odds ratio might not correspond to bias in the risk ratio calculated for the same data.
- ❖ There is no guarantee that the presence or absence of bias in the odds ratio will always correspond to the same degree of bias in the risk ratio.

Quiz (Q8.9)

Fill in the Blanks

Table of selection probabilities

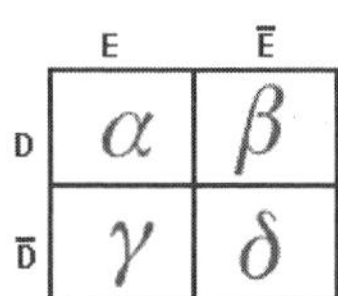

Critics of case-control studies that found an association between estrogen use (E) and endometrial cancer (D) claimed that estrogen users with the disease are more likely to be selected as cases than nonusers of estrogen with the disease.

1. In the table of selection probabilities shown above, how do the critics claim translate into a statement about selection probabilities? **???**

These critics also claimed that if gynecological patients with benign tumors were chosen for controls, then estrogen users without the disease would be more likely to be selected as controls than nonusers of estrogen without the disease.

2. In the table above, how does this latter claim translate into a statement about selection probabilities? **???**

The critics in the previous question were countered with the argument that nearly all women with invasive endometrial cancer will ultimately have the disease diagnosed, so that estrogen users were not likely to be over-selected in previous studies.

3. In the table of selection probabilities shown above, how does this counterclaim translate into a statement about selection probabilities? **???**

It was also argued here that if women with other gynecological cancers are chosen for controls, this latter group would not be subject to selective screening and would nearly always be detected without screening.

4. In the table above, how does this latter claim translate into a statement about selection probabilities? **???**

Choices
alpha = beta **alpha > beta** **alpha > delta** **alpha > gamma** **gamma = delta** **gamma > delta**

8-3 Selection Bias (continued)

Selection Bias Due to Inappropriate Choice of Controls

The use of condoms is widely recommended for prevention of sexually transmitted diseases. Nevertheless, a number of studies have found surprisingly little evidence of condom effectiveness, even for STDs where high condom efficacy would be expected. This apparent lack of effect of condom use found in STD studies is typically attributed to misclassification of exposure from inaccurate reporting of condom usage. However, another explanation might be selection bias arising from the choice of the comparison group used.

The study population in condom effectiveness studies is typically all individuals examined at STD clinics. Correspondingly, in case-control studies of condom effectiveness, the control group is chosen to be a random sample of persons without STDs from this clinic population. A criticism of such studies is that the correct source population consists of individuals who have infected sex partners, since one cannot acquire an STD without having an infected partner. This source population is more restrictive than a population of all individuals attending STD clinics, but ensures that an individual eligible for study is actually at risk for an STD.

If controls are selected from ***all*** persons without STDs, and not from only those with infected partners, then unexposed controls could include persons who do not have infected partners and perhaps exhibit no high-risk behavior leading to STDs. In contrast, exposed controls may include a much higher proportion of persons who have infected partners.

Study Questions (Q8.10)

Suppose controls are selected from all persons examined at STD clinics who do not have an STD. Are the following questions **True** or **False**?

1. Unexposed controls are likely to be over-represented in the study when compared to unexposed non-cases in the source (i.e., only those at risk) population.
2. Exposed controls are likely to be over-represented in the study when compared to exposed non-cases in the source population.

The tables below consider a hypothetical case-control study in which controls are sampled from all clinic patients with***out*** STDs, even though the source population is restricted to patients at risk for being a case.

SOURCE POPULATION

	Condom Use Yes	No	Total
Case	100	600	700
Noncase	200	600	800
Total	300	1200	1500

OR = 0.5

STUDY POPULATION

	Condom Use Yes	No	Total
Case	100	600	700
Control	200	1200	1500
Total	300	1800	2100

$OR^0 = 1.0$

3. Is there selection bias in the odds ratio?
4. Are the unexposed non-cases over-represented in the study population when compared to unexposed non-cases in the source population?
5. What are α, β, γ, and δ for the hypothetical study?
6. Using the selection ratios just calculated, how does this example illustrate selection bias when controls are allowed to include persons who have sexual partners without STDs?
7. Would selection bias described in the previous question be towards or away from the null?

One way to minimize selection bias in these studies would be to re-define the control group to uninfected individuals known to have sex partners infected with STD's. For example, one could restrict controls to be individuals who have been notified by their local health department as being a recent sexual contact of persons diagnosed with STD. Such restriction of the control group would help to ensure that all controls, regardless of condom use, were recently exposed to infected partners and thus equally at risk for acquiring STD.

Study Questions (Q8.10) continued

8. True or False. If controls were restricted to persons without STD's who had sexual partners with STD's, unexposed controls are likely to be over-represented in the study when compared to unexposed non-cases in the source population.
9. True or False. Exposed controls are likely to be over-represented in the study when compared to exposed non-cases in the source population.
10. What do the answers to the above questions imply about relationships between the selection ratios γ and δ?
11. If it is assumed that the selection ratios for exposed and unexposed cases are equal, why would there likely not be selection bias in studies that restrict controls to only persons who have sexual partners with STD's?

Summary

- In case-control studies of STDs, the typical control group has been a random sample of persons without STDs from a clinic population.
- A criticism of such studies is that the correct source population consists of individuals from these clinics who have sexual partners with STDs.

Summary continued on next page

- ❖ In such studies, unexposed controls are likely to be over-represented when compared to unexposed non-cases in the source (i.e., clinic) population.
- ❖ Selection bias is likely when controls are allowed to include persons who have sexual partners without STDs.
- ❖ One way to minimize selection bias would be to re-define the control group to be uninfected individuals known to have sexual partners with STD's.

Selection Bias in Hospital Case-Control Studies

In 1946, Berkson demonstrated that case-control studies carried out using hospital patients are subject to a type of "selection" bias called **Berkson's bias**. Berkson's bias arises because patients with two disease conditions or high-risk behaviors are more likely to be hospitalized than those with a single condition. Such patients will tend to be over-represented in the study population when compared to the community population.

For example, because cigarette smoking is strongly associated with several cancers, we would expect a hospital study of the relationship between cigarette smoking and almost any cancer to demonstrate a stronger relationship than would exist in a community population. To illustrate Berkson's bias, consider the two by two table describing exposure and disease for a certain community population shown here.

Community Population

	Smoke?		
Disease	Yes	No	Total
Yes	400	200	600
No	100,000	200,000	300,000
Total	100,400	200,200	300,600

$\hat{OR}$ =(400 x 200,000) / (200 x 100,000) = 4.0

Hospital Case-Control Sample

	Smoke?		
	Yes	No	Total
Case	225	75	300
Control	100	200	300
Total	325	275	600

Suppose of the 600 cases, 300, or 50%, are selected for a case-control study from several hospitals. Suppose further that 225, or 75%, of these 300 cases are smokers. Assume that there are 300 controls selected as a 0.1% sample of all non-cases from the community. Assuming no bias or random error in choosing these controls, we would therefore expect 100 exposed and 200 unexposed controls, as shown above.

Study Questions (Q8.11)

1. What are the source and study populations for this study?
2. What are α, β, γ, and δ for this study?
3. Are α, β, γ, and δ selection probabilities?
4. Why is there selection bias in the odds ratio estimate? Explain in terms of selection ratios.
5. What is the direction of the bias?
6. Why is this an illustration of Berkson's bias?

In the previous example, cases came from the hospital but controls came from the community. To compensate for choosing hospitalized cases, the tables below describe a study in which 300 hospitalized non-cases are chosen as controls.

Community Population

Disease	Smoke? Yes	No	Total
Yes	400	200	600
No	100,000	200,000	300,000
Total	100,400	200,200	300,600

Hospital Case-Control Sample

	Smoke? Yes	No	Total
Case	225	75	300
Control	150	150	300
Total	375	225	600

7. Are the controls in the above study over-representative of the non-cases in the community?
8. What are α, β, γ, and δ for this study?
9. In terms of selection probabilities, why is there selection bias in the odds ratio estimate?
10. What is the direction of the bias?
11. Why is this an illustration of Berkson's bias?

Summary

- Berkson demonstrated that case-controls studies carried out using hospitalized patients are subject to a type of selection bias called Berkson's bias.
- Patients with two disease conditions or high-risk behaviors are more likely to be hospitalized than those with a single condition.
- Hospital patients in a study population tend to be over-represented when compared to a community population.
- Berkson's bias can either be towards or away from the null depending on how hospital cases are more or less over-represented than hospital controls when compared to the community.

Selection Bias in Cross-Sectional Studies

In cross-sectional studies, there are two distinct ways to consider selection bias. If the study objective is primarily to survey an assumed stable population at a point or short interval of time, then the source population is that subset of an assumed stable cross-sectional population who were at risk for the disease prior to the study onset. Selection bias may then occur when some aspect of the selection process, such as non-response, distorts the study population enough to bias the estimate.

Here, the study sample is a subset of the source population, so the key parameters of interest are selection probabilities. Alpha (α) denotes the probability that an exposed case in the cross-sectional source population is eligible to be sampled in the study population. It is given by A^o divided by A, the respective cell frequencies in the source and study population. β, γ, and δ are defined similarly.

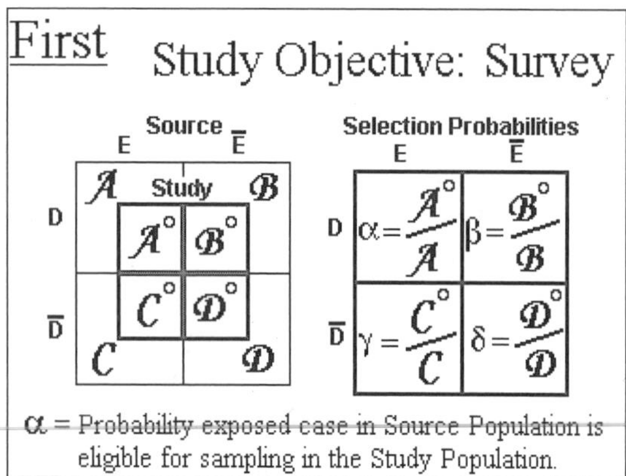

Selection bias in the odds ratio, (the effect measure of interest in a cross-sectional study), can occur when the cross product of the selection probabilities [(αδ)/(βγ)] meaningfully differs from unity.

If the study objective, however, is to determine whether there is an etiologic relationship between exposure and disease, then the source population of interest is no longer the cross-sectional population being surveyed. The appropriate source population is instead given by a cohort identified at an earlier time point that contains the persons at risk for developing the disease by the time of the cross-sectional study. In other words, a cross-sectional study provides prevalence data from which we wish to make inferences about incidence data. In this situation, the primary source of selection bias is what is called **selective survival**. (Note that selective survival is an important issue with diseases with high mortality risk, such as coronary heart disease, but less of an issue with diseases with low mortality risk, such as arthritis.) Subjects in cross-sectional studies are survivors, particularly those with the disease, whether exposed or unexposed.

The selection probabilities α, β, γ, and δ give probabilities of surviving from time T_0 to time T_1. Selection bias can occur here if the probability of surviving long enough to be included in the cross-sectional study is different for the four cells of the source population cohort.

To illustrate selective survival, the table shown below (upper left table) describes the incidence of disease from time T_0 to time T_1 for the follow-up of a fixed cohort of 1000 persons who were disease-free at time T_0. We compare the results from the above incidence data to expected results from a cross-sectional study carried out on survivors. The study population of survivors at time T_1 is shown in the lower right table below.

Incidence for Fixed Cohort from T_0 to T_1

	E	not E	Total
D	50	80	130
not D	150	720	870
Total	200	800	1000

RR = 2.5, ROR = 3

Expected Results at time T1

	E	Not E	Total
D	25	40	65
not D	142.5	540	682.5
Total	162.5	580	747.5

PR = 2.2, POR = 2.4

Study Question (Q8.12)

1. For exposed and unexposed separately, what proportion of persons who develop the disease are expected to survive by time T_1?
2. What proportion of exposed persons who do not develop the disease are expected to survive to time T_1?
3. What proportion of unexposed persons who do not develop the disease are expected to survive by time T_1?
4. What are the 4 selection probabilities for selective survival?
5. What is the cross-product of selection probabilities?
6. Is there selection bias in both the odds ratio and the risk ratio?

Summary

- In cross-sectional studies, there are two distinct ways to consider selection bias, depending on the objective of the study.
- If the study objective is to survey an assumed stable population at a point or short interval of time, then selection bias may occur because of non-response and/or some other selective distortion of the study population.
- If the study objective is to determine whether there is an etiologic relationship between exposure and disease, then the primary source of selection bias is selective survival.
- Selective survival can occur if the probability of surviving long enough to be included in the cross-sectional study is different for the four cells of the source population cohort.

Quiz (Q8.13)

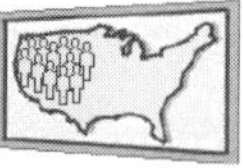

1. If the study objective is primarily to survey an assumed stable population at a point or short interval of time, then the source population is that subset of an assumed stable cross-sectional population who were at risk for the disease **???** to the study onset.
2. Selection bias may then occur when some aspect of the selection process such as **???** distorts the **???** enough to bias the estimate.
3. Selection bias in the odds ratio, an effect measure of interest in a cross-sectional study, can occur when the cross-product of the selection probabilities meaningfully differs from **???**.

Choices
100 **after** **non-response** **one** **prior to** **selective survival** **source population** **study population** **zero**

If the study objective, however, is to determine whether there is an etiologic relationship between exposure and disease, then the source population of interest is no longer the cross-sectional population being surveyed.

4. The appropriate source population is instead given by a cohort identified at a(n) **???** time point that contains the persons at risk for developing the disease **???** the time of the cross-sectional study.
5. In other words, a cross-sectional study provides **???** from which we wish to make inferences about **???**.

Choices
after **by** **earlier** **incidence data** **later** **prevalence data**

Consider again a cross-sectional study to determine whether there is an etiologic relationship between exposure and disease.

6. In this situation, the primary source of selection bias is what is called **???**.
7. Subjects in cross-sectional studies are **???**, particularly those with the disease, whether exposed or unexposed.
8. Those that do not **???** are not included in the study.

Choices
cases **die** **non-response** **selective survival** **survive** **survivors**

8-4 Selection Bias (continued)

What Can Be Done About Selection Bias Quantitatively?

Suppose you are convinced that a study already designed and carried out is flawed by selection bias. Is it possible to quantitatively correct for such bias? For example, can the value of an estimated odds ratio that is biased be adjusted to provide a modified value that is free from selection bias?

The answer here depends on whether reliable estimates of the underlying selection ratio parameters can be determined. Actually, even if these parameters cannot be individually estimated, correcting for selection bias is possible if we can estimate ratios of these selection parameters, such as α over γ, β over δ, or just the cross-product ratio of the selection parameters.

$$\hat{\alpha} \quad \hat{\beta} \quad \hat{\gamma} \quad \hat{\delta}$$

$$\frac{\hat{\alpha}}{\hat{\gamma}} \quad \frac{\hat{\beta}}{\hat{\delta}} \quad \text{or} \quad \frac{\hat{\alpha}\hat{\delta}}{\hat{\beta}\hat{\gamma}}$$

Here are three 2x2 tables that describe the study data, estimated selection parameters, and adjusted study data that we need to consider if we want to adjust for selection bias.

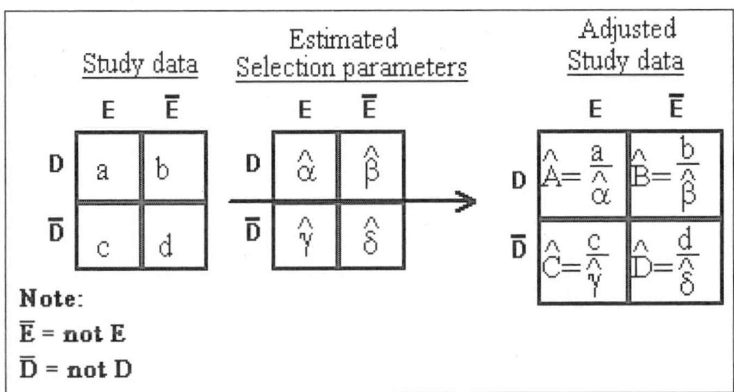

The table of adjusted study data is derived using our previous definitions of the selection parameters as ratios of corresponding cell frequencies from the study and source populations.

Selection Parameters

$$\alpha = \frac{A^{o}}{A} \quad \beta = \frac{B^{o}}{B}$$

$$\gamma = \frac{C^{o}}{C} \quad \delta = \frac{D^{o}}{D}$$

This table is obtained by substituting a for A^{o}, b for B^{o}, and so on and then solving for capital A, B, C and D in terms of lower-case a, b, c, and d divided by their corresponding selection parameter estimates.

Selection Parameters

$$\hat{\alpha} = \frac{a}{\mathcal{A}} \quad \hat{\beta} = \frac{b}{\mathcal{B}}$$

$$\hat{\gamma} = \frac{c}{\mathcal{C}} \quad \hat{\delta} = \frac{d}{\mathcal{D}}$$

An adjusted estimate of the odds ratio that corrects for selection bias is then given by the cross product of the adjusted cell frequencies, which can be written in terms of the study data and selection parameter estimates as shown here.

Adjusted Study data

	E	$\bar{E}$
D	$\hat{A} = \frac{a}{\hat{\alpha}}$	$\hat{B} = \frac{b}{\hat{\beta}}$
$\bar{D}$	$\hat{C} = \frac{c}{\hat{\gamma}}$	$\hat{D} = \frac{d}{\hat{\delta}}$

$$\hat{OR}(adj) = \frac{\hat{A}\hat{D}}{\hat{B}\hat{C}} = \frac{\frac{a}{\hat{\alpha}}\frac{d}{\hat{\delta}}}{\frac{b}{\hat{\beta}}\frac{c}{\hat{\gamma}}}$$

Thus, the odds ratio that adjusts for selection bias is given by the estimated odds ratio from the study data divided by the estimated cross product of selection parameters.

$$\hat{OR}_{adj} = \frac{\hat{OR}}{\left(\frac{\hat{\alpha}\hat{\delta}}{\hat{\beta}\hat{\gamma}}\right)}$$

This adjustment formula can alternatively be written using ratios of the selection parameters as shown here.

$$= \hat{OR} \times \frac{\hat{r}_{\bar{D}}}{\hat{r}_D} = \hat{OR} \times \frac{\hat{r}_{\bar{E}}}{\hat{r}_E}$$

where

$$\hat{r}_D = \frac{\hat{\alpha}}{\hat{\beta}} \qquad \hat{r}_E = \frac{\hat{\alpha}}{\hat{\gamma}}$$

$$\hat{r}_{\bar{D}} = \frac{\hat{\gamma}}{\hat{\delta}} \qquad \hat{r}_{\bar{E}} = \frac{\hat{\beta}}{\hat{\delta}}$$

The alternative formulas allow the correction for bias even if only the ratios of selection parameters can be determined rather than the individual selection parameters themselves.

Study Questions (Q8.14)

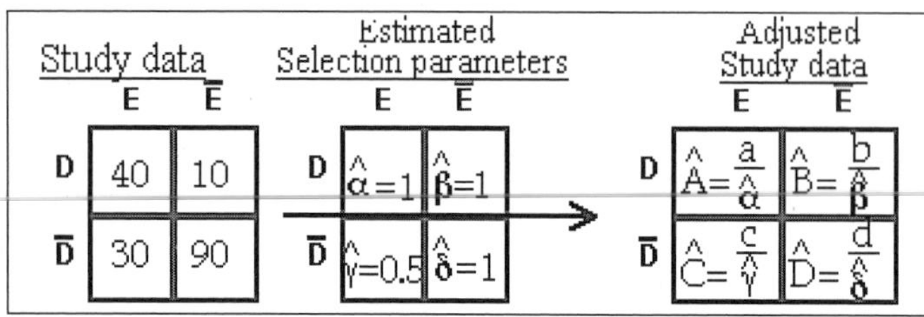

Consider the above three tables of **study data**, **estimated selection parameters**, and **adjusted study data**.

1. Compute the adjusted cell frequencies $\hat{A}, \hat{B}, \hat{C}, \text{and } \hat{D}$.
2. What is the adjusted odds ratio that corrects for selection bias?
3. Compute the ratios of selection parameters $\hat{r}_D = \hat{\alpha}/\hat{\beta}$ and $\hat{r}_{\bar{D}} = \hat{\lambda}/\hat{\delta}$
4. Based on $\hat{r}_D$ and $\hat{r}_{\bar{D}}$, use the alternative correction formula to compute the odds ratio that corrects for selection bias.

An adjusted estimate of the risk ratio is also possible if estimates of the selection parameters can be identified. From the 2x2 table of adjusted cell frequencies we can obtain an adjusted risk ratio as shown here.

$$\hat{RR}(adj) = \frac{\dfrac{\hat{A}}{\hat{A}+\hat{C}}}{\dfrac{\hat{B}}{\hat{B}+\hat{D}}}$$

If we then substitute for $\hat{A}, \hat{B}, \hat{C}, \text{and } \hat{D}$ their corresponding expressions involving the study data divided by the estimated selection parameter, and carry out a few algebraic manipulations, we obtain the adjusted risk ratio formula shown here.

$$RR_{adj} = \frac{\left(\dfrac{a}{a + c\dfrac{\hat{\alpha}}{\hat{\gamma}}}\right)}{\left(\dfrac{b}{b + d\dfrac{\hat{\beta}}{\hat{\delta}}}\right)}$$

An alternative formula can be stated involving the ratio of selection parameters $\hat{r}_E$ and $\hat{r}_{\bar{E}}$ as shown here.

$$\hat{RR}(adj) = \frac{\dfrac{a}{a + c\hat{r}_E}}{\dfrac{b}{b + d\hat{r}_{\bar{E}}}}$$

Study Questions (Q8.14) continued

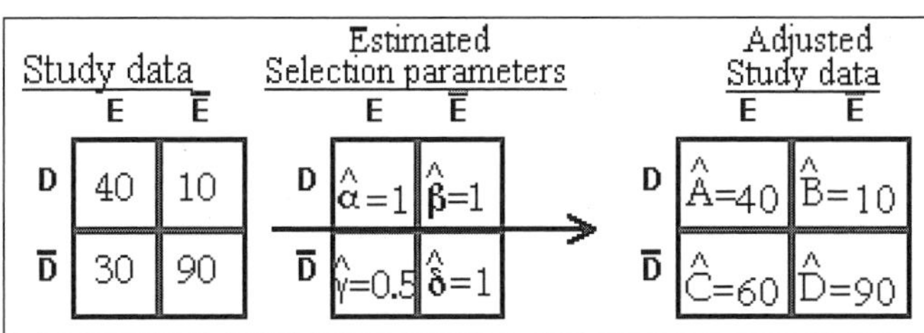

Study data

	E	$\bar{E}$
D	40	10
$\bar{D}$	30	90

Estimated Selection parameters

	E	$\bar{E}$
D	$\hat{\alpha}=1$	$\hat{\beta}=1$
$\bar{D}$	$\hat{\gamma}=0.5$	$\hat{\delta}=1$

Adjusted Study data

	E	$\bar{E}$
D	$\hat{A}=40$	$\hat{B}=10$
$\bar{D}$	$\hat{C}=60$	$\hat{D}=90$

Consider again the above three tables of **study data**, **estimated selection parameters**, and **adjusted study data**.

5. Using the above information, what is the adjusted risk ratio that corrects for selection bias?
6. What are the values of ratios $\hat{\alpha}/\hat{\gamma}$ and $\hat{\beta}/\hat{\delta}$?
7. Use the formula for correcting selection bias in the risk ratio to obtain the adjusted risk ratio.

Summary

- Selection bias can be corrected quantitatively depending on whether the four selection parameters or ratios of these parameters can be estimated.
- An adjusted odds ratio formula is:

$$\hat{OR}_{adj} = \frac{\hat{OR}}{\left(\frac{\hat{\alpha}\,\hat{\delta}}{\hat{\beta}\,\hat{\gamma}}\right)}$$

- Alternative formulae are:

* Alternative formulae are

$$\hat{OR}_{adj} = \hat{OR}\frac{\hat{r}_{\bar{D}}}{\hat{r}_D} = \hat{OR}\frac{\hat{r}_{\bar{E}}}{\hat{r}_E}$$

- Selection bias can be corrected quantitatively depending on whether the four selection parameters can be estimated.
- An adjusted risk ratio formula is at left and an alternative formula for RR_{adj} is at right:

$$RR_{adj} = \frac{\left(\frac{a}{a + c\frac{\hat{\alpha}}{\hat{\gamma}}}\right)}{\left(\frac{b}{b + d\frac{\hat{\beta}}{\hat{\delta}}}\right)} \qquad RR_{adj} = \frac{\left(\frac{a}{a + c\hat{r}_E}\right)}{\left(\frac{b}{b + d\hat{r}_{\bar{E}}}\right)}$$

What Can Be Done About Selection Bias Qualitatively?

The previous activity provided quantitative formulas for correcting for selection bias. Unfortunately the information required to apply these formulas, namely the selection ratio parameters or their ratios, is conceptually complicated, rarely available, and not easily quantified. The ideal way to address selection bias is to prevent or at least minimize such bias when designing a study rather than to attempt to correct for the bias once the data have been collected.

In case-control studies, the controls should be carefully chosen to represent the source population that produced the cases. The use of two or more control groups should also be considered.

Study Questions (Q8.15)

1. Suppose two control groups are used in a case-control study and the resulting estimated effects differ considerably depending on which control group is used. How can it be determined which of the two estimates is correct?
2. Suppose two control groups are used and the resulting estimated effects are essentially the same regardless of which control group is used. Does this mean that there is no selection bias?

Case-control studies using incident cases and nested case-control studies are preferable to studies that use prevalent cases or to hospital-based studies. Selection bias may also be avoided by assuring equal opportunity for disease detection among exposed and unexposed subjects.

In cohort studies, efforts should be made to achieve high response and low loss-to-follow-up.

Study Questions (Q8.15) continued

3. Suppose only 1% of a study cohort is lost to follow-up. Is selection bias still possible because of loss to follow-up?
4. Suppose 20% of exposed subjects and 30% of unexposed subjects are lost to follow-up in a cohort study. Will there be selection bias?

Observational studies involving volunteers should be avoided, although clinical trials involving volunteers are possible because of randomization. Occupational cohort studies should avoid a healthy worker bias by ensuring that unexposed subjects are as healthy as exposed subjects.

At the analysis stage, it may be possible to determine the direction of the bias, even if the magnitude of the bias cannot be estimated or a numerical correction for bias is not feasible.

Study Questions (Q8.15) continued

Suppose a reasonable argument could be made that alpha is less than beta, whereas gamma equals delta, even though specific questimates of each selection parameter or ratios of selection parameters were not possible. Answer the following questions based on this information:

5. Is it possible to assess the magnitude of selection bias in the odds ratio?
6. Is it possible to quantitatively correct for selection bias in the odds ratio?
7. Is it possible to determine the direction of the bias?

Suppose it can be argued that α is greater than γ, but that β is also greater than γ, even though specific guestimates of each selection parameter or ratios of selection parameters were not possible.

8. Is it possible to assess whether or not there is selection bias?
9. Suppose it could be argued that α / γ is equal to β / δ. Is it possible to assess whether or not there is selection bias in the odds ratio?
10. Suppose the correct OR is greater than 1 and it could be argued that α / γ is greater than β / δ. Is it possible to determine the direction of the bias?

A final approach to addressing selection bias as well as information and confounding biases is in the write-up of the study. A description of the potential biases of the study is typically provided in the "Discussion" section of a publication. This discussion, particularly when it concerns possible selection bias, is quite subjective, but such judgment is required because of the inherent difficulty in quantifying such biases.

Summary

- ❖ The information required to assess selection bias is conceptually complicated, rarely available, nor is easily quantifiable.
- ❖ At the study design stage, decisions may be made to avoid selection bias in the study's results.
- ❖ At the analysis stage, it may be possible to determine the direction of selection bias without being able to quantitatively correct for the bias.
- ❖ At the publication stage, potential biases are typically addressed qualitatively in the **Discussion** section of a paper.

The "Worst-Case Scenario" Approach.

This is a practical approach for assessing the **direction** of **selection bias** that considers the most extreme changes in the estimate of effect that are realistically possible as a result of the way subjects are selected. Through such an approach it may be possible to show that the worst amount of bias possible will have a negligible effect on the conclusions of one's study.

For example, consider a cohort study involving lung-cancer-free 1000 smokers and 1000 non-smokers all over 40 years of age that are followed for 10 years. Suppose further that over the follow-up period, 200 smokers and 100 non-smokers are lost-to-follow-up. Also, suppose, that among those 800 smokers and 900 non-smokers remaining in the study, the 2x2 table relating smoking status at the start of the study to the development of lung cancer (LC) over the 10 year follow-up is shown as follows:

	Smokers	**Non-smokers**
LC	80	10
No LC	720	890
Total	800	900

The estimated risk ratio from these data is (80/800)/(10/900) = 9, which suggests a very strong relationship between smoking status and the development of lung cancer. A worst-case scenario might determine what the risk ratio estimates would be if either all 200 smokers lost-to-follow-up did not develop lung cancer and/or all 100 non-smokers lost to follow-up did develop lung. Here are comparison of estimates for "worst-case" scenarios:

Scenario	**Risk Ratio**
1. Actual observed data	9
2. 1/10th of 200 lost-to-follow-up smokers get LC and 1/90th of the 100 lost-to-follow-up non-smokers get LC	9
3. 1/10th of the 200 lost-to-follow-up smokers get LC and 2/90th of the 100 lost-to-follow-up non-smokers get LC	8.2
4. None of the 200 lost-to-follow-up smokers get LC and 1/90th of the 100 lost-to-follow-up non-smokers get LC	7.3
5. None of the 200 lost-to-follow-up smokers get LC and 2/90th of the 100 lost-to-follow-up non-smokers get LC	6.6
6. None of the 200 lost-to-follow-up smokers get LC and all 100 lost-to-follow-up non-smokers get LC	0.7

Notice that scenario #6 above changes smoking from being harmful to being protective. Yet this is not a very realistic scenario. Scenario #2 is not really a "worst case" type of scenario because it assumes that those lost to follow-up have the same risk as those actually followed over the entire 10 years. The other three scenarios, i.e., #'s 3, 4, and 5, are "realistic" and of a "worst-case" type; all of these show that the risk ratio is reduced, but is still high. The difficulty with this approach, as illustrated above, concerns the extent to which the investigator can identify a "worst-case" scenario that is the most realistic among all possible scenarios.

Quiz (Q8.16)

1. An adjusted estimate of the odds ratio that corrects for selection bias is given by the cross product of the **???**, which can be written in terms of the study data and **???** estimates.
2. Thus, the odds ratio that adjusts for selection bias is given by the **???** from the study data divided by the **???**.

Choices
adjusted cell frequencies **estimated cross-product of selection parameters**
estimated odds ratio **population data** **selection bias** **selection parameter**

3. Case-control studies using **???** are preferable to studies that use **???**.
4. And **???** studies are preferable to **???** studies.
5. In cohort studies, efforts should be made to achieve **???** and to avoid **???**.

Choices
cases **controls** **high response** **hospital-based** **incident cases** **loss-to-follow-up** **nested case-control** **prevalent cases** **selection bias**

At the analysis stage, the extent of possible selection bias may be assessed using what are often referred to as "worst-case" analyses. Such analyses consider the most extreme changes in the estimated effect that are possible as a result of selection bias. Determine whether each of the following is **True** or **False**.

6. This approach is useful since it could demonstrate that the worst amount of bias possible will have a negligible effect on the conclusions of the study. **???**
7. This approach can rule out selection bias, but it cannot confirm selection bias. **???**

The tables below show the observed results and a 'worst-case' scenario for a clinical trial. Ten subjects receiving standard treatment and 15 subjects receiving a new treatment were lost-to-follow-up. The outcome was whether or not a subject went out of remission (D = out, not D = in) by the end of the trial. In the 'worst-case' scenario, all 10 subjects on standard treatment who were lost-to-follow-up remained in remission but all 15 subjects on the new treatment who were lost-to-follow-up went out of remission.

	Observed			Worst-Case	
	Standard Rx	New Rx		Standard Rx	New Rx
D=out	32	7	D=out	a*	b*
Not D = in	100	106	Not D = in	c*	d*

8. What are the values of a*? **???**, b*? **???**, c*? **???**, d*? **???**

Choices
106 **110** **121** **22** **32** **42** **90** **91**

Quiz continued on next page

9. Refer to the data in the tables above. What is the risk ratio estimate based on the observed data? **???**
10. What is the risk ratio estimate based on the 'worst-case' scenario? **???**
11. In the worst-case scenario, is there selection bias? **???**
12. Does a "worst-case" assessment such as illustrated here "prove" that there is selection bias? **???**

Choices
1.3 **1.4** **3.9** **4.8** **No** **Yes**

Nomenclature

Source vs. study population

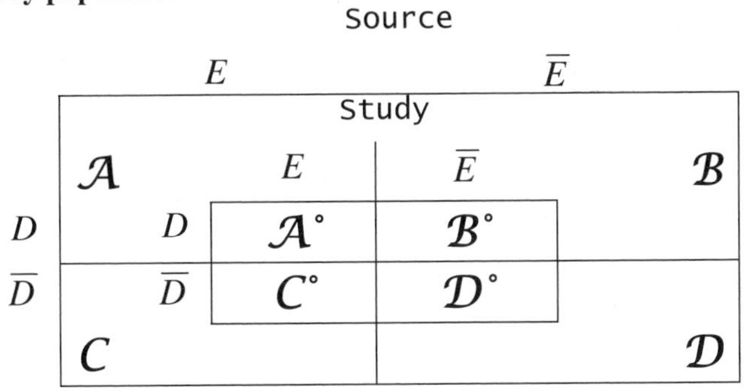

Source

	E	$\overline{E}$
D	$\mathcal{A}$	$\mathcal{B}$
$\overline{D}$	$\mathcal{C}$	$\mathcal{D}$

Study

	E	$\overline{E}$
D	$\mathcal{A}^\circ$	$\mathcal{B}^\circ$
$\overline{D}$	$\mathcal{C}^\circ$	$\mathcal{D}^\circ$

Selection Probabilities

	E	$\overline{E}$
D	$\alpha = \mathcal{A}^\circ / \mathcal{A}$	$\beta = \mathcal{B}^\circ / \mathcal{B}$
$\overline{D}$	$\gamma = \mathcal{C}^\circ / \mathcal{C}$	$\delta = \mathcal{D}^\circ / \mathcal{D}$

Study Data

	E	$\overline{E}$
D	a	b
$\overline{D}$	c	d

Estimated Selection parameters

	E	$\overline{E}$
D	$\hat{\alpha}$	$\hat{\beta}$
$\overline{D}$	$\hat{\gamma}$	$\hat{\delta}$

Adjusted Study Data

	E	$\overline{E}$
D	$\hat{A} = \frac{a}{\hat{\alpha}}$	$\hat{B} = \frac{b}{\hat{\beta}}$
$\overline{D}$	$\hat{C} = \frac{c}{\hat{\lambda}}$	$\hat{D} = \frac{d}{\hat{\gamma}}$

$$\text{Bias}(\hat{OR}, OR) = OR^\circ - OR = \frac{A^\circ D^\circ}{B^\circ C^\circ} - \frac{AD}{BC} = \frac{\alpha A \delta d}{\beta B \gamma C} - \frac{AD}{BC}$$

$$\hat{OR}_{adj} = \frac{\hat{A}\hat{D}}{\hat{B}\hat{C}} = \frac{\frac{a}{\hat{\alpha}}\frac{d}{\hat{\delta}}}{\frac{b}{\hat{\beta}}\frac{c}{\hat{\gamma}}} = \frac{\hat{OR}}{\frac{\hat{\alpha}\hat{\delta}}{\hat{\beta}\hat{\gamma}}}$$

$$\hat{RR}_{adj} = \frac{\frac{\hat{A}}{\hat{A}+\hat{C}}}{\frac{\hat{B}}{\hat{B}+\hat{D}}} = \frac{\frac{a}{a+c\frac{\hat{\alpha}}{\hat{\gamma}}}}{\frac{b}{b+d\frac{\hat{\beta}}{\hat{\delta}}}}$$

D	Truly has disease
E	Truly exposed
Not D *or* $\overline{D}$	Truly does not have disease
Not E *or* $\overline{E}$	Truly not exposed
$\hat{OR}$	Odds ratio from observed data
OR^o	Odds Ratio from Study Population
$\hat{OR}_{adj}$	Odds ratio from corrected or adjusted data
$\hat{RR}$	Risk ratio from observed
OR^o	Risk Ratio from Study Population
$\hat{RR}_{adj}$	Risk ratio from corrected or adjusted

References

References on Selection Bias

Berkson J. Limitations of the application of fourfold table analysis to hospital data. Biometrics Bulletin 1946;2: 47-53.

Greenberg RS, Daniels SR, Flanders WD, Eley JW, Boring JR. Medical Epidemiology (3rd Ed). Lange Medical Books, New York, 2001.

Hill H, Kleinbaum DG. Bias in Observational Studies. In Encyclopedia of Biostatistics, pp 323-329, Oxford University Press, 1999.

Horwitz RI, Feinstein AR. Alternative analytic methods for case-control studies of estrogens and endometrial cancer. N Engl J Med 1978;299(20):1089-94.

Kleinbaum DG, Kupper LL, Morgenstern H. Epidemiologic Research: Principles and Quantitative Methods. John Wiley and Sons Publishers, New York, 1982.

References on Condom Effectiveness Studies

Crosby RA. Condom use as a dependent variable: measurement issues relevant to HIV prevention programs. AIDS Educ Prev 1998;10(6): 548-57.

Fishbein M, Pequegnat W. Evaluating AIDS prevention interventions using behavioral and biological outcome measures. Sex Transm Dis 2000;37 (2):101-10.

Peterman TA, Lin LS, Newman DR, Kamb ML, Bolan G, Zenilman J, Douglas JM Jr, Rogers J, Malotte CK. Does measured behavior reflect STD risk? An analysis of data from a randomized controlled behavioral intervention study. Project RESPECT Study Group. Sex Transm Dis 2000;27(8):446-51.

National Institute of Allergy and Infectious Disease, Workshop Summary: Scientific Evidence on Condom Effectiveness for Sexually Transmitted Disease (SID) Prevention, 2001.

Stone KM, Thomas E, Timyan J. Barrier methods for the prevention of sexually transmitted diseases. In: Holmes KK, Sparling PF, Mardh P-A,eds. Sexually Transmitted Diseases, 3rd ed., McGraw-Hill, NewYork,1998.

Warner L, Clay-Warner J, Boles J, Williamson J. Assessing condom use practices. Implications for evaluating method and user effectiveness. Sex Transm Dis 1998; 25(6):273-7.

Warner DL, Hatcher RA. A meta-analysis of condom effectiveness in reducing sexually transmitted HIV. Soc Sci Med 1994;38(8):1169-70.

Homework

ACE-1. Selection Bias: Dietary Fat vs. Colon Cancer

Suppose an investigator is interested in whether dietary fat is associated with the development of colon cancer. S/he identifies several hundred incident cases of colon cancer and a comparable number of community controls. Each subject fills out a food frequency questionnaire describing his/her dietary habits in the previous year; information from the questionnaire is used to divide subjects into those who consumed a high fat diet (exposed) and those who consumed a low fat diet (unexposed).

Assume the following:

- Among cases, the likelihood of agreeing to participate in the study is NOT related to exposure.
- Among controls, those who consume a high fat diet are half as likely to participate, compared to those who consume a low fat diet.

a. Express each of the assumptions above in terms of selection probabilities.
b. Calculation the selection odds for the diseased (r_D) and the selection odds for the non-diseased ($r_{\text{not D}}$).
c. Based on your answers to a and b above, do you expect bias in the estimated odds ratio (OR) from this study. Justify your answer.

Suppose the following table reflects the data that one would have observed if there had been 100% participation among both cases and controls (i.e., these are unbiased data):

	High Fat	**Low Fat**	**Total**
Case	200	175	375
Control	150	225	375
Total	350	400	750

d. Based on the information about selection probabilities described earlier, describe the 2x2 table that gives the study population for this study.
e. Compute and compare the odds ratio in the source population (i.e., 100% participation) with the odds ratio in the study population. Is there bias, and if so, what is the direction of the bias?

ACE-2. Selection Bias: Food Allergies vs. CHD Development

Consider a ten-year follow-up of a fixed cohort free of CHD at the start of follow-up. The study objective is to determine whether persons experiencing food allergies (FA) prior to the start of follow-up are at increased risk for CHD development. Suppose that a simple analysis of the study results yields the following study information.

	FA		
CHD	**Yes**	**No**	**Total**
Yes	49	286	355
No	1378	5816	7194
Total	1447	6102	7549

RR = 1.02, OR = 1.02

Which of the following statements about possible selection bias is appropriate for this study design (Choose only one answer):

a. If persons who had high blood pressure (a known risk factor for CHD) prior to the start of follow-up were excluded from the comparison (no allergy) group, but NOT from the FA (allergy) group, then there should be concern about selection bias that would be AWAY FROM THE NULL.
b. If the likelihood of being not lost to follow-up for persons with food allergies is exactly the same as the likelihood of not being lost to follow-up for persons without food allergies, then there can be no selection bias due to follow-up loss.

c. Bias from selective survival should be a concern in this study.
d. Berkson's bias should be a concern in this study.
e. If persons who were free of any kind of illness (except FA) prior to start of follow-up were excluded from both the exposed group and the comparison (no allergy) group, then there should be concern about selection bias that would be TOWARDS THE NULL.

ACE-3. Selective Survival

a. What is meant by "bias due to selective survival" in cross-sectional studies? (In your answer, make sure to define appropriate selection probability parameters.)
b. Under what circumstances might there be no selective survival bias even if the selection probabilities are not all equal?
c. Suppose that you could assess that the direction of possible selective survival bias in your study was towards the null. If your study data yielded a non-statistically significant odds ratio of 1.04, would it be correct to conclude that there was no exposure-disease association in your source population? Explain.

ACE-4. Choice of Cases and Controls

You are asked to advise a clinician interested in conducting two case-control studies examining the relationship between coffee consumption and:

i. Primary ovarian cancer.
ii. Coronary artery disease

The clinician wishes to use hospital admissions as the source of cases. Advise how valid the proposal is, particularly in light of possible selection biases. How would you advise him about choice of cases and controls?

ACE-5. Mortality and Clofibrate: Selection Bias?

A randomized placebo-controlled trial was carried out to estimate the effect of Clofibrate (a cholesterol-lowering drug) on 5-year mortality in men aged 30 to 64 years who had a recent myocardial infarct. The mortality in the Clofibrate group was 18.2% and that in the control group was 19.4%. This difference was not statistically significant. Subsequent analysis showed that about one-third of men given Clofibrate did not take their medication as directed. If these men are omitted from the analysis, there is a statistically significant effect of Clofibrate, with mortality in the Clofibrate group being 15.0%.

a. Which of the above two alternative analyses is more appropriate? Why?
b. Is this a selection bias issue? Explain.

ACE-6. Identifying Study Subjects

A study entitled "Antidepressant Medication and Breast Cancer Risk" (Amer. J. of Epi, late 1990's) stated in the methods section of the paper that "Cases were an age-stratified (< 50 and ≥ 50 years of age) random sample of women aged 25-74 years diagnosed with primary breast cancer during 1995 and 1996 (pathology report confirmed) and recorded in the population-based Ontario Cancer Registry. As the 1-year survival for breast cancer is 90%, surrogate respondents were not used. Population controls, aged 25-74 years, were randomly sampled from the property assessment rolls of the Ontario Ministry of Finance; this database includes all home owners and tenants and lists age, sex, and address."

a. Discuss the authors' approach to the identification of cases with respect to the potential for selection bias.
b. Discuss the authors' approach to the identification of controls with respect to the potential for selection bias.
c. Does the decision not to allow surrogate respondents potentially affect selection bias?

ACE-7. Selection Bias in Observational Studies

The following questions concern the assessment of selection bias in observational studies where the putative association between salt intake and hypertension is being investigated. It is hypothesized that persons who have a greater intake of salt are at greater risk of developing chronic hypertension. It is assumed that both variables are dichotomized: high and low salt intake; hypertensive (DBP≥95 and SPB≥160) and normotensive. In considering the following questions, ignore the possibility of either information bias or confounding. Also, assume that the measure of effect in the source population is greater than one.

a. Consider a case-control study in which equal numbers of hypertensives (cases) and normotensives (controls) are enrolled. Suppose that hypertensives from the study are gathered from an outpatient clinic designed to screen for hypertensives in the population, using the clinic on a volunteer basis. Normotensives are selected randomly from the surrounding population served by the clinic. Suppose that high-salt-intake hypertensives are four times as likely to be screened as low-salt-intake hypertensives in the clinic and that the sample of normotensives from the surrounding community is indeed a representative sample. What can be said about selection bias in estimating the exposure odds ratio (EOR)? (Formulate your answer in terms of the selection probabilities □, □, □, and □, or the selection odds r_D and $r_{not\ D}$.)

b. In a type of case-control study similar to that described in part a, suppose that high- and low-salt-intake hypertensives are equally likely to be screened but that high-salt-intake normotensives in the community are more likely to cooperate in the study than are low-salt-intake normotensives in the community. What can be said about selection bias in estimating the EOR? (Formulate your answer in terms of the selection probabilities α, β, γ, and δ, or the selection odds r_D and $r_{not\ D}$.)

Answers to Study Questions and Quizzes

Q8.1

1. 2
2. 1; exposed cases are likely to be over-represented in the study when compared to unexposed cases.

Q8.2

1. Withdrawals from the study can distort the final sample to be analyzed as compared to the random sample obtained at the start of the trial.
2. Those who withdraw from the study have an unknown outcome and therefore cannot be analyzed.
3. A general population of patients with the specified cancer and eligible to receive either the standard or the new treatments.
4. The (expected) sample ignoring random error that would be obtained after the withdrawals from the random sample initially selected for the trial.
5. The source population is the population cohort from which the cases would be derived if an appropriate cohort study had been carried out.
6. The study population is the expected sample obtained from the cross-sectional sample that is retained for analysis. Alternatively, the study population is the stable population from which the cross-sectional sample is obtained for study.

Q8.3

1. It is unlikely that women in this control group (e.g., with cervical or ovarian cancers) would be selectively screened for their cancer from vaginal bleeding caused by estrogen use.
2. Away from the null because of selective screening of cases but not controls. This would yield too high a proportion of estrogen users among cases but a correct estimate of the proportion of estrogen users among controls.
3. Because those who have vaginal bleeding from using estrogen will be more likely to have their benign endometrial tumor detected than those non-users with benign endometrial tumors.
4. Using benign endometrial tumors as the control group would hopefully compensate for the selective screening of cases. However, it is not clear that the extent of selective screening would be the "same" for both cases and controls.
5. Having a benign tumor in the endometrium is not readily detected without vaginal bleeding. Therefore, controls with benign endometrial tumors who use estrogen are more likely to have their tumor detected than would nonuser controls.
6. Towards the null because there would be selective screening of controls but not cases. This would yield too high a proportion of estrogen users among controls but a correct estimate of the proportion of estrogen users among cases.
7. Because there is unlikely to be selective screening in the detection of control cases (with other gynecological cancers) when comparing estrogen users to nonusers.

Q8.4

1. No, since the risk ratio for the expected sample is 3, which equals the risk ratio in the source population.
2. Study population, because the sample just described gives the expected number of subjects obtained in the final sample.
3. The source is the population from which the initial 10% sample obtained prior to follow-up was selected. This is the population of subjects from which cases were derived.
4. There is no selection bias because the RR=3 in the study population, the same as in the source population.
5. Yes, because the estimated risk ratio in the study population of 3.9 is somewhat higher than (3) in the source population as a result of subjects being lost to follow-up.
6. No, the risk for exposed persons is 150/10,000 or .0150 in the source population and is 14/800 = .0175 in the sample.
7. No, the risk for unexposed persons is 50/10,000 or .0050 in the source population and 4/900 or .0044 in the sample.
8. No. There will only be selection bias if loss to follow-up results in risks for disease in the exposed and/or unexposed groups that are different in the final sample than in the original cohort.

Q8.5

1. Workers tend to be healthier than those in the general population and may therefore have a more favorable outcome regardless of exposure status.
2. Volunteers may have different characteristics from person who do not volunteer. The study population here is restricted to volunteers, whereas the source population is population-based, e.g., a community.
3. There is lack of external validity in drawing conclusions from a source population of volunteers to an external population that is population-based.

4. Yes. Clinic-based studies may lead to spurious conclusions because patients from clinics tend to have more severe illness than persons in a population-based sample.

Q8.6

1. $\boldsymbol{\alpha}$ = 12/150 = .08; $\boldsymbol{\beta}$ = 4.5/50 = .09; $\boldsymbol{\gamma}$ = 788/9850 = .08; and $\boldsymbol{\delta}$ = 895.5/9950 = .09.
2. No. The odds ratio in both source and study populations is 4.03. The risk ratio in both target and study populations is 3.0.
3. (α x δ) / (β x γ) = (.08 x .09) / (.09 x .08) = 1. This result illustrates a rule (described in the next activity) that says that the cross-product ratio of selection probabilities equal 1 if there is no bias in the odds ratio.
4. $\boldsymbol{\alpha}$ = 14/150 = .093; $\boldsymbol{\beta}$ = 4/50 = .08; $\boldsymbol{\gamma}$ = 786/9850 = .08; and $\boldsymbol{\delta}$ = 896/9950 = .09.
5. Yes. The odds ratio in the source population is 3.03 and in the study population is 4.00. The risk ratio in the source population is 3.0 and in the study population 3.9.
6. (α x δ) / (β x γ) = (.093 x .09) / (.08 x .08) = 1.3. This result illustrates a rule (described in the next activity) that says that the cross-product ratio of selection probabilities will differ from 1 if there is bias in the odds ratio.

Q8.7

1. No. The cross product ratio of selection ratios equals 1.
2. Yes. The cross product ratio of selection ratios equals (.093 x .09)/(.08 x .08) = 1.3, which is larger than 1.
3. Since the bias is defined as OR^o – OR, this means that the bias is away from the null (i.e., OR^o is further away from the null than is OR).

Q8.8

1. α = 100/100 = 1.0; β = 50/50 = 1.0; γ = 7920/9900 = .8; and δ = 7960/9950 = .8
2. Yes, the study population is a subset of the source population.
3. Yes, more or less; the incidence of disease in the overall source population is 150/20,000=.0075 and 150/16,030=.0093 in the overall study population. Within each exposure group, the risks are 100/10,000=.01 and 50/110,000 =.005, respectively, in the source population; and 100/8,020 =.0124 and 50/8,010=.0062, respectively, in the study population.
4. No. The cross product ratio of selection ratios equal 1.
5. No. The risk ratio in the study population is 1.998, which is not identical but essentially equal to the risk ratio in the source population.
6. α = 100/100 = 1.0; β = 50/50 = 1.0; γ = 99/9,900 = .01; and δ = 99.5/9950 = .01.
7. No. The cross-product ratio of selection ratios equals 1.
8. Yes. The risk ratio in the study population is 1.503, which is smaller than the risk ratio of 2.0 in the source population. The bias is towards the null.
9. No. The disease incidence is not at all rare in the study population. Overall the disease incidence is 150/348.5 = .4304 in the study population, and is 100/199 = .5025 and 50/149.5 = .3344 within exposed and unexposed groups, respectively.
10. α = 40/40 = 1.0; β = 10/10 = 1.0; γ = 30/60 = .5; and δ = 90/90 = 1.0.
11. Yes. The cross product ratio of selection ratios equals (1 x 1)/(1 x .5) = 2, which is clearly different from 1. The incorrect OR = 12, which is greater than the correct OR = 6, indicating that the bias is away from the null.
12. Yes. The risk ratio in the study population is 5.71, which is larger than the risk ratio of 4.0 in the source population. The bias is away from the null.
13. The disease is not rare in either source or study populations. It is particularly not rare for exposed subjects in the study population.
14. No. The bias in the odds ratio (OR^o =12 whereas OR = 6) is much larger than the bias in the risk ratio (RR^o =5.71 whereas RR =4.0).

Q8.9

1. $\alpha > \beta$
2. $\gamma > \delta$
3. $\alpha = \beta$
4. $\gamma = \delta$

Q8.10

1. True. Unexposed controls in the study will include persons with sexual partners with and without STD's. Unexposed non-cases in the source population will not include sexual partners without STDs.
2. False. Exposed controls will go to STD clinics primarily for reasons of disease prevention rather than contraception. They are more likely to have sex partners with STDs than unexposed controls, and therefore, are more likely to represent the source population.
3. Yes. In the source population the odds for condom users in one-half the odds for non-users, whereas the odds ratio is one in the study population.

4. Yes. The ratio (i.e., odds) of unexposed to exposed non-cases in the study population is 6 to 1 whereas the odds of unexposed to exposed non-cases in the source population is 3 to 1.
5. $\alpha = 100/100 = 1$; $\beta = 600/600 = 1$; $\gamma = 200/200 = 1$; and $\delta = 1200/600 = 2$.
6. In this example, α (1) equals β (1), but γ (1) is less than δ (2). Consequently, the cross product ratio of selection ratios should be greater than 1. Thus, there is selection bias in the odds ratio.
7. The bias would be towards the null, since the (incorrect) OR in the study population is equal to 1 whereas the correct OR is different (in this case, smaller) than 1 in the source population.
8. False. Both unexposed controls in the study and unexposed non-cases will be restricted to have sexual partners with STD's.
9. False. Exposed controls and exposed non-cases will also be restricted to have sexual partners with STD's. Consequently, exposed controls in the study will reflect the source population of exposed non-cases, all those partners have STD's.
10. γ, the selection ratio for exposed controls, is equal to δ, the selection ratio of unexposed controls.
11. The cross product of selection ratios should be equal to 1.

Q8.11

1. The source population is the community population and the study population is the expected case-control sample under the selection conditions described for the study.
2. $\alpha = 225/400 = .5625$; $\beta = 75/200 = .375$; $\gamma = 100/100{,}000 = .001$; and $\delta = 200/200{,}000 = .001$.
3. Yes, because each is derived as a ratio in which the numerator is a subset of the denominator.
4. $\alpha = .5625$ is greater than $\beta = .375$, whereas $\gamma = \delta$. Consequently, the cross-product ratio of selection probabilities = (.5625 x .001)/(.375 x .001) = 1.5, which is different from 1.
5. The OR in the study population is 6 whereas the OR in the source population is 4. The incorrect (i.e., biased) odds ratio of 6 is an overestimate of the correct odds ratio, and is therefore away from the null.
6. Patients who are both cases and smokers are over-represented in the study when compared to the community population.
7. Yes. Controls are split 50:50 among exposed and unexposed, whereas non-cases in the community are split 1:2.
8. $\alpha = 225/400 = .5625$; $\beta = 75/200 = .375$; $\gamma = 150/100{,}000 = .0015$; and $\delta = 150/200{,}000 = .00075$.
9. $\alpha = .5625$ is greater than $\beta = .375$, and $\gamma = .0015$ is less than $\delta = .00075$. The cross-product ratio of selection probabilities = (.5625 x .00075)/(.375 x .0015) = 3.0, which is different from 1.
10. The OR in the study population is 3 whereas the OR in the source population is 4. The incorrect (i.e., biased) odds ratio of 3 is an underestimate of the correct odds ratio, and is therefore towards the null.
11. Hospital patients who are both cases and smokers are more over-represented than are hospital patients who are both controls and smokers when compared to the community population.

Q8.12

1. 25/50 = .5 for exposed and 40/80 = .5 for unexposed.
2. 142.5/150 = .95
3. 540/720 = .75
4. $\alpha = \beta = .5$; $\gamma = .95$, and $\delta = .75$.
5. (.5 x .75)/(.95 x 5) = .79
6. Yes, a slight bias. For the odds ratio, ROR = 3 whereas POR (= OR in the study population) = 2.4. For the risk ratio, RR 2.5 whereas PR (= effect in the study population) = 2.2. The bias is towards the null.

Q8.13

1. prior to
2. non-response, study population
3. one
4. earlier, by
5. prevalence data, incidence data
6. selective survival
7. survivors
8. survive

Q8.14

1. $\hat{A} = 40, \hat{B} = 10, \hat{C} = 60, \hat{D} = 90$
2. $OR_{adj} = 6$
3. $\hat{r}_D = 1,\ \hat{r}_{\bar{D}} = .5$
4. $OR_{adj} = O\hat{R} \times (\hat{r}_{\bar{D}} / \hat{r}_D) = 12 \times (.5/1) = 6$
   $RR_{adj} = (40/100)/(10/100) = 4$
5. $\hat{r}_E = \hat{\alpha}/\hat{\gamma} = 2, \hat{r}_{\bar{E}} = \hat{\beta}/\hat{\delta} = 1$
6. RR_{adj} = [40/(40 + [30 x 2])]/{10/(10 + [90 x 1])} = 4

Q8.15

1. A decision will have to be made as to which of the two control groups is the most suitable, e.g., which control group is more representative of the source population.
2. Either of the following is possible: 1.) There is no selection bias because both estimated effects are correct; or 2.) There is selection bias because both estimated effects are biased.
3. Yes. Since the presence of selection bias depends on the selection parameters within the 2x2 table, selection bias may still occur even with excellent response rates and minimal loss-to-follow-up.
4. Not necessarily. As with the previous question, the presence of selection bias depends on the selection parameters within the 2x2 table, whereas the information provided only considers follow-up loss on the marginals (i.e., total exposed and total unexposed) of the 2x2 table.
5. No. The magnitude of selection bias cannot be determined without guestimates of the selection parameters or their ratios.
6. No. A corrected odds ratio would require guestimates of the selection parameters or their ratios.
7. Since α is less than β and gamma equals delta, the cross-product $(\alpha \times \delta) / (\beta \times \gamma)$ must be less than one. Thus, OR^o must be less than OR, so that the bias would be towards the null if the OR is greater than 1 and away from the null if the OR is less than 1.
8. No. Without knowing whether α over γ is either equal to, less than, or greater than β over δ, it is not possible to determine whether the cross-product $(\alpha \times \delta)/(\beta \times \gamma) = (\alpha / \gamma)/(\beta / \delta)$ is equal to, less than, or greater than 1.
9. Yes. There is no selection bias in the odds ratio because the cross-product $(\alpha \times \delta)/(\beta \times \gamma)$ equals 1.
10. Yes. The bias must be away from the null because the cross-product $(\alpha \times \delta)/(\beta \times \gamma)$ is greater than 1 so that the biased OR^o is greater than the correct OR, which is greater than 1.

Q8.16

1. adjusted cell frequencies, selection parameter
2. estimated odds ratio, estimated cross-product of selection parameters
3. incident cases, prevalent cases
4. nested case-control, hospital-based
5. high response, loss-to-follow-up
6. True
7. True – Since the "worst-case" analysis can demonstrate that the worst amount of bias will have a negligible effect on the conclusions, one could rule out selection bias. However, since the "worst-case" analysis gives us the "worst possible" results, we cannot confirm selection bias. We cannot be sure that our results will be as extreme as "worst-case" results.
8. 32, 22, 110, 106
9. 3.9
10. 1.3
11. Yes
12. No - A worst-case analysis gives the "worst possible" results. Therefore, we cannot be sure that the lost-to-follow-up results that "actually" occur are as extreme as the worst-case "possible".

LESSON 9

INFORMATION BIAS

9-1 Information Bias

***Information bias** is a **systematic error** in a study that arises because of incorrect information obtained on one or more variables measured in the study. The focus here is on the consequences of having inaccurate information about exposure and disease variables that are dichotomous, that is, when there is **misclassification** of exposure and disease that leads to a **bias** in the resulting **measure of effect**. We consider exposure and disease variables that are **dichotomous**. More general situations, such as several categories of exposure or disease, continuous exposure or disease, adjusting for covariates, matched data, and mathematical modeling approaches, are beyond the scope of the activities provided below.*

What is Misclassification Bias?

The two-way table below shows the correct classification of 16 subjects according to their true exposure and disease status. Let's see what might happen to this table and its corresponding odds ratio if some of these subjects were misclassified.

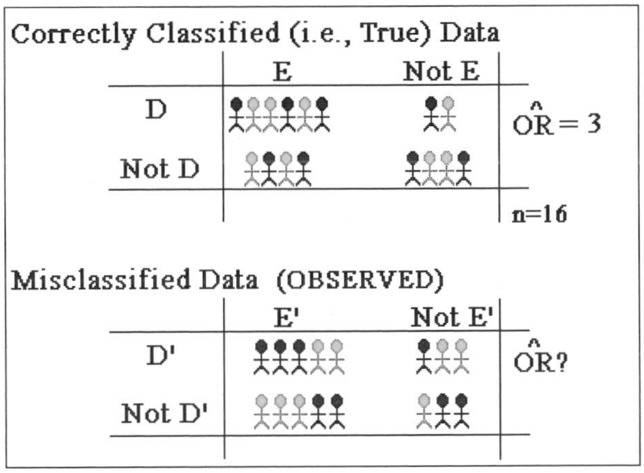

Correctly Classified (i.e., True) Data

	E	Not E	
D			$\hat{OR} = 3$
Not D			
			n=16

Misclassified Data (OBSERVED)

	E'	Not E'	
D'			$\hat{OR}$?
Not D'			

Suppose that 3 of the 6 exposed cases, shown here in the lighter shade, were actually misdiagnosed as exposed non-cases. Suppose further that one of the two unexposed cases was also misclassified as an unexposed non-case. To complete the misclassification picture, we assume that two of the four truly exposed non-cases and two of the four truly unexposed non-cases were misclassified as cases.

The misclassified data are the data that would actually be analyzed because these data are what is observed in the study. So, what is the odds ratio for these data and how does it compare to the correct odds ratio? The odds ratio for the misclassified data is 1; the correct odds ratio is 3. Clearly, there is a bias due to misclassification. The misclassified data suggests no effect of exposure on disease, but the true effect of exposure is quite strong.

Summary

- ❖ If subjects are misclassified by exposure or disease status, the effect measure, e.g., the OR, may become biased
- ❖ Bias from misclassification can occur if the effect measure for the correctly classified data is meaningfully different from the estimated effect actually observed in the misclassified data.
- ❖ Subjects are misclassified if their location in one of the four cells of the correctly classified data changes to a different cell location in the (misclassified) data that is actually observed.

General Formulation of Misclassification Bias

The general framework used in Lesson 7 (Validity) to describe what is meant by bias incorporates misclassification bias as well as selection bias. For misclassification bias, the target population is represented by a 2x2 table (Table 1 below) that assumes no misclassification of any kind; whereas the study population is represented by a 2x2 table (Table 2 below) that presents a rearrangement (rather than a subset, as in selection bias) of the target population that results from misclassification.

Table 1. Target Population (No Misclassification)

	E	Not E
D	**A** $(=A^o_{11}+B^o_{11}+C^o_{11}+D^o_{11})$	**B** $(=A^o_{12}+B^o_{12}+C^o_{12}+D^o_{12})$
Not D	**C** $(A^o_{21}+B^o_{21}+C^o_{21}+D^o_{21})$	**D** $(A^o_{22}+B^o_{22}+C^o_{22}+D^o_{22})$

Table 2. Study Population (Misclassified)

	E′	Not E′
D′	$\mathbf{A^o}$ $(=A^o_{11}+A^o_{12}+A^o_{21}+A^o_{22})$	$\mathbf{B^o}$ $(=B^o_{11}+B^o_{12}+B^o_{21}+B^o_{22})$
Not D′	$\mathbf{C^o}$ $(=C^o_{11}+C^o_{12}+C^o_{21}+C^o_{22})$	$\mathbf{D^o}$ $(=D^o_{11}+D^o_{12}+D^o_{21}+D^o_{22})$

In Table 1, the **A** persons who are truly diseased and exposed (**DE**) are classified into each of the four cells of Table 2 as follows:

A^o_{11} are classified as diseased and exposed (**D'E'**);

B^o_{11} are classified as diseased and unexposed (**D', not E'**);

C^o_{11} are classified as nondiseased and exposed (**not D', E'**); and

D^o_{11} are classified as nondiseased and unexposed (**not D', not E'**).

In other words, $A = A^o_{11} + B^o_{11} + C^o_{11} + D^o_{11}$

A similar rearrangement is shown for the **B** persons who are truly diseased and unexposed (**D, not E**), the **C** persons who are truly nondiseased and exposed (**not D, E**), and the **D** persons who are truly nondiseased and unexposed (**not D, not E**). Consequently, the $\mathbf{A^o}$ persons in Table 2 who are classified as diseased and exposed (D'E') are derived from the four cells of the target population (Table 1), as expressed by:

$$A^o = A^o_{11} + A^o_{12} + A^o_{21} + A^o_{22}$$

Similar statements apply to $\mathbf{B^o}$, $\mathbf{C^o}$, and $\mathbf{D^o}$, as shown in Table 2 above.

Misclassifying Disease Status

What are the reasons why a subject, like an exposed case, might be misclassified on exposure or disease status? In particular, why might subjects be misclassified from diseased to non-diseased or from non-diseased to diseased status? First, a subject may be incorrectly diagnosed. This can occur because of limited knowledge about the disease, because the diagnostic process is complex, because of inadequate access to state-of-the-art diagnostic technology, or because of a laboratory error in the measurement of biologic markers for the disease. In addition, the presence of disease may be not be detected if the disease is sub-clinical at the time of physical exam. Misdiagnosis can occur because of a detection bias if a physician gives a more thorough exam to patients who are exposed or have symptoms related to exposure.

Why misclassification of disease status ?

* Incorrect Diagnosis
- Limited knowledge
- Diagnostic process complex
- Inadequate access to technology
- Laboratory error
- Disease subclinical
- Detection Bias (e.g., more thorough exam if exposed)

* Subject Self-report
- Incorrect recall
- Reluctant to be truthful

* Records Incorrectly Coded in Data-Base

Another source of error occurs when disease status is obtained solely by self-report of subjects rather than by physician examination. In particular, a subject may incorrectly recall illness status, such as respiratory or other infectious illness, that may have occurred at an earlier time period. A subject may be reluctant to be truthful about an illness he or she considers socially or personally unacceptable. Finally, patient records may be inaccurate or coded incorrectly in a database.

The table below summarizes misclassification of disease status. The columns of the table show true disease status. The rows of the table show classified disease status. We call this table a **misclassification table**.

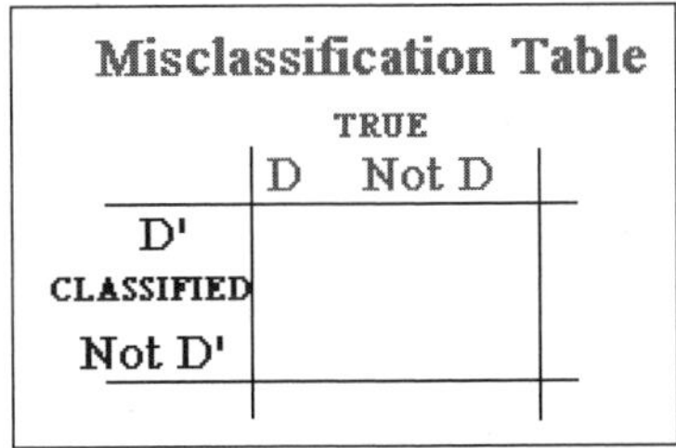

Misclassification Table

CLASSIFIED	TRUE: D	TRUE: Not D
D'		
Not D'		

Suppose the following numbers appear in the table:

Misclassification Table

CLASSIFIED	TRUE: D	TRUE: Not D	
D'	6	3	9
Not D'	2	7	9
	8	10	

Study Questions (Q9.1)

The following questions refer to the table above.

1. How many truly diseased persons are misclassified?
2. How many truly non-diseased persons are correctly classified?
3. The percentage of truly diseased persons correctly classified?
4. The percentage of truly non-diseased persons correctly classified?

Summary

- Misclassification of disease status may occur from any of the following sources:
  - Incorrect diagnosis
  - Subject self-report
  - Coding errors
- A misclassification table provides a convenient summary of how disease status can be misclassified from true disease status to observed disease status.

Misclassifying Exposure Status

How can subjects be misclassified from exposed to unexposed or from unexposed to exposed? Misclassification of exposure status can occur because of imprecise measurement of exposure. This can result from a poorly constructed questionnaire or survey process that doesn't ask the right questions, or from a faulty measuring device or observation technique.

Why misclassification of exposure status?
* Imprecise measurement
* Subject's Self-report
* Interviewer bias
* Incorrect Coding of Exposure Data

Study Questions (Q9.2)

A primary criticism of studies evaluating whether living near power lines increases one's risk for cancer is the quality of measurements of personal exposure to electromagnetic fields (EMFs). Which of the following "reasons" for imprecise measurement of personal EMF exposure do you think are **True** or **False**?

1. Measurements are usually made at only one time point and/or in one location of a residence.
2. Instruments for measuring EMF exposure are not available.
3. Methods for measuring distances and configuration of transmission lines near residences are poorly developed.
4. Better methods for monitoring measurements over time are needed.
5. Measuring EMF exposure from intermittent use of appliances or tools is difficult to measure.
6. Mobility patterns of individual related to EMF exposure are difficult to measure.

Exposure error may occur when exposure is determined solely from self-report by subjects, particularly when recalling prior exposure status. This is typically a problem in case-control studies, since cases may be more motivated to recall past exposures than controls. Recall error can also occur in cohort studies. For example, in the Sydney Beach Users study described in Lesson 2, subjects were asked to report their swimming status seven days after swimming may have occurred.

Subject self-report may also be incorrect because of reluctance of subjects to be truthful in reporting exposures relating to behaviors considered socially unacceptable. This problem often occurs in studies that measure food intake, sexual behavior, and illegal drug-use.

A third source of error in classifying exposure is interviewer bias. In particular, an interviewer may probe more thoroughly about exposure for cases than for controls.

Study Questions (Q9.2) continued

Consider a case-control study of the effect of oral contraceptive use on the development of venous (i.e., in the vein) thrombosis (i.e., clotting). (Note: there are no questions numbered 7 to 9.)

10. Why might there be misclassification of exposure, i.e., oral contraceptive use, in such a study?
11. What you expect to be the direction of such misclassification bias?
12. How might you avoid such bias?

Finally, exposure data can be coded incorrectly in a database. The table below summarizes exposure status misclassifications. The columns of the table show true exposure status. The rows of the table show classified exposure status.

Misclassification Table

CLASSIFIED	TRUE: E	TRUE: Not E
E'		
Not E'		

Suppose the following numbers appear in this table:

Misclassification Table

CLASSIFIED	TRUE: E	TRUE: Not E	
E'	95	10	105
Not E'	5	70	75
	100	80	

Study Questions (Q9.2) continued

The questions are based on the table above.

13. How many truly exposed persons are misclassified?
14. How many truly unexposed persons are correctly classified?
15. The percentage of truly exposed persons correctly classified?
16. The percentage of truly unexposed persons correctly classified?

Summary

- ❖ Misclassification of exposure status may occur from any of the following sources:
  - o Imprecise measurement
  - o Subject self-report
  - o Interviewer bias
  - o Incorrect coding of exposure data
- ❖ A misclassification table provides a convenient summary of how exposure can be misclassified from true exposure to observed exposure.

Misclassifying Both Exposure and Disease Status – An Example

Misclassification can sometimes occur for both exposure and disease in the same study. For example, the table below considers hypothetical cohort data from subjects surveyed on the beaches of Sydney, Australia during the summer months of a recent year. The study objective was to determine if those who swam at the beach were more likely to become ill than those who did not swim.

Correctly Classified Data	Swam **(E)**	Not Swam **(Not E)**	
Ill **(D)**	367	233	
not Ill **(Not D)**	300	1100	$\hat{RR} = 3.14$
Total	667	1333	

Subjects were asked to recall one week later whether they had swum for at least a half an hour on the day they were interviewed on the beach and whether they had developed a cold, cough or flu during the subsequent week. Since both exposure and illness information were obtained by subjects' recall, it is reasonable to expect some subjects may incorrectly report either swimming or illness status or both.

Suppose of the 367 subjects who got ill and swam, only 264 reported that they got ill and swam, and 30 subjects reported that they got ill but didn't swim, 66 subjects reported that they did not get ill but swam, and 7 subjects reported that they did not get ill and did not swim.

Suppose, further, that of the 233 subjects who truly got ill but did not swim, 130 reported this correctly, but 56 reported that they got ill and swam, 14 reported that they did not get ill but swam, and 33 reported that they did not swim and did not get ill.

Continuing in this way, the table can be further revised to describe how the 300 subjects who truly did not get ill and swam were misclassified. The table can also be revised to describe the misclassification of the 1100 subjects who truly did not get ill and did not swim.

Observed Data

	Swam'		Not Swam'	
Ill'	**264**	**56**	**30**	**130**
	27	**33**	**3**	**77**
not Ill'	**66**	**14**	**7**	**33**
	243	**297**	**27**	**693**

We can now separately sum up the 4 frequencies within each of the 4 cells in the table of observed data to obtain a summarized table of the observed data as shown here:

Observed Data

	Swam'	Not Swam'
Ill'	**380**	**240**
not Ill'	**620**	**760**
Total	**1000**	**1000**

Study Questions (Q9.3)

1. What is the estimated risk ratio for the observed data?
2. Why is there misclassification bias? (Hint: RR=3.14 for true data)
3. What is the direction of the bias?

4. If there is misclassification of both exposure and disease, will the bias always be towards the null?

Summary

- Misclassification can sometimes occur for both exposure and disease in the same study.
- An example of such misclassification is likely if both the exposure variable and the disease variable are determined by subject recall.
- When there is misclassification of both exposure and disease, the observed data results from how the cell frequencies in each of the four cells of the 2x2 table for the true data get split up into the four cells of the 2x2 table for the observed data.

9-2 Information Bias (continued)

Misclassification Probabilities – Sensitivity and Specificity

The misclassification table that follows describes how a disease **D** may be misdiagnosed. Twelve subjects who were truly diseased were misclassified as non-diseased and 14 subjects who were truly not diseased were misclassified as diseased. 48 subjects who were truly diseased and 126 subjects who were truly non-diseased were correctly classified.

Misclassification of Disease Status

	TRUE D	TRUE Not D	
CLASSIFIED D'	48	14	62
CLASSIFIED Not D'	12	126	138
	60	140	

In a perfect world, we would hope that no one was misclassified, that is, we would want our table to look like the table below. Then the proportion correctly classified as diseased would be 1 and the proportion correctly classified as non-diseased is also 1.

Perfect World	TRUE D	TRUE Not D		Proportion
CLASSIFIED D'	60	0	60	$\frac{60}{60} = 1$
CLASSIFIED Not D'	0	140	140	$\frac{140}{140} = 1$
	60	140		

In the real world, however, these proportions are not equal to one. In our example, the proportion of truly diseased correctly classified as diseased is .80 (48/60). The proportion of truly non-diseased correctly classified as non-diseased is .90 (126/140).

$\frac{48}{60} = .80$ → **Sensitivity = Prob(D'| D)** (| = given)

$\frac{126}{140} = .90$ → **Specificity = Prob(not D'|not D)**

The first of these proportions is called the **sensitivity**. Generally, the sensitivity for misclassification of disease status is the probability that a subject is classified as diseased given that he or she is truly diseased.

The second of these proportions is called the **specificity**. Generally the specificity for misclassification of disease is the probability that a subject is classified as not diseased given that he or she is truly not diseased.

The ideal value for both sensitivity and specificity is 1.0 or 100%. We can also make use the misclassification table for **exposure** status to define sensitivity and specificity parameters.

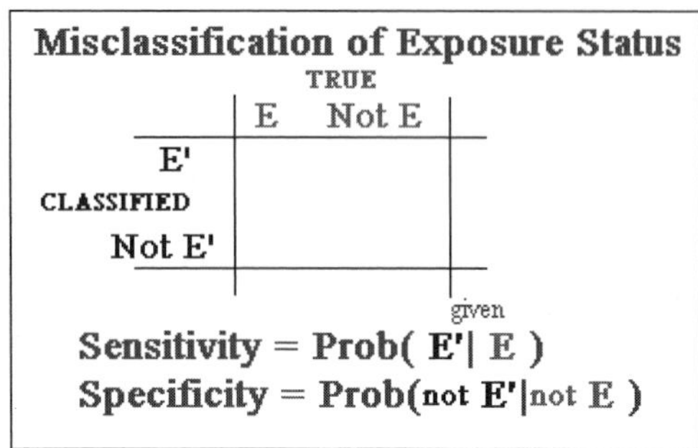

Study Questions (Q9.4)

Consider the numbers in the following misclassification table for exposure.

Misclassification of Exposure Status

		TRUE E	TRUE Not E	
CLASSIFIED	E'	720	190	910
	Not E'	180	910	1090
		900	1100	

Sensitivity = Prob(E'| E)
Specificity = Prob(not E'|not E)

1. What is the sensitivity for misclassifying exposure?
2. What is the specificity for misclassifying exposure?
3. Do your answers to the previous questions suggest that there should be some concern about misclassification bias?

Summary

- The underlying parameters that must be considered when assessing information bias are called **sensitivity** and **specificity**.
- **Sensitivity** gives the probability that a subject who is truly diseased (or exposed) will be classified as diseased (or exposed) in one's study.
- **Specificity** gives the probability that a subject who is truly non-diseased (or unexposed) is classified as non-diseased (or unexposed) in one's study.
- The ideal value for both sensitivity and specificity is 1, or 100%, which means there is no misclassification.

Sensitivity and Specificity Parameters Are Conditional Probabilities

If we allow for possible misclassification of both disease and exposure in a 2x2 table, then there are several sensitivity and specificity parameters that can be defined in terms of **conditional probabilities**. These include:

Sensitivity (Se)

Disease misclassification: **Se_D=Pr(D′|D)**
Exposure misclassification: **Se_E=Pr(E′|E)**

Specificity (Sp)

Disease misclassification: **Sp_D=Pr(not D'|not D)**
Exposure misclassification: **Sp_E=Pr(not E′|not E)**

where **D** and **E** denote truly diseased and exposed, respectively, and **D'** and **E'** denote (possibly mis-) classified as diseased and exposed, respectively.

Moreover, if we allow for the possibility that misclassification of disease might differ depending on exposure status and that misclassification of exposure status might differ depending on disease status, we can further categorize the sensitivity and specificity parameters as follows:

Sensitivity

1. Disease misclassification given exposed: **$Se_{D|E}$=Pr(D′|DE)**
2. Disease misclassification given unexposed: **$Se_{D|not\ E}$=Pr(D′|D, not E)**
3. Exposure misclassification given diseased: **$Se_{E|D}$=Pr(E′|DE)**
4. Exposure misclassification given nondiseased: **$Se_{E|not\ D}$=Pr(E′not D, E)**

Specificity

5. Disease misclassification given exposed: **$Sp_{D|E}$=Pr(not D′|not D, E)**
6. Disease misclassification given unexposed: **$Sp_{D|not\ E}$=Pr(not D′|not D, not E)**
7. Exposure misclassification given diseased: **$Sp_{E|D}$=Pr(not D′|D, not E)**
8. Exposure misclassification given nondiseased: **$Sp_{E|not\ D}$=Pr(not E′|not D, not E)**

All eight parameters above are **conditional probabilities** and they all are potentially different if both exposure and disease misclassifications are present. The tables shown below illustrate where these sensitivity and specificity probabilities belong within misclassification tables that allow for misclassification of disease status (given exposure status) and exposure status (given disease status):

Probabilities for Misclassification of Disease (Given Exposure Status)

Classified	True E: D	True E: Not D
D′	$Se_{D\|E}$	1-$Sp_{D\|E}$
Not D′	1-$Se_{D\|E}$	$Sp_{D\|E}$

Classified	True Not E: D	True Not E: Not D
D′	$Se_{D\|not\ E}$	1-$Sp_{D\|not\ E}$
Not D′	1-$Se_{D\|not\ E}$	$Sp_{D\|not\ E}$

Continued on next page

Sensitivity and Specificity Parameters Are Conditional Probabilities (continued)

Probabilities for Misclassification of Exposure (Given Disease Status)

True D	E	Not E
Classified E′	$Se_{E\|D}$	$1\text{-}Sp_{E\|D}$
Not E′	$1\text{-}Se_{E\|D}$	$Sp_{E\|D}$

True Not D	E	Not E
Classified E′	$Se_{E\|not D}$	$1\text{-}Sp_{E\|not D}$
Not E′	$1\text{-}Se_{E\|not D}$	$Sp_{D\|not D}$

Nondifferential Misclassification

The table that follows describes the true exposure and disease status for 2000 subjects in a hypothetical case-control study of the relationship between diet and coronary heart disease (CHD):

True Exposure and Disease Status

Fruit/Vegetable Consumption

	Low	High	Total
CHD	600	400	1000
not CHD	300	700	1000

$$\hat{OR} = \frac{600 \times 700}{400 \times 300} = 3.5$$

The exposure variable is the amount of fruits and vegetables eaten in an average week, categorized as low or high, as recalled by the study subjects. The disease variable is the presence or absence of CHD. Suppose there is no misclassification of disease status, but that most subjects over-report their intake of fruits and vegetables because they think that diets with high amounts of fruits and vegetables are more acceptable to the investigator. In other words, there is misclassification of exposure status.

The two tables that follow describe how exposure is misclassified separately for both the CHD cases and the non-cases.

Misclassifying Exposure

CHD cases	True Low	True High	Total
Classified Low'	480	20	500
Classified High'	120	380	500
Total	600	400	1000

CHD non-cases	True Low	True High	Total
Classified Low'	240	35	275
Classified High'	60	665	725
Total	300	700	1000

Study Questions (Q9.5)

1. What are the sensitivity and specificity for the CHD cases?
2. What are the sensitivity and specificity for the non-cases?
3. What do these two misclassification tables have in common?

This example illustrates **non-differential misclassification of exposure**. This occurs whenever the sensitivities and specificities do **not** vary with disease status. We have assumed that CHD status is not misclassified in this example. The sensitivities and specificities for misclassifying disease are all equal to 1 regardless of exposure group. Thus, in this example there is no misclassification of disease.

Study Questions (Q9.5) continued

4. Use the column total in both misclassification tables (i.e., 600, 400, 300, and 700) to determine the odds ratio for the correctly (i.e., true) classified data.

The row totals from each of the misclassification tables for exposure allow us to determine the study data that would actually be observed as a result of misclassification.

Observed (Classified) Data

	Low	High	Total
CHD	500	500	1000
not CHD	275	725	1000

$$\hat{OR} = \frac{500 \times 725}{500 \times 275} = 2.6$$

Study Questions (Q9.5) continued

5. Why is there bias due to misclassifying exposure (Note: the correct odds ratio is 3.5)?
6. What is the direction of the bias?

This example illustrates a general rule about non-differential misclassification. Whenever there is non-differential misclassification of both exposure and disease, the bias is always towards the null, provided that there are no other variables being controlled that might also be misclassified.

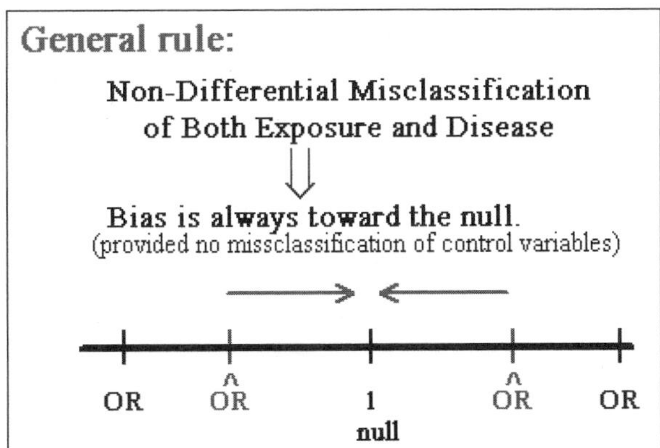

Summary

- **Nondifferential misclassification of disease**: the sensitivities and specificities for misclassifying disease do not differ by exposure.
- **Nondifferential misclassification of exposure**: the sensitivities and specificities for misclassifying exposure do not differ by disease.
- **Nondifferential misclassification** of **both** disease and exposure leads to a **bias towards the null**.

Mathematical Definition of Nondifferential Misclassification

In order to give a general definition of nondifferential misclassification, we need to consider the sensitivity and specificity parameters in the misclassification tables that follow:

Probabilities for Misclassification of Disease (Given Exposure Status)

Classified	True E: D	True E: Not D
D′	$Se_{D \mid E}$	$1\text{-}Sp_{D \mid E}$
Not D′	$1\text{-}Se_{D \mid E}$	$Sp_{D \mid E}$

Classified	True Not E: D	True Not E: Not D
D′	$Se_{D \mid \text{not } E}$	$1\text{-}Sp_{D \mid \text{not } E}$
Not D′	$1\text{-}Se_{D \mid \text{not } E}$	$Sp_{D \mid \text{not } E}$

Probabilities for Misclassification of Exposure (Given Disease Status)

Classified	True D: E	True D: Not E
E′	$Se_{E \mid D}$	$1\text{-}Sp_{E \mid D}$
Not E′	$1\text{-}Se_{E \mid D}$	$Sp_{E \mid D}$

Classified	True Not D: E	True Not D: Not E
E′	$Se_{E \mid \text{not } D}$	$1\text{-}Sp_{E \mid \text{not } D}$
Not E′	$1\text{-}Se_{E \mid \text{not } D}$	$Sp_{D \mid \text{not } D}$

In general, there is **non-differential misclassification** of **both** exposure and disease if the following four equations are satisfied:

Disease Misclassification:

$\mathbf{Se_{D|E}=Se_{D|not\ E}\ (=Se_D)}$ and $\mathbf{Sp_{D|E}=Sp_{D|not\ E}\ (=Sp_D)}$

Exposure Misclassification:

$\mathbf{Se_{E|D}=Se_{E|not\ D}\ (=Se_E)}$ and $\mathbf{Sp_{E|D}=Sp_{E|not\ D}\ (=Sp_E)}$

In words, this definition means that when classifying disease status, the sensitivities are the same among both exposed (**E**) and unexposed (**not E**) groups, as are the specificities. Similarly, when classifying exposure status, the sensitivities are the same among both diseased (**D**) and nondiseased (**not D**) groups, as are the specificities.

Thus, when there is non-differential misclassification of both exposure and disease, the misclassification tables presented above will be simplified as follows:

Probabilities for Non-differential Misclassification of Disease

Classified	True E: D	True E: Not D
D′	Se_D	$1\text{-}Sp_D$
Not D′	$1\text{-}Se_D$	Sp_D

Classified	True Not E: D	True Not E: Not D
D′	Se_D	$1\text{-}Sp_D$
Not D′	$1\text{-}Se_D$	Sp_D

Continued on next page

Mathematical Definition of Nondifferential Misclassification (continued)

Probabilities for Non-differential Misclassification of Exposure

Classified	True D: E	True D: Not E
E′	Se_E	$1\text{-}Sp_E$
Not E′	$1\text{-}Se_E$	Sp_E

Classified	True Not D: E	True Not D: Not E
E′	Se_E	$1\text{-}Sp_E$
Not E′	$1\text{-}Se_E$	Sp_D

Note that if there is no misclassification of disease status, then:

$\mathbf{Se_{D|E}=Se_{D|not\ E}\ (=Se_D) = 1}$ and $\mathbf{Sp_{D|E}=Sp_{D|not\ E}\ (=Sp_D) = 1}$

which means that there is nondifferential misclassification of disease, though not necessarily nondifferential misclassification of exposure.

Also, if there is no misclassification of exposure status, then

$\mathbf{Se_{E|D}=Se_{E|not\ D}\ (=Se_E) = 1}$ and $\mathbf{Sp_{E|D}=Sp_{E|not\ D}\ (=Sp_E) = 1}$

which means that there is non-differential misclassification of exposure, though not necessarily non-differential misclassification of disease.

What Happens if a Variable <u>other</u> than Exposure or Disease Gets Misclassified?

Greenland (1980) showed that if there is non-differential misclassification of exposure and disease, but also misclassification of a covariate, then there is no guarantee that there will be a bias towards the null. However, if misclassification of exposure and disease is non-differential and a covariate that is not misclassified is controlled in the analysis (say, by stratification), then both stratum-specific and summary measures that adjust for the covariate will be biased towards the null.

Other issues about misclassification of covariates were also addressed as follows:

- Misclassification of exposure can spuriously introduce effect modification (described in Lesson 10) by a covariate.
- Misclassification of a confounder (also described in Lesson 10) can reintroduce confounding in a summary estimate that controls for confounding using misclassified data.

Differential Misclassification

The table that follows describes the true exposure and disease status for the same 2000 subjects described in the previous activity for a hypothetical case-control study of the relationship between diet and coronary heart disease:

True Exposure and Disease Status

	Fruit/Vegetable Consumption: Low	High	Total
CHD	600	400	1000
not CHD	300	700	1000

$$\hat{OR} = \frac{600 \times 700}{400 \times 300} = 3.50$$

Suppose, as before, there is no misclassification of disease status, but that subjects over-report their intake of fruits and vegetables because they think that diets with high amounts of fruits and vegetables are more acceptable to the investigator. Suppose also that a CHD case, who is concerned about the reasons for his or her illness, is not as likely to over-estimate his or her intake of fruits and vegetables as is a control. Here are the two tables that describe how exposure is misclassified for both the CHD cases and controls:

Disease Status: No Misclassification
Exposure Status: Misclassification

CHD cases	True Low	True High	Total
Classified Low'	580	20	600
Classified High'	20	380	400
Total	600	400	1000

CHD non-cases	True Low	True High	Total
Classified Low'	240	35	275
Classified High'	60	665	725
Total	300	700	1000

Study Questions (Q9.6)

The following questions are based on the previous two tables.

1. What are the sensitivity and specificity for the CHD cases?
2. What are the sensitivity and specificity for the non-cases?
3. Is there non-differential misclassification of exposure?

This example illustrates what is called **differential misclassification of exposure**. This occurs because the sensitivities and specificities for misclassifying exposure vary with disease status.

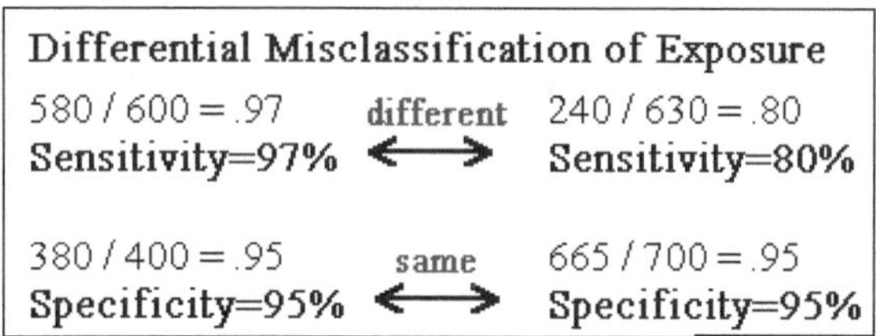

Differential Misclassification of Exposure
580 / 600 = .97
different
240 / 630 = .80
Sensitivity=97%
Sensitivity=80%
380 / 400 = .95
same
665 / 700 = .95
Specificity=95%
Specificity=95%

The row totals from each of the misclassification tables for exposure allow us to determine the study data that would actually be observed as a result of misclassification:

Observed (Classified) Data

	Low	High	Total
CHD	600	400	1000
not CHD	275	725	1000

$$\hat{OR} = \frac{600 \times 725}{400 \times 275} = 3.95$$

Study Questions (Q9.6) continued

The following questions refer to the previous table and the table with the true exposure information shown previously.

4. Is there a bias due to misclassifying exposure? (Note, the correct OR is 3.5.)
5. What is the direction of the bias, if any?

In general, when there is differential misclassification of either exposure or disease, the bias can be either **towards the null** or **away from the null** (see below for an example of bias away from the null).

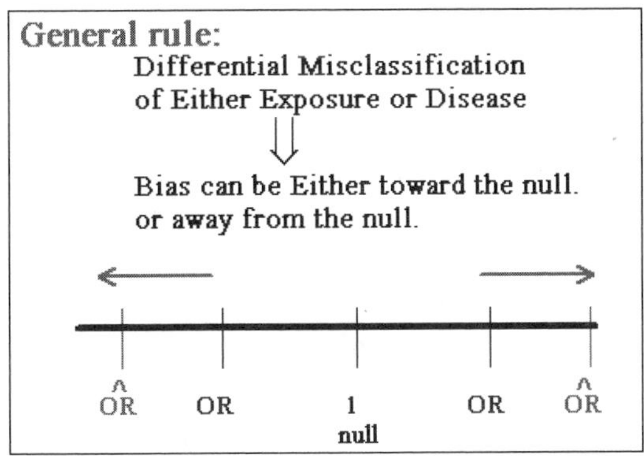
General rule:
Differential Misclassification of Either Exposure or Disease
Bias can be Either toward the null. or away from the null.
OR
OR
1
null
OR
OR

<u>Summary</u>

- ❖ With **differential** misclassification, either the sensitivities and specificities for misclassifying **D** differ by **E** or the sensitivities and specificities for misclassifying **E** differ by **D**.
- ❖ Differential misclassification of either **D** or **E** can lead to bias either towards the null or away from the null.

Mathematical Definition of Differential Misclassification

In general, **differential misclassification** is defined to be misclassification of disease and/or exposure in which there are differences in sensitivities or specificities of one variable (say, disease status) over the categories of the other variable (say, exposure status). To clarify this definition, we once again present below the misclassification tables containing the sensitivity and specificity parameters we are considering:

Probabilities for Misclassification of Disease (Given Exposure Status)

	True E	
Classified	**D**	**Not D**
D′	$Se_{D\|E}$	$1\text{-}Sp_{D\|E}$
Not D′	$1\text{-}Se_{D\|E}$	$Sp_{D\|E}$

	True Not E	
Classified	**D**	**Not D**
D′	$Se_{D\|not\ E}$	$1\text{-}Sp_{D\|not\ E}$
Not D′	$1\text{-}Se_{D\|not\ E}$	$Sp_{D\|not\ E}$

Probabilities for Misclassification of Exposure (Given Disease Status)

	True D	
Classified	**E**	**Not E**
E′	$Se_{E\|D}$	$1\text{-}Sp_{E\|D}$
Not E′	$1\text{-}Se_{E\|D}$	$Sp_{E\|D}$

	True Not D	
Classified	**E**	**Not E**
E′	$Se_{E\|not\ D}$	$1\text{-}Sp_{E\|not\ D}$
Not E′	$1\text{-}Se_{E\|not\ D}$	$Sp_{D\|not\ D}$

Using the notation given in these tables, **differential misclassification** occurs if any of the following four inequalities hold:

Disease Misclassification:

$Se_{D|E} \neq Se_{D|not\ E}$ $\qquad$ $Sp_{D|E} \neq Sp_{D|not\ E}$

Exposure Misclassification:

$Se_{E|D} \neq Se_{E|not\ D}$ $\qquad$ $Sp_{E|D} \neq Sp_{E|not\ D}$

Independent Misclassification of Both Exposure and Disease

Two events A and B are independent if the probability that both events occur is equal to the product of their individual probabilities:

Events A and B are independent if:
P(A and B) = P(A) * P(B)

When considering misclassification of exposure and disease, the two events we are concerned about are **A**, the way we classify disease status, and **B**, the way we classify exposure status. These two events are independent if the probability of classifying exposure AND disease status equals the product of the probabilities of classifying disease status and exposure status separately.

Independent misclassification of E and D
A = (classified D status) B = (classified E status)
Pr(classified D status and classified E status) =
Pr(classified D status)×Pr(classified E status)

In many epidemiologic studies, it makes sense to assume that exposure and disease status determinations are more or less independent. Typically, disease status is determined by physical exam or medical records; exposure status is determined by a separate process, often at a different time, such as an interview, occupational history, or measurement technique. For example, the way a certain cancer is diagnosed does not typically influence or depend on how occupational history, smoking history, or diet history is determined.

On the other hand, if both disease and exposure status are determined at the same time and by subject self-report or recall, then they might not be independent. Such a situation occurred in the Sydney Beach Users Study of 1989, where subjects were asked to recall previous swimming exposure and illness status one week after they were interviewed on the beach.

So, why is the assumption of independent misclassification important? We need this assumption to conveniently assess or correct for misclassification bias whenever *both* disease *and* exposure are misclassified in the same study. If only one of exposure or disease is misclassified, but not the other, the independence assumption is not needed.

Study Questions (Q9.7)

1. Is the assumption of **independent misclassification** of disease and exposure equivalent to assuming **nondifferential misclassification** of disease or exposure status?

For the following questions, assume disease and exposure are independently misclassified. Suppose also:

(Sensitivity D|E) = .8, (Sensitivity D|not E) = .75
(Sensitivity E|D) = .9, (Specificity E|D) = .95

2. Is misclassification nondifferential or differential?
3. Are there any sensitivities and specificities missing from the above information?
4. How can you express Pr(D′E′|DE) as the product of sensitivity and/or specificity parameters?
5. Calculate Pr(D′E′|DE) using the above information.
6. How can you express Pr(D′E′|not D not E) as the product of sensitivity and/or specificity parameters?
7. Calculate Pr(D′E′|D not E).

Summary

- Another important way to characterize misclassification concerns whether or not there is **independent misclassification** of both exposure and disease.
- Misclassification of exposure and disease is defined to be independent if:
  Pr(classifying D and E| true D and E) =
  Pr(classifying D| true D and E) x Pr(classifying E | true D and E)
- The assumption of independent misclassification allows for the assessment of information bias from misclassification of both exposure and disease simultaneously.

More on Independent Misclassification

This box provides a mathematical description of how the cell frequencies in the misclassified 2x2 table can be expressed in terms of the true cell frequencies under the assumption that there is independent misclassification of disease and exposure. You may not wish to read what follows if you are not interested in the mathematical underpinnings of misclassification bias.

We have defined **independent misclassification of disease and exposure** as follows:

Pr(classifying D and E|true D and E)=
Pr(classifying D|true D and E) x Pr(classifying E|true D and E)

If we assume independent misclassification as defined above, we can then express the cell frequencies $\mathbf{A^o}$, $\mathbf{B^o}$, $\mathbf{C^o}$ and $\mathbf{D^o}$ in the 2x2 table for the misclassified population in terms of the true cell frequencies **A**, **B**, **C**, and **D** in the corresponding 2x2 table for the correctly classified (i.e., true) population.

Here are the equations:

$$\mathbf{A^o}=Se_{D|E}Se_{E|D}\mathbf{A} + Se_{D|not\ E}(1-Sp_{E|D})\mathbf{B} + (1-Sp_{D|E})Se_{E|not\ D}\mathbf{C} + (1-Sp_{D|not\ E})(1-Sp_{E|not\ D})\mathbf{D}$$

$$\mathbf{B^o}=Se_{D|E}(1-Se_{E|D})\mathbf{A} + Se_{D|not\ E}Sp_{E|D}\mathbf{B} + (1-Sp_{D|E})(1-Se_{E|not\ D}\mathbf{C} + (1-Sp_{D|not\ E})Sp_{E|not\ D}\mathbf{D}$$

$$\mathbf{C^o}=(1-Se_{D|E})Se_{E|D}\mathbf{A} + (1-Se_{D|not\ E})\ (1-Sp_{E|D})\mathbf{B} + Sp_{D|E}Se_{E|not\ D}\mathbf{C} +Sp_{D|not\ E}(1-Sp_{E|not\ D})\mathbf{D}$$

$$\mathbf{D^o}=(1-Se_{D|E})\ (1-Se_{E|D})\mathbf{A} + (1-Se_{D|not\ E})1-Sp_{E|D}\mathbf{B} + Sp_{D|E}\ (1-Se_{E|not\ D})\mathbf{C} + Sp_{D|not\ E}Sp_{E|not\ D}\mathbf{D}$$

The first of these equations is derived by applying the independence definition given above to the probability terms in the following expression for $\mathbf{A}^\circ$:

$$A^o = A^o_{11} + A^o_{12} + A^o_{21} + A^o_{22}$$
=**A** Pr(D′E′|DE) + **B** Pr(D′E′|D, not E) + **C** Pr(D′E′|not D, E) + **D** Pr(D′E′|not D, not E)

The other three equations (i.e., for $\mathbf{B^o}$, $\mathbf{C^o}$ and $\mathbf{D^o}$) are derived similarly. These equations represent a system of four equations for the misclassified cell frequencies $\mathbf{A^o}$, $\mathbf{B^o}$, $\mathbf{C^o}$ and $\mathbf{D^o}$ in terms of four unknown true cell frequencies **A**, **B**, **C**, and **D**. By solving this system of equations to obtain expressions for **A**, **B**, **C**, and **D** in terms of $\mathbf{A^o}$, $\mathbf{B^o}$, $\mathbf{C^o}$ and $\mathbf{D^o}$, we can derive formulae for adjusting the observed estimate an effect measure (e.g., an odds ratio) to correct for misclassification. Such correction formulae are described in later expositions on misclassification.

Quiz (Q9.8)

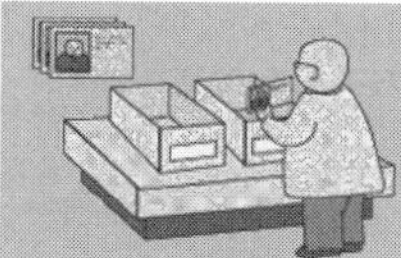

Label the following statement as **True** or **False**.

1. If there is misclassification of disease status but not of exposure status in a follow-up study, and if the sensitivity probability is the same for exposed and unexposed groups, then whatever bias exists (due to misclassification) must be towards the null. **???**
2. If there is differential misclassification of disease status but not of exposure status in a case-control study, then there may be bias in estimating the odds ratio, which is either away from or toward the null. **???**
3. Suppose there is independent misclassification of both exposure status and disease status. Then:
   P(D', not E'ID, not E) = $(Se_{DInot\ E})*(Sp_{EID})$. **???**

9-3 Information Bias (continued)

Quantitative Assessment of Misclassification Bias

Let's assume that we know there is a likely misclassification of exposure or disease that could bias our study results. How can we quantitatively correct for such bias to obtain an adjusted effect measure that is no longer biased?

To quantify bias, we need to adjust the cell frequencies for the observed data to obtain a two-way table of corrected cell frequencies from which we can compute a corrected effect measure. We can then compare the observed and possibly biased estimate with our corrected estimate to determine the extent of the possible bias and its direction.

Observed (i.e., misclassified) Data

	E'	Not E'
D'	a	b
Not D'	c	d

$$\hat{OR} = \frac{ad}{bc}$$

$$\hat{RR} = \frac{a/(a+c)}{b/(b+d)}$$

Corrected (i.e., adjusted) Data

	E	Not E
D	A	B
Not D	C	D

$$\hat{OR}_{adj} = \frac{AD}{BC}$$

$$\hat{RR}_{adj} = \frac{A/(A+C)}{B/(B+D)}$$

Study Questions (Q9.9)

Use the information in the table below about the observed and corrected effect measures to determine whether the observed bias is towards or away from the null.

Question Number	Observed Effect	Corrected Effect	Towards the null?	Away from the null?
a.	2.2	1.7		
b.	2.5	3.8		
c.	4.0	6.1		
d.	4.1	1.2		
e.	0.5	0.9		
f.	0.8	0.9		
g.	0.3	0.2		
h.	0.7	0.1		

Suppose we have determined that whatever bias that exists results from nondifferential misclassification of exposure or disease.

1. Do we need to obtain an adjusted effect measure that corrects for such bias?
2. How can we determine whether or not misclassification is nondifferential?

Suppose we have determined that there is differential misclassification of exposure or disease.

3. Do we need to obtain an adjusted effect measure that corrects for possible misclassification bias?

In the presentations that follow, we give formulas for obtaining corrected estimates. These formulas require reliable estimates of sensitivity and specificity. How can we determine the sensitivity and specificity if all we have are the observed data? One option is to take a small sample of the observed data for which true disease and exposure status is determined, so that sensitivities and specificities can be estimated (a drawback to this approach is the sample might be too small to give reliable values for sensitivity and specificity). Another option is to determine sensitivity and specificity parameters from separate data obtained in a previous study involving the same variables. A third option is simply to make an educated guess of the sensitivity and specificity parameters from your clinical or other knowledge about the study variables. A fourth option is to carry out what is often referred to as a sensitivity analysis with several educated guesses to determine a range of possible biases.

Need estimates of sensitivity and specificity.
How?

1. Estimate from a small sample (gold standard).
2. Determine from previous study involving same variables.
3. Make an educated guess from your clinical or other knowledge.
4. Carry out sensitivity analysis with several educated guesses to determine range of possible biases.

Study Questions (Q9.9) continued

Consider the following results of a sensitivity analysis for correcting for misclassification that is assumed to be nondifferential:

Observed OR	Sensitivity?	Specificity?	Corrected OR
1.5	80%	80%	3.5
1.5	80%	90%	2.5
1.5	90%	80%	2.5
1.5	90%	90%	1.8

4. Why do the observed results compared to the corresponding corrected results illustrate a bias that is towards the null?
5. Which values of sensitivity and specificity are associated with the most bias?
6. Which values of sensitivity and specificity are associated with the least bias?
7. Based on the sensitivity analysis described above, how might you decide which corrected OR is "best"?

Summary

- ❖ We can correct for misclassification bias by computing an adjusted effect measure from a two-way table whose cell frequencies are corrected from the misclassification found in observed cell frequencies.
- ❖ The correction requires accurate estimation of sensitivity and specificity parameters.
- ❖ Options for estimating sensitivity and specificity:
  - o A sub-sample of the study data
  - o A separate sample from another study
  - o A questimate based on clinical or other theory/experience
  - o A sensitivity analysis that considers several guestimates

Correcting for Nondifferential Misclassification of Disease

This table gives the observed data from a hypothetical cohort study of the relationship between gender and peptic ulcer disease:

Observed (i.e., Classified) Data

	Male'	Female'
PEU'	380	240
not PEU'	620	760
Total	1000	1000

PEU = Peptic Ulcer Disease

$$\hat{RR} = \frac{(380/1000)}{(240/1000)} = 1.58$$

Assume that gender, the exposure variable, has not been misclassified, but that diagnosing peptic ulcer involves some amount of misclassification. How can we correct for possible misclassification of disease to obtain an adjusted risk ratio estimate that gives an accurate estimate of the exposure-disease relationship?

We must first obtain reliable estimates of the sensitivities and specificities for misclassifying the disease. Suppose we carried out a more extensive physical examination of a sub-sample of 200 subjects to obtain a "gold standard" determination of peptic ulcer status. Suppose, that separate misclassification tables for males and females of these 200 subjects are obtained as shown here.

Misclassfying Disease (subsample)

Males:

	Gold Standard PEU	not PEU	Total
PEU'	32	6	38
not PEU'	8	54	62
Total	40	60	100

Females:

	Gold Standard PEU	not PEU	Total
PEU'	16	8	24
not PEU'	4	72	76
Total	20	80	100

Study Questions (Q9.10)

The following questions refer to the "Misclassifying Disease (sub sample)" table.

1. What are the sensitivity and specificity estimates for males?
2. What are the sensitivity and specificity estimates for females?
3. Is misclassification nondifferential or differential?

When, as in our example, misclassification is nondifferential and we have reliable estimates of the sensitivity and specificity parameters, we can transform misclassified cell frequencies of our observed data into 'corrected' cell frequencies from which an adjusted, corrected, effect measure can be calculated.

We can write a relatively simple formula for the corrected cell frequencies if, as in our example, only the disease variable is misclassified.

Observed (i.e., Classified) Data

	Male	Female
PEU'	a= 380	b= 240
not PEU'	c= 620	d= 760
Total	1000	1000

Corrected Data

	Male	Female
PEU	A	B
not PEU	C	D
Total	1000	1000

$$A=\left[a\ SpD\ -\ c\ \left(1-SpD\right)\right]/q$$
$$B=\left[b\ SpD\ -\ d\ \left(1-SpD\right)\right]/q$$
$$C=\left[c\ SeD\ -\ a\ \left(1-SeD\right)\right]/q$$
$$D=\left[d\ SeD\ -\ b\ \left(1-SeD\right)\right]/q$$
$$q=SeD+SpD-1$$

To compute the corrected cell frequencies, we substitute the observed cell frequencies into the formula as shown below. We also substitute our estimated sensitivity and specificity values into this formula as follows:

$$A=\left[380\ 0.90\ -\ 620\left(1-0.90\right)\right]/0.70$$
$$B=\left[240\ 0.90\ -\ 760\left(1-0.90\right)\right]/0.70$$
$$C=\left[620\ 0.80\ -\ 380\left(1-0.80\right)\right]/0.70$$
$$D=\left[760\ 0.80\ -\ 240\left(1-0.80\right)\right]/0.70$$
$$q=0.70$$
$$SeD=0.80\quad SpD=0.90$$

The resulting corrected cell frequencies are shown here.

Corrected Data

	Male	Female
PEU	400	200
not PEU	600	800
Total	1000	1000

Study Questions (Q9.10) continued

4. Using the corrected cell frequencies, compute the adjusted risk ratio that corrects for misclassification of disease.
5. What is the direction of the bias?
6. Although there was no problem in transforming the cell frequencies in this example, what should bother you about the denominator q = Sensitivity(D) + Specificity(D) – 1, in the formulae for A, B, C, and D?

Summary

- ❖ Nondifferential misclassification of disease without misclassification of exposure can be corrected when estimates of sensitivity and specificity are available.
- ❖ This procedure involves transforming the misclassified cell frequencies into "corrected" cell frequencies from which a corrected effect measure can be calculated that is no longer biased.
- ❖ The procedure requires as input estimates of sensitivities and specificities for misclassification of disease as well as the cell frequencies in the observed 2x2 table that relates exposure and disease.
- ❖ An explicit algebraic formula for the correction procedure is available although these calculations are most efficiently done using a computer program.

Correcting for Nondifferential Misclassification of Disease by Computer

A computer program is available in **DataDesk**, the software package that is accessible using ActivEpi. This is a general program that allows for differential misclassification of both exposure and disease, so that nondifferential misclassification of disease is a special case of the program in which the sensitivities and specificities for exposure given disease status are all specified to equal unity, corresponding sensitivities for disease given exposure status are specified as equal, and corresponding specificities for disease given exposure status are also specified as equal. That is, the input into the program are:

1. Observed cell frequencies **a**, **b**, **c**, and **d**
2. Sensitivities and specificities for exposure given disease:

$$Se_{E|D} = Se_{E|not\ D} \ (= Se_E) = 1$$
$$Sp_{E|D} = Sp_{E|not\ D} \ (= Sp_E) = 1$$

3. Equal sensitivities for disease given exposure:

$$Se_{D|E} = Se_{D|not\ E} \ (= Se_D)$$

4. Equal specificities for disease given exposure:

$$Sp_{D|E} = Sp_{D|not\ E} \ (= Sp_D)$$

Problems with Correcting for Nondifferential Misclassification of Disease

The formulae for correcting for nondifferential misclassification of disease when there is no misclassification of exposure, may yield inappropriate results for the corrected cell frequencies **A**, **B**, **C**, and **D** under the following circumstances:

1. **Se + Sp = 1** (i.e., **q** = Se + Sp - 1 = 0, where **q** is the denominator of the correction formula for each of the four cells and **Se** and **Sp** are the sensitivity and specificity for misclassification of **disease**)
2. **Sp < c / (a + c)**, in which case the corrected value for **A** is negative
3. **Sp < d / (b + d)**, in which case the corrected value for **B** is negative
4. **Se < a / (a + c)**, in which case the corrected value for **C** is negative.
5. **Se < b / (b + d)**, in which case the corrected value for **D** is negative.

Correcting for Nondifferential Misclassification of Disease in DataDesk

An exercise in DataDesk is provided to correct for nondifferential misclassification of disease.

Correcting for Nondifferential Misclassification of Exposure

The following table describes the observed data from a hypothetical case-control study of the relationship between reported history of peptic ulcer, the exposure variable, and stomach cancer. The controls were persons with colon cancer, obtained from the same hospital population as were the cases of stomach cancer. Assume that case-control status has been correctly diagnosed, but that subjects' reported history of peptic ulcer does involves some misclassification.

Observed (i.e., Classified) Data

	PEU'	not PEU'	Total
Cases	75	225	300
Controls	30	270	300

$$\hat{OR} = \frac{(75 \times 270)}{(225 \times 30)} = 3.0$$

PEU=peptic ulcer disease

As in the previous activity, only one of the disease and exposure variables is misclassified. This time, it's the exposure variable. Once again, we ask, "how can we correct for possible misclassification in this situation?" We can obtain a corrected answer provided we have reliable estimates of the sensitivity and specificity for misclassifying the exposure.

Suppose a separate study previously conducted on risk factors for peptic ulcer disease had evaluated the extent of misclassification of reported disease by obtaining a gold standard determination of peptic ulcer status on a sample of 200 subjects. Suppose, further, that the misclassification table is as shown here:

Exposure Misclassification Table: (**Separate Study**)

	Gold Standard PEU	Not PEU	Total
PEU'	48	7	55
Not PEU'	12	133	145
Total	60	140	200

Study Questions (Q9.11)

1. What are the sensitivity and specificity estimates for misclassifying PEU status?
2. How can you determine whether misclassification is nondifferential or differential?

It is reasonable to assume that misclassification of exposure is nondifferential in this example. We can now transform the misclassified cell frequencies of our observed data into corrected cell frequencies from which a corrected odds ratio estimate can be estimated. Here are the expressions for the corrected cell frequencies if only the exposure variable is misclassified.

$$A=\left[a\ SpE - b\left(1 - SpE\right)\right]/q$$
$$B=\left[b\ SeE - a\left(1 - SeE\right)\right]/q$$
$$C=\left[c\ SpE - d\left(1 - SpE\right)\right]/q$$
$$D=\left[d\ SeE - c\left(1 - SeE\right)\right]/q$$
$$q = SeE + SpE - 1 \quad SeE = 0.80 = \text{Sensitivity E}$$
$$SpE = 0.95 = \text{Specificity E}$$

To compute the corrected cell frequencies, we substitute the observed cell frequencies into the formula as shown below. We also substitute our estimated sensitivity and specificity values into this formula as follows:

$$A=\left[75\ 0.95 - 225\left(1 - 0.95\right)\right]/0.75 = 80$$
$$B=\left[225\ 0.80 - 75\left(1 - 0.80\right)\right]/0.75 = 220$$
$$C=\left[30\ 0.95 - 270\left(1 - 0.95\right)\right]/0.75 = 20$$
$$D=\left[270\ 0.80 - 30\left(1 - 0.80\right)\right]/0.75 = 280$$
$$q = 0.75 \quad SeE = 0.80 = \text{Sensitivity E}$$
$$SpE = 0.95 = \text{Specificity E}$$

The resulting corrected cell frequencies are shown here:

Corrected Data

	PEU	not PEU'	Total
Cases	80	220	300
Controls	20	280	300

Study Questions (Q9.11) continued

(Note: there is no question 3 in Study Questions Q9.11)

4. Using the corrected cell frequencies, compute the adjusted odds ratio that corrects for misclassification of disease.
5. What is the direction of the bias?

Summary

- Nondifferential misclassification of exposure without misclassification of disease can be corrected when estimates of sensitivity and specificity are available.
- This procedure involves transforming the misclassified cell frequencies into "corrected" cell frequencies from which a corrected effect measure can be calculated.
- The procedure requires as input estimates of sensitivities and specificities for misclassification of exposure as well as the cell frequencies in the observed 2x2 table that relates exposure to disease.
- An explicit algebraic formula for the correction procedure is available although these calculations are most efficiently done using a computer program.

Correcting for Nondifferential Misclassification of Exposure by Computer

A computer program is available in **DataDesk**, the software package that is accessible using ActivEpi. This is a general program that allows for differential misclassification of both exposure and disease, so that nondifferential misclassification of exposure is a special case of the program in which the sensitivities and specificities for disease given exposure status are all specified to equal unity, corresponding sensitivities for exposure given disease status are specified as equal, and corresponding specificities for exposure given disease status are also specified as equal. That is, the inputs into the program are:

1. Observed cell frequencies **a**, **b**, **c**, and **d**
2. Sensitivities and specificities for disease given exposure:

$$Se_{D|E} = Se_{D|\text{not } E} \; (= Se_D) = 1$$
$$Sp_{D|E} = Sp_{D|\text{not } E} \; (= Sp_D) = 1$$

3. Equal sensitivities for exposure given disease:

$$Se_{E|D} = Se_{E|\text{not } D} \; (= Se_E)$$

4. Equal specificities for exposure given disease:

$$Sp_{E|D} = Sp_{E|\text{not } D} \; (= Sp_E)$$

Problems with Correcting for Nondifferential Misclassification of Exposure

The formulae for correcting for nondifferential misclassification of exposure when there is no misclassification of disease, may yield inappropriate results for the corrected cell frequencies **A**, **B**, **C**, and **D** under the following circumstances:

1. **Se + Sp = 1** (i.e., **q** = Se + Sp - 1 = 0, where **q** is the denominator of the correction formula for each of the four cells and **Se** and **Sp** are the sensitivity and specificity for misclassification of **exposure**)
2. **Sp < c / (a + c)**, in which case the corrected value for **A** is negative
3. **Sp < d / (b + d)**, in which case the corrected value for **B** is negative
4. **Se < a / (a + c)**, in which case the corrected value for **C** is negative.
5. **Se < b / (b + d)**, in which case the corrected value for **D** is negative.

Correct for Nondifferential Misclassification of Exposure in DataDesk

An exercise in DataDesk is provided to correct for nondifferential misclassification of exposure.

Correcting for Nondifferential Misclassification of Exposure and Disease

This table (below) considers hypothetical cohort data from subjects surveyed on the beaches of a certain Caribbean island. This study closely resembles the Beach User's Study described in Lesson 2.

Observed (i.e., Recall) Data

	Pollution Level		
	High'	Low'	Total
Ill'	400	375	775
not Ill'	600	1125	1725
Total	1000	1500	2500

$$\hat{RR} = \frac{(400/1000)}{(375/1500)} = 1.6$$

The table describes 2500 subjects who were observed swimming on the beaches on the day of interview. Pollution level for each subject was based on average water quality counts taken from three pre-specified samples on the day the subject was interviewed on the beach. Subjects were asked by telephone one week later to recall whether they had developed a cough, flu or gastro-intestinal illness since their beach interview.

Because illness information was obtained by subjects' recall, it is reasonable to expect some subjects may incorrectly report illness status. Also, since pollution level per subject might not correspond to the time of day that a subject actually swam, and since the measurement technique was subject to error, it is also likely there was misclassification of pollution level.

Suppose that a sub-sample of study subjects were thoroughly probed to determine their true illness status, and it was found that the sensitivity was 80% and, the specificity was 90% for misclassifying illness. Suppose, further, that a previous study of beach water quality measurements indicated the sensitivity and specificity for misclassifying pollution level at 90% and 75%, respectively.

Illness Information Misclassified	Pollution Level Misclassified
study subsample reveals:	Previous study revealed:
Sensitivity = 80%	Sensitivity = 90%
Specificity = 90%	Specificity = 70%

Study Questions (Q9.12)

1. If appropriate information were available for this study, how would you determine whether misclassifying either disease or exposure was nondifferential or differential?
2. Based on the information provided for this example, is it possible to explicitly determine whether misclassification of either E or D is nondifferential or differential?
3. Based on the information provided, is it necessary to assume nondifferential or differential misclassification in order to correct for misclassification bias?
4. Do you think that there was independent classification of pollution level and illness status?

The answers to the study questions indicate that it is reasonable to assume that the misclassification of both exposure and disease is nondifferential and the classification probabilities are independent. Under these assumptions, the observed cell frequencies can be transformed to corrected cell frequencies using the expressions shown here:

$$A=\{SpD[(a+b)SpE-b]-(1-SpD)[(c+d)SpE-d]\}/q^*$$
$$B=\{SpD[(a+b)SeE-a]-(1-SpD)[(c+d)SeE-c]\}/q^*$$
$$C=\{SeD[(c+d)SpE-d]-(1-SeD)[(a+b)SpE-b]\}/q^*$$
$$D=\{SeD[(c+d)SeE-c]-(1-SeD)[(a+b)SeE-a]\}/q^*$$
$$q^*=(SeD+SpD-1)(SeE+SpE-1)$$

These formulas are more complicated than those when only one of exposure or disease was misclassified. To compute the corrected cell frequencies, we substitute the observed cell frequencies into the formulae as shown below. We also substitute our estimated sensitivity and specificity values into these formulae as follows:

$$A=\{0.90[(775)0.70-375]-(1-0.90)[(1725)0.70-1125]\}/0.42$$
$$B=\{0.90[(775)0.90-400]-(1-0.90)[(1725)0.90-600]\}/0.42$$
$$C=\{0.80[(1725)0.70-1125]-(1-0.80)[(775)0.70-375]\}/0.42$$
$$D=\{0.80[(1725)0.90-600]-(1-0.80)[(775)0.90-400]\}/0.42$$
$$q^*=0.42$$

The resulting corrected cell frequencies are shown here:

Corrected Table

	High	Low
Ill	**339.3**	**410.8**
not Ill	**77.4**	**1672.6**
Total	**416.7**	**2083.4**

Study Questions (Q9.12) continued

5. Using the corrected cell frequencies, compute the adjusted risk ratio that corrects for misclassification of disease and exposure.
6. What is the direction of the bias?
7. What do you notice about the extent of the bias when both exposure and disease are misclassified?
8. Even though there is no problem with q* in this example, what should bother you about the use of q* in the calculation formula?

Summary

- Nondifferential misclassification of both exposure and disease can be corrected when estimates of sensitivity and specificity are available.
- The formulae used to calculate corrected estimates assume that misclassification of exposure and diseases are independent.
- The procedure requires as input estimates of sensitivities and specificities for misclassification of exposure and disease as well as the cell frequencies in the observed 2x2 table that relates exposure to disease.
- An explicit algebraic formula for the correction procedure is available although these calculations are most efficiently done using a computer program.

Problems with Correcting for Nondifferential Misclassification of Both Disease and Exposure

The formulae for correcting for nondifferential misclassification of both disease and exposure are given as follows:

A = {SpD[(a + b) SpE – b] – (1 – SpD)[(c + d) SpE – d]} / q*
B = {SpD[(a + b) SeE – a] – (1 – SpD)[(c + d) SeE – c]} / q*
C = {SeD[(c + d) SpE – d] – (1 – SeD)[(a + b) SpE – b]} / q*
D = {SeD[(c + d) SeE – c] – (1 – SeD)[(a + b) SeE – a]} / q*

where **q* = (SeD + SpD – 1) (SeE + SpE – 1)**

These formulae may yield inappropriate results for the corrected cell frequencies **A**, **B**, **C**, and **D** under the following circumstances:

1. SeD + SpD = 1 or SeE + SpE = 1 (i.e., q* = 0)
2. (a+b+c+d)SpDSpE - (b+d)SpD - (c+d)SpE + d < 0, in which case the corrected value for **A** is negative.
3. (a+b+c+d)SpDSeE - (a+c)SpD - (c+d)SeE + c < 0, in which case the corrected value for **B** is negative.
4. (a+b+c+d)SeDSpE - (b+d)SeD - (a+b)SpE + b < 0, in which case the corrected value for **C** is negative.
5. (a+b+c+d)SeDSeE - (a+c)SeD - (a+b)SeE + a < 0, in which case the corrected value for **D** is negative.

An example in which negative values for the corrected cell frequencies is now given. Suppose the observed 2x2 table of cell frequencies is given as follows:

Observed (i.e., Classified) Data

	Swam	**Not Swam**
Ill′	a= 500	b= 150
Not ill′	c= 1500	d= 750
Total	2000	900

Suppose, further, that nondifferential misclassification of both exposure and disease is assumed, and that estimated values for the sensitivities and specificities are given as follows:

SeD = .90, SpD = .80, SeE =.85, SpE = .95.

We now use the correction formulae given earlier to obtain corrected cell frequencies for the following 2x2 table:

Corrected Table

	Swam	**Not Swam**
Ill	A	B
Not ill	C	D

We now check the five conditions (1-5) described above that determine whether any of the corrected cell frequencies will be either undefined or negative:

1. SeD + SpD = .90 + .80 = 1.70 > 1, and SeE + SpE = .85 + .95 > 1, so that q* = (SeD + SpD - 1) x (SeE + SpE - 1) = .70 x .80 = .56 > 0, therefore, there is no problem associated with the denominator, q*.
2. (a+b+c+d)SpDSpE - (b+d)SpD - (c+d)SpE + d = 2900 x .80 x .95 – 900 x .80 – 2250 x .95 + 750 = 2204 - 2857 + 750 = 96.5 > 0, in which case the corrected value for **A** should be positive.
3. (a+b+c+d)SpDSeE - (a+c)SpD - (c+d)SeE + c = 2900 x .80 x .85 – 2000 x .80 – 2250 x .85 + 1500 = 1972 - 3512.5 + 1500 = -40.5 < 0, in which case the corrected value for **B** should be negative.
4. (a+b+c+d)SeDSpE - (b+d)SeD - (a+b)SpE + b = 2900 x .90 x .95 – 900 x .90 – 650 x .95 + 150 = 2479.5 - 1427.5 + 150 = 1202 > 0, in which case the corrected value for **C** should be positive.
5. (a+b+c+d)SeDSeE - (a+c)SeD - (a+b)SeE + a = 2900 x .90 x .85 – 2000 x .90 – 650 x .85 + 500 = 2218.5 - 2352.5 + 500 = -366 > 0, in which case the corrected value for **D** should be positive.

Continued on next page

Problems with Correcting for Nondifferential Misclassification of Both Disease and Exposure (continued)

We now use the actual correction formula to demonstrate that the above checks actually lead to the positive values for A, C, and D, and a negative value for B:

A = { SpD [(a+b)SpE - b] - (I-SpD) [(c+d)SpE - d] }/q*
= {.80 [(650).95 - 150] - (1-.80) [(2250).95 - 750] }/.56
= 96.5/.56= 172.3
B = { SpD [(a+b)SeE - a] - {1-SpD) [(c+d)SeE - c] }/q*
= { .80 [(650).85 - 500] - {1-.80) [(2250).85 - 1500 }/.56
= -40.5/.56 = -72.3
C = { SeD [(c+d)SpE - d] - (I-SeD) [(a+b)SpE - b] }/q*
= { .90 [(2250).95 - 750] - (1-.90) [(650).95 - 150] }/.56
= 1202.0/.56 = 2146.4
D = { SeD [(c+d)SeE - c] - (I-SeD) [(a+b)SeE - a] }/q*
= { .90 [(2250).85 - 1500] - (1-.90) [(650).85 - 500] }/.56
= 366/.56 = 653.6

The corrected table is therefore inappropriate because the B cell is negative:

(Inappropriately) Corrected Table

	Swam	Not Swam
Ill	172.3	-72.3
Not ill	2146.4	653.6
Total	2318.7	581.3

Correct for Nondifferential Misclassification of Exposure and Disease in DataDesk

An exercise in DataDesk is provided to correct for nondifferential misclassification of exposure and disease.

Quiz (Q9.13)

Use the "observed" data in the table below to answer the following questions. Assume there is nondifferential misclassification of disease, SeD=0.8, SpD=0.7 and that there is no misclassification of exposure.

1. Calculate an Odds ratio that is corrected for misclassification ???.
2. The bias due to misclassification is ??? the null.

Choices
1.4 2.3 5.1 7.4 away from towards

	E	Not E	
D	70	30	100
Not D	50	50	100
Total	120	80	200

Quiz continued on next page

Label the following statements as **True** or **False**.

3. It is possible to obtain indeterminate results when trying to correct for nondifferential misclassification, even if good estimates of the sensitivity and specificity probabilities are available. . **???**
4. If misclassification is nondifferential and independent, with SpD=SeD=SeE=SpE=0.5, then each person in the target population has an equal chance of being misclassified into anyone of the four cells of the 2x2 table. . . . . **???**
5. For the situation described above, it is possible to obtain an adjusted estimate of the odds ratio that will correct for misclassification. . . . . **???**

9-4 Information Bias (continued)

Correcting for Differential Misclassification of Exposure and/or Disease

We previously illustrated differential misclassification of exposure using hypothetical case-control data on the relationship between diet and coronary heart disease. We assumed that we knew the 2 x 2 table describing **true** exposure and disease status and then determined the 2 x 2 table for the observed data based on differential misclassification of exposure but no misclassification of disease.

Observed (Classified) Data

	Low'	High'	Total
CHD	600	400	1000
not CHD	275	725	1000

$$\hat{OR} = \frac{600 \times 725}{400 \times 275} = 2.6$$

Here, we more realistically start the other way around by assuming that we have the observed data, as shown below. We wish to correct for differential misclassification by transforming the observed cell frequencies to corrected cell frequencies from which we can obtain a corrected odds ratio. Suppose, as before, that a CHD case, who is concerned about the reasons for his or her illness, is not as likely to over-estimate his or her intake of fruits and vegetables as is a control. Assume that from either a sub-sample or a separate study, the sensitivities and specificities for misclassifying exposure are known.

Observed (i.e., Classified) Data

	Low'	High'	Total
CHD	a= 600	b= 400	1000
not CHD	c= 275	d= 725	1000

$$\hat{OR} = \frac{(600 \times 725)}{(400 \times 275)} = 3.95$$

Corrected Cell Frequencies

	Low	High	Total
CHD	A	B	1000
not CHD	C	D	1000

$$\hat{OR}_{adj} = \frac{(A \times D)}{(B \times C)}$$

Suppose:

CHD Case is less likely to over-estimate fruit/vegetable intake as Control.

Cases: Sensitivity = 96.7%, Specificity = 95%

Controls: Sensitivity = 80%, Specificity = 95%

Study Questions (Q9.14)

The following questions are based on the sensitivities and specificities described above.

1. Is there nondifferential or differential misclassification of exposure?
2. What does the sensitivity information say about the likelihood that a case will over-estimate his or her intake of fruits and vegetables when compared to a control?

When misclassification is differential, general formulas can be derived for calculating corrected cell frequencies, capital **A**, **B**, **C**, and **D**, in terms of the observed cell frequencies, little **a**, **b**, **c**, and **d**. (Note: in the box at the end of this activity are the general formula correcting for differential misclassification of both exposure and disease.) The resulting corrected cell frequencies are shown here.

Observed (i.e., Classified) Data

	Low'	High'	Total
CHD	a= 600	b= 400	1000
not CHD	c= 275	d= 725	1000

$$\hat{OR} = \frac{(600 \times 725)}{(400 \times 275)} = 3.95$$

Corrected Cell Frequencies

	Low	High	Total
CHD	599.7	400.2	1000
not CHD	300.0	700	1000

$$\hat{OR}_{adj} = \frac{(A \times D)}{(B \times C)}$$

Study Questions (Q9.14) continued

The following questions are based on the tables above.

3. Using the corrected cell frequencies, compute the odds ratio that corrects for misclassification of disease.
4. What is the direction of the bias?

Summary

- As in the nondifferential case, differential misclassification can be corrected when estimates of sensitivity and specificity are available.
- The procedure involves transforming the misclassified cell frequencies into "corrected" cell frequencies and requires the assumption that misclassification of exposure and disease are independent.
- The procedure requires as input estimates of sensitivities and specificities for misclassification of exposure and disease as well as the cell frequencies in the observed 2x2 table that relates exposure to disease.
- An explicit algebraic formula for the correction procedure is not available although calculations can be carried out using a computer program.

Correcting for Differential Misclassification and Associated Problems

The general formulae for correcting for differential misclassification are given as follows:

A={Sp(D|E)[(a + b)Sp(E|D) - b] – [1 – Sp(D|E)][(c + d)Sp(E|D) – d]} / q_A

B={Sp(D|not E)[(a + b)Se(E|D) – a] – [1 – Sp(D|not E)][(c + d)Se(E|D) – c]} / q_B

C={Se(D|E)[(c + d)Sp(E|not D) – d] – [1 – Se(D|E)][(a + b)Sp(E|not D) – b]} / q_C

D={Se(D|not E)[(c + d)Se(E|not D) – c] – [1 – Se(D|not E)][(a + b)Se(E|not D) – a]}/q_D

Continued on next page

Correcting for Differential Misclassification and Associated Problems (continued)

where

q_A= [Se(D|E) + Sp(D|E) - 1] x [Se(E|D) + Sp(E|D) - 1]

q_B= [Se(D|not E) + Sp(D|not E) - 1] x [Se(E|D) + Sp(E|D) - 1]

q_C= [Se(D|E) + Sp(D|E) - 1] x [Se(E|not D) + Sp(E|not D) - 1]

q_D= [Se(D|not E) + Sp(D|not E) - 1] x [Se(E|not D) + Sp(E|not D) - 1]

Se(D|E) = Sensitivity **D** given **E**

Se(D|not E) = Sensitivity **D** given not **E**

Sp(D|E) = Specificity **D** given **E**

Sp(D|not E) = Specificity **D** given not **E**

Se(E|D) = Sensitivity **E** given **D**

Se(E|not D) = Sensitivity **E** given not **D**

Sp(E|D) = Specificity **E** given **D**

Sp(E|not D) = Specificity **E** given not **D**

These formulae may yield inappropriate results for the corrected cell frequencies **A**, **B**, **C**, and **D** under the following circumstances:

1. Either $q_A = 0$, $q_B = 0$, $q_C = 0$, or $q_D = 0$
2. The corrected value for **A** is negative
3. The corrected value for **B** is negative
4. The corrected value for **C** is negative
5. The corrected value for **D** is negative.

Correct for Differential Misclassification in DataDesk

An exercise in DataDesk is provided to correct for differential misclassification.

Quiz (Q9.15)

Use the data in the table below to answer the following questions. Assume there is differential misclassification of Exposure, Se(E|D)=0.9, Se(E|not D) = 0.6, Sp(E|D) = 0.7, Sp(E|not D) = 0.9, and there is no misclassification of disease.

1. Calculate an Odds ratio that is corrected for misclassification. ???
2. The bias due to misclassification is ??? the null.

Choices
3.5 **4.0** **5.1** **5.8** **away from** **towards**

	E	Not E	Total
D	660	340	1000
Not D	250	750	1000
Total	910	1090	2000

Diagnostic Testing and Its Relationship to Misclassification

An Example of Clinical Diagnosis

A 43 year-old man with an acutely swollen leg is examined at a medical clinic. The examining physician is concerned that the patient might have deep vein thrombosis, or DVT, a blood clot in the deep vein system of the leg. How does the physician make the diagnosis for this patient? The **gold standard** diagnostic test for DVT is a contrast venogram, a test that requires that a radiographic contrast dye be injected into a vein on the top of the foot. A filling defect seen in the dye column on the x-ray would correctly diagnose a deep vein clot. However, the radiographic contrast itself can inflame the veins and cause DVT. Moreover, allergies to the dye are not uncommon and may prevent the patient from undergoing such a procedure. Also, a venogram always involves a needle stick; a less painful test would be preferred if it were available and sufficiently accurate. The patient's true DVT status might still be assessed by prolonged follow-up since DVT should declare itself eventually by causing persistent symptoms. Nevertheless, the clinician prefers to make a diagnosis quickly without prolonged follow-up, not only to avoid the negative aspects of using a venogram but primarily because of the immediate risk of pulmonary embolism, when a deep leg vein clot migrates upstream to the lungs.

An alternative diagnostic test for DVT is ultrasound, which can be quickly scheduled and is not invasive to the patient. An ultrasound, however, may be less accurate in diagnosing DVT than a venogram. So, how do we choose between the gold standard procedure with its complications and the more simple procedure with its potential for misclassification? Diagnostic test studies may provide some answers. Such studies are described in the next activity.

Summary

- The gold standard test procedure for diagnosing a specific disease condition may have problems (e.g., invasive, risky, costly) associated with its use.
- An example of such a procedure is the use of a contract venogram to diagnose deep vein thrombosis (DVT).
- An alternative diagnostic test for DVT is ultrasound, which may not be as accurate as a venogram.
- A diagnostic test study can help a clinician assess the performance of an alternative diagnostic procedure in comparison with using the gold standard procedure.

Diagnostic Test Studies

In clinical medicine, studies concerned with misclassification of disease are usually called **diagnostic test studies**. The primary goal of such a study is to evaluate the performance of a test for diagnosing a disease condition of interest. Suppose, for example, that the disease condition is deep vein thrombosis, or DVT. In a diagnostic study for DVT, the clinician targets only patients with a specific symptom, for example "acute leg swelling" and then performs both the diagnostic test, typically an ultrasound, and the gold standard procedure, typically an x-ray venogram, on these patients. Here are the results of such a diagnostic test study in the form of a misclassification table:

Misclassification Table - Diagnostic test

		GOLD STANDARD		
		+	-	Total
TEST RESULT	+	48	14	62
	-	12	126	138
	Total	60	140	200

Using the diagnostic test, the disease classification status that is determined for a given patient is called the **test result**, and is labeled as positive (+) or negative (-) on the rows of the table. The procedure used to define true disease is called the **gold standard**, however imperfect it may be. In the misclassification table, the results from using the gold standard are labeled on the columns of the table. Typically, the gold standard is a test that is more detailed, expensive, or risky than the diagnostic test used by the physician. The gold standard might even require prolonged follow-up of the patient if the disease is expected to eventually declare itself, post-mortem examination, or a measure combining more than one strategy, sometimes in complex ways tailored to the specific disease.

Using the information in the misclassification table, the performance of a diagnostic test can be evaluated using several important measures, including the **sensitivity**, the **specificity**, and the **prevalence**. Recall that **sensitivity** describes the test's performance in patients who truly have the disease, and is defined as the conditional probability of a positive test result given true disease, P(Test + | True +).

Study Questions (Q9.16)

1. What is the sensitivity of the test in the above table?
2. If the sensitivity had been 0.99, what could you conclude about a truly diseased patient who had a negative test result?

Specificity describes the tests performance among patients who are truly without the disease. It is defined as the conditional probability of a negative test result given the absence of disease, P(Test - | True -).

Study Questions (Q9.16) continued

3. What is the specificity of the test?
4. If the specificity had been .99, what would you conclude about a truly nondiseased patient who had a positive test result?

Prevalence is calculated as the proportion of patients in the study sample who truly have the disease, P(True +). If little is known about a patient, disease prevalence in a diagnostic test study is the best estimate of pre-test probability that the patient has the disease.

Study Questions (Q9.16) continued

5. What is the prevalence of true disease from these data?
6. Based on the sensitivity, specificity, and prevalence calculations above, do you think that the test is a good diagnostic tool for DVT? Explain briefly.

Although the three measures, sensitivity, specificity, and prevalence provide important summary information about the performance of the diagnostic test, a more useful measure of overall performance is called the **predictive value**. It is described in the next activity.

Summary

- In clinical medicine, studies concerned with misclassification of disease are usually called **diagnostic test studies**.
- The purpose of a diagnostic test study is to evaluate test performance rather than to adjust for information bias.
- The procedure used to define true disease is called the **gold standard**.
- In a diagnostic study, the clinician targets patients with a specific symptom and then performs both the diagnostic test and the gold standard procedure on these patients.
- The performance of a diagnostic test can be evaluated using several important measures, including **sensitivity**, **specificity**, and **prevalence**.
- A more useful measure of the performance of a diagnostic test is provided by the **predictive value**.

Screening Tests

A second type of clinical study concerned with misclassification is called a **screening test**. In contrast to a diagnostic test, a screening test targets a broad population of asymptomatic subjects to identify those subjects that may require more detailed diagnostic evaluation. The subjects in a screening test have not gone to a physician for a specific complaint.

Members of the general public are typically invited to undergo screening tests of various sorts to separate them into those with higher and lower probabilities of disease. Those with higher probabilities are then urged to seek medical attention for definitive diagnosis. Those with lower probabilities receive no direct health benefit because they do not have the disease condition being screened. Also, depending on invasiveness of the screening test and/or the disease condition being targeted, persons under going screening may suffer risks as well as face some inconvenience, anxiety, personal cost, and sometimes discomfort, e.g., as with the use of a colonoscopy to screen for bowel cancer.

The Predictive Value of a Diagnostic Test

The probability of true disease status for an individual patient given the result of a diagnostic test is called the test's **predictive value**. The predictive value is particularly useful to the clinician for individual patient diagnosis because it directly estimates the probability that the patient truly does or does not have the disease depending on the results of the diagnostic test. That's what the clinician wants to know.

Study Questions (Q9.17)

Suppose T+ denotes the event that a patient truly has a disease condition of interest, whereas D+ denotes the event of a positive diagnostic test result on the same patient.

1. Which of the following two probability statements describes sensitivity and which describes predictive value?
   A. P(T+|D+) B. P(D+|T+)

The predictive value can be obtained directly from the misclassification table generated by a diagnostic test study. Because there are two possible results for a test, there are two different predictive values. The probability of actually having the disease when the test is positive is called the **positive predictive value**, and is denoted as **PV+**. The probability of actually not having the disease if the test is negative is the **negative predictive value**, and is denoted as **PV-**. Both **PV+** and **PV-** are proportions. The closer these proportions are to 1, the better the test's predictive performance.

Misclassification Table: diagnostic test study

		TRUE DISEASE STATUS (T) +	-	
DIAGNOSTIC TEST RESULT (D)	+	48	14	62
	-	12	126	138
		60	140	200

PV+ : P(T+ | D+) = **Positive Predictive Value**

PV- : P(T- | D-) = **Negative Predictive Value**

The positive predictive value is calculated as the number of true positive results divided by all positive results.

$$PV+ = \frac{\text{\# true positive results}}{\text{all positive results}}$$

Study Questions (Q9.17) continued

2. What is PV+ for the above table?

The positive predictive value is often referred to as the **post-test probability** of having disease. It contrasts with prevalence, which gives the average patient's **pre-test probability** of having disease.

Study Questions (Q9.17) continued

3. Based on the data in the misclassification table, what is the estimate of the average patient's probability of having DVT prior to performing an ultrasound?
4. Has the use of an ultrasound improved disease diagnosis for persons with positive ultrasound results? Explain briefly.

The **negative predictive value** is the number of true negative results divided by the total number of subjects with negative test results.

$$PV- = \frac{\text{\# true negative results}}{\text{all negative results}}$$

Study Questions (Q9.17) continued

5. What is PV- for these data?
6. Based on the data in the misclassification table, what is the estimate of the average patient's probability of not having DVT prior to performing an ultrasound?
7. Has the use of an ultrasound improved disease diagnosis for persons with negative ultrasound results? Explain briefly.

The prevalence of true disease in a diagnostic test study can greatly influence the size of the predictive values obtained. To illustrate this, we now consider a second misclassification table for a different group of patients who have presented to their clinician with pain but without swelling in their leg.

Misclassification Table: diagnostic test study

		Venogram for DVT +	Venogram for DVT -	
Ultrasound	+	16	18	34
	-	4	162	166
		20	180	200

Study Questions (Q9.17) continued

8. What are the sensitivity, specificity, and prevalence in the table?
9. In the previously considered misclassification table, the sensitivity, specificity, and prevalence were 0.80, 0.90, and 0.30, respectively. How do these values compare with the corresponding values computed in the previous question?
10. What are the values of PV+ and PV- in the above table?
11. In the previously considered misclassification table, PV+ and PV- were .77 and .91, respectively, and the prevalence was .30. How do these values compare with the corresponding predicted values computed in the previous question?
12. What is the moral of this story relating predictive value to disease prevalence?

Summary

- ❖ The predictive value (or post-test probability) is the probability of true disease status given the result of a diagnostic test.
- ❖ The predictive value can be obtained directly from the misclassification table generated by a diagnostic test study.
- ❖ The probability of disease when the test is positive is called the positive predictive value, and is denoted as PV+
- ❖ The probability of disease when the test is negative is called the negative predictive value, and is denoted as PV-.
- ❖ PV+ = # true positives / all positive
- ❖ PV- = # true negatives / all negatives
- ❖ The closer PV+ and PV- are to 1, the better the test.
- ❖ The prevalence of true disease in a diagnostic test study can greatly influence the size of the predictive value.

Expressing Predicted Value in Terms of Sensitivity, Specificity, and Prevalence

The prevalence of true disease in a diagnostic test study can greatly influence the size of the predictive value. We first illustrate this statement using an example. We then describe mathematically how we computed the numerical values in our example using formulae that express the positive and negative predictive values in terms of the sensitivity, specificity and prevalence parameters.

Suppose we fix the sensitivity and specificity values at .80 and .90, respectively, and we consider what the predictive values would be for different prevalences, say .10, .30, .50, .70, and .90. The following results will be obtained:

Prevalence of True Disease	Sensitivity D	Specificity D	PV+	PV-
.01	.80	.90	.08	.998
.10	.80	.90	.47	.98
.30	.80	.90	.77	.91
.50	.80	.90	.89	.82
.70	.80	.90	.95	.66
.90	.80	.90	.99	.33

Continued on next page

Expressing Predicted Value in Terms of Sensitivity, Specificity, and Prevalence (*continued*)

From the above table, we can see that for fixed values of sensitivity and specificity the predicted value positive (**PV +**) increases from .08 to .99 as the prevalence increases from .01 to .90; moreover, even though both sensitivities and specificities are relatively high, the predictive value positive can be very low (e.g., 8%) if the prevalence of the disease is small (e.g., 1 %). A reverse trend is seen for predicted value negative (**PV-**), which decreases as the prevalence of disease increases.

We now describe how the formulae for **PV+** and **PV-** can be expressed in terms of sensitivity, specificity, and prevalence parameters. To do this, we need to use a famous theorem about conditional probabilities, called **Bayes Theorem**, which allows us to express conditional probabilities of the form:

P(A|B) and P(A|not B)

in terms of conditional probabilities of the form:

P(B|A) and P(B|not A)

in which **A** and **B** have been switched into opposite sides of the given sign. In diagnostic test studies, the event **A** refers to a subject's true disease status being positive, which we denote as:

A = T+

The event **B** refers to a subject's diagnostic test result being positive, which we denote as:

B = D+

It follows then that

not A = T- and
not B = D-

denote being a true negative and having a negative diagnostic test result, respectively. Thus:

P(A|B) = P(T+|D+)

defines the **positive predictive value** (**PV+**) of a diagnostic test,

P(not A|not B) = P(T -|D-)
defines the **negative predictive value** (**PV -**) of a diagnostic test,

P(B|A) = P(D+|T+)

denotes the **sensitivity** of the diagnostic test,

(not B|not A) = P(D-|T-)

denotes the **specificity** of the diagnostic test, and

P(A)=P(T+)

denotes the **prevalence** of true disease in the population under study.

Using Bayes Theorem, we can express **P(A|B)** and as follows:

Continued on next page

Expressing Predicted Value in Terms of Sensitivity, Specificity, and Prevalence (continued)

$$P(A \mid B) = \frac{P(A)P(B \mid A)}{P(A)P(B \mid A) + [1 - P(A)]P(B \mid \text{not } A)}$$

Substituting **T+**, **T-**, **D+** into this formula for **A**, not A, **B**, respectively, we get the following result

$$P(T+ \mid D+) = \frac{P(T+)P(D+ \mid T+)}{P(T+)P(D+ \mid T+) + [1 - P(T+)]P(D+ \mid T-)}$$

which can be re-written as

$$PV+ = \frac{\text{PrevD} \times \text{SensD}}{[\text{PrevD} \times \text{SensD}] + [(1 - \text{PrevD}) \times (1 - \text{SpecD})]}$$

Similarly, we can use **Bayes Theorem** to describe **PV-** in terms of sensitivity, specificity and prevalence of disease:

$$PV- = \frac{(1 - \text{PrevD}) \times \text{SpecD}}{[\text{PrevD} \times (1 - \text{SensD})] + [(1 - \text{PrevD}) \times \text{SpecD}]}$$

Quiz (Q9.18)

For the classification table shown on the right, determine each of the following:

1. What is the sensitivity? . . ???
2. What is the specificity? . . ???
3. What is the prevalence of the disease? ???
4. What is the positive predictive value? ???

		Truth		
		D	not D	
Classified	D	90	20	110
	not D	10	180	190
		100	200	300

Choices
10% 33.3% 36.7% 81.8% 90%

The sensitivity and specificity for the classification table shown on the right are still 90% as in the previous questions. For this table, answer each of the following:

5. What is the prevalence of the disease? ???
6. What is the positive predictive value? ???
7. The prevalence in this table is smaller than in the previous table; therefore, the positive predictive value is ??? than in the previous table.

		Truth		
		D	not D	
Classified	D	90	50	140
	not D	10	450	460
		100	900	600

Choices
16.7% 64.3% 90% larger smaller

Quiz continued on next page

Once again, the sensitivity and specificity for the classification table shown on the right are 90%. For this table, answer each of the following:

		Truth		
		D	not D	
Classified	D	90	90	180
	not D	10	810	820
		100	900	1000

8. What is the prevalence of the disease? **???**
9. What is the positive predictive value? **???**
10. The prevalence in this table is smaller than in the previous two tables, therefore, the positive predictive value is **???** than in the previous two tables.
11. These results illustrate the fact that if the prevalence is small, the predictive value can be quite **???** even if the sensitivity and specificity parameters are quite **???**

Choices
10% **36.7%** **50%** **high** **larger** **small** **smaller**

Nomenclature

Misclassification tables for disease and exposure

Misclassification table for disease:

		Truth		
		D	Not D	
Classified	D′			
	Not D′			

Misclassification table for exposure:

		Truth		
		E	Not E	
Classified	E′			
	Not E′			

Observed (misclassified) and corrected tables

Observed (i.e., misclassified) Data

	E′	Not E′	
D′	a	b	
Not D′	c	d	

Corrected (i.e., adjusted) Data

	E	Not E	
D	A	B	
Not D	C	D	

Formulae for correcting for nondifferential misclassification of disease (no misclassification of exposure)

$A = [a\ SpD - c(1 - SpD)] / q$
$B = [b\ SpD - d(1 - SpD)] / q$
$C = [c\ SeD - a(1 - SeD)] / q$
$D = [d\ SeD - b(1 - SeD)] / q$

Where $q = SeD + SpD - 1$

Formulae for correcting for nondifferential misclassification of exposure (no misclassification of disease)

$A = [a\ SpE - b(1 - SpE)] / q$
$B = [b\ SeE - a(1 - SeE)] / q$
$C = [c\ SpE - d(1 - SpE)] / q$
$D = [d\ SeE - c(1 - SeE)] / q$

Where $q = SeE + SpE - 1$

Formulae for correcting for nondifferential misclassification of exposure and disease

$A = \{SpD[(a + b)\ SpE - b] - (1 - SpD)[(c + d)\ SpE - d]\} / q^*$
$B = \{SpD[(a + b)\ SeE - a] - (1 - SpD)[(c + d)\ SeE - c]\} / q^*$
$C = \{SeD[(c + d)\ SpE - d] - (1 - SeD)[(a + b)\ SpE - b]\} / q^*$
$D = \{SeD[(c + d)\ SeE - c] - (1 - SeD)[(a + b)\ SeE - a]\} / q^*$

Where $q^* = (SeD + SpD - 1)\ (SeE + SpE - 1)$

D	Truly has disease
D′	Classified as having disease
E	Truly exposed
E′	Classified as exposed
Not D	Truly does not have disease
Not D′	Classified as not having disease
Not E	Truly not exposed
Not E′	Classified as not exposed
$\hat{OR}$	Odds ratio from *observed* data = ad/bc
$\hat{OR}_{adj}$	Odds ratio from *corrected* or *adjusted* data = AD/BC
$\hat{RR}$	Risk ratio from *observed* data = [a/(a+c)]/[b/(b+d)]
$\hat{RR}_{adj}$	Risk ratio from *corrected* or *adjusted* data = [A/(A+C)]/[B/(B+D)]
Sensitivity	Of those truly with the characteristic, the proportion that will be correctly classified as having the characteristic; for disease, Pr(D′|D); for exposure, Pr(E′|E)
SeD	Sensitivity of disease misclassification
SeE	Sensitivity of exposure misclassification
SpD	Specificity of disease misclassification
SpE	Specificity of exposure misclassification
Specificity	Of those truly without the characteristic, the proportion that will be correctly classified as not having the characteristic; for disease, P(not D′|not D); for exposure, P(not E′|not E).

References on Information/Misclassification Bias

Overviews

Greenberg RS, Daniels SR, Flanders WD, Eley JW, Boring JR. Medical Epidemiology (3rd Ed). Lange Medical Books, New York, 2001.

Hill H, Kleinbaum DG. Bias in Observational Studies. In Encyclopedia of Biostatistics, pp 323-329, Oxford University Press, 1999.

Kleinbaum DG, Kupper LL, Morgenstern H. Epidemiologic Research: Principles and Quantitative Methods. John Wiley and Sons Publishers, New York, 1982.

Special Issues

Barron, BA. The effects of misclassification on the estimation of relative risk. Biometrics 1977;33(2):414-8.

Copeland KT, Checkoway H, McMichael AJ and Holbrook RH. Bias due to misclassification in the estimation of relative risk. Am J Epidemiol 1977;105(5):488-95.

Dosemeci M, Wacholder S, and Lubin JH. Does nondifferential misclassification of exposure always bias a true effect toward the null value? Am J Epidemiol 1990:132(4):746-8.

Espeland MA, Hui SL. A general approach to analyzing epidemiologic data that contain misclassification errors. Biometrics 1987;43(4):1001-12.

Greenland S. The effect of misclassification in the presence of covariates. Am J Epidemiol 1980;112(4):554-69.
Greenland S, Kleinbaum DG. Correcting for misclassification in two-way tables and matched-pair studies. Int J Epidemiol 1983;12(1):93-7.
Reade-Christopher SJ, Kupper LL. Effects of exposure misclassification on regression analyses of epidemiologic follow-up study data. Biometrics 1991;47(2):535-48.
Satten GA and Kupper LL. Inferences about exposure-disease associations using probability-of-exposure information. J Am Stat Assoc 1993;88:200-8.
Wynder EL. Investigator bias and interviewer bias: the problem of reporting systematic error in epidemiology. J Clin Epidemiol 1994;47(8):825-7.

Homework

ACE-1. Radiation Exposure vs. GI Tumors

An investigator is interested in studying the relationship between radiation exposure and development of gastrointestinal (GI) tumors. He assembles a cohort of cancer-free subjects, divides them into "high" and "low" radiation exposure groups, and follows them over 10 years for evidence of GI tumors.

a. Suppose that the investigator uses a diagnostic test that fails to identify 20% of <u>all</u> GI tumors but never registers a false positive result (i.e. no one without the cancer is ever misdiagnosed as having cancer). Describe this situation in terms of sensitivities and specificities for the exposed and unexposed subjects:

Sensitivity (D | E) =   Sensitivity (D | not E) =

Specificity (D | E) =   Specificity (D | not E) =

The following summarizes the data from the SOURCE population (no misclassification):

	High Radiation	Low Radiation	Total
GI Tumor	600	200	800
No Tumor	1400	1300	2700
Total	2000	1500	3500

b. What is the unbiased estimate of the risk ratio (RR)?
c. Use the sensitivities and specificities from part a to show the data that would have been observed by the investigator:

	High Radiation	Low Radiation	Total
GI Tumor			
No Tumor			
Total			

d. Calculate the observed RR. Is there no bias, bias toward the null, bias away from the null, or switchover bias?
e. The following table summarizes the OBSERVED data from a different study of radiation and GI tumors:

	High Radiation	Low Radiation	Total
GI Tumor	476	376	852
No Tumor	524	624	1148
Total	1000	1000	2000

Assuming the scenario described in part "a" above, what is the corrected (unbiased) estimate of the RR for the relationship between radiation exposure and GI tumors for this study? Show your calculations. What is the nature of the bias, if any?

f. Calculate an odds ratio (rather than a RR) for both the observed and corrected data in part e. Is your conclusion regarding bias the same?

ACE-2. CHD and Behavior

The following data represent the SOURCE population in a case-control study of coronary heart disease (CHD) and type A behavior:

	Type A	Non-Type A	Total
CHD	80	25	105
No CHD	50	55	105
Total	130	80	210

a. Calculate the unbiased estimate of the exposure odds ratio (EOR).
b. Suppose that exposure was misclassified with the following sensitivities and specificities:

Cases:	Sensitivity = .8	Specificity = 1
Controls:	Sensitivity = .9	Specificity = .6

Show the data that would have been observed, given the misclassification:

	Type A	Non-Type A	Total
CHD			
No CHD			
Total			

c. Calculate the EOR from the observed data. Is there no bias, bias toward the null, bias away from the null, or switchover bias?

ACE-3. Sleep Disturbance vs. Clinical Depression

The Johns Hopkins Precursors Study, a long-term prospective cohort study, was used to evaluate the relation between self-reported sleep disturbances and subsequent clinical depression. A total of 1,053 men at several U.S. universities provided information on sleep habits during medical school and then were followed for 20 years for development of depression. Subjects underwent extensive psychological testing and lengthy structured interviews conducted by psychiatrists trained in the diagnosis of depression. Results of the study are summarized below (you may assume that these data are correctly classified):

	Sleep Disturbance	
	Yes	**No**
Depression	168	93
No Depression	258	534
	426	627

Now suppose that the investigators had limited funds and were not able to use the sophisticated diagnostic tools described above. They had two options available to them for diagnosing depression among the study subjects: (1) an interview or (2) a self-administered questionnaire.

The interview was able to classify depression status with the following sensitivities and specificities:

Exposed: Sensitivity = 0.84 and Specificity = 0.90
Unexposed: Sensitivity = 0.84 and Specificity = 0.90

The questionnaire was able to classify depression status with the following sensitivities and specificities:

Exposed: Sensitivity = 0.80 and Specificity = 0.88
Unexposed: Sensitivity = 0.90 and Specificity = 0.95

a. What is the true relative risk (RR) that describes the association between sleep disturbance and depression?
b. What is the observed RR when the interview is used to classify depression?
c. What is the observed RR when the questionnaire is used to classify depression?
d. Taking only issues of validity into account, which tool for diagnosing depression is preferable - the interview or the questionnaire? Justify your answer.

ACE-4. Case-Control Studies of Coffee Consumption

An epidemiologist was interested in determining whether use of a new aspirin-containing pain reliever was associated with an increased risk of gastrointestinal bleeding. S/he identified 600 patients who were taking the drug on a regular basis and 600 unexposed subjects. Subjects were followed for one year to detect the occurrence of gastrointestinal bleeding. Due to publicity about the potential hazards of the new drug, physicians participating in the study followed their exposed subjects more closely than unexposed subjects, and were thus more likely to diagnose gastrointestinal bleeds when they occurred in this group of study subjects.

a. Which of the following describe(s) the situation? [You may choose more than one]:

   i. Nondifferential misclassification of exposure
   ii. Differential misclassification of exposure
   iii. Nondifferential misclassification of disease
   iv. Differential misclassification of disease
   v. Recall bias
   vi. Detection bias
   vii. Berkson's bias

Suppose that the following table represents the SOURCE population (i.e. correctly classified) for the study of the new drug and gastrointestinal bleeding:

	Use of New Drug	
	Yes	**No**
GI bleed	400	300
no GI bleed	200	300

b. Assume that when the study was carried out, GI bleeds were classified with sensitivity = 0.9 and specificity = 0.75 for subjects who were using the drug. For subjects not using the drug, GI bleeds were classified with sensitivity = 0.6 and specificity = 0.75. Use this information to fill in the table below:

	Use of New Drug	
	Yes	**No**
GI bleed		
no GI bleed		

c. Calculate the appropriate ratio measure of association for the OBSERVED (misclassified) data. Indicate whether there is bias, and if so, in what direction.

d. Suppose the study were repeated, with GI bleeds detected by physicians who were blinded as to the exposure status of the subjects. If the sensitivity for unexposed subjects were increased to 0.9, while all other sensitivities and specificities remained the same, what impact would this have on the bias?

i. The bias would be eliminated.
ii. The magnitude of the bias would decrease, but the direction would remain the same.
iii. The magnitude of the bias would increase, but the direction would remain the same.
iv. The direction of the bias would change.
v. None of the above.

ACE-5. Antidepressant Medication and Breast Cancer Risk

A study entitled "Antidepressant Medication and Breast Cancer Risk" was published in a recent issue of the *American Journal of Epidemiology*. According to the methods section of the paper, "Cases were an age-stratified (<50 and >=50 years of age) random sample of women aged 25-74 years, diagnosed with primary breast cancer during 1995 and 1996 (pathology report confirmed) and recorded in the population-based Ontario Cancer Registry. As the 1-year survival for breast cancer is 90 percent, surrogate respondents were not used. Population controls, aged 25-74 years, were randomly sampled from the property assessment rolls of the Ontario Ministry of Finance; this database includes all home owners and tenants and lists age, sex, and address."

a. Discuss the authors' approach to the identification of cases with respect to the potential misclassification bias.
b. Discuss the authors' approach to the identification of controls with respect to the potential for (a) selection and (b) misclassification bias.
c. Discuss the pros and cons of the decision not to allow surrogate respondents.

The methods section of the paper goes on to say: "Data were collected through mailed, self-administered, structured questionnaires that included information on (1) sociodemographic data; (2) duration, dosage, timing, and type of antidepressant medications used; and (3) potential confounders. Subjects were asked "have you ever taken antidepressants for at least 2 weeks at any time in your life?" (a list of 11 antidepressants was given to provide examples)."

d. Discuss the pros and cons of the authors' approach to exposure assessment. List and describe how information bias could result from this approach.

ACE-6. Validation Study

You have conducted:

i) a cohort study of a disease and dichotomous exposure; and

ii) a separate validation study to assess sensitivity and specificity of the measure used in the cohort study, since disease status was misclassified differentially with respect to exposure.

Results of the validation study:

Exposed:

	True Disease	
Measured Disease	+	-
+	95	10
-	5	90

Unexposed:

	True Disease	
Measured Disease	+	-
+	97	5
-	3	95

Results of the cohort study (observed data):

	Exposure Status	
Disease Status	+	-
+	540	284
-	1460	1716

Using the above information, estimate the true risk ratio adjusted for misclassification.
(To answer this question, you may wish to use the Data Desk template for correcting for misclassification., Specificity.ise)

ACE-7. Diagnostic Testing: Baseball Fever

A researcher at Javier Lopez University developed a new test for detecting baseball fever. The new test was evaluated in a population of 10,000 people, 21% of whom were definitely known to have baseball fever. The number of negative tests was 8,162. The positive predictive value was discovered to be 91.4%.

a. Use the information provided above to fill in the following 2 x 2 table:

	Baseball Fever	
	Present	Absent
Test Result		
Positive	______	______
Negative	______	______

b. What are the test's sensitivity and specificity? Provide an interpretation of the sensitivity, using value you just calculated.

c. If you were to apply this test to a patient and the result came back negative, what would you advise that patient regarding his/her chances of having baseball fever?

ACE-8. Predictive Value: Diabetes

In a certain community, eight percent of all adults over age 50 have diabetes. If a health service in this community correctly diagnosis 95% of all persons with diabetes as having the disease and incorrectly diagnoses ten percent of all persons without diabetes as having the disease, find the probabilities that:

a. The health service will diagnose an adult over age 50 as having diabetes.
b. A person over 50 diagnosed by the health service as having diabetes actually has the disease.

ACE-9. Diagnostic Testing: Prostate Cancer

A group of 50,000 men over 60 years of age were tested for prostate cancer using the PSA test with the following results:

	Prostate Cancer	
PSA Test Result	**Yes**	**No**
Positive	2,900	20,000
Negative	100	27,000
Total	3,000	47,000

a. What are the sensitivity and the specificity of this test?
b. What is the positive predictive value of this test?
c. If a person gets a PSA test result that is negative, should he worry about having prostate cancer? Explain.
d. If a person gets a PSA test result that is positive, should he worry about having prostate cancer? Explain.
e. Suppose a new test was developed that had the same sensitivity as the PSA test, but had a specificity that was 95%. Assuming the same numbers subjects with and without (true) prostate cancer, describe the misclassification table that would result using the new test.

	Prostate Cancer	
New Test Result	**Yes**	**No**
Positive		
Negative		
Total	3,000	47,000

f. What is the positive predictive value of the new test?
g. If a person gets a new test result that is negative, should he worry about having prostate cancer? Explain.
h. If a person gets a new test result that is positive, should he worry about having prostate cancer? Explain.

ACE-10. Prostate Cancer Screening

Suppose that a new screening test for prostate cancer has been under development and is almost ready to be put on the market. Subjects undergoing screening are required to provide a small blood sample that is then tested to determine the level of a certain factor (Factor P). The higher the level of Factor P, the more likely the presence of prostate cancer. However, Factor P may be elevated as a result of other non-cancerous conditions involving the prostate.

a. Scientists in Europe and those in the United States have disagreed on the appropriate cut-point for determining whether the screening test is to be considered positive. Discuss the implications of the choice of cut-point for this test. Think in terms of evaluation of the test's performance as well as ramifications for patient care.
b. Design an epidemiologic study that would be appropriate for evaluating the effectiveness of the new screening test. Be sure to comment on the study design, study subjects, exposure(s) of interest, outcome(s) of interest, analytic plan, potential biases, and any other important aspects of your study.
c. As soon as the new screening test becomes available for use, it receives a great deal of media attention. It is quickly endorsed by the American Medical Association, the American Urological Association, and the American College of Surgeons. A community-based health advocacy group founded by prostate cancer patients and their families begins to call for widespread screening using the new test. Their goal is see that every adult male in the United States is screened each year for prostate. Discuss the pros and cons of such a plan for widespread screening of the general public for prostate cancer.

Answers to Study Questions and Quizzes

Q9.1

1. 2
2. 7
3. 100 x (6/8) = 75%
4. 100 x (7/10) = 79%

Q9.2

1. False. Typically, several times and locations are used in the same residence, and a time-weighted average (TWA) is often calculated.
2. False. Good instrumentation for measuring time-weighted average has been available for some time.
3. False. A system of wire codes to measure distance and configuration has been used consistently since 1979 to rank homes crudely according to EMF intensity. However, the usefulness of this system for predicting past exposure remains an open question.
4. True
5. True
6. True

(Note: there are no questions numbered 7 to 9)

10. Interviewer bias. Subjects known to have experienced a venous thrombosis might be probed more extensively than controls for a history of oral contraceptive use.
11. Away from the null. The proportion of exposed among controls would be less than it should have been if both cases and controls were probed to the same extent. Consequently, the odds ratio in the misclassified data would be higher than it should be.
12. Keep the interviewers blind to case-control status of the study subject.
13. 5
14. 70
15. 100 x (95/100) = 95%
16. 100 x (70/80) = 87.5%

Q9.3

1. The estimated risk ratio for the observed data is (380/1000)/(240/1000) = 1.58.
2. Because the observed risk ratio of 1.58 is meaningfully different than the true (i.e., correct) risk ratio of 3.14.
3. Towards the null, since the biased estimate of 1.58 is closer to the null value than is the correct estimate.
4. No way to tell from one example, but the answer is no, the bias might be either towards the null or away from the null.
5. The observed OR of 2.6 that results from misclassifying exposure is meaningfully different than the true odds ratio of 3.5.
6. The bias is towards the null. The biased OR estimated of 2.6 is closer to the null value of 1 than is the correct OR.

Q9.4

1. Sensitivity = 720 / 900 = .8 or 80%
2. Specificity = 910 / 1100 = .83 or 83%
3. Yes. Both sensitivity and specificity are smaller than one. However, without correcting for the bias, it is not clear that the amount of bias will be large.

Q9.5

1. Sensitivity = 480/600 = .80 or 80% and Specificity = 380/400 = .95 or 95%.
2. Sensitivity = 240/300 = .80 or 80% and Specificity = 665/700 = .95 or 95%.
3. The sensitivities for CHD cases and non-cases are equal. Also, the specificities for CHD cases and non-cases are equal. The sensitivity information indicates that 20% of both cases and non-cases with low intake of fruits and vegetables tend to over-estimate their intake. The specificity information indicates that only 5% of both cases and non-cases with high intake tend to under-estimate their intake.
4. In the correctly classified 2x2 table, a=600, b=400, c=300, and d=700, so the estimated odds ratio is ad/bc = (600 x 700) / (400 x 300) = 3.5.
5. The observed OR of 2.6 that results from misclassifying exposure is meaningfully different than the true odds ratio of 3.5.
6. The bias is towards the null. The biased OR of 2.6 is closer to the null value of 1 than the correct OR.

Q9.6

1. Sensitivity = 580/600 = .97 or 97% and Specificity = 380/400 = .95 or 95%.
2. Sensitivity = 240/300 = .80 or 80% and Specificity = 665/700 = .95 or 95%.
3. No. Although the specificities for cases and non-cases are equal (i.e., 95%), the sensitivity for the cases (97%) is quite different from the sensitivity for the non-cases (80%). This difference in sensitivities indicates that cases with low intake of

fruits and vegetables are less likely to over-estimate their intake than non-cases.

4. Not much. The observed OR of 3.95 that results from misclassifying exposure is slightly higher than the true odds ratio of 3.50.
5. The bias is slightly away from the null. The biased OR of 3.95 is further away from the null value of 1 than is the correct OR of 3.5.

Q9.7

1. No. Assuming **independent misclassification** is <u>not</u> equivalent to assuming **nondifferential misclassification**. The latter assumes that how a subject classifies exposure will not vary with their true disease status, i.e., Pr(classifying disease status|truly E) = Pr(classifying disease status|truly not E) or Pr(classifying exposure status|truly D) = Pr(classifying exposure status|truly not D).
2. Differential because (Sensitivity D | E) is not equal to (Sensitivity D | not E)=.75.
3. Yes, the following are missing:
   (Specificity D | E)
   (Specificity D | not E)
   (Specificity E | not D)
4. Pr(D′E′ | D E) = Pr(D′ | D E) x Pr(E′ | D E) = (Sensitivity D | E) x (Sensitivity E | D)
5. Pr(D′ E′ | D E) = .8 x .9 = .72.
6. Pr(D′ E′ | D not E) = Pr(D′ | D not E) x Pr(E′ | D not E) = (Sensitivity D | not E) x (1 – (Specificity E | D)).
7. Pr(D′ E′ | D not E) = .75 x (1 - .95) = .0375.

Q9.8

1. False – in addition to the sensitivities, if the specificities for both exposed and unexposed are the same, then the bias must be towards the null.
2. True
3. True

Q9.9

a) Away
b) Towards
c) Towards
d) Away
e) Away
f) Away
g) Towards
h) Towards

1. It depends. We know that the bias must be towards the null. If the direction of the bias is all that we are interested in, then we do not need to correct for the bias. However, if we want to determine the extent of the bias and to obtain a quantitative measure of the true effect, then we need to correct for the bias.
2. We can either reason that misclassification is nondifferential from our knowledge or experience with the exposure and disease variables of our study, or we can base our decision on reliable estimates of the sensitivity and specificity parameters.
3. It depends. The bias may be either towards the null or away from the null. We might be able to determine the direction of the bias by logical reasoning about study characteristics. Otherwise, the only way we can determine either the extent or direction of the bias is to compare a corrected estimate with an observed estimate.
4. The biased (i.e., misclassified) observed odds ratio is closer to the null than the corrected odds ratio.
5. The greatest amount of bias is seen with the observed OR is 1.5 compared to the corrected OR of 3.5, which occurs when both the sensitivity and specificity are 80%.
6. The bias is smallest when the correct OR is 1.8, which results when both sensitivity and specificity are 90%.
7. One way to decide is to choose the corrected OR corresponding to the most realistic set of values for sensitivity and specificity. Another way is to choose the corrected OR (here, 3.5) that is most distant from the observed OR. A third alternative is to choose the corrected OR that changes least (here, 1.8) from the observed OR.

Q9.10

1. For males, SeD = 32/40 = 80% and SpD = 54/60 = 90%.
2. For females, SeD = 16/20 = 80% and SpD = 72/80 = 90%.
3. Nondifferential: The SeD for males and females are equal at 80% and the SpD for males and females are equal at 90%.
4. RR(adjusted) = (A/1000)/(B/1000) = (400/1000)/(200/1000) = 2.0.
5. The bias is towards the null because the biased risk ratio estimate of 1.58 is closer to the null value than is the corrected risk ratio.
6. q will be zero if both the SeD and SpD add up to 1. For example, if both the SeD and SpD equal .5, then q = 0. In this case there would be an equal chance of being misclassified into any one of the four cells of the 2x2 tables. There would be no point in computing corrected effect estimates for such a situation, since misclassification would have completely invalidated one's study results.

Q9.11

1. SeE = 48/60 = 80% and SpE = 133/140 = 95%.
2. You would need to stratify the misclassification table into two tables, one for cases and the other for controls, and then determine whether corresponding sensitivities and specificities for cases and controls were equal. Without such information, you might be able to reason that since both cases and controls have cancer that is gastrointestinal, they may tend to have similar reporting tendencies about PEU history. Such an argument suggests that misclassification is likely to be nondifferential.
3. Note: there is no question 3 in this section.
4. OR(adjusted) = (A x D)/(B x C) = (80 x 280)/(220 x 20) = 5.1.
5. The bias is towards the null, because the biased risk ratio estimated of 3.0 is closer to the null than the corrected risk ratio of 5.1.

Q9.12

1. Stratify the classification information on pollution level by true illness status, and stratify the classification information on illness by true pollution level.
2. Not really because the true stratified misclassification information involving the true illness status and pollution level is not provided.
3. Non-differential misclassification is assumed since no stratum-specific sensitivity or specificity values are provided in the sub-sample or the previous pollution study to be applied here. The non-differential misclassification assumption does appear reasonable. It is unlikely that illness would be misreported one week later according to pollution level, or that water quality was measured incorrectly or misreported according to illness level.
4. It is reasonable to assume independent classification since the exposure and disease variables were measured at different times and likely by different investigators.
5. RR(corrected) = (339.3/416.7)/(410.8/2083.4) = 4.1
6. The bias is towards the null because the biased risk ratio estimate of 1.6 is closer to the null value than is the corrected risk ratio of 4.1.
7. The biased estimate of 1.6 jumps quite a lot to 4.1 when corrected. Such a large jump from biased to correct estimate often occurs when both disease and exposure are misclassified, even when the sensitivity and specificity parameters are close to 100%.
8. If either the sensitivity and specificity for disease sums to 1 or the sensitivity and specificity for exposure sums to 1, then the corrected cell frequencies are undefined because q* = 0.

Q9.13

1. 7.4 – To answer this question, you will need to use the appropriate formula to correct for nondifferential misclassification of disease.
2. towards
3. true – If the sum of the specificity and sensitivity equals 1, you will obtain indeterminate results.
4. true
5. false – Since the sum of either the specificities and sensitivities is 1, the results will be indeterminate.

Q9.14

1. Differential because the sensitivities of 96.7% and 80% are different, even though the specificities are the same.
2. A CHD case, who might be concerned about the reasons for his or her illness, is not as likely to over-estimate his or her intake of fruits and vegetables as is a control.
3. OR(corrected) = (599.7 x 700)/(400.2 x 300) = 3.5. This is the same value that we previously obtained for the true odds ratio in our previous presentation about differential misclassification that showed how to obtain observed cell frequencies when starting out with the true cell frequencies.
4. The bias is away from the null because the biased odds ratio estimate of 3.95 is further away from the null value than the corrected odds ratio of 3.5.

Q9.15

1. 3.5 – See the appropriate formulas required to calculate this estimate. A=600, B=400, C=300, D=700.
2. Away from – Since the observed OR is further from the null than the adjusted estimate, the observed estimate must be away from the null.

Q9.16

1. Sensitivity = 48 / 60 = 0.80.
2. The patient is very unlikely to have the disease, since the probability of getting a negative test result for a patient with the disease is.01, which is very small.
3. Specificity = 126 / 140 = 0.90
4. The patient is very likely to have the disease, because the probability of getting a positive result for a patient without the disease is .01, which is very small.

5. Prevalence of true disease = 60 / 200 = 0.30.
6. Cannot fully answer this question. Both the sensitivity and specificity are relatively high at .80 and .90, but the prevalence is only 30%. What is required is the proportion of total ultrasound positives that truly have DVT, which in this study is 48 / 62 = 0.77, which is high but not over .90 or .95.

Q9.17

1. Choice A is the predictive value and Choice B is sensitivity.
2. PV+ = 48 / 62 = 0.77
3. Based on the table, the prior probability of developing DVT is 60 / 200 = 0.30, which is the estimated prevalence of disease among patients studied.
4. Yes, the prior probability was 0.30, whereas the (post-test) probability using an ultrasound increased to 0.77 given a positive result on the test.
5. PV- = 126 / 138 = 0.91
6. Based on the table, the prior probability of **not** developing DVT is 140 / 200 = 0.70, which is 1 minus the estimated prevalence of disease among patients studied.
7. Yes, the prior probability of not developing DVT was 0.70 whereas the (post-test) probability of not developing DVT using an ultrasound increased to 0.91 given a negative test result.
8. Sensitivity = 16 / 20 = 0.80, specificity = 162 / 180 = 0.90, prevalence = 20 / 200 = .10.
9. Corresponding sensitivity and specificity values are identical in both tables, but prevalence computed for this data is much lower at 0.10 than computed for the previous table (.30).
10. PV+ = 16 / 34 = 0.47 and PV- = 162 / 166 = 0.98.
11. PV+ has decreased from 0.77 to 0.47 and PV- has increased from 0.91 to 0.98 whereas the prevalence has dropped from 0.30 to 0.10 while sensitivity and specificity has remained the same and high.
12. If the prevalence decreases, the predictive value positive will decrease and may be quite low even if sensitivity and specificity are high. Similarly, the predictive value negative will increase and may be very high, even if the sensitivity and specificity are not very high.

Q9.18

1. 90%
2. 90%
3. 33.3%
4. 81.8%
5. 16.7%
6. 64.3%
7. smaller
8. 10%
9. 50%
10. smaller
11. small, high

LESSON 10

CONFOUNDING

***Confounding** is a form of bias that concerns how a measure of association may change in value depending on whether variables other than the exposure variable are controlled in the analysis.*

10-1 Concept and Definition

Simpson's Paradox

Simpson's Paradox is illustrated by a hat shopping story in which a different conclusion is made about which color hat fits better when the same collection of hats are moved from separate tables to a single table.

A man enters a store to buy a hat. He sees two tables on which there are hats of only two colors, green and blue. The first table holds 6 green and 10 blue hats. He finds that 5 of the 6 green hats fit and 8 of the 10 blue hats fit, so 83% of the green hats fit and 80% of the blue hats fit.

The other table holds 10 green and 6 blue hats. Only 2 of the 10, or 20%, of the green hats fit, and 1 of the 6, or 17%, of the blue hats fit. But he goes home without picking a hat.

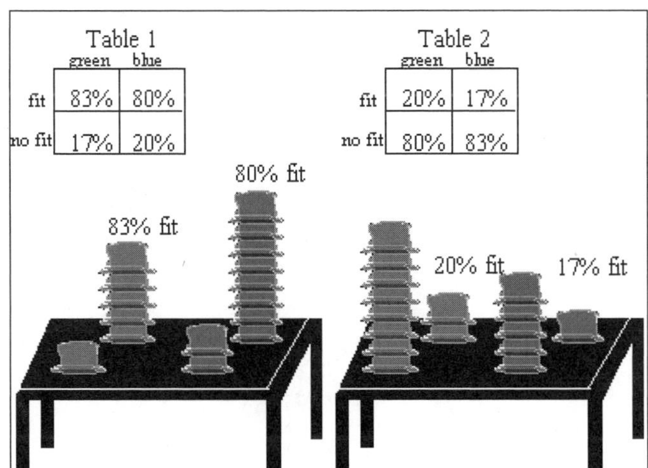

(Note: green hats are on the left side of the tables, blue on the right)

The next day he returns to the store just as it opens. He remembers that yesterday the percentage of green hats that fit was greater than the corresponding percentage of blue hats that fit at each table. But now the same 32 hats have been mixed together on one big table. Being a very thorough shopper, he tries on all the hats again. He finds that 7 of the 16 green hats, or 44%, fit, and 9 of the 16 blue hats, or 56%, fit. How can he explain such a reversal? Now blue hats seem more likely to fit than green hats! This situation illustrates a surprising result known as Simpson's Paradox.

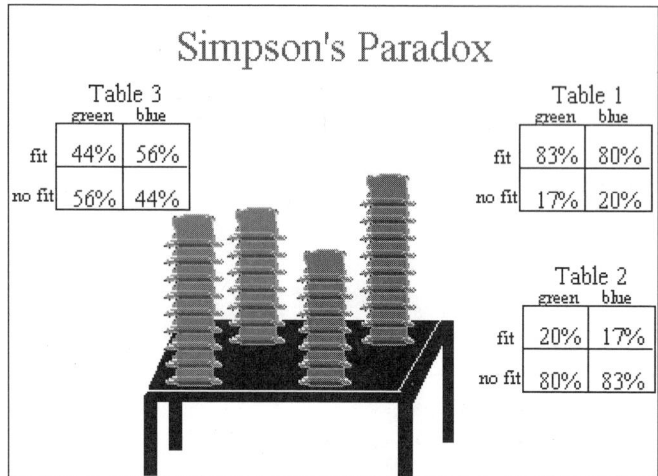

(Note: green hats are on the left side of the table, blue on the right)

Simpson's Paradox Explanation

In our hat shopping story, we found that even though green hats fit better than blue hats on each table separately, the combined table shows a reversal, blue hats fit better than green hats overall. The apparent paradox arises here because there exist two important underlying relationships involving the three variables considered in this story. These variables are table number, hat fitting outcome, and hat color. Table number is related to both the fit of the hat and to the color of the hat. The first relationship relating table number to hat fitting means that: for each color separately, the hats on table 1 fit better than the hats on table 2. Yet the second relationship relating table number to hat color says that there are more blue hats than green hats on table 1, where hats fit better, and there are more green hats than blue hats on table 2, where hats fit worse. The observed reversal is not really a paradox at all because such a reversal can actually happen if, as in our story, table number is related to both hat fitting and hat color. We call this principle **data-based confounding**. We will describe it in more detail in the lessons to follow.

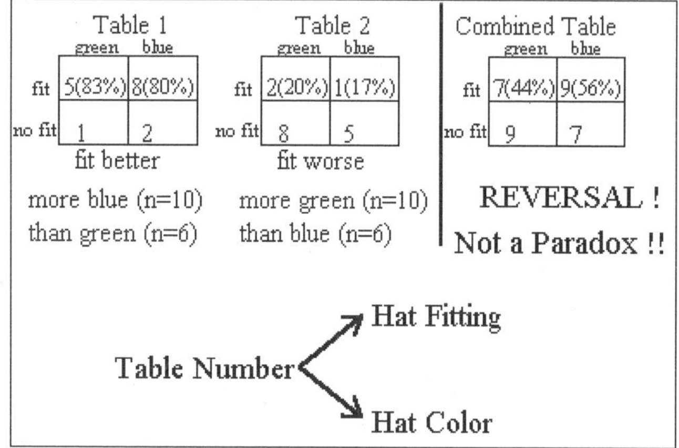

(Note: "Table Number" is the confounder (C); "Hat Fitting" the outcome (D), and "Hat Color" the exposure (E).)

Summary

- Simpson's Paradox illustrates a general principle about the relationship between two variables E and D when a third variable C is also considered.
- The principle: If the variable C is both strongly related to E and to D, then the relationship between E and D for the combined data, which ignores C, can be different from the relationship between E and D when we stratify on categories of C.
- The principle is the basis of the **data-based criterion** for confounding, which is described in the activities that follow.

The Concept of Confounding

Simpson's Paradox is an example of the confusion that can result from **confounding**. Confounding is an important problem for health and medical researchers whenever they conduct studies to assess a relationship between an exposure, **E**, and some health outcome or disease of interest, **D**. Confounding is a type of bias that may occur when we fail to take into account other variables, like age, gender, or smoking status, in attempting to assess an E→D relationship.

To illustrate confounding consider the results from a hypothetical retrospective cohort study to determine the effect of exposure to a suspected toxic chemical on the development of lung cancer for workers in a chemical industry. We will call the chemical TCX. The ten-year risks for lung cancer are estimated to be 0.36 for those who were exposed to TCX and 0.17 for those who were not exposed to TCX. The estimated risk ratio is 2.1, which indicates that those exposed to TCX have twice the risk for lung cancer as those unexposed. So far, we have considered two variables, exposure to TCX, and lung cancer status, the health outcome.

Chemical Workers	TCX	no TCX	Total
LC	27	14	41
No LC	48	67	115
Total	75	81	156

Ten-year risks for LC

TCX: 27/75 = 0.36

no TCX: 14/81 = 0.17

$\hat{RR} = 0.36/0.17 = 2.1$

We haven't yet considered any other variables that might also have been measured or observed on the patients in this study. For example, we might wonder whether there were relatively more smokers among those who were exposed to TCX than those unexposed to TCX. If so, that may explain why workers exposed to TCX were found to have an increased risk of 2.1 compared to unexposed workers. Those exposed to TCX may simply have been heavier smokers and, therefore, more likely to develop lung cancer than among those not exposed to TCX, regardless of exposure to TCX. Perhaps TCX exposure is a determinant of some other form of cancer or another disease, but not necessarily lung cancer.

Suppose, that we categorize our study data into two smoking history categories, non-smokers and smokers. For these tables, the estimated risk ratio is computed to be 1.0 for non-smokers and 1.3 for smokers. Notice that these two stratum-specific risk ratios suggest no association between exposure to TCX and the development of lung cancer.

Chemical Workers	TCX	no TCX	Total
LC	27	14	41
No LC	48	67	115
Total	75	81	156

$\hat{RR} = 2.1$

non-smokers

	TCX	no TCX	Total
LC	1	2	3
No LC	24	48	72
Total	25	50	75

smokers

	TCX	no TCX	Total
LC	26	12	38
No LC	24	19	43
Total	50	31	81

$\hat{RR} = 1.0$ No Association $\hat{RR} = 1.3$

Study Question (Q10.1)

1. Actually, concluding from these data that there is no association after controlling for smoking is debatable. Can you explain why?

When we form strata by categorizing the entire dataset according to one or more variables, like smoking history in our example here, we say that we are **controlling** for these variables, which we often refer to as **control variables**. Thus, what looks like a twofold increase in risk when we ignore smoking history, changes to no association when controlling for smoking history. This suggests that the reason why workers exposed to TCX had a twofold increase in risk compared to unexposed workers might be explained simply by noting that there were relatively more smokers among those exposed to TCX. This is an example of what we call **confounding**, and we say that smoking history is a confounder of the relationship between TCX exposure status and ten-year risk for lung-cancer. In general, confounding may be described as a distortion in a measure of association, like a risk ratio, that may arise because we fail to control for other variables, for example, smoking history, that might be risk factors for the health outcome being studied. If we fail to control the confounder we will obtain an incorrect, or biased, estimate of the measure of effect.

Summary

- ❖ Confounding is a distortion in a measure of effect, e.g., RR, that may arise because we fail to control for other variables, for example, smoking history, that are previously known risk factors for the health outcome being studied.
- ❖ If we ignore the effect of a confounder, we will obtain an incorrect, or biased, estimate of the measure of effect.

Quiz (Q10.2)

1. Confounding is a **???** in a **???** that may arise because we fail to **???** other variables that are previously known **???** for the health outcome being studied.

Choices
case-control study **control for** **distortion** **effect modifiers** **eliminate**
measure of effect **risk factors**

A study finds that alcohol consumption is associated with lung cancer, crude OR = 3.5. Using the data below, determine whether smoking could be confounding this relationship.

2. What is the OR among smokers? . . . **???**
3. What is the OR among non-smokers? . . . **???**
4. Does the OR change when we control for smoking status? **???**
5. Is there evidence from this data that smoking is a confounder of the relationship between alcohol consumption and lung cancer? . . . **???**

Choices
0.01 **1.0** **3.5** **5.4** **no** **yes**

	Smokers			Non-Smokers	
	Alcohol	No Alc		Alcohol	No Alc
Cases	560	140	Cases	40	160
Controls	240	60	Controls	160	640

Crude versus Adjusted Estimates

Confounding is assessed in epidemiologic studies by comparing the **crude estimate of effect** (e.g., $c\hat{RR}$) in which no variables are controlled, with an **adjusted estimate of effect** (e.g., $a\hat{RR}$), in which one or more variables is controlled. The adjusted estimate is typically computed by combining stratum-specific estimates into a single number.

For example, to assess confounding by smoking history in the previously described retrospective cohort study of the effects of exposure to the chemical TCX on the development of lung cancer, we can compare the crude risk ratio of 2.1 to an adjusted risk ratio that combines the risk ratios of 1.0 and 1.3 for the two smoking history categories. The method for combining these estimates into a single summary measure is the topic of the next activity. Once we have combined these stratum specific estimates, how do we decide if there is confounding? The **data-based criterion** for confounding requires the crude estimate of effect to be different from the adjusted estimate of effect. How different must these two estimates be to conclude that there is confounding? To answer this question, the investigator must decide whether or not there is a clinically important difference.

In our retrospective cohort study the adjusted estimate would be some number between 1.0 and 1.3, which suggests a much weaker relationship than indicated by the crude estimate of 2.1. Most investigators would consider this a clinically important difference. Suppose the crude estimate had been 4.2 instead of 2.1, the difference between crude and adjusted estimates would indicate even much stronger confounding.

We can compare other crude and adjusted estimates. Suppose, for example that an estimated crude risk ratio was 4.2 and the estimated adjusted risk ratio was 3.8. Both these values indicate an association that is about equally strong, so there is no clinically important difference between crude and adjusted estimates. Similarly, if the estimated crude risk ratio is 1.2 and the estimated adjusted risk ratio is 1.3, both these values indicate about the same very weak or no association. So, here again, there is no clinically important difference between these crude and adjusted estimates.

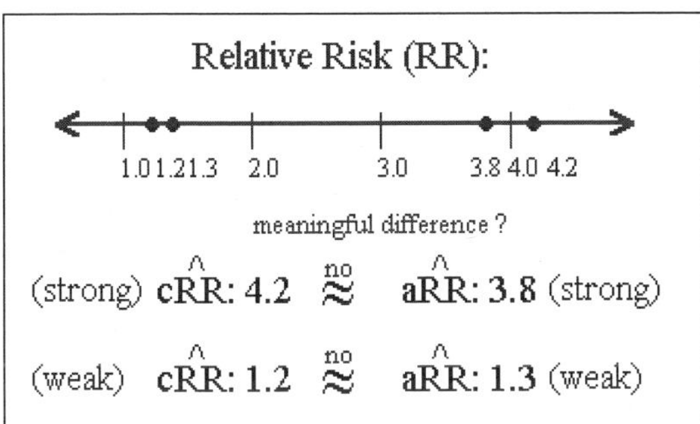

Risk Ratio (RR)

Clearly, deciding on what is clinically important requires a subjective decision by the investigators. One investigator might conclude, for example, that the difference between a 1.2 and a 1.3 is clinically important whereas another investigator might conclude otherwise. This problem may lead one to want to use a test of statistical significance to decide on whether there is a difference between the crude and adjusted estimate. However, because confounding is a validity issue, it should not be evaluated using a statistical test, but rather by looking for a meaningful difference, however imprecise.

A commonly used approach for assessing confounding is to specify, prior to looking at one's data, how much of a change in going from the crude to the adjusted estimate is required. Typically, a 10 per cent change is specified, so that if the crude risk ratio estimate is say, 4, then a 10% change in this estimate either up or down would be obtained for an adjusted risk ratio of either 3.6 or 4.4. Thus, if the adjusted risk ratio were found to be below 3.6 or above 4.4, we would say that confounding has occurred with at least a 10% change in the estimated association.

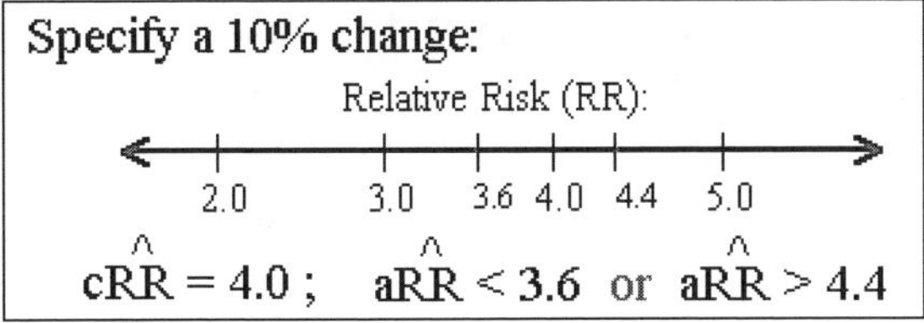

As another example, if a 20% change is specified and the crude risk ratio is, say, 2.5, the adjusted risk ratio would have to be either below 2 or above 3 to conclude that there is confounding.

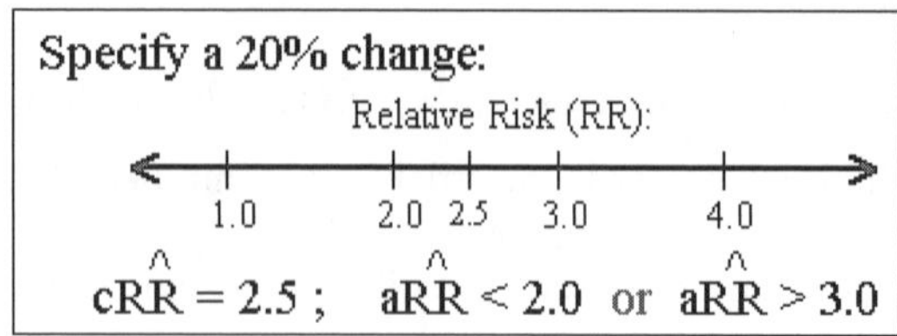

Summary

- ❖ Confounding is assessed by comparing the crude estimate of effect, in which no variables are controlled, with an adjusted estimate of effect, in which one or more variables are controlled.
- ❖ Confounding is present if we conclude that there is a clinically important or meaningful difference between crude and adjusted estimates.
- ❖ We do not use statistical testing to evaluate confounding.
- ❖ A commonly used approach for assessing confounding is to specify, prior to looking at one's data, how much of a change in going from the crude to the adjusted estimate is required.

Quiz (Q10.3)

A case-control study was conducted to study the relationship between oral contraceptive use and ovarian cancer. The crude OR was calculated as 0.77. Since age was considered a potential confounder in this study, the data were stratified into 3 age groups as shown below.

1. The OR for the 20-39 year age group is **???**
2. The OR for the 40-49 year age group is **???**
3. The OR for the 50-54 year age group is **???**
4. Do you think that this data provides some evidence that age is a confounder of the relationship between oral contraceptive use and ovarian cancer? **???**

Choices
0.58 **0.61** **0.65** **0.69** **0.77** **1.45** **no** **yes**

	Ages 20-39		Ages 40-49		Ages 50-54	
	OCs	No OCs	OCs	No OCs	OCs	No OCs
Case	46	12	30	30	17	44
Control	285	51	463	301	211	331

In this study described in the previous question, the crude OR was 0.77, and the adjusted OR controlling for age was 0.64.

5. If a 15% change in the crude versus adjusted OR is specified by the investigator as a meaningful difference, an adjusted OR less than **???** or greater than **???** provides evidence of confounding.
6. Is there evidence of confounding? **???**
7. Suppose the investigators determined that a 20% change was a meaningful difference. Is there evidence of confounding? **???**

Choices
0.15 **0.63** **0.65** **0.85** **0.89** **0.92** **no** **yes**

10-2 Adjusted Estimates

Characteristics of Adjusted Estimates

Consider again the results of a ten-year retrospective cohort study to determine whether exposure to the chemical TCX is a determinant of lung cancer for workers in a chemical industry. The estimated ten-year risk ratio is 2.1. This value is our crude estimate of effect because its computation ignores the control of other variables like smoking history.

Retrospective Cohort Study

	TCX	No TCX	Total
LC	27	14	41
No LC	48	67	115
Total	75	81	156

$c\hat{RR} = 2.1$

We compute an adjusted estimate by combining the stratum specific estimates to obtain a single summary measure. The typical summary measure used as an adjusted estimate is a suitably chosen weighted average of the stratum specific estimates.

In our retrospective cohort example, where the estimated risk ratios for two smoking categories were 1.0 and 1.3, a simple way to obtain this adjusted risk ratio as a weighted average is to use this formula:

Adjusted Estimate ($a\hat{RR}$): Weighted Average

Estimated Risk Ratios : 1.0 and 1.3

$$a\hat{RR} = (w_1 \times 1.0 + w_2 \times 1.3) / (w_1 + w_2)$$

Arithmetic Average

$$w_1 = w_2$$

$$a\hat{RR} = (1.0 + 1.3) / 2 = 1.15$$

The terms w_1 and w_2 represent weights that are respectively multiplied by the corresponding stratum-specific risk ratio estimates and then combined to form a weighted average. (Note: a more complicated version of a weighted average that is typically used for adjusted risk ratio estimates is described on page 14-3 in Lesson 14 when we return to discuss weighted average formulae used in a stratified analysis.) What weights should we use? In a simple arithmetic average, w_1 equals w_2, and that the adjusted estimate simplifies to the sum of the two risk ratio estimates divided by 2. The value 1.15 is a reasonable number to use as our adjusted estimate, because it lies halfway between the two estimated risk ratios. However, most epidemiologists prefer to give unequal weights to each stratum-specific estimate, especially if the sample sizes are very different between different strata. If we again look at the tables for the two smoking history categories in our example, the sample sizes for each group are 75 and 81, which though not exactly equal, are not very different. So maybe an arithmetic average isn't so bad for this example.

Retrospective Cohort Study

Stratum 1: Non-smokers	TCX	No TCX	Total	Stratum 2: Smokers	TCX	No TCX	Total
LC	1	2	3	LC	26	12	38
No LC	24	48	72	No LC	24	19	43
Total	25	50	75	Total	50	31	81

$\hat{RR}_1 = 1.0$ $\hat{RR}_2 = 1.3$

On the other hand, if there were 10 times as many non-smokers as there were smokers, we might want to give more weight to the non-smokers, for which we had more information. In fact, we might consider giving the first stratum ten times the weight of the second stratum, thus yielding the following weighted average, which comes out to be 1.03. Notice that this value is much closer to the estimated risk ratio of 1.0 for stratum 1 because we are giving more weight to that stratum.

$$a\hat{RR} = (10 \times 1.0 + 1 \times 1.3) / (10 + 1) = 1.03$$

Summary

- The typical adjusted estimate is a suitably chosen weighted average of stratum-specific estimates of the measure of effect.
- Epidemiologists prefer to use unequal weights, particularly if the sample sizes are different in different strata.
- When weights are chosen according to sample size, larger strata receive larger weights.
- A more complicated version of a weighted average typically used for adjusted RR estimates is described on page 14-3 in Lesson 14 on stratified analysis. This more complicated version is called a **precision-based adjusted RR**.

Quiz (Q10.4)

Determine whether each of the following is **True** or **False**.

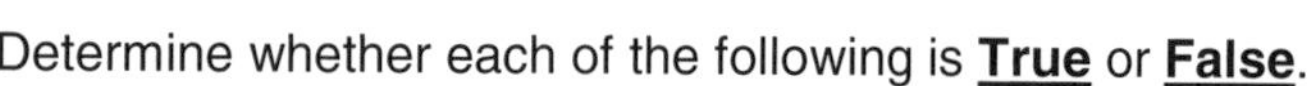

1. An adjusted estimate is a suitably chosen weighted average of the stratum specific estimates. ???. T
2. An adjusted estimate is always less than the corresponding crude estimate. . . ??? F
3. Most epidemiologists prefer to give equal weight to each stratum specific estimate in case-control studies. . . . . ???. F
4. Confounding is a validity issue and therefore, requires the use of a statistical test to determine its significance. . . . . ???. F

Use the formula below to calculate the adjusted RR for the following examples whose stratum specific estimates are given. (Note: Although the formula below gives a weighted average, the usual formula for aRR is a more complicated "precision-based" weighted average described in Lesson 14.)

$$a\hat{RR} = (w_1 \times \hat{RR}_1 + w_2 \times \hat{RR}_2)/(w_1 + w_2)$$

5. Stratum 1: RR=1.13, w=13.1; Stratum 2: RR=1.00, w=7.7. The adjusted RR is . . ???. 1.08
6. Stratum 1: RR=2.25, w=31.3; Stratum 2: RR=1.75, w=5.6. The adjusted RR is . . ??? 2.17

Choices
1.06 1.07 1.08 1.09 1.98 2.08 2.17

Criteria for Confounding

In addition to the data-based criterion for confounding, we must assess several **a priori criteria**. These are conditions to consider at the study design stage, prior to data collection, to identify variables to be measured for possible control in the data analysis.

The first a priori criterion is that a **confounder must be a risk factor for the health outcome**. This criterion ensures that a crude association between exposure and disease cannot be explained away by other variables already known to predict the disease. Such variables are called **risk factors**. For example, suppose we are studying the link between exposure to a toxic chemical and the development of lung cancer in a chemical industry. Based on the epidemiologic literature on the determinants of lung cancer, we would want to control for age and smoking status, two known risk factors. Our goal is to determine whether exposure to the chemical contributes anything over and above the effects of age and smoking on the development of lung cancer.

The second criterion is that a confounder cannot be an **intervening variable** between the exposure and the disease. A pure intervening variable (**V**) is any variable whose relationship to exposure and disease lies entirely within the causal pathway between exposure and disease.

Study Question (Q10.5)

1. Given a hypothetical scenario where saturated fat levels are measured to determine their effects on CHD, would we want to control for HDL levels?

If we control for HDL level, we essentially control for the saturated fat level, and we would likely find an adjusted risk ratio or odds ratio relating saturated fat to coronary heart disease status to be close to the null value. The intervening variable here is HDL level, and we should not control for it.

E → V → D

Saturated Fat → HDL → Coronary Heart Disease

The third criterion is that a **confounder must be associated with the exposure in the source population being studied**. By source population we mean the underlying population cohort that gives rise to the cases used in the study. Consider a study to assess whether a particular genetic factor, BRCAl, is a determinant of breast cancer. Age is a well-known risk factor for breast cancer, but is clearly not associated with the presence or absence of the gene in whatever source population is being studied.

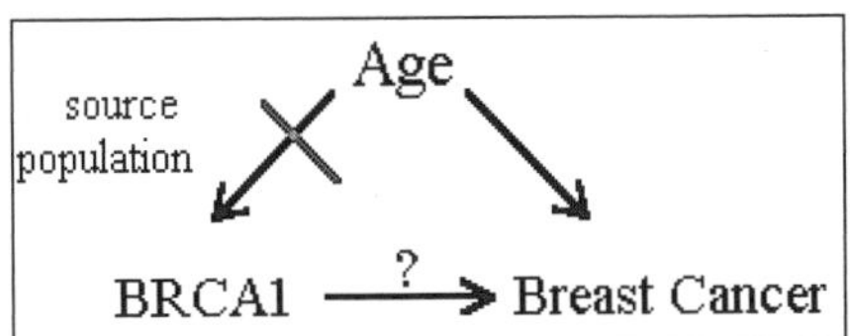

The third a priori criterion is therefore not satisfied. Consequently, even if by some chance, age turned out to be associated with the gene in the study data, we would not control for age, even though there is data-based confounding, because age does not satisfy all a priori criteria.

All three a priori criteria plus the data-based criterion are required for a variable to be considered a true confounder.

A priori criteria

A confounder

✓ 1. must be a risk factor.

✓ 2. cannot be an intervening variable

✓ 3. must be associated with the exposure in the source population.

Data-based criterion

✓ adjusted estimate $\neq$ crude estimate

Summary

A confounder must satisfy 3 a priori criteria in addition to the data-based criterion for confounding. These are:

- A confounder must be a **risk factor** for the health outcome.
- A confounder **cannot be an intervening variable** in the causal pathway between exposure and disease.
- A confounder must be **related to exposure in the source population** from which the cases are derived.

Some "fine points" about risk factors

The decision regarding which variables to include in the list of risk factors is, in practice, rarely a clear-cut matter. Such is the case when only a small amount of literature is available on a given study subject. On the other hand, a large literature may be controversial in terms of which previously studied variables are truly predictive of the disease.

Also, after data collection, but prior to the primary data analysis, the list of risk factors may need to be re-evaluated to allow for the addition or deletion of variables already measured but not explicitly considered for control. Variables measured for other purposes, say in a broad study to evaluate several etiologic questions, may be added to the list of risk factors if they were previously overlooked.

Furthermore, a **surrogate** of a risk factor may have to be used when the latter is difficult to measure. For example, the number of years spent in a given job in a particular industry is often used as a surrogate measure for the actual amount of exposure to a toxic substance suspected of being an occupationally related carcinogen.

Some "fine points" about a priori criterion 3

The third a priori criterion for confounding is of particular concern in case-control studies, where the controls are usually selected into the study after the cases have already occurred. In such studies, it is possible that the study data are not representative of the source population with regard to the exposure as well as other variables.

Therefore, a variable, say **C**, that may be not associated with exposure in the source population may still be associated with the exposure in the actual study data. In such a case, criterion 3 says that the variable **C** cannot be considered a confounder, and should not be controlled, even if there is data-based confounding.

In cohort studies, the exposure status is determined before disease status has occurred, so that the source population is the study cohort. In such studies, a variable, **C**, that is not associated with the exposure in the study data, does not satisfy condition 3 and therefore should not be considered a confounder.

The main difficulty in assessing the third a priori criterion concerns how to determine the association of the suspected confounder, **C**, with the exposure, **E**, in the in the source population. This requires some knowledge of the epidemiologic literature about the relationship between **C** and **E** and about the source population being studied

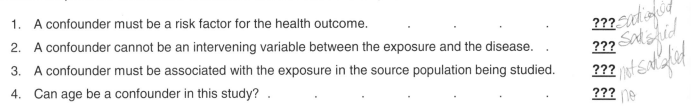

Quiz (Q10.6)

A study was conducted to assess the relationship between blood type (O-positive and O-negative) and a particular disease. Since age is often related to disease outcome, it was considered a potential confounder. Determine whether each of the apriori criteria below is satisfied.

1. A confounder must be a risk factor for the health outcome. . . . . . ???
2. A confounder cannot be an intervening variable between the exposure and the disease. . ???
3. A confounder must be associated with the exposure in the source population being studied. ???
4. Can age be a confounder in this study? . . . . ???

Choices

no not satisfied satisfied yes

10-3 A priori Criteria/Study Designs

Confounding in Different Study Designs

We have seen that the assessment of confounding requires both data-based and apriori criteria. The data-based criterion requires that the crude estimate of effect be meaningfully different from the adjusted estimate of effect. The adjusted estimate of effect is computed as a weighted average of stratum-specific estimates obtained over different categories of the potential confounder. The measure of effect used for this comparison changes with the study design but it always compares a crude estimate with an adjusted one.

We have thus far only considered follow-up studies where the measure of effect of interest is the risk ratio. For case-control studies, we compare crude and adjusted estimates of the exposure odds ratio. In cross-sectional studies, we compare crude and adjusted estimates of the prevalence odds ratio or the prevalence ratio.

Data-based Confounding	
	Crude vs. Adjusted
Follow-up:	$\text{c}\widehat{\text{RR}}$ vs. $\text{a}\widehat{\text{RR}}$
Case-control:	$\text{c}\widehat{\text{EOR}}$ vs. $\text{a}\widehat{\text{EOR}}$
Cross-sectional:	$\text{c}\widehat{\text{POR}}$ vs. $\text{a}\widehat{\text{POR}}$ or $\text{c}\widehat{\text{PR}}$ vs. $\text{a}\widehat{\text{PR}}$

Summary

- The measure of association used to assess confounding will depend on the study design.
- In a follow-up study, we typically compare a crude risk ratio with an adjusted risk ratio.
- In a case-control study, we typically compare a crude exposure odds ratio with an adjusted exposure odds ratio.
- In a cross-sectional study, we typically compare a crude prevalence odds ratio with an adjusted prevalence odds ratio, or we might use a prevalence ratio instead of a prevalence odds ratio.
- Regardless of the study design, data based confounding is assessed by comparing a crude estimated of effect with an appropriate adjusted estimate of effect.

Assessing Confounding in Case-Control Studies

These tables show results from a case-control study to assess the relationship of alcohol consumption to oral cancer. The tables describe the crude data when age is ignored and the stratified data when age has been categorized into three groups. The investigators wanted to evaluate whether there was possible confounding due to age.

Case-control Study
Alcohol consumption and oral cancer

Crude:

	Alc	no Alc
OCa	27	47
no OCa	90	443

n = 607

Stratified:

	Ages 40-49 Alc	Ages 40-49 no Alc	Ages 50-59 Alc	Ages 50-59 no Alc	Ages 60+ Alc	Ages 60+ no Alc
OCa	4	25	12	10	11	12
no OCa	22	309	37	67	31	67
	n = 360		n = 126		n = 131	

Study Questions (Q10.7)

1. Which expression should be used to calculate the crude odds ratio relating alcohol consumption to oral cancers?
   A. (27*90)/(47*443)
   B. 27*443/47*90
   C. (27*443)/(47*90)
3. What are the stratum-specific odds ratios?
   D. 2.2, 2.2, 2.0
   E. 0.45, 0.45, 0.5
   F. 0.6, 0.2, 0.2
3. Is there data-based confounding due to age? (Yes or No)
4. Assuming age satisfied all three a priori conditions for confounding, why is age a confounding of the relationships between alcohol consumption and oral cancer? a. all 3 apriori conditions for confounding are satisfied b. data based citerion is satisfied

Summary

- As an example to assess confounding involving the exposure odds ratio, we consider a case-control study of the relationship between alcohol consumption and oral cancer.
- Age, a possible confounder, has been categorized into three groups.
- The crude estimate of 2.8 indicates a threefold excess risk whereas the adjusted estimate of 2.1 indicates a twofold excess risk for drinkers over non-drinkers.
- The results indicate that there is confounding due to age.

Quiz (Q10.8)

The data below are from a cross-sectional seroprevalence survey of HIV among prostitutes in relation to IV drug use. The crude prevalence odds ratio is 3.59.

	Black or Hispanic		White	
	IV Drug Use	No IV Drug Use	IV Drug Use	No IV Drug Use
HIV +	31	12	16	3
HIV -	93	144	141	124

1. What is the estimated POR among the black or Hispanic group? . . . . . **???**
2. What is the estimated POR among the whites? . . . . **???**
3. Which table do you think should receive more weight when computing an adjusted odds ratio? **???**

Choices
3.25 **3.59** **4.00** **4.31** **4.69** **Black or Hispanic** **White**

In the study described in the previous questions, the estimated POR for the Black or Hispanic group was 4.00 and the estimated POR for the Whites was 4.69. The "precision-based" adjusted POR for this study is 4.16. Recall that the crude POR was 3.59.

4. Is there confounding? . . . . **???**

Choices
1.00 **4.35** **4.97** **maybe** **no** **yes**

10-4 Confounding, Interaction, and Effect Modification

Confounding versus Interaction

Another reason to control variables in an epidemiologic study is to assess for interaction. To assess interaction, we need to determine whether the estimate of the effect measure differs at different levels of the control variable. Consider the results from a case-control study to assess the potential relationship between alcohol consumption and bladder cancer. These data are stratified on race in three categories.

Case-control study

Alcohol → ? → Bladder cancer

	White		Black		Asian		Combined	
	Alc	no Alc	Alc	no Alc	Alc	no Alc	Alc	no Alc
Case	72	41	93	54	68	33	233	128
Control	106	105	113	113	78	142	297	360
	$\hat{OR} = 1.74$		$\hat{OR} = 1.72$		$\hat{OR} = 3.75$		$c\hat{OR} = 1.73$	

The estimated odds ratios for the three race strata and the combined strata are computed to be 1.74, 1.72, 3.75, and 1.73. There is clear evidence of interaction here because the effect is strong in Asians, but less so in Whites and Blacks. It is not clear whether there is confounding, since the value of the adjusted estimate could vary between 1.72 and 3.75, depending on the weights assigned to the strata. The precision-based adjusted odds ratio is computed to be 2.16, which is not very different from the crude odds ratio of 1.73, suggesting that there is little evidence of confounding.

Confounding and interaction are different concepts. Confounding compares the estimated effects before and after control whereas interaction compares estimated effects after control. When assessing confounding and interaction in the same study, it is possible to find one with or without the other.

Consider the table shown here giving stratum specific and crude risk ratio estimates from several hypothetical data sets in which one dichotomous variable is being considered for control. For each data set in this table, do you think there is interaction or confounding? Think about your answer for a few minutes, and then continue to see the answers below.

Confounding versus Interaction

Data Set	Stratum 1 RR	Stratum 2 RR	Crude RR	Interaction?	Confounding?
1	1.02	3.50	6.00		
2	1.02	3.50	2.00		
3	0.03	3.50	1.70		
4	1.00	1.00	4.10		
5	4.00	4.10	4.20		

Let us look at the data sets, one at a time. For dataset 1, there is clearly interaction, because the estimate for stratum 1 indicates no association but the estimate for stratum 2 indicates a reasonably strong association. There is clearly confounding, because any weighted average of the values 1.02 and 3.50 will be meaningfully different from the crude estimate of 6.0.

For data set 2, again there is clearly interaction, as in data set 1. However, it is not clear whether or not there is confounding. The value of an adjusted estimate will depend on the weights assigned to each stratum. If all the weight is given to either stratum 1 or stratum 2, then the crude estimate of 2.0 will differ considerably from the adjusted estimate, but if equal weight is given to each stratum, the adjusted estimate will be much closer to the crude estimate. Nevertheless, the use of an adjusted estimate here is not as important as the conclusion that the E→D association is different for different strata.

Dataset 3 also shows interaction, although this time the nature of the interaction is different from what we observed in datasets 1 and 2. Here, the two stratum specific estimates are on opposite sides of the null risk ratio value of 1. It appears there is a protective effect of exposure on disease in stratum 1, but a harmful effect of exposure on disease in stratum 2. In this situation, the assessment of confounding is questionable and potentially very misleading, since the important finding here is the interaction effect, especially if this strong interaction holds up after performing a statistical test for interaction.

In dataset 4, the two stratum specific estimates are identically equal to one, so there is no interaction. However, there is clear evidence of confounding, since the crude estimate of 4.0 is meaningfully different from both stratum-specific estimates.

In dataset 5, there is no interaction, because the stratum-specific estimates are both equal to 4. There is also no confounding because the crude estimate is essentially equal to both stratum specific estimates.

Data Set	Stratum 1 RR	Stratum 2 RR	Crude RR	Interaction?	Confounding?
1	1.02	3.50	6.00	yes	yes
2	1.02	3.50	2.00	yes	?
3	0.03	3.50	1.70	yes	??
4	1.00	1.00	4.10	no	yes
5	4.00 ↖	4.10 ↖	4.20 ↖	no	no

Summary

- Confounding and interaction are different concepts.
- Interaction considers what happens after we control for another variable.
- Interaction is present if the estimate of the measure of association differs at different levels of a variable being controlled.
- When assessing confounding and interaction in the same study, it is possible to find one with or without the other.
- In the presence of strong interaction, the assessment of confounding may be irrelevant or misleading.

Two Types of Interaction - Additive and Multiplicative

Consider the following table that gives the cell frequencies for the categorization of three dichotomous variables **e**, **d** and **f**. Assume that these data derive from a cohort study that estimates risk.

	e=1, f=1	e=1, f=0	e=0, f=1	e=0, f=0
d=1	a1	a2	b1	b2
d=0	c1	c2	d1	d2

Note that this table equivalently represents two 2 x 2 tables that describes the relationship between e and d after stratifying on **f**.

We now define three risk measures from this table that we will use to provide formulae that distinguish additive from multiplicative interaction.

$\mathbf{R_{11}}$ **= Risk for (e = 1, f = 1) = a1 / (a1 + c1)**

$\mathbf{R_{10}}$ **= Risk for (e=1, f = 0) = a2 / (a2 + c2)**

$\mathbf{R_{01}}$**= Risk for (e=0, f= 1) = b1 / (b1 + d1)**

$\mathbf{R_{00}}$ **= Risk for (e=0, f = 0) = b2 / (b2 + d2)** (background risk)

We then define **no interaction on an additive scale** if: $R_{11} - R_{10} - R_{01} + R_{00} = 0$
We also define **no interaction on an multiplicative scale** if: $RR_{11} = RR_{10} \times RR_{01}$

where $RR_{11} = R_{11} / R_{00}$, $RR_{10} = R_{10} / R_{00}$, and $RR_{01} = R_{01} / R_{00}$

To illustrate these two definitions consider the following risk data that consider the possible effect modification of genetic make-up (**f**) on the relationship between the presence of athlerosclerosis (**e**) and Alzheimer's disease (**d**) in an elderly population in Rotterdam, The Netherlands.

$\mathbf{R_{11}}$ **= Risk for (e = 1, f = 1) = 13.3%**

$\mathbf{R_{10}}$ **= Risk for (e = 1, f = 0) = 4.4%**

$\mathbf{R_{01}}$ **= Risk for (e = 0, f = 1) = 4.8%**

$\mathbf{R_{00}}$ **= Risk for (e = 0, f = 0) = 3.4%**

To evaluate interaction on an additive scale, we calculate: $R_{11} - R_{10} - R_{01} + R_{00} = 13.3 - 4.4 - 4.8 + 3.4 = 7.5$

which is meaningfully different from zero, thus indicating a "departure from no interaction on an additive scale."

To evaluate interaction on an multiplicative scale, we evaluate whether:

$RR_{11} - (RR_{10} \times RR_{01}) = (13.3/3.4) - (4.4/3.4) \times (4.8/3.4) = 3.912 - 1.827 = 2.1$

Continued on next page

Two Types of Interaction - Additive and Multiplicative (continued)

which is also meaningfully different from zero, thus indicating a "departure from no interaction on a multiplicative scale."

It turns out from algebra that "no interaction on an additive scale" is equivalent to obtaining "equal risk difference effect measures" when comparing the **e-d** effect stratified by **f**.

It also turns out from algebra that "no interaction on a multiplicative scale" is equivalent to obtaining "equal risk ratio effect measures" when comparing the **e-d** effect stratified by **f**.

Reference: Kleinbaum et al., Epidemiologic Research: Principles and Quantitative Methods, Chapter 19, John Wiley and Sons, 1982.

Quiz (Q10.9)

1. In contrast to **???** when interaction is present, the estimates of the **???** differ at various levels of the control variable.
2. When assessing confounding and interaction in the same study, it is **???** to find one without the other.

Choices

confounding **effect measure** **effect modification** **not possible** **possible** **precision** **variance**

For datasets 1-3 in the table below, select the best answer from the following:

A. Confounding
B. Interaction
C. No confounding or interaction
D. Calculation error (not possible)

3. Data set 1 ? **???**
4. Data set 2? **???**
5. Data set 3? **???**

	Crude OR	Stratum 1 OR	Stratum 2 OR	Adjusted OR
Dataset 1	4.0	1.0	6.0	4.0
Dataset 2	4.0	5.0	5.0	5.0
Dataset 3	4.0	4.0	4.0	3.0

Interaction versus Effect Modification

The term **effect modification** is often used interchangeably with the term **interaction**. We use effect modification from an epidemiologic point of view to emphasize that the effect of exposure on the health outcome is modified depending on the value of one or more control variables. Such control variables are called **effect modifiers** of the relationship between exposure and outcome. We use interaction from a statistical point of view to emphasize that the exposure variable and the control variable are interacting in some way within a mathematical model for determining the health outcome.

To illustrate effect modification, consider the case-control data to assess the relationship between alcohol consumption and bladder cancer. The data showed clear evidence of interaction, since the estimated effect was much stronger in Asians than in either Blacks or Whites. This evidence suggests that race is an effect modifier of the relationship between alcohol consumption and bladder cancer. Such a conclusion is supported by the epidemiologic literature, which indicates that alcohol is metabolized in Asians differently than in other racial groupings.

	White		Black		Asian		Combined	
	Alc	no Alc	Alc	no Alc	Alc	no Alc	Alc	no Alc
Case	72	41	93	54	68	33	233	128
Control	106	105	113	113	78	142	297	360
	$\hat{OR} = 1.74$		$\hat{OR} = 1.72$		$\hat{OR} = 3.75$		$c\hat{OR} = 1.73$	

The assessment of interaction or effect modification is typically supported using the results of statistical testing. Recall that confounding does not involve significance testing because it is a validity issue. Nevertheless, statistical testing of interaction is considered appropriate in epidemiologic studies because effect modification concerns understanding the underlying causal mechanisms involved in the E→D relationship, which is not considered a validity issue. One such test for stratified data that has been incorporated into available computer software, is called the **Breslow-Day test**.

Summary

- Effect modification and interaction are often used interchangeably.
- If there is effect modification, then the control variable or variables involved are called effect modifiers.
- The assessment of effect modification is typically supported by statistical testing for significant interaction.
- One popular statistical test for interaction is called the **Breslow-Day** test.

Is There Really a Difference between Effect Modification and Interaction?

Although the terms effect modification and interaction are often used interchangeably, there is some controversy in the epidemiologic literature about the precise definitions of effect modification and interaction (see Kleinbaum et al., Epidemiologic Research: Principles and Quantitative Methods, Chapter 19, John Wiley and Sons, 1982).

One distinction frequently made is that effect modification describes a non-quantitative clinical or biological attribute of a population, whereas interaction is typically quantitative and data-specific, and in particular, depends on the scale on which the "interacting" variables are measured. Nevertheless, this conceptual distinction is often overlooked in the applied research studies.

Why Do Epidemiologists Statistically Test for Interaction but not for Confounding?

We have previously (lesson page 10-1) pointed out that the assessment of confounding should not involve statistical testing, essentially because **confounding is a validity issue** involving systematic rather than random error. Moreover, if there is a meaningful difference between estimated crude and adjusted effects, then a decision has to be made as to which of these estimates to report; consequently, the adjusted effect must be used, without consideration of a statistical test, because it controls for the variables (i.e., risk factors) designated for adjustment.

Furthermore, it is not obvious, even if we wanted to statistically test for confounding, exactly how to properly perform a test for confounding. What is typically done, though incorrect, is to test whether the potential confounder, e.g., age, is significantly associated with the health outcome, possibly also controlling for exposure status. Such a test does not really assess confounding, since it concerns random error (i.e., variability) rather than whether or not the crude and adjusted estimates are different in the data!

As to whether or not one should do a statistical test for assessing **interaction/effect modification**, the answer is not as clear-cut. If we consider interaction as a data-based manifestation of a population-based phenomenon (i.e., effect modification), then a statistical test can be justified to account for the random error associated with a data-based result. Moreover, in contrast, to the assessment of confounding, there are several 'legitimate' approaches to testing for interaction, one of which is the Breslow-Day test for stratified data (described in Lesson 14) and another is a test for the significance of product terms in a logistic model.

Continued on next page

Why Do Epidemiologists Statistically Test for Interaction but not for Confounding? (continued)

Furthermore, it may be argued that the assessment of interaction/effect modification isn't a validity issue, but rather concerns the conceptual understanding/explanation of the relationships among variables designated for control. The latter argument, in this author's opinion, is a little too esoteric to accept at face value. In fact, a counter argument can be made that the presence of interaction/effect modification implies that the most "valid" estimates are obtained by stratifying on effect modifiers, provided that one can determine which variables are the "true" effect modifiers.

As in many issues like this one that arise in the undertaking of epidemiologic research, the best answer is probably, "it depends!" That is, it depends on the researcher's point of view whether or not a statistical test for interaction/effect modification is appropriate. Nevertheless, this author tends to weigh in with the opinion that effect modification is a population phenomenon that can be assessed using 'legitimate' statistical testing of a data-based measure of interaction.

Effect Modification – An Example

In the early 1990's, investigators of the Rotterdam Study screened 8,000 elderly men and women for the presence of Alzheimer's disease. One of the research questions was whether the presence of atherosclerosis increased the risk of this disease. In a cross-sectional study, the investigators found that patients with high levels of atherosclerosis had a three times increased risk of having Alzheimer's disease compared to participants with only very little atherosclerosis. These results were suggestive of a link between cardiovascular disease and neurodegenerative disease.

The investigators knew from previous research that one of the genes involved in lipid metabolism influences the risk of Alzheimer's disease. For this gene, there are two alternative forms, allele A and allele B. Each person's genetic make-up consists of two of these alleles. Persons with at least one B-allele have a higher risk of Alzheimer's disease than persons with two A-alleles.

The investigators hypothesized that a person's genetic make-up might modify the association they found between atherosclerosis and Alzheimer's disease. Therefore, they divided the study population into a group of participants who had at least one B-allele, and a group of participants with two A-alleles. They found the following results: Among those with at least one B allele, the prevalence of Alzheimer's disease for those with high levels of atherosclerosis was three times the prevalence of those with low levels. This result is the same as the crude. However, among those with two A-alleles, the prevalence for those with high levels of atherosclerosis was only 1.4 times the prevalence of those with low levels.

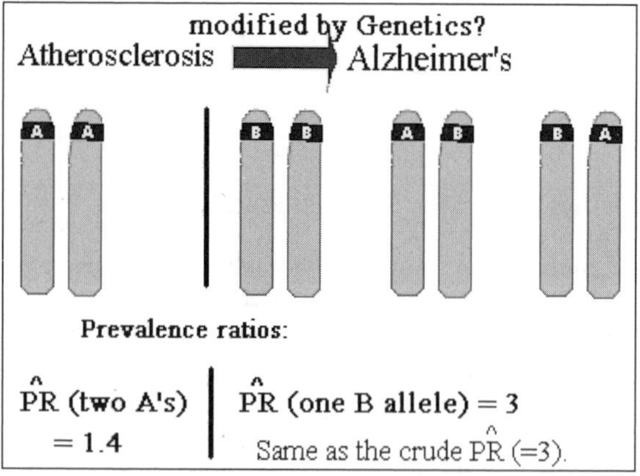

These results provide an example of effect modification. Genetic make-up is the effect modifier. The investigators showed that the extent of atherosclerosis is associated with Alzheimer's disease, but only in those whose genetic make-up predisposes them to the development of this disease.

Study Questions (Q10.10)

Prevalence of Alzheimer's Disease
High ATH, 1 B-allele: 13.3%
Low ATH, 1 B-allele: 4.4%
} $\hat{PR} = 3.0$
High ATH, 2 A-alleles: 4.8%
Low ATH, 2 A-alleles: 3.4%
} $\hat{PR} = 1.4$

The prevalence of Alzheimer's and the prevalence ratios for each gene group are listed above. Answer the following assuming the group with low ATH and two A-alleles is the reference.

1. What is the prevalence ratio comparing those with high ATH and two A-alleles to the reference group?
2. What is the prevalence ratio comparing those with low ATH and at least one B-allele to the reference group?
3. What is the prevalence ratio comparing those with both high ATH and at least one B-allele to the reference group?
4. What is the difference between the three prevalence ratios you just calculated and the two listed above?

Quiz (Q10.11)

True or **False**

1. The term effect modification emphasizes that the effect of exposure on the health outcome is modified depending on the value of one or more control variables. . . . . . . **???**
2. Evidence for effect modification is present when the stratum-specific measures of association are approximately the same. . . . . **???**
3. This assessment can be supported by a statistical test known as the Breslow-Day test. . **???**

A measles vaccine may be highly effective in preventing disease if given after a child is 15 months of age, but less effective if given before 15 months.

4. This example illustrates **???**, where the exposure is **???**, the outcome is **???**, and the effect modifier is **???**.

Choices
age at vaccination **confounding** **effect modification** **measles** **measles vaccine**

Nomenclature

$aE\hat{O}R$	Estimate of the adjusted exposure odds ratio
$aP\hat{O}R$	Estimate of the adjusted prevalence odds ratio
$a\hat{P}R$	Estimate of the prevalence ratio
$a\hat{R}R$	Estimate of an adjusted risk ratio
C	Confounding variable
$cE\hat{O}R$	Estimate of the crude exposure odds ratio
$cP\hat{O}R$	Estimate of the prevalence odds ratio
$c\hat{P}R$	Estimate of the prevalence ratio
$c\hat{R}R$	Estimate of the crude risk ratio
D	Disease
E	Exposure
V	Intervening variable
w or w_i	Weight; with a subscript i, denotes the weight for a stratum

References

Hofman A, Ott A, Breteler MM, Bots ML, Slooter AJ, van Harksamp F, van Duijn CN, Van Broeckhoven C, Grobbee DE. Atherosclerosis, apolipoprotein E, and prevalence of dementia and Alzheimer's disease in the Rotterdam Study. Lancet 1997;349 (9046):151-4.

Greenland S, Morgenstern H. Confounding in health research. Annu Rev Public Health 2001;22:189-212.

Kleinbaum DG, Kupper LL, Morgenstern H. Epidemiologic Research: Principles and Quantitative Methods. John Wiley and Sons Publishers, New York, 1982.

Miettinen O. Confounding and effect modification. Am J Epidemiol 1974;100(5):350-3.

Mundt KA, Shy CM. Interaction: An epidemiological perspective for risk assessment. In: Fan AM, Chang LW (eds.). Toxicology and risk assessment: Principles, methods, and applications. Marcel Dekker, Inc., New York, NY, p.p. 329-351, 1996

Whittemore AS. Collapsibility of multidimensional contingency tables. J R Stat Soc B 1978;40:328-40.

Homework

ACE-1. Confounding: Smoking

A case-control study was conducted to assess whether coffee consumption (high vs. low) is associated with peptic ulcer disease. The results, stratified on smoking status, are summarized below:

Smokers	High Coffee	Low Coffee
Ulcer	40	65
No Ulcer	45	150

Non-smokers	High Coffee	Low Coffee
Ulcer	150	45
No Ulcer	65	40

a. Calculate the stratum-specific odds ratios and the crude odds ratio. Show your calculations.
b. Based upon these calculations, is smoking status a confounder in these data? Justify your answer.
c. Is there evidence of interaction?
d. Assess whether

   i. smoking is related to coffee consumption among those without peptic ulcer disease
   ii. smoking is related to peptic ulcer disease among subjects with low coffee consumption.

e. What do your answers in part d tell you about whether smoking is a confounder in this study?
f. Is your conclusion in part e about whether smoking status is a confounder the same as in part a?

ACE-2. Confounding and Interaction

Because of an unusually high occurrence of endogenous anxiety syndrome among undergraduate students at University X, a case-control study was carried out to determine whether taking an introductory statistics course (a 0,1 exposure variable, where 1 = exposed and 0 = unexposed) might be a cause. The study involved 57 students diagnosed with the anxiety syndrome, and these 'cases' were compared with a sample of 750 'normal' controls. Consider the following 2 x 2 tables that stratify for previous history of mental disorder (MD):

MD = 1	E = 1	E = 0
D = 1	25	15
D = 0	50	50

MD = 0	E = 1	E = 0
D = 1	2	15
D = 0	50	600

a. What is the estimated measure of effect that describes the E-D relationship that ignores the control of the variable MD (i.e., what is the estimated crude effect for these data?)? In answering this question, show your calculations.
b. What are the estimated effect measures for each MD group? Again, show your calculations.
c. Based on your calculations for either or both of the above questions, should MD be controlled because there is meaning interaction? Explain.
d. Based on your calculations for parts a and b, should the variable MD be controlled because it is a confounder? (Assume that the a priori conditions for confounding are already satisfied for the variable MD.) Explain your answer with appropriate information and logic.
e. What is the estimated odds ratio that describes the association of MD with exposure? What does this odds ratio say about the distribution of previous history of mental disorder (i.e., MD) when comparing students taking introductory

statistics with those not taking introductory statistics. Does this result support the conclusion that MD is a confounder in these data?

ACE-3. Exposure and Potential Confounders

A cohort study was conducted to examine the relation between use of estrogen replacement therapy and ovarian cancer mortality among peri- and postmenopausal women. The following information relating the exposure variable to each of several possible confounders was provided:

Distribution of potential ovarian cancer risk factors and their association with use of estrogen replacement therapy (ERT).

Potential risk factor	Odds Ratio	95% Confidence Interval
Age at menarche		
< 12 yrs	1.00	Referent
12 yrs	0.98	0.95-1.00
13 yrs	0.94	0.83-1.05
Number of live births		
0	1.00	Referent
1	1.06	1.01-1.10
2-3	1.08	1.05-1.11
4	0.86	0.82-0.89
Education		
< High school	1.00	Referent
High school	1.92	1.89-1.95
College	3.48	2.26-5.59

Indicate which of the following statements is TRUE [Choose one best answer]:

i. Age at menarche is unlikely to be a strong confounder of the relationship between ERT and ovarian cancer.
ii. Number of live births is unlikely to be a strong confounder of the relationship between ERT and ovarian cancer.
iii. Education is definitely a confounder of the relationship between ERT and ovarian cancer.

ACE-4. Error type

For each of the situations described below, indicate the type of error that would most likely occur. Each of the numbered options can be used once, more than once, or not at all.

i. Selection bias
ii. Nondifferential misclassification
iii. Differential misclassification
iv. Confounding
v. Ecologic fallacy
vi. Random error

a. In a cohort study of hormone replacement therapy (HRT) and risk of atherosclerotic coronary artery disease (CAD), high income level is associated with both HRT use and risk of CAD.
b. In a case-control study of the relation between stressful daily events and asthma attacks, cases are more likely than controls to over-report the amount of stress.
c. In a cohort study of use of video display terminals (VDTs) and risk of carpal tunnel syndrome, the users of VDTs are more difficult to trace than nonusers, resulting in a greater loss to follow-up of VDT users.
d. In a case-control study of beta carotene and risk of esophageal cancer, serum specimens frozen and stored 20 years earlier are compared between cases and controls. Later it is found that the specimens deteriorated while in storage.

ACE-5. External Validity

Which one of the following approaches to control of confounding is most likely to affect the external validity of a study?

a. Randomization
b. Restriction
c. Stratified Analysis
d. Regression Analysis

ACE-6. Categorizing Exposure

A hospital-based case-control study was conducted to determine whether residential exposure to magnetic fields was associated with occurrence of lymphoma. The exposure variable (strength of magnetic field in the home) was originally measured as a continuous variable but was categorized for purposes of data analysis. As illustrated in the table below, the investigators presented results for three different categorization schemes:

Field Strength	Crude OR	95% CI	Adjusted OR*	95% CI
Scheme 1				
Low	1.0 (referent)		1.0 (referent)	
High	1.59	0.87-2.93	1.77	0.91-3.41
Scheme 2				
Low	1.0 (referent)		1.0 (referent)	
Medium	1.07	0.68-1.99	1.20	0.62-2.32
High	1.64	0.84-3.20	2.06	0.96-4.42
Scheme 3				
Very Low	1.0 (referent)		1.0 (referent)	
Low	1.22	0.49-3.02	1.46	0.53-4.04
Medium	1.01	0.50-2.04	1.19	0.56-2.64
High	1.24	0.61-2.56	1.48	0.66-3.32
Very High	6.81	1.63-28.5	13.43	1.76-102.7

- Adjusted for age, sex, socioeconomic status, geographic area, and years lived in home

a. Which categorization scheme produces results that are most consistent with a dose-response effect?

_____ Scheme 1 _____ Scheme 2 _____ Scheme 3

b. For each of the following statements, indicate whether the statement is true or false.

_____ Bias due to confounding (by the factors listed below the table) was consistently away from the null.

_____ Exposure categorization scheme #3 is least susceptible to residual confounding, since it divides the study subjects into the largest number of categories.

_____ Among subjects exposed to very high magnetic fields in the home, adjustment for confounding factors led to an increase in validity but a decrease in precision.

ACE-7. Confounding

Confounding has been defined in the literature as being comprised of both data-based criteria and apriori (to the data) criteria.

a. State the data-based criteria for confounding.
b. Give a numerical example of data-based confounding such that the bias is away from the null. (Note: you can answer this question without completely specifying all the cell frequencies in the 2x2 tables used for your example)
c. Give a numerical example of data-based confounding such that the bias is away from the null. (Note: you can answer this question without completely specifying all the cell frequencies in the 2x2 tables used for your example)
d. When assessing interaction, should you test for significant interaction? Why or why not?
e. When assessing confounding, should you test for confounding? Why or why not?

Three apriori conditions that have been defined for confounding are:

I. The potential confounder must be a risk factor for the disease.

II. The potential confounder cannot be an intervening variable between the exposure and the disease.

III. The potential confounder must be associated with the exposure in the source population being studied (note: this condition is the most controversial)

f. Give an example of a potential confounder that is a risk factor for some disease.
g. Give an example of an intervening variable between an exposure and a disease.
h. Give an example of a risk factor for a disease that is not associated with the exposure in the source population under study.
i. Would you control for a risk factor that is not an intervening variable and does not satisfy apriori condition III, but is, nevertheless, associated with both the exposure and the disease in the data? Explain.
j. For each of the conditions I-III above, provide a justification (either conceptual or numerical) for the condition if you agree that the condition is appropriate.

If, on the other hand, you believe that a condition is not appropriate (or needs further qualification), explain (either conceptually or numerically) why you think it is not appropriate.

ACE-8. Effect Modification and Confounding

A cohort study of physical activity (PA) and incidence of diabetes was conducted over a six-year period among Japanese-American men in Honolulu. Data from that study are summarized below, stratified on body mass index (BMI):

	High BMI		Low BMI	
	High PA	Low PA	High PA	Low PA
Diabetes	48	62	54	71
Person-Yrs	1050	1067	1132	1134

a. Based upon these data, what is the observed rate ratio (IDR) for someone with high BMI and high physical activity (i.e. relative to the background rate)?
b. Is there evidence of effect modification by BMI? (Show any calculations and justify your answer.)
c. Is there evidence of confounding by BMI in these data? (Show any calculations and justify your answer.)

ACE-9. Interaction: Epidemiologists and Headaches

The following data are from a cumulative incidence type cohort study of exposure to manuscripts written by a certain epidemiologist and intractable headaches (d) among epidemiology graduate students. The data are stratified by degree of previous experience with foreign languages (f).

	f = 1 (yes)		f = 0 (no)	
	e = 1	e = 0	e = 1	e = 0
d = 1	25	20	10	5
d = 0	975	980	990	995
	1000	1000	1000	1000

a. Assess whether there is interaction when comparing risk ratios and when comparing risk differences between the two categories of the variable f.
b. Fill in the following table:

	e=1, f = 1	e=1, f=0	e=0, f=1	e=0, f = 0
d = 1				
d = 0				

c. Calculate the following:

R_{11} = Risk for (e=1, f = 1) =

R_{10} = Risk for (e=1, f = 0) =

R_{01} = Risk for (e=0, f = 1) =

R_{00} = Risk for (e=0, f = 0) = (background risk)

d. We say that there is "no interaction on an additive scale if

$$R_{11} - R_{10} - R_{01} + R_{00} = 0$$

Based on the data, is there interaction on an additive scale?

e. We say there is "no interaction on an multiplicative scale" if

$$RR_{11} = RR_{10} \times RR_{01} \quad \text{where}$$

$$RR_{11} = R_{11} / R_{00} \qquad RR_{10} = R_{10} / R_{00} \text{ and } RR_{01} = R_{01} / R_{00}$$

Based on the data, is there interaction on a multiplicative scale?

f. Based on these results, which of the following would be appropriate as part of your overall analytic plan? (You may choose more than one.)

1. Report stratum-specific RRs (relative risks)
2. Report stratum-specific RDs (risk differences)
3. Calculate an overall summary estimate of the RR
4. Calculate an overall summary estimate of the RD
5. Calculate the crude RR and assess confounding
6. Calculate the crude RD and assess confounding

ACE-10. Misclassification Bias

The following table summarizes data from a case-control study:

	C = 1		C = 0	
	E = 1	E = 0	E = 1	E = 0
Cases	120	100	45	200
Controls	80	200	30	400

E = exposure status C= potential confounder/effect modifier

a. Assume that all subjects in the study have been correctly classified and that there is no bias due to selection. Is there evidence of effect modification and/or confounding due to C in these data? Justify your answer.
b. Now assume that the above observed data had not been correctly classified and that, moreover, the exposure was misclassified with a sensitivity of 0.8 and a specificity of 0.75. The misclassification was non-differential with respect to both outcome status and category of C.
c. Is there evidence of bias in the misclassified data? If so can you correct for the bias? Justify your answer.
d. Is there evidence of effect modification in the misclassified data? Justify your answer.
e. How does exposure prevalence affect the magnitude of bias due to misclassification in this example? Explain.

Answers to Study Questions and Quizzes

Q10.1

1. The 1.3 estimated RR for smokers says that there is a 30% increased risk for those exposed to TCX compared to unexposed. This amount of increase might be viewed by some researchers as indicating a moderate, rather than no association.

Q10.2

1. distortion, measure of association, control for, risk factors
2. 1.0
3. 1.0
4. yes
5. yes – The data-based assessment of confounding is made by determining whether the crude estimate of the measure of association is meaningfully different from an adjusted measure of association.

Q10.3

1. 0.69
2. 0.65
3. 0.61
4. yes – The data-based assessment of confounding is made by determining whether the crude estimate of the measure of association is meaningfully different from an adjusted measure of association.
5. 0.65, 0.89 – To determine whether there is a meaningful difference in the crude and adjusted estimates based on a specified percent change required between the crude and adjusted estimates, multiply the crude estimate by the specified percent change, and then add and subtract that value to the crude estimated. If the interval obtained contains the adjusted estimate, then there is no meaningful difference.
6. yes
7. no

Q10.4

1. True
2. False – The adjusted estimate can be greater or less than the corresponding crude estimate.
3. False – Most epidemiologists prefer to give unequal weight to each stratum specific estimate. Weights are usually determined based on sample size or precision.
4. False – Since confounding is a validity issue, it should not be evaluated by statistical testing, but by looking for a meaningful difference in the crude and adjusted estimates.
5. 1.08
6. 2.17

Q10.5

1. No

Q10.6

1. satisfied
2. satisfied
3. not satisfied – Age cannot possibly be associated with blood type.
4. no – Age cannot possibly be associated with blood type.

Q10.7

1. C; the crude odds ratio relating alcohol to oral cancer is calculated by using the formula (a x d)/(b x c) which in this case equals 2.8
2. A; the stratum-specific odds ratio relating alcohol to oral cancer can be calculated for each strata by using the formula (a x d)/(b x c). The stratum-specific odds ratios are 2.2 for the 40 to 49 age group, 2.2 for the 50 to 59 age group and 2.0 for the 60 and higher age group.
3. Yes; there is data-based confounding because the crude odds ratio of 2.8 is meaningfully different than any weighted average of stratum-specific odds ratios, all of which are about 2.
4. Age is a confounder because all three a priori conditions for confounding are assumed to be satisfied and the data-based criterion is also satisfied.

Q10.8

1. 4.00
2. 4.69
3. Black or Hispanic – You might think that the table for the white group should receive more weight since it has a slightly larger sample size, however, the table for the Black or Hispanic group is actually more balanced. See Lesson 14 for a more complete explanation on balanced data.
4. maybe – It depends on whether the investigator considers the difference between 3.59 and 4.16 a meaningful difference.

Q10.9

1. confounding, effect measure
2. possible

3. B
4. A
5. D – Data set 3: Recall that the adjusted estimate is a weighted average of the two stratum specific estimates and therefore, must lie between them.

Q10.10

1. 4.8% / 3.4% = 1.4
2. 4.4% / 3.4% = 1.3
3. 13.3% / 3.4% = 3.9
4. The two PRs above are the stratum-specific PRs for each of the two gene groups. The three calculated here use one group as a reference and compare the other three to that group. In this example, having low ATH and 2 A-alleles is the reference group compared to those having either one or both of the risk factors (high ATH, 1 B-allele). To see how these 3 PRs can be used to define two different types of interaction, see the first asterisk on this lesson page (10-4) or the box labeled **Two Types of Interaction - Additive and Multiplicative.**

Q10.11

1. True
2. False – Evidence for effect modification is present when the stratum-specific measures of association are different.
3. True
4. effect modification, measles vaccine, measles, age at vaccination

LESSON 11

CONFOUNDING INVOLVING SEVERAL RISK FACTORS

11-1 Confounding Involving Several Risk Factors

This lesson considers how the assessment of ***confounding*** *gets somewhat more complicated when controlling for more than one risk factor. In particular, when several* ***risk factors*** *are being controlled, we may find that considering all risk factors simultaneously may not lead to the same conclusion as when considering risk factors separately. We have previously (Lesson 10) argued that the assessment of confounding is not appropriate for variables that are* ***effect modifiers*** *of the exposure-disease relationship under study. Consequently, throughout this lesson, our discussion of confounding will assume that* ***none*** *of the variables being considered for control are effect modifiers (i.e., there is no interaction between exposure and any variable being controlled).*

Assessing Confounding in the Presence of Interaction

We have restricted our discussion of confounding involving several variables to the situation where **none** of the variables considered for control are effect modifiers of the exposure-disease relationship under study. This restriction has been made primarily for pedagogical reasons, since it is easier to discuss the confounding among several variables when there is no effect modification.

Nevertheless, **it is often quite appropriate to consider confounding even when interaction is present**. For example, if we are only controlling for one variable, say gender, and we find that the odds ratio for males is 1.3 whereas the odds ratio for females is 3.6 and the crude odds ratio is 10.1, then both confounding and interaction are present and each may be addressed. A similar situation may present itself when two or more variables are being controlled.

Moreover, when several variables are being controlled and there is interaction of, say, only one of these variables with the exposure variable, then the remaining variables considered for control may be assessed as potential confounders. For example, if in a cohort study of risk factors for coronary heart disease (CHD), it was determined that cholesterol level (CHL) was the only effect modifier of the exposure variable (say, physical activity level) among risk factors that included age, smoking status, gender and blood pressure, then these latter variables may still be assessed for possible confounding.

In the latter situation, one method for carrying out confounding assessment involves **stratifying on the effect modifier** (CHL) and assessing confounding involving the other variables separately within different categories of CHL.

Another approach is to use a **mathematical model** (e.g., using logistic regression) that contains all risk factors considered as main effects and also contains a product term of exposure with cholesterol. Those risk factors other than CHL can then be assessed for confounding provided the main effect of cholesterol, the exposure variable, and the product of exposure with CHL remains in the model throughout the assessment.

Two Important Principles

We have thus far considered only the control of a single confounder in an epidemiologic study. But usually several risk factors are identified and measured for possible control. Recall the a priori criteria for confounding. When several factors meet these criteria, how do we determine which to control for in the analysis?

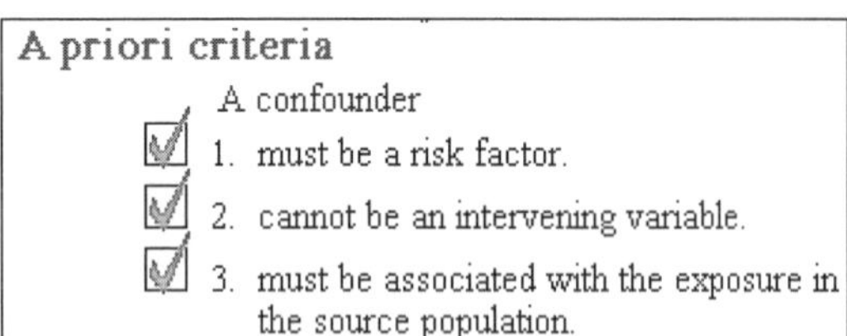

A priori criteria
A confounder
1. must be a risk factor.
2. cannot be an intervening variable.
3. must be associated with the exposure in the source population.

Suppose age and race are two risk factors identified and measured for possible control in a case-control study to assess an exposure disease relationship. It is possible that the adjusted odds ratio, which simultaneously controls for both age and race to give different results from those obtained by controlling for each variable separately.

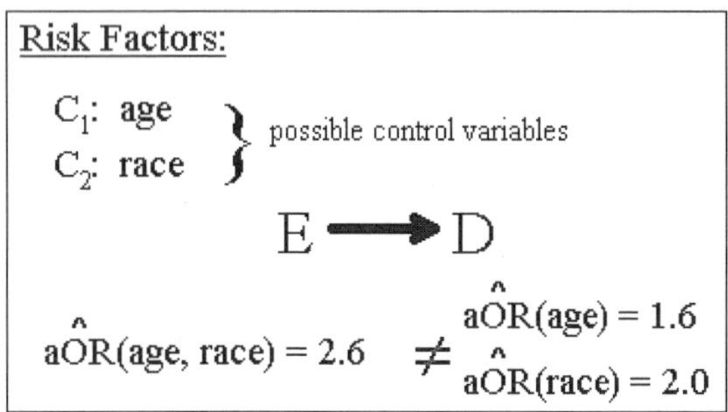

Risk Factors:
C_1: age
C_2: race
possible control variables
E → D
$a\hat{OR}(\text{age, race}) = 2.6$
≠
$a\hat{OR}(\text{age}) = 1.6$
$a\hat{OR}(\text{race}) = 2.0$

If the odds ratio that controls for all potential risk factors is our standard, then should we **always** control for **all** risk factors? Not necessarily. It is possible that only a subset of these factors needs to be controlled to obtain valid results.

Suppose these results were obtained from our case-control study:

$a\hat{OR}(\text{age, race}) = 2.6$ ↓ **the standard**

$a\hat{OR}(\text{age}) = 2.6$

$a\hat{OR}(\text{race}) = 2.0$

Here, the odds ratio controlling for age alone is equivalent to the odds ratio controlling for both age and race. In this case, we would not lose anything with regards to validity by selecting only age for control.

These examples illustrate two fundamental principles about the control of confounding when several risk factors have been identified and measured. First, the joint (or simultaneous) control of two or more variables may give different results from those obtained by controlling for each variable separately. The adjusted estimate (denoted here as a theta hat, $\hat{\theta}$) that simultaneously controls for all risk factors under consideration should be the standard on which all conclusions about confounding and the identification of specific confounders must be based.

Second, not all the variables in a given list of risk factors may need to be controlled; it is possible that different subsets of such variables can correct for confounding. We will discuss these two principles in the activities that follow.

Confounding With Several Covariates
Two Fundamental Principles

1. Joint (i.e., simultaneous) control of two or more variables may give different results from controlling for each variable separately.
   $a\hat{\theta}\,(C_1, C_2, \ldots, C_p)$ → standard
2. Not all the variables in a given list of risk factors may need to be controlled; it is possible that different subsets of such variables can correct for confounding.

Note: θ denotes any effect measure of interest.

Study Questions (Q11.1)

Suppose age, race, gender, and smoking status are the only risk factors considered for control in assessing an exposure-disease relationship.

1. Describe the adjusted estimate that should be the standard on which all conclusions about confounding and the identification of specific confounders be based.
2. If one fails to consider all potential confounders simultaneously, what might be some of the problems to arise?
3. If the gold standard adjusted estimate controls for all risk factors, is it possible that a subset of such risk factors may also control for confounding?
4. Why might the use of such a subset of variables be advantageous over the use of the adjusted estimate that controls for all potential confounders?

Summary

- There are two fundamental principles about the control of confounding when two or more risk factors have been identified and measured for possible control.
  1. The joint or simultaneous control of two or more variables can give different results from those obtained by controlling for each variable separately.
  2. Not all variables in a given list of risk factors may need to be controlled.

- Moreover, depending on the relationships among these risk factors, it is possible that confounding can be corrected by using different subsets of risk factors on the list.

Joint Versus Marginal Confounding

We defined **data-based confounding** involving a single potential confounder to mean that there is a meaningful difference between the estimated crude effect (which completely ignores a potential confounder) and the **estimated adjusted effect** (which controls for a potential confounder). We now define **data-based joint confounding** in the presence of 2 or more potential confounders. This occurs when there is a meaningful difference between the estimated crude effect and the estimated adjusted effect, which simultaneously controls for all the potential confounding.

Data-Based Joint Confounding

(2 or more C's)

Crude estimate		Adjusted estimate
$c\hat{\theta}$	$\neq$ (meaningful)	$a\hat{\theta}(C_1, C_2, \ldots, C_p)$

Study Questions (Q11.2)

Suppose a follow-up study was conducted to evaluate an E→D relationship. Age and smoking status were determined as possible control variables. Suppose further that:

aRR(age, smoking) =	2.4
aRR(age) =	1.7
aRR(smoking) =	1.9
cRR =	1.5

Study questions continue on next page

1. Is this evidence of joint confounding? Why or why not?

Suppose for a different follow-up study of the same E→D relationship that once again age and smoking status were possible control variables. Suppose further that:

aRR(age, smoking) =	1.4
aRR(age) =	2.4
aRR(smoking) =	2.4
cRR =	1.5

2. Is this evidence of joint confounding? Why or why not?

In contrast, we define **data-based marginal confounding** to mean that there is a meaningful difference between the estimated crude effect and the estimated adjusted effect that controls for only one of several potential confounders.

Data-Based Marginal Confounding

$$c\hat{\theta} \underset{\text{meaningful}}{\neq} a\hat{\theta}(C_j)$$

where C_j is one of p potential confounders.

Study Questions (Q11.2) continued

Suppose a follow-up study was conducted to evaluate an E→D relationship. Age and smoking status were determined as possible control variables. Suppose that:

aRR(age, smoking) =	2.4
cRR =	1.5

3. Is there evidence of marginal confounding? Why or why not?
4. If the aRR(age) = 1.4, does this provide evidence of marginal confounding?
5. Does this mean that we should not control for age as a confounder?

Joint confounding is the primary criterion for determining the presence of data-based confounding when all are eligible for control. Nevertheless, data-based marginal confounding can help determine whether **some** potential confounders **need not** be controlled.

Study Questions (Q11.2) continued

6. In the follow-up study described in the previous study question, the:

aRR(age, smoking) =	2.4
aRR(age) =	1.5
aRR(smoking) =	2.4
cRR =	1.5

   Does this mean that we do not have to control for age?
7. What problem might there be in practice that could prevent the estimate of the effect that controls for all risk factors (e.g., C_1, C_2, …, C_k)?
8. What should we do if there are too many potential confounders in our list and we are unable to determine the appropriate adjusted estimate?
9. What if the choice of such a subset becomes difficult?

Summary

- **Data-based joint confounding** occurs when there is a meaningful difference between the estimated crude effect and the estimated adjusted effect that simultaneously controls for *all* the potential confounders.
- **Data-based marginal confounding** occurs when there is a meaningful difference between the estimated crude effect and the estimated adjusted effect, which controls for only *one* of the several potential confounders.
- Our conclusions regarding confounding should be based on joint confounding whenever possible.

Joint Versus Marginal Confounding – An Example

Suppose that a follow-up study is conducted to assess an exposure disease relationship and that the crude risk ratio for these data is 2.

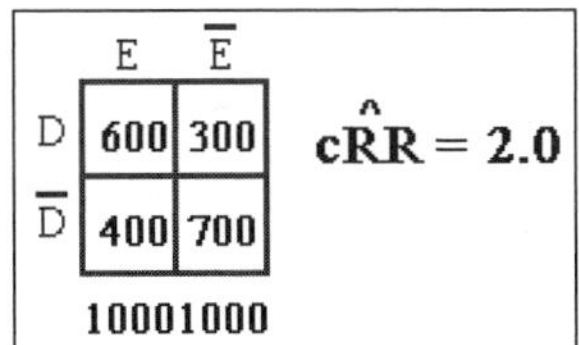

Suppose also that two dichotomous variables **F** and **G** have been identified and measured for possible control. We would like to know whether we can control for either **F** or **G** separately or whether we must simultaneously control for both **F** and **G** in order to properly control for confounding.

Recall the first fundamental principle regarding confounding with several variables. The adjusted estimate that simultaneously controls for all risk factors under consideration is the standard on which all conclusions about confounding must be based. In our example, the adjusted risk ratio controlling for both F and G is the standard. Stratifying the data by both **F** and **G**, we find that each stratum specific estimated risk ratio equals 1.0. It thus follows that the adjusted risk ratio controlling for both **F** and **G** is 1.0. This differs from the crude risk ratio of 2.0. So, we can conclude there is data-based joint confounding due to **F** and **G**.

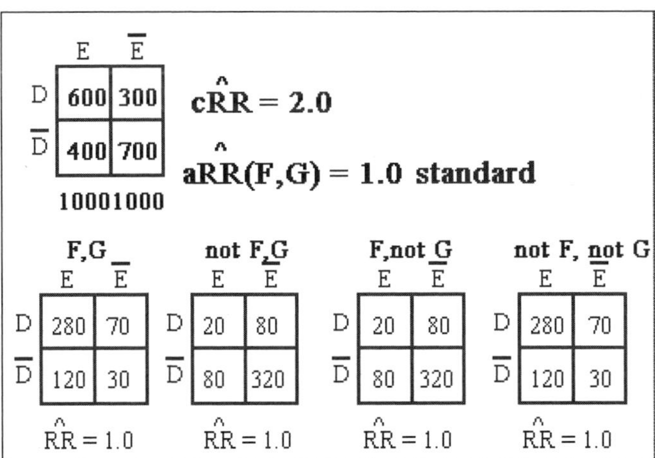

Suppose that we ignore the joint control of these two variables and evaluate each variable separately. When we stratify the data by **F**, we see the stratum specific risk ratios are 2.0 for when **F** is both present and absent and hence the adjusted risk ratio controlling for **F** is 2.0.

	E	$\bar{E}$
D	600	300
$\bar{D}$	400	700
	1000	1000

$c\hat{RR} = 2.0$

$a\hat{RR}(F) = 2.0$

F

	E	$\bar{E}$
D	300	150
$\bar{D}$	200	350

$\hat{RR} = 2.0$

not F

	E	$\bar{E}$
D	300	150
$\bar{D}$	200	350

$\hat{RR} = 2.0$

Likewise, if we stratify the data by **G**, we see the stratum specific risk ratios are 2.0 when **G** is both present and absent, and therefore the adjusted risk ratio is 2.0.

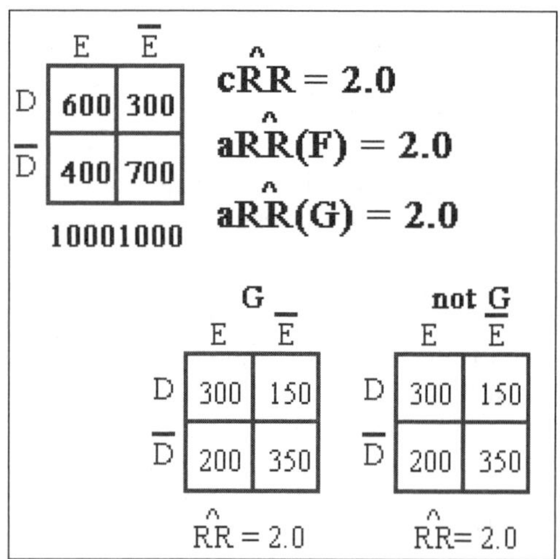

Because the adjusted estimates that control for both **F** and **G** separately are equal to the crude risk ratio of 2.0, we can conclude there is no data-based marginal confounding due to either **F** or **G**. This example illustrates that incorrect conclusions about confounding may result if one fails to consider the conditions for joint confounding.

Study Questions (Q11.3)

1. In general, which is a better indicator of data-based confounding in the presence of two or more risk factors: marginal or joint confounding?
2. In the example described in this presentation, was there any interaction between **F** and **E** or between G and **E**?
3. When two or more risk factors are being considered for possible control, is it possible that not all of the risk factors may need to be controlled?
4. In this example, is it possible that either **F** or **G** alone would be an appropriate subset to control for confounding?

Summary

- ❖ The adjusted estimate that simultaneously controls for all risk factors under study is the standard on which all conclusions about confounding must be based.
- ❖ Incorrect conclusions about joint confounding may result if one considers only the conditions for marginal confounding.

Quiz (Q11.4)

Suppose **F** and **G** are two distinct risk factors for some disease with dichotomous levels $\mathbf{F_1}$, $\mathbf{F_0}$, and $\mathbf{G_1}$, $\mathbf{G_0}$, respectively. The estimated risk ratios describing the association between the disease and some exposure are listed below for various combinations of levels of **F** and **G**. Assume that the risk ratio estimates are of high precision (i.e., are based on large sample sizes).

$\hat{RR}(F_1G_1) = 3.0$	$\hat{RR}(F_1) = 1.0$	$c\hat{RR} = 1.0$
$\hat{RR}(F_1G_0) = 3.0$	$\hat{RR}(F_0) = 1.0$	
$\hat{RR}(F_0G_1) = 0.3$	$\hat{RR}(G_1) = 1.0$	
$\hat{RR}(F_0G_0) = 3.0$	$\hat{RR}(G_0) = 1.0$	

Determine whether the following statements are **True** or **False**.

1. There is evidence of interaction in the data. . . . ???
2. There is evidence of confounding in the data. . . . ???
3. At level $\mathbf{G_0}$, there is no confounding due to factor **F**. . . ???
4. At level $\mathbf{F_1}$, there is no interaction due to factor **G**. . . . ???
5. At level $\mathbf{F_0}$, there is no interaction and no confounding due to factor **G** ???
6. At level $\mathbf{G_0}$, there is confounding but no interaction due to factor **F** . ???.
7. It is not necessary to control for either **F** or **G** (or both) in order to understand the relationship between D and E. . . . . ???.

11-2 Confounding Involving Several Risk Factors

Confounding Due to One of Two Potential Risk Factors

The second fundamental principle of confounding regarding several risk factors says that it may not be necessary to control for all the risk factors in a particular study. A subset of these variables may control for the confounding. How do we determine candidate subsets?

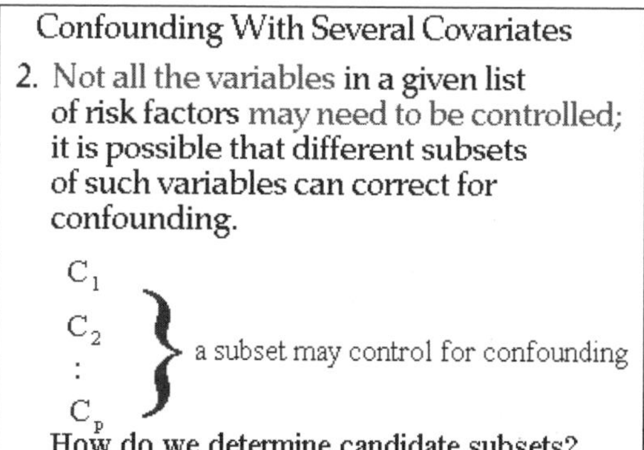

Confounding With Several Covariates
2. Not all the variables in a given list of risk factors may need to be controlled; it is possible that different subsets of such variables can correct for confounding.
C1
C2
:
Cp
a subset may control for confounding
How do we determine candidate subsets?

Suppose that a follow-up study is conducted to evaluate an exposure-disease relationship. The crude risk ratio equals 2. Two dichotomous variables **K** and **L** have been identified and measured for possible control. The adjusted risk ratio controlling for both **K** and **L** is the standard on which conclusions about confounding should be based.

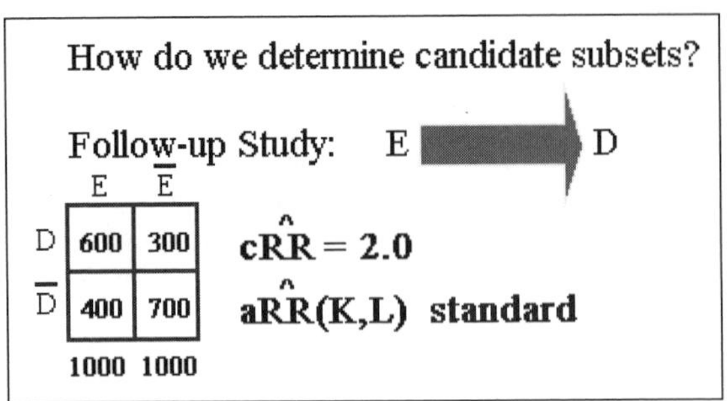

Stratifying the data by both **K** and **L**, we find that each stratum specific estimated risk ratio is approximately 1. It thus follows that the adjusted risk ratio controlling for both **K** and **L** is approximately 1. This differs from the crude risk ratio of 2. So, we can conclude there is data-based joint confounding due to **K** and **L**.

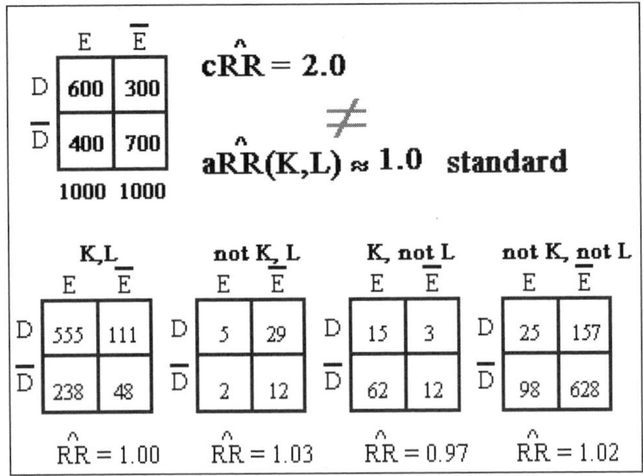

Can we control for a subset of these two risk factors, either **K** or **L** alone? If we stratify the data by **K**, we see the stratum specific risk ratios are 1.0 when **K** is both present and absent, and hence the adjusted risk ratio controlling for **K** is 1.0.

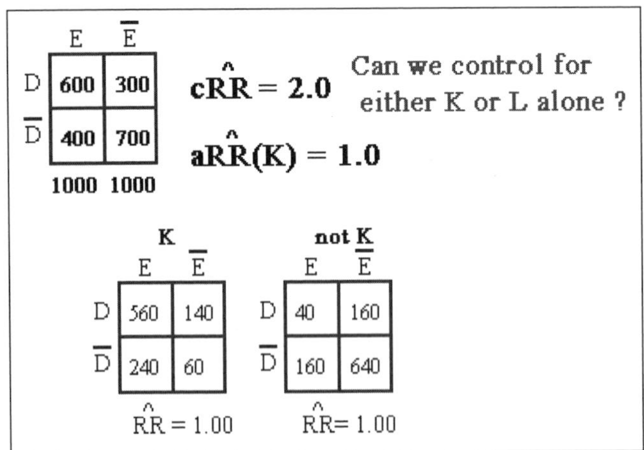

Likewise, if we stratify the data by **L**, we see the stratum specific ratios are approximately 1 when **L** is both present and absent, and therefore the adjusted risk ratio is 1.0.

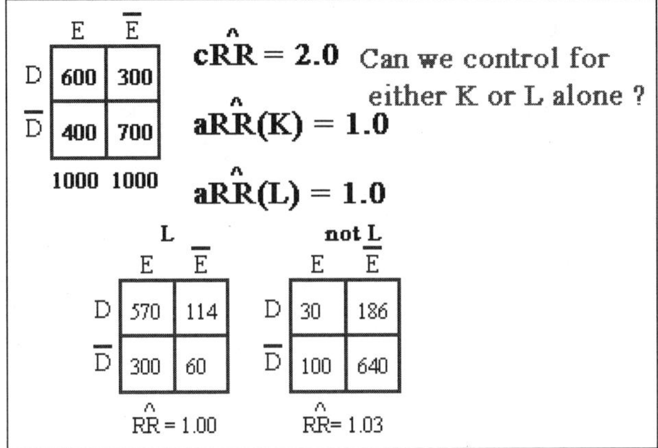

Because the adjusted estimates that control for both **K** and **L** separately are different from the crude risk ratio, these results illustrate marginal confounding. In this example, it would not be necessary to control for both **K** and **L**, since the control of either of these two variables provides the same results as the standard. It is sufficient to control for either one of these two variables separately.

<u>Study Questions (Q11.5)</u>

1. How would you determine which variable, **K** or **L** or both, to control for in the above mentioned study?
2. If the results are the same, is there any advantage to controlling for both variables?

Suppose now that **K** and **L** are the only risk factors eligible for control and that these are the results. Here, the adjusted risk ratio controlling for both **K** and **L** is not equal to the adjusted risk ratio controlling for **L** alone. In this case, **K** is the only candidate subset of confounders eligible for control.

$\bar{D}$ = not D
$\bar{E}$ = not E

	E	$\bar{E}$
D	600	300
$\bar{D}$	400	700

$c\hat{RR} = 2.0$

$\neq$

$a\hat{RR}(K) = 1.0$

$a\hat{RR}(L) = 1.7$

$a\hat{RR}(K,L) = 1.0$ standard

- K is the only eligible subset

<u>Study Questions (Q11.5) continued</u>

3. Is it appropriate to control for only **L** in the study above?
4. Why is it appropriate to control for only **K** in this study?
5. Is it appropriate to control for both **K** and **L** in this study?

Summary

- It may not be necessary to control for all the risk factors in a particular study.
- A candidate subset of these variables may control for the confounding.

Variable Selection and Control of Confounding

The adjusted estimate that controls for all potential risk factors is the standard on which conclusions about confounding should be based. However, if an adjusted estimate that controls for only a subset of risk factors is equivalent to this standard, we may then choose such a subset for control.

$a\hat{\theta}(C_1, C_2, \ldots, C_p)$ → standard

=

$a\hat{\theta}(C_1, C_2)$

$a\hat{\theta}(C_2, C_3, C_7)$ Maybe this subset is 'best'

$a\hat{\theta}(C_3)$

Consider a case-control study in which three risk factors, **F**, **G**, and **H** are being considered for control. The crude odds ratio differs from the adjusted odds ratio that controls for all three factors. Because the crude and adjusted estimates differ, we have evidence of data-based joint confounding in these data.

Case-control study

F, G, and H eligible for control

$c\hat{OR} = 3.1$

$\neq$

$a\hat{OR}(F,G,H) = 1.6$ **(standard)**

Joint Confounding

Suppose now that controlling for any two of these factors provides the same results as controlling for all three. **F**, **G**, and **H** do not **all** need to be controlled simultaneously. Controlling for any two of the three risk factors will provide the same results as the standard.

Case-control study

F, G, and H eligible for control

$c\hat{OR} = 3.1$

$\neq$

$a\hat{OR}(F,G,H) = 1.6$ **(standard)**

$a\hat{\theta}(F, G) = 1.6$

$a\hat{\theta}(F, H) = 1.6$ **Do not need to control for all 3!**

$a\hat{\theta}(G, H) = 1.6$

We may also wish to consider marginal confounding to see if any single variable is an appropriate subgroup. These results indicate that there is no marginal confounding because each of these results differs from the standard estimate, not one of these variables alone would be an appropriate subgroup for control.

Case-control study

F, G, and H eligible for control

$c\hat{OR} = 3.1$ $\neq$ $a\hat{OR}(F,G,H) = 1.6$ (standard)

Marginal Confounding? NO!

$a\hat{OR}(F) = 3.0$

$a\hat{OR}(G) = 2.7$

$a\hat{OR}(H) = 3.2$

Study Questions (Q11.6)

$a\hat{OR}(F,G,H) = 1.6$ (standard)

$a\hat{OR}(F,G) = 1.6$	$a\hat{OR}(F) = 3.0$
$a\hat{OR}(F,H) = 1.6$	$a\hat{OR}(G) = 2.7$
$a\hat{OR}(G,H) = 1.6$	$a\hat{OR}(H) = 3.2$

1. What might be the advantage to controlling for only two of these risk factors rather than all three even though it is the standard?
2. How might you determine which two variables to controls?
3. Why can't we control for **F**, **G**, or **H** separately?

Assume that F, G, H, I, and J are the only risk factors in a case-control study. Suppose further that:

cOR $\neq$ aOR(F, G, H, I, J)

but

aOR(F, G, H, I) = aOR(G, J) = aOR(I) = aOR(F, G, H) = aOR(F, G, H, I, J)

and that

aOR(any other subset of risk factors) $\neq$ aOR(F, G, H, I, J)

Determine whether each of the following is a proper subset of confounders that controls for (joint) confounding in this study by answering **Yes** or **No**:

4. {G, J}?
5. {I}?
6. {G, H, I}?
7. {F, G, H}?

Summary

- When two or more risk factors are considered for control, we can select an appropriate subset of confounders for control.
- When the results from controlling for various subsets of risk factors are equivalent to the joint control of all risk factors, we can select any one of which provides valid and precise results.

Confounding: Validity versus Precision

The fully adjusted estimate that controls for all factors simultaneously is the standard on which all decisions should be based. Why then would we want to go through the process of seeking out candidate subsets of confounders? If we cannot improve on the validity of the effect measure, why not just use the fully adjusted estimate?

$a\hat{\theta}(C_1, C_2, C_3, \cdots, C_k)$ → Standard

controls all risk factors simultaneously

$a\hat{\theta}(C_1, C_2)$

$a\hat{\theta}(C_2, C_3, C_7)$ } Why bother?

$a\hat{\theta}(C_3)$

The precision of the estimate may justify such an effort. Controlling for a smaller number of variables may yield a more precise estimate of effect. The identification of the subset of confounders giving the most precise estimate is important enough to make such examination worthwhile. Consider the following exercise to illustrate this point.

Study Questions (Q11.7)

A clinical trial was conducted to determine the effectiveness of a particular treatment on the survival of stage 3 cancer patients. The following variables were considered in the analysis:

RX = exposure
SERH = serum hemoglobin level
INSG = combined index that measures tumor stage and grade.
AGE = age at trial entry
TSZ = size of primary tumor

1. The cRR = 6.28 and the aRR(AGE, SERH, TSZ, INSG) = 8.24. Does this provide evidence of joint confounding in the study? (Assume all quantities above are estimates.)

We calculated the aRR for all possible subsets of the four potential confounders. Excluding the crude results and the gold standard, there are 14 possible subsets of these 4 confounders.

2. What criteria may we use to reduce the number of candidate subsets?

Below are the results from the gold standard and the 4 candidate subsets whose aRR is within 10% of the gold standard:

Covariates	aRR	95% CI	CI Width
INSG, AGE, SEHR, TSZ	8.24	(3.59, 18.91)	15.32
AGE, SERH, TSZ	8.26	(3.64, 18.75)	15.11
INSG, AGE	7.63	(3.75, 15.56)	11.81
INSG, SERH	7.63	(3.64, 15.97)	12.33
SERH	8.25	(4.06, 16.76)	12.70

3. The most valid estimate results from controlling which covariates?
4. The most precise estimate results from controlling which covariates?
5. Which covariates do you think are most appropriate to control?

This exercise has illustrated that we need to consider both validity and precision when assessing an exposure-disease relationship. Getting a valid estimate of effect is most important. Nevertheless, you must also consider the trade-off between controlling for enough risk factors to maintain validity and the possible loss in precision from the control of too many variables.

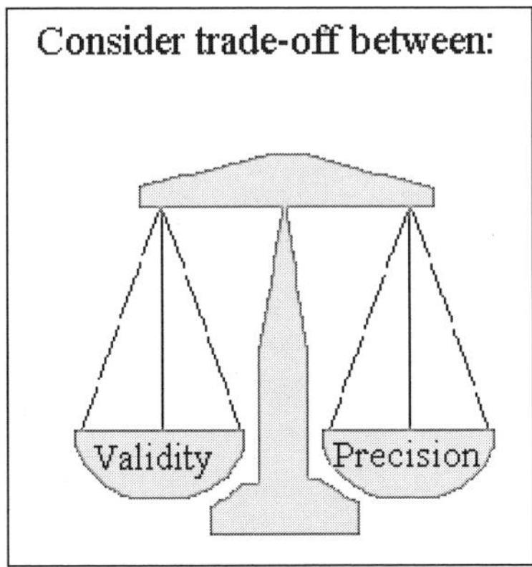

Summary

- ❖ The reason for seeking candidate subsets of all potential confounders is the possibility of improving the precision of the estimated effect.
- ❖ Controlling for fewer variables may (or may not) lead to a more precise estimate of effect.
- ❖ When controlling for several potential confounders, you should consider the possible trade-offs between:
  - o Controlling for enough risk factors to maintain validity

    versus
  - o Possible loss in precision from the control of too many variables.

Quiz (Q11.8)

Suppose that variables **F**, **G**, and **H** have been measured in a certain study and that only **F** and **G** are considered to be risk factors for some disease (D). Suppose that it is of interest to describe the relationship between this disease and some study factor (E), and that there is no interaction of any kind present in the data. Finally, suppose that the following relationships hold among various odds ratios computed from the data:

$c\hat{OR} \neq a\hat{OR}\,(F)$	$a\hat{OR}\,(F,G) = a\hat{OR}\,(F)$
$c\hat{OR} = a\hat{OR}\,(G)$	$a\hat{OR}\,(F,G) \neq a\hat{OR}\,(G)$
$c\hat{OR} = a\hat{OR}\,(H)$	$a\hat{OR}\,(F,H) = a\hat{OR}\,(H)$
$c\hat{OR} = a\hat{OR}\,(F,G,H)$	$a\hat{OR}\,(F,G,H) = a\hat{OR}\,(G,H)$
$c\hat{OR} \neq a\hat{OR}\,(F,G)$	

Determine whether the following statements are **True** or **False**.

1. There is confounding in the data. . . . . . **???**
2. Variable **F** needs to be controlled to avoid confounding. . . . . **???**
3. Variable **G** needs to be controlled to avoid confounding. . . . . **???**
4. Variable **H** needs to be controlled to avoid confounding. . . . . **???**
5. Both variables **F** and **G** do not need to be controlled simultaneously in order to avoid confounding. **???**

Quiz continued on next page

A ten-year follow-up study was conducted to determine if someone experiencing food allergies is at increased risk of coronary heart disease. The following covariates were considered in the analysis: **AGE** = age at enrollment, **BMI** = body mass index, and **SMK** = smoking status. The results for the standard and all possible subsets are listed below. The crude risk ratio = 1.02.

	Covariates	aRR	95%CI	
1.	AGE, BMI, SMK	4.10	(1.7, 8.8)	(standard)
2.	AGE, BMI	2.70	(1.6, 7.8)	
3.	AGE, SMK	4.12	(1.4, 9.5)	
4.	BMI, SMK	2.50	(1.8, 6.2)	
5.	AGE	4.01	(1.6, 8.7)	
6.	BMI	2.50	(1.4, 7.5)	
7.	SMK	4.16	(1.5, 8.6)	

6. Is there evidence of confounding? . . . ???.
7. Besides the standard, which are candidate subgroups for control? ???, ???,. ???.
8. Which of the candidate subgroups corresponds to the most valid estimate (including the standard)? . . . . ???.
9. Is there more than one candidate subgroup that is the most precise? ???.
10. Which estimate should be used? . . ???.

Choices
#2 #3 #4 #5 #6 #7 no yes

Nomenclature

$\hat{\theta}$	Estimated measure of effect
aOR	Adjusted odds ratio
aRR	Adjusted risk ratio
C_i	Confounding variable
CI	Confidence interval
cOR	Crude odds ratio
cRR	Crude risk ratio
D	Disease
E	Exposure

Reference

Kleinbaum DG, Kupper LL, Morgenstern H. Epidemiologic Research: Principles and Quantitative Methods. John Wiley and Sons Publishers, New York, 1982 (Chapter 14).

Homework

ACE-Identifying Risk Factors for Control

Suppose you carry out a cohort study to assess the relationship between a dichotomous exposure variable and a dichotomous disease variable. You identify and measure three risk factors f1, f2 and f3 that you want to control for in assessing the E-D relationship. Suppose further that the analysis of your data gives the following estimates of effect:

aRR(f1, f2, f3) = 4.1	cRR = 1.3
aRR(f1, f2) = 2.5	aRR(f1) = 2.1
aRR(f2, f3) = 4.0	aRR(f2) = 4.0
aRR(f1, f3) = 2.9	aRR(f3) = 2.8

Assuming no interaction of any kind and that all of the above estimates are very precise, answer the following questions:

a. Is there confounding? Explain
b. How should you decide on which of the above variables need to be controlled? (You are being asked for a strategy, not a conclusion.)
c. Which of the variables are not confounders. Explain
d. What conclusions can you make about which variables should be controlled in this study?
e. If you strictly use a 10% change rule to determine whether an adjusted estimate differs from the gold standard adjusted estimate, what conclusions do you draw about which variables should be controlled in this study?
f. If you strictly use a 20% change rule to determine whether an adjusted estimate differs from the gold standard adjusted estimate, what conclusions do you draw about which variables should be controlled in this study?

ACE-2. Variable Selection

Suppose you carry out a case-control study to assess the relationship between a dichotomous exposure variable and a dichotomous disease variable. You identify and measure four risk factors f1 (previous history of mental disorder), f2 (personality type), f3 (age), and f4 (gender) that you want to control for in assessing the E-D relationship. Suppose further that the analysis of your data gives the following estimates of effect:

a0R(f1, f2, f3, f4) = 1.68
aOR(f2, f3, f4) = 1.20
aOR(f1, f3, f4) = 1.20
aOR(f1, f2, f4)
aOR(f1, f2, f3) = 1.18
aOR(f3, f4) = 1.69
aOR(f1, f4) = 2.78
aOR(f1, f3) = 1.68
aOR(f2, f4) = 1.70
aOR(f2, f3) = 5.88
aOR (f1, f2) = 5.95
aOR (f4) = 2.75
aOR (f3) = 5.85
aOR (f2) = 1.20
aOR (f1) = 1.69

Given the above information and assuming no interaction of any kind, use a 10% change rule to determine which variables should be included to correct for confounding. How might precision play a role in terms of which variables are selected for control?

ACE-3. Joint and Marginal Confounding

A case-control study was carried out to evaluate whether alcohol consumption was a risk factor for the development of breast cancer in women. The exposure variable was denoted as ALC and categorized into 3 groups (1= no alcohol intake, 2= small to moderate alcohol intake, and 3=high alcohol intake. Three risk factors were considered as control variables, AGE (1= under 50, 2 = race 50), SMK status(1= ever, 2=never), and OBESITY (0=No, 1=Yes). The following adjusted odds ratio were obtained comparing moderate drinkers (ALC=1) to non-drinkers (ALC=0) and heavy drinkers (ALC=2) to non-drinkers (ALC=0):

Variables Controlled	aOR(1 to 0)	aOR(2 to 0)
None	4.5	6.0
AGE	3.4	5.1
SMK	2.0	2.8
OBESITY	2.9	4.1
AGE, SMK	1.8	3.3
AGE, OBESITY	4.6	5.6
SMK, OBESITY	3.2	5.4
AGE, SMK, OBESITY	1.9	3.1

Assuming no interaction of any of the control variables with ALC and using a 10% change rule to determine a meaningful difference in adjusted odds ratios, answer the following questions:

a. Is there confounding? Justify your answer.
b. What subsets of variables give the same adjusted odds ratio as the gold standard adjusted odds ratio? Justify your answer.
c. How would you consider precision to determine which subset of variables to control?
d. Suppose no meaningful gain in precision is made when controlling for a proper subset of all three control variables. Which variables would you control? Justify your answer.

ACE-4. Marginal Confounding

The accompanying Table provides the results of a stratified analysis of data collected in a case-control study. The outcome variable is dichotomous and is labeled A. The predictor variables are labeled 1 through 10 with variable 1 the exposure variable of interest. Variables 2 through 10 are control variables.

Results of Stratified Analysis in Examination of the Association Between Variable 1 and the Outcome (Variable A)

Risk Variable Controlled	Sub-Strata estimated OR		Adjusted $OR_{M\text{-}H}$	95% Confidence Interval	Breslow-Day Test of Homogeneity (p-value)
	Stratum 1	Stratum 2			
Variable 2	1.550	2.360	1.998	(1.070, 3.730)	0.522
Variable 3	5.758	1.840	3.083	(1.517, 6.268)	0.136
Variable 4	3.300	1.875	2.040	(1.088, 3.822)	0.534
Variable 5	1.250	3.829	2.711	(1.357, 5.415)	0.151
Variable 6	0.563	2.972	1.813	(.984, 3.341)	0.022
Variable 7	2.000	1.648	1.711	(0.889, 3.293)	0.819
Variable 8	1.125	2.134	2.032	(1.082, 3.819)	0.603
Variable 9	1.333	2.146	1.964	(1.044, 3.693)	0.570
Variable 10	1.9028	1.950	1.931	(1.026, 3.632)	0.970

cOR for Variable 1 vs. Variable A (outcome) = 2.022 (1.084 - 3.772)

a. Assuming no interaction of any kind, how would you assess whether or not there is confounding? Has enough information been provided in the above table to allow you to answer this question?
b. Again, assuming no interaction, is there marginal confounding due to any of the control variables? Explain.
c. Again, assuming no interaction, how would you determine which variables to control? What is the primary reason why you can't answer this question based on the data provided above (assuming no interaction)?
d. Assuming that there is interaction of variables 9 and 10 with the exposure (variable 1), how would you modify your answer to part c to determine which variables to control?

Answers to Study Questions and Quizzes

Q11.1

1. The adjusted estimate that simultaneously controls for all 4 risk factors under consideration.
2. Confounding might not be controlled if there is not a subset of potential confounders that yields (essentially) the same adjusted estimate as obtained when all confounders are controlled.
3. Yes, provided the subset yields essentially the same adjusted estimate as the gold standard.
4. Adjusting for a smaller number of variables may increase precision. Also, such a subset provides a more parsimonious description of the exposure-disease relationship.

Q11.2

1. Yes, the cRR of 1.5 differs from the aRR(age, smoking) of 2.4 that controls for both potential confounders.
2. No, the cRR of 1.5 is essentially equal to the aRR(age, smoking) of 1.4 that controls for both potential confounders.
3. No, the cRR of 1.5 differs from the aRR(age, smoking) of 2.4, which controls for all potential confounders. This is evidence of joint confounding.
4. No, since the cRR of 1.5 is approximately equal to the aRR(age) of 1.4, there is no evidence of marginal confounding due to age.
5. Not necessarily. Our conclusions regarding confounding should be based on the joint control of all risk factors.
6. Yes. Controlling for smoking alone gives us the same result as controlling for both risk factors. We might still wish to evaluate the precision of the estimates before making a final conclusion.
7. There may be so many risk factors in our list relative to the amount of data available that the adjusted estimate cannot be estimated with any precision at all.
8. Then we may be forced to make decisions by using a subset of this large initial set of risk factors.
9. The use of marginal confounding may be the only alternative.

Q11.3

1. Joint confounding should be used, whenever possible, as the baseline from which all other confounding issues should be examined.
2. No, the two stratum-specific RRs that compare **F** with **not F** are equal and the two stratum-specific RRs that compare **G** with **not G** are equal.
3. Yes, the second fundamental principal of confounding states that "not all variables in a given list of risk factors my need to be controlled"; it is possible that different subsets of such variables can alternatively correct for confounding.
4. No, controlling for either of these risk factors separately yields the same results as the crude data. It is only in the joint control of these factors that we observe confounding.

Q11.4

1. True – There is interaction because the risk ratio estimated in one stratum (F_0G_1) is 0.3, which is quite different from the stratum-specific risk ratios of 3.0 in the other strata.
2. True – The presence of strong interaction may preclude the assessment of confounding. Also, the value of an adjusted estimate may vary depending on the weights chosen for the different strata.
3. False – The RR for F_1 and F_0 at level G_0 are both 3.0. These differ from the overall RR at level G_0 of 1.0. Therefore, at level G_0, there is confounding due to factor F.
4. True
5. False – There is interaction and possibly confounding. At level F_0, the RR for G_1 and G_0 are very different, and both are very different from the overall risk ratio at level F_0.
6. True
7. False – Both confounding and interaction are present and each should be addressed.

Q11.5

1. You may wish to control for the variable(s) that yield(s) the most precise estimate.
2. Yes, you may wish to control for both if you do not gain anything regarding precision by dropping one. Although these results may be the same, if you drop a variable from analysis, it is not clear to a reviewer that you controlled for both.
3. No; the results when controlling for L alone differ from the results controlling for both K and L, the standard on which all conclusions about confounding must be based.
4. Controlling for K yields the same results as controlling for both factors K and L simultaneously.

5. Yes, if you do not gain anything regarding precision from dropping L.

Q11.6

1. Controlling for fewer variables will likely increase the precision of the results.
2. The two that provide the most precise adjusted estimate.
3. Controlling for any of these three factors alone yields different results than controlling for all three, which is the standard on which our conclusions should be based.
4. Yes
5. Yes
6. No
7. Yes

Q11.7

1. Yes, the cRR differs from the aRR controlling for all potential confounders, which is the gold standard.
2. We may choose to only consider those results within 10% of the gold standard. In this case, that would be 8.24 ± 0.82 which is a range of values between 7.42 and 9.06.
3. Controlling for all the covariates provides the most valid estimate. It is the gold standard.
4. Controlling for both INSG and AGE provides the narrowest confidence interval and hence is the most precise.
5. Debatable: Controlling for SERH alone yields an almost identical aRR as the gold standard, increases precision, and is the stingiest subset. Controlling for INSG and AGE provides a slightly larger increase in precision (than controlling for SERH only) and its aRR is within 10% of the standard. Consider the trade-off between parsimony and political/scientific implications of not controlling for all risk factors, and more precision from controlling for fewer risk factors.

Q11.8

1. True
2. True
3. False – Variable G does not need to be controlled since aOR(F,G)=aOR(F). In other words, controlling for F alone yields the same results as the gold standard, controlling for both F and G.
4. False – Variable H is not a risk factor in this study, and therefore should not be considered a confounding.
5. True
6. Yes – The cRR of 1.02 differs from the standard RR of 4.10 that controls for all potential confounders.
7. #3, #5, #7
8. #1 – The most valid estimate controls for all risk factors measured.
9. Yes – Candidate subgroups 1 and 7 are equally precise.
10. #1 – The gold standard is the most valid estimate; has the same precision as obtained for candidate 7. No precision is gained by dropping any risk factors so it can be argued the gold standard is the 'political' choice for it controls for all considered risk factors. Controlling only for SMK is the best choice for it gives the smallest, most precise subset of variables.

LESSON 12

SIMPLE ANALYSES

12-1 Statistical Inference for Simple Analyses

This lesson discusses methods for carrying out ***statistical inference*** *procedures for epidemiologic data given in a simple two-way table. We call such procedures* ***simple analyses*** *because we are restricting the discussion here to dichotomous disease and exposure variables only and we are ignoring the typical analysis situation that considers the control of* ***other*** *variables when studying the effect of an exposure on disease.*

WHAT IS SIMPLE ANALYSIS?

When analyzing the crude data that describes the relationship between a dichotomous exposure and dichotomous disease variable, we typically want to make statistical inferences about this relationship. That is, we would like to determine whether the measure of effect being estimated is statistically significant and we would like to obtain an interval estimate that considers the sample variability of the measure of effect.

The tables shown here have been described in previous lessons to illustrate data from three different studies, a cohort study to assess whether quitting smoking after a heart attack will reduce one's risk for dying, a case-control study to determine the source of an outbreak of diarrhea at a resort in Haiti, and a person-time cohort study to assess the relationship between serum cholesterol level and mortality.

Heart Attack Patients	Smoke	Quit	Total
Death	27	14	41
Survival	48	67	115
Total	75	81	156

COHORT

Quit Smoking → Mortality

$\hat{RR} = 2.1$

Resort Study	Raw Hamburger: Ate	Did not Eat	Total
Cases	17	20	37
Controls	7	26	33
Total	24	46	70

CASE-CONTROL

Source → Diarrhea

$\hat{OR} = 3.2$

Cholestrol Study	Cholestrol Level: Borderline High	Normal
Deaths	26	14
P-Years	36,581	68,239

PERSON-TIME COHORT

Cholesterol → Mortality

$\hat{IDR} = 3.5$

In each study, a measure of effect was computed to estimate the extent of the relationship between the exposure variable and the health outcome variable. In the quit smoking cohort study, the effect measure was a risk ratio and its estimate was 2.1. In the outbreak study, the effect measure was an odds ratio and its estimate was 3.2. And in the cholesterol mortality study, the effect measure was a rate ratio, also called an incidence density ratio, and its estimate was 3.5.

We have discussed how to interpret each of these estimates in terms of the exposure disease-relationship being studied. All three estimates, even though dealing with different study types and different study questions, are similar in that they are all larger than the **null value of one**, and they all indicate that there is a moderately large effect from exposure.

Nevertheless, we must be careful to realize that each of these three estimates is based on sample data. If a different sample had been drawn, any of these estimates might have resulted in a different value, maybe a lot larger, maybe closer to the null value of one. That is, there is always **random error** associated with any sample estimate.

We call these estimates **point estimates** because they each represent a single number or point from the possibly wide range of numbers that might have been obtained if different samples had been drawn. So, we might wonder, given the inherent variability in a point estimate, how can we draw conclusions about the **population parameters** being estimated?

For example, in the first cohort study, what can we say about the **population risk ratio** based on the **estimated risk ratio**? Or, in the case-control study, what can we conclude about the **population odds ratio** based on the **estimated odds ratio**? In the person-time study what can we conclude about the **population rate ratio** based on the **estimated rate ratio**? In answering these questions, we typically have one of two objectives. We may want to determine whether we have evidence from the sample that the population risk ratio, odds ratio or rate ratio being estimated is different from the null value of one. For this objective, we use **hypothesis testing**. Or, we may want to determine the precision of our point estimate by accounting for its **sampling variability**. For this objective, we use **interval estimation**.

The methods used to achieve these objectives comprise the general subject matter of **statistical inference**. When our attention is focused on the relatively simple situation involving only one dichotomous exposure variable and one dichotomous disease variable, as illustrated by these three studies, we call these methods **simple analyses**.

Summary

- Estimates of measures of effect such as RR, OR, and IDR are **point estimates**, since each estimate represents a single number that may vary from sample to sample.
- **Statistical inference** involves drawing conclusions about the value of a measure of effect in a population, based on its estimate obtained from a sample.
- The two types of **statistical inference** procedures are **hypothesis testing** and **interval estimation**.

STATISTICAL INFERENCES – A REVIEW

The activities in this section review fundamental concepts and methods of statistics. Our primary focus concerns how to draw conclusions about populations based on data obtained in a sample. We assume that you already have some previous exposure to basic statistical concepts, including the distinction between a sample and a population, a sample statistic and a population parameter, some important distributions like the normal, binomial, Student's t, and chi square distributions. We also assume that you have some previous exposure to the concepts underlying statistical tests of hypothesis and confidence intervals, which are the two types of statistical inferences possible. Our focus here will be to review statistical inference concepts in the context of the statistical questions that apply to the analysis of a 2 x 2 table, which is the kind of data we are considering in a simple analysis. You may wish to skip this entire review section and proceed to the next section, Cohort Studies, on page 12-4. We begin by using data from a famous "incident" to distinguish between the two types of statistical inference procedures: hypothesis (significance) testing and confidence interval estimation. See if you can guess what "incident" we illustrate.

Statistical Inference Overview

Here are some data from an incident in which a group of persons were at risk of dying. From these data, we can find the proportions who died for men and women, separately. We can see that 79.7% of the men died, but only 25.6% of the women died.

AN INCIDENT

Persons at Risk

	Men	Women	Total
Died	1329	109	1438
Lived	338	316	654
Total	1667	425	2092
	.797	.256	Meaningfully
	79.7 %	25.6 %	different !

Clearly these two percentages are meaningfully different since the men had a much higher risk for dying than the women. But can we also claim that there is a difference in the risk for dying among men and women in the population from which these samples came? In other words, is the difference in the risks for men and women **statistically significant**?

If we wish to draw conclusions about a population from data collected from a sample, we must consider the methods of **statistical inference**. In particular, we must view the two proportions or percentages as estimates obtained from a sample. Let's focus on the two sample proportions, which we denote $\hat{p}_M$ and $\hat{p}_W$. The corresponding population proportions are denoted p_M and p_W, without "hats".

Statistical inference draws conclusions about a population parameter based on information obtained from a sample statistic. So, what is the population parameter considered for these data and what is its corresponding sample statistic?

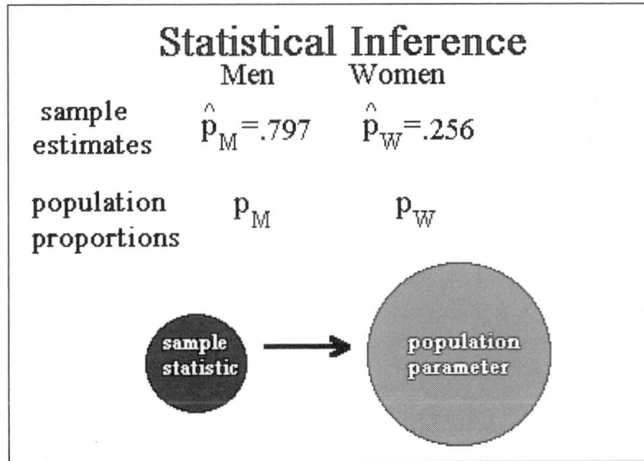

Since our focus here is to **compare** the proportions for males and females, one logical choice for our parameter of interest is the difference between the two population proportions. The corresponding sample statistic is the difference between the two estimated proportions.

Statistical Inference

	Men - Women	
sample statistic	$\hat{p}_M - \hat{p}_W$	= **.541**
population parameter	$p_M - p_W$	

<u>Study Questions (Q12.1)</u>

1. What other (epidemiologic) parameters could also be considered as alternatives to the difference in the two proportions?

Hypothesis testing can be used here to determine whether the difference in the two proportions is statistically significant. This is one of the two types of statistical inference questions we may ask. Our hypothesis in this case, usually stated as what we want to disprove, is that the true difference in the two population proportions is zero. This is called the **null hypothesis**. In hypothesis testing, we seek evidence from the sample to disprove the null hypothesis in favor of the **alternative hypothesis** that there is a difference in the population proportions.

Statistical Inference - Two Types:
1. Hypothesis Testing

$$\hat{p}_M - \hat{p}_W = .541$$

$$p_M - p_W$$

Statistically significant ? Hypothesis Testing → $H_0: p_M - p_W = 0$

$H_A: p_M - p_W \neq 0$

__Study Questions (Q12.1) continued__

2. If the parameter of interest is the risk ratio (RR), how would you state the null hypothesis?
3. If the parameter of interest is the odds ratio (OR), how would you state the null hypothesis?

We can use **interval estimation** to determine the precision of our point estimate. Here, our goal is to use our sample information to compute two numbers, say, **L** and **U**, that define a confidence interval for the difference between the two population proportions. Using a confidence interval, we can predict with a certain amount of confidence, say 95%, that the limits, **L** and **U**, bound the true value of the parameter. For our data, it turns out, that the lower and upper limits for the difference in the two proportions are .407 and .675, respectively.

$$L = .407 < p_M - p_W < U = .675$$

$$\hat{p}_M - \hat{p}_W = .541$$

It may appear from these two numbers that an interval estimate is less precise than a point estimate. The opposite is actually true. The range of values specified by the interval estimate actually takes into account the unreliability or variance of the point estimate. It is therefore more precise, since it uses more information to describe the point estimate.

In general, interval estimation and hypothesis testing can be contrasted by their different approaches to answering questions. A test of hypothesis arrives at an answer by looking for rare or unlikely sample results. In contrast, interval estimation arrives at its answer by looking at the most likely results, that is, those values that we are confident lie close to the parameter under investigation.

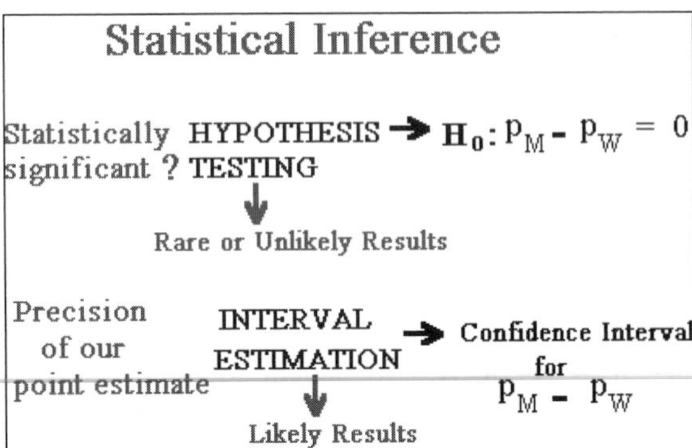

Summary

- ❖ **Statistical inference** concerns drawing conclusions about a population parameter based on information obtained from a sample statistic.
- ❖ The two types of statistical inference are **hypothesis testing** and **interval estimation**.
- ❖ When we test a hypothesis, we typically want to disprove a null hypothesis.
- ❖ When doing interval estimation, we want to obtain a confidence interval that provides upper and lower limits that we can be, say 95%, confident covers the population parameter of interest.
- ❖ A test of hypothesis looks for rare or unlikely results from our sample, where a confidence interval looks for likely results.

The Incident

The story of the Titanic is well known. The largest ship that had ever been built up to that time, she left Southampton, England on her maiden voyage to New York on Wednesday, April 10, 1912, carrying many of the rich and famous of England and the United States, but also many of more modest means. Because the Titanic was so large and so modern, many thought that she could not sink.

After a stop at Cherbourg France, where she took on many 3rd class passengers seeking new lives in the New World, and a brief stop off Queenstown, Ireland, she set out across the Atlantic. At 11:40 on the evening of April 14th, the Titanic struck an iceberg and, by 2:15 the next morning, sank.

Of 2,201 passengers and crew, only 710 survived. Some facts can be gleamed about the passengers, about who survived and who did not. One underlining question of interest in any disaster of this sort is did everyone have an equal chance of surviving? Or, stated in statistics terms, was the probability of surviving independent of other factors.

12-2 Statistical Inference for Simple Analyses (continued)

Hypothesis Testing

Hypothesis Testing, Part 1

We illustrate how to carry out a statistical test of hypothesis to compare survival of men and women passengers on the Titanic. We want to assess whether the difference in sample proportions for men and women is statistically significant.

A **test of hypotheses**, also called a **test of significance**, can be described as a seven-step procedure. In **step one** we state the information available, the statistical assumptions, and the population parameter being considered. In our example, the information includes the numbers of men and women passengers and the sample proportions. We need to assume that the data represents a sample from a larger population and that each sample proportion is approximately normally distributed. The **population parameter** of interest is the difference between the two population proportions being estimated (P_M - P_W). The corresponding **sample statistic** is the difference between the two estimated proportions $(\hat{P}_M - \hat{P}_W)$.

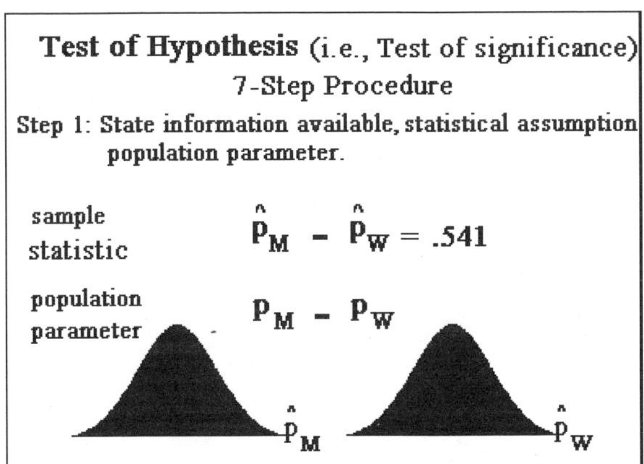

In **step 2**, we specify the **null** and **alternative hypotheses**. A hypothesis is a claim about a value of a population parameter. The hypothesis that we plan to test is commonly called the null hypothesis. This is often the accepted state of knowledge that we want to question. The null hypothesis in our example is that the difference in the two population proportions is zero. $\mathbf{H_0}$ is essentially treated like the defendant in a trial. It is assumed true, or innocent, until the evidence from the data makes it highly unlikely to have occurred by chance. Our testing goal here is to see if we have evidence to disprove this null hypothesis. The alternative hypothesis, typically called $\mathbf{H_A}$, gives the values the parameter may take if the null is false. In our example, the alternative hypothesis is that the difference in population proportions is not equal to zero. This is called a **two-sided alternative** because it states that we are interested in values both above and below the null. The alternative hypothesis would be called **one-sided** if, before looking at our data, we were interested in determining only whether men were at greater risk for dying than women. To avoid biasing one's analysis, both the null and alternative hypothesis should be made without looking at the study data and be based only on the a priori objectives of the study.

Step 2: Specify the null and alternative hypotheses.

$H_0: p_M - p_W = 0$

$H_A: p_M - p_W \neq 0$ Two-sided

APRIORI

$p_M - p_W > 0$ One-sided

In **step 3**, we specify the **significance level, alpha**. We will set the significance level for our example at .05 or 5 percent. This means that in carrying out our procedure, we are willing to take a 5 percent risk of rejecting the null hypothesis even if it is actually true. Equivalently, the significance level tells us how rare or unlikely our study results have to be under the null hypothesis in order for us to reject the null hypothesis in favor of the alternative hypothesis. An alpha of 5 percent means that, if the null hypothesis is actually true, we will have a 5% chance of rejecting it.

Step 3: Specify the significance level α.

$\alpha = .05$ (i.e., 5%)

Even if H_0 true, 5% risk of rejecting H_0

How unlikely must our study results be (under H_0) in order to reject H_0.

<u>Summary</u>

- A statistical test of hypothesis can be described as a seven-step procedure. The first three steps are:
  - Step 1: State the information available, the statistical assumptions, and the population parameter being considered.
  - Step 2: Specify the null and alternative hypotheses.
  - Step 3: Specify the significance level alpha (α).

Hypothesis Testing, Part 2

In **step 4** of our hypothesis testing procedure, we must select the **test statistic** to use, and we must state its **sampling distribution** under the assumption that the null hypothesis is true. Because the parameter of interest is the difference between two proportions, the test statistic T is given by the difference in the two sample proportions divided by the estimated standard error of this sample difference under the null hypothesis. The denominator here is computed using an expression involving the pooled estimate, $\hat{p}$, of the common proportion for both groups that would result under the null hypothesis.

Step 4: Select the test statistic and state its sampling distribution under H_0.

$$T = \frac{\hat{p}_M - \hat{p}_W}{\sqrt{\hat{p}(1-\hat{p})\left(\frac{1}{n_M}+\frac{1}{n_W}\right)}} \qquad \hat{p} = \frac{n_M\hat{p}_M + n_W\hat{p}_M}{n_M + n_W}$$

Study Question (Q12.2)

The pooled estimate of the common proportion for two groups is a weighted average of the two sample proportions, where the weights are the sample sizes used to compute each proportion. The sample size proportions and their corresponding sample sizes are .7972 and 1667, respectively for men, and .256 and 425 for women.

1. Compute the pooled estimate from the above information. (You will need a calculator to obtain your answer.)

The sampling distribution of this test statistic is approximately the standard normal distribution, with zero mean and unit standard deviation, under the null hypothesis.

In **step 5**, we formulate the decision rule that partitions the possible outcomes of the test statistic into acceptance and rejection regions. Because our test statistic has approximately the standard normal or Z distribution under the null hypothesis, the acceptance and rejection regions will be specified as intervals along the Z-axis under the curve of this distribution. In particular, because our alternative hypothesis is two-tailed and since our significance level is .05, these two regions turn out to be as shown here by the red and green lines. (Note: the red lines are in the tail areas < -1.96 and > 1.96; the green line between –1.96 and 1.96.) The area under the standard normal curve above the interval described as the acceptance region is .95 .The area under the curve in each tail of the distribution identified as rejection regions is .025. The sum of these two areas is .05, which is our chosen significance level. The -1.96 on the left side under the curve is the 2.5 percentage point of the standard normal distribution, and the 1.96 on the right side under the curve is the 97.5 percentage point.

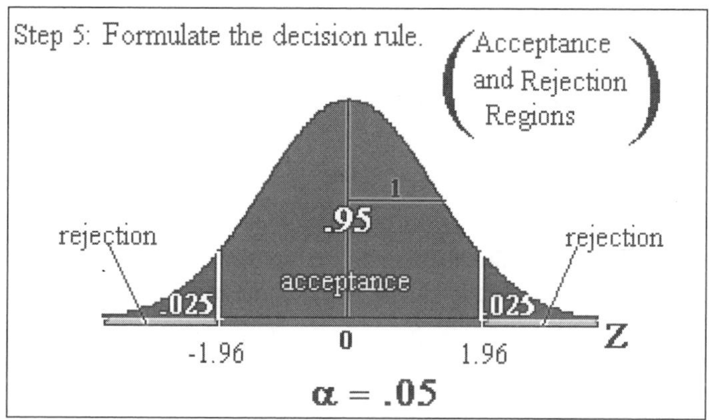

Our decision rule can now be described as follows. If the value of the **test statistic T** computed from our data falls into the rejection region, we reject the null hypothesis in favor of the alternative hypothesis. However, if the observed study value falls into the acceptance region, we do not reject the null hypothesis.

Step 6 of our process simply requires us to compute the value of the test statistic T from the observed data. We will call the computed value **T*** to distinguish it from the test statistic T. Here again are the sample results:

Step 6: Compute the test statistic T from the data.

$$T = \frac{\hat{p}_M - \hat{p}_W}{\sqrt{\hat{p}(1-\hat{p})\left(\frac{1}{n_M}+\frac{1}{n_W}\right)}} \qquad \hat{p} = \frac{n_M\hat{p}_M + n_W\hat{p}_M}{n_M + n_W}$$

Computed T = T*

$\hat{p}$ = .6874 (men and women combined)

$\hat{p}_M$ = .7972 n_M = 1667 (men)

$\hat{p}_W$ = .2565 n_W = 425 (women)

Substituting the sample information into the formula for T, our computed value T* turns out to be 12.0.

$$T^* = \frac{.7972 \quad .2565}{\sqrt{.6874\,(1 - .6874)\,\left(\frac{1}{1667} + \frac{1}{425}\right)}}$$

$$= \mathbf{12.00}$$

Finally, in **Step 7**, we use our computed test statistic to draw conclusions about our test of significance. In this example, the computed test statistic falls into the extreme right tail of the rejection region because it is much larger than 1.96.

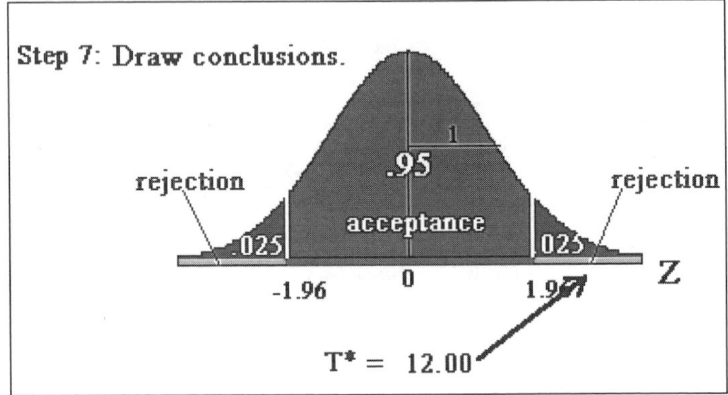

Consequently, we reject the null hypothesis and conclude that we have a statistically significant difference between the two proportions at the .05 significance level. We therefore conclude that of all those that could have been aboard the Titanic, men were more likely to die than women.

Summary

- ❖ A statistical test of hypothesis can be described as a seven-step procedure. The first three steps are:
  - o Step 1: State the information available, the statistical assumptions, and the population parameter being considered.
  - o Step 2: Specify the null and alternative hypotheses.
  - o Step 3: Specify the significance level alpha (α).
  - o Step 4: Select the test statistic and state its sampling distribution under the null hypothesis.
  - o Step 5: Formulate the decision rule in terms of rejection and acceptance regions under the null hypothesis.
  - o Step 6: Compute the test statistic using the observed data.
  - o Step 7: Draw conclusions, i.e., reject or do not reject the null hypothesis at the alpha significance level.

Hypothesis Testing – The P-value

We have found that the difference in the sample proportions of the men and women who died on the Titanic was statistically significant at the 0.05 significance level. In particular, the computed value of the test statistic fell into the extreme right tail of the rejection region. This tells us that if the null hypothesis were true, the observed results had less than a 5% chance of occurring. That is, the results were quite unlikely under the null hypothesis.

We may wonder, then, exactly how unlikely, or how rare, were the observed results under the null hypothesis? Were they also less than 1% likely, or less than 0.1% likely, or even rarer? The answer to these questions is given by the **P-value**. The P-value gives the probability of obtaining the value of the test statistic we have computed or a more extreme value if the null hypothesis is true.

Let's assume, as in our example, that the test statistic has the standard normal distribution under the null hypothesis. To obtain the P-value, we must determine an area under this curve. Here, we show four different areas that correspond to where the computed value T* falls under the curve and to whether the alternative hypothesis is one-sided or two-sided.

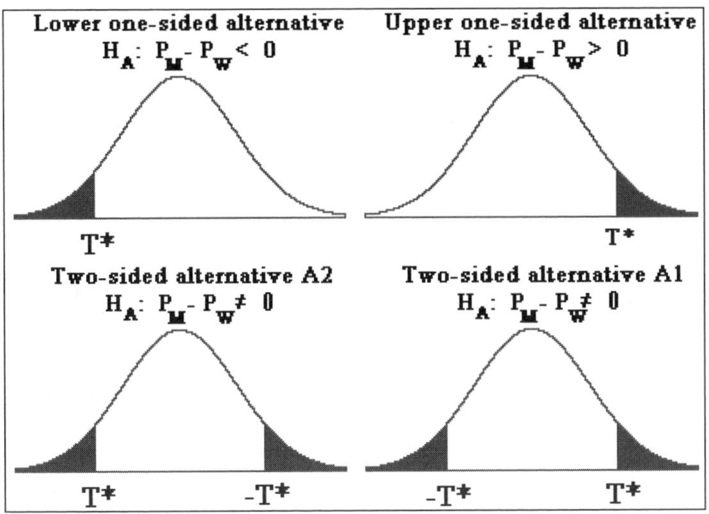

If the alternative hypothesis is an upper one-sided hypothesis, then the P-value is the area under the curve to the right of T* (upper right distribution in the above figure). If the alternative hypothesis is a lower one-sided hypothesis, then the P-value is the area under the curve to the left of T* (upper left distribution in the above figure). If the alternative hypothesis is two-sided and T* falls in the right tail under the curve, then the P-value is the sum of the areas under the curve to the right of T* and to the left of -T*. If the alternative hypothesis is two-sided and T* falls in the left tail under the curve, then the P-value is the sum of the areas under the curve to the left of T* and to the right of -T*. The P-value gives the area under the curve that shows the probability of the study results under the null hypothesis.

Study Questions (Q12.3)

1. Which of the four scenarios above correspond to the P-value for our Titanic example? (Hint: T* = 12, H_A is two-sided.)
2. To obtain the P-value for a 2-sided H_A, why is it not necessary to compute 2 areas under the normal curve?

Now, let's see how rare our computed test statistic is under the null hypothesis. The computed test statistic is 12.0, so we need to find the area under the normal curve to the right of the value 12.0, and to the left of -12.0. One way to determine this area is to use a table of the percentage points of the standard normal or Z distribution. In one such table, as illustrated in the figure that follows this paragraph, the highest percentage point is 3.8, corresponding to the 99.99 percentage point. Although we can't find the area to the right of 12.0 under the normal curve exactly, we can say this area is less than .0001, clearly a very small value.

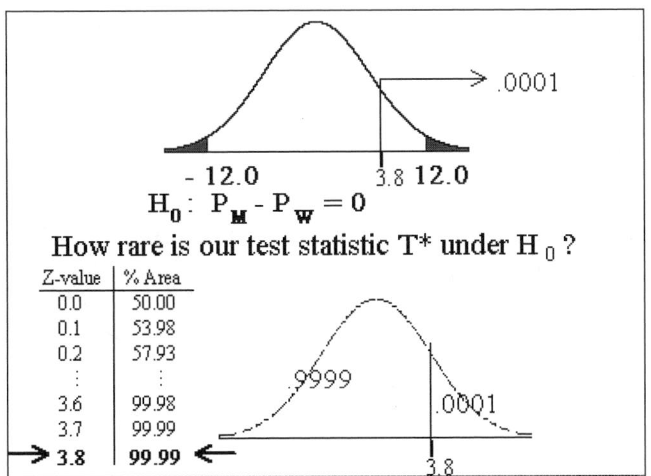

Z-value	% Area
0.0	50.00
0.1	53.98
0.2	57.93
⋮	⋮
3.6	99.98
3.7	99.99
3.8	**99.99**

Study Questions (Q12.3) continued

3. If our alternative hypothesis had been one-tailed, what would be our P-value?
4. Since our alternative hypothesis was actually two-tailed, what is our P-value?
5. Based on your answer to question 2, has a rare event occurred under the null hypothesis?
6. Based on the P-value here, what should you conclude about whether or not the test of hypothesis is significant?

P-values are often used as an alternative way to draw conclusions about a test of hypothesis rather than specifying a fixed significance level in advance of computing the test statistic. If the P-value is small enough, so that a rare event has occurred, then we reject the null hypothesis. If the P-value is not small, then we would not reject the null hypothesis.

So, how small must the P-value be for our results to be considered rare? The answer here essentially depends on the alpha (α) significance level we wish to use. A conventional choice for alpha is 0.05, although a frequent alternative choice is 0.01. Thus if the P-value is <0.05 or <0.01, then the test results are typically considered rare enough to reject the null hypothesis in favor of the alternative hypothesis.

Study Questions (Q12.3) continued

If your significance level was .05, what conclusions would you draw about the null hypothesis for the following P-values?

7. a) P > .01? b) P = .023?
8. c) P < .001? d) P = .54? e) P = .0002?

If your significance level was .001, what conclusions would you draw about H_0 for the following P-values?

9. a) P > .01? b) P = .023?
10. c) P < .001? d) .01 < P = .05? e) P = .0002?

Summary

- The P-value describes how unlikely, or how rare, are the observed results of one's study under the null hypothesis.
- For one-tailed alternative hypotheses, the P-value is determined by the area in the tail of the distribution, beyond the computed test statistics (i.e., to the right or left), under the null hypothesis.
- For two-tail alternative hypotheses, the P-value is twice the area in the tail of the distribution, beyond the computed test statistic, under the null hypothesis.
- The P-value is often used as an alternative way to draw conclusions about a null hypothesis rather than specifying a significance level prior to computing the test statistics.
- If the P-value is considered small by the investigators, say, less than .05, .01, we reject the null hypothesis in favor of the alternative hypothesis.
- If the P-value is not considered small, usually greater than .10, we do not reject the null hypothesis.

Z-scores and Relative Frequencies – The Normal Density Function

Most statistics texts include a table that lets you relate z-scores and relative frequencies in a **normal density**. The tables always give this information for the **Standard Normal Density**, so that the x-axis of the density is marked out in z-scores. The normal density tool we have been working with provides the same information more easily. For example, to find the relative frequency of values with z-scores below -1.5, just drag the left flag to the z-score value -1.5 and read the relative frequency in the lower left box, 0.067.

(Note: please use the ActivEpi CD Rom to use the **normal density tool** to answer a number of example questions.)

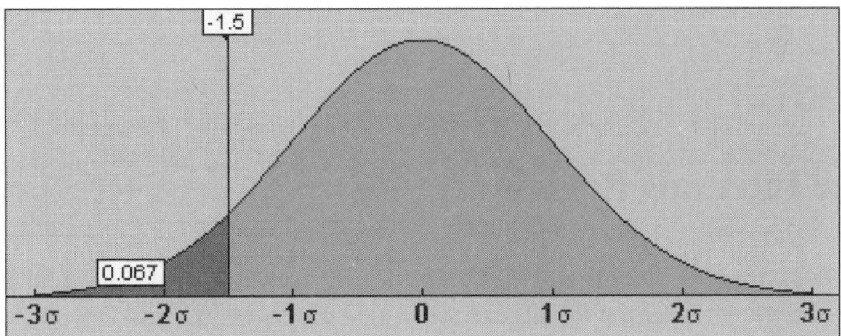

Summary

- The Standard Normal Density curve, for which tables are usually given in statistics textbooks, relates z-scores and relative frequencies.
- The Normal Density tool provides the same information directly.

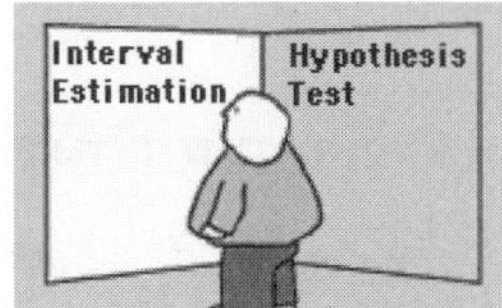

Quiz (Q12.4)

Fill in the Blanks.

1. We use ??? to assess whether the population parameter is different from the null value.
2. When determining the precision of a point-estimate, ??? accounts for sampling variability.
3. ??? looks for rare or unlikely results.
4. By looking at the most likely results, ??? finds those values that we are confident lie close to the population parameter.

Choices
hypothesis testing interval estimation

5. The ??? gives the risk we are willing to take for rejecting the null hypothesis when the null hypothesis is false.
6. The ??? can be either upper-one-sided, lower one-sided or two-sided.
7. If the computed value of the test statistic falls into the ???, we reject the ??? and conclude that the results are ??? significant.

Choices
acceptance region alternative hypothesis meaningfully null hypothesis
rejection region significance levels statistically

8. The ??? describes how rare or how unlikely are the observed results of one's study under the ???.
9. If the P-value satisfies the inequality P>.30, we should ??? the null hypothesis.
10. If the P-value satisfies the inequality P<.005, we should reject the null hypothesis at the ???. significance level, but not at the ??? level.

Choices
.001 .01 P-value alternative hypothesis not reject
null hypothesis reject significance level

12-3 Simple Analyses (continued)

Confidence Intervals Review

*A **confidence interval (CI)** provides two numbers **L** and **U** between which the population parameter lies with a specified level of confidence. Here we describe how to compute a large-sample 95% CI for the difference in two proportions.*

Confidence Interval for Comparing Two Proportions

We now show how to calculate a confidence interval for the difference in two proportions using the Titanic data. Our goal is to use our sample information to compute two numbers, **L** and **U**, about which we can claim with a certain amount of confidence, say 95%, that they surround the true value of the parameter. Here is the formula for this 95 percent confidence interval:

CONFIDENCE INTERVAL for $p_M - p_W$

The data: $\hat{p}_M = .797 \quad \hat{p}_W = .256$

$n_M = 1667 \quad n_W = 425$

The goal: $L < p_M - p_W < U$

e.g., 95% Confidence

$$(\hat{p}_M - \hat{p}_W) \pm 1.96 \sqrt{\frac{\hat{p}_M(1-\hat{p}_M)}{n_M} + \frac{\hat{p}_W(1-\hat{p}_W)}{n_W}}$$

The standard error of the difference is the square root of the sum of the variances of the proportions, where each variance is of the form $(\hat{p})(1-\hat{p})$ / (sample size). The value 1.96 is the 97.5 percent point of the standard normal distribution. This percent point is chosen because the area between -1.96 and +1.96 under the normal curve is .95, corresponding to the 95% confidence level we specified. The normal distribution is used here because the difference in the two sample proportions has approximately the normal distribution if the sample sizes in both groups are reasonably large, which they are for these data. This is why the confidence interval formula described here is often referred to as a **large-sample** confidence interval.

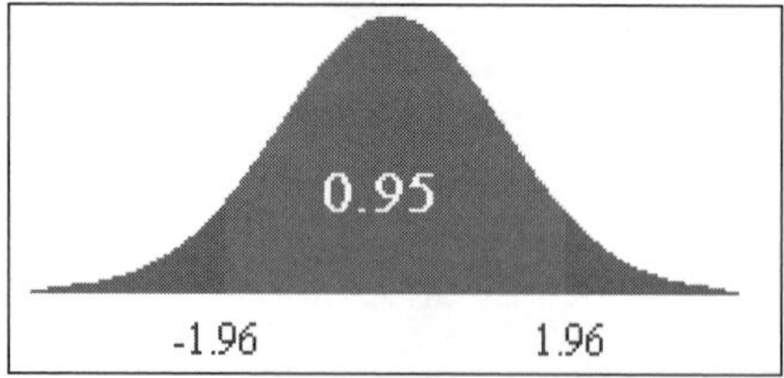

Study Question (Q12.5)

1. Why is the standard error formula used here different from the standard error formula used when testing the null hypothesis of no difference in the two proportions?

$$\text{S.E. for testing: } \sqrt{\hat{p}(1-\hat{p})\left(\frac{1}{n_M}+\frac{1}{n_W}\right)}$$

We can calculate the confidence interval for our data by substituting into the formula the values for $\hat{p}_M$, $\hat{p}_W$, n_M, and n_W:

CONFIDENCE INTERVAL for $p_M - p_W$

The data: $\hat{p}_M = .797$ $\hat{p}_W = .256$

$n_M = 1667$ $n_W = 425$

$$(\hat{p}_M - \hat{p}_W) \pm 1.96 \sqrt{\frac{.797(1-.797)}{1667} + \frac{.256(1-.256)}{425}}$$

The standard error turns out to be .0234. The lower and upper limits of the 95% interval are then .495 and .587, respectively. Thus, the 95 percent confidence interval for the difference in proportions of men and women who died on the Titanic is given by the range of values between .495 and .587.

$$(\hat{p}_M - \hat{p}_W) \pm (1.96)\ (.0234)$$

$$.495 < p_M - p_W < .587$$

Summary

- A **confidence interval** (**CI**) provides two numbers **L** and **U** between which the population parameter lies with a specified level of confidence.
- A large-sample 95% CI for the difference in two proportions is given by the difference ± 1.96 times the estimated standard error of the estimated difference.
- The estimated standard error is given by the square root of the sum of the estimated variances of each proportion.

Interpretation of a Confidence Interval

How do we interpret this confidence interval? A proper interpretation requires that we consider what might happen if we were able to repeat the study, in this example, the sailing and sinking of the Titanic, several times. If we computed 95 percent confidence intervals for the data resulting from each repeat, then we would expect that about 95 percent of these confidence intervals would cover the true population difference in proportions.

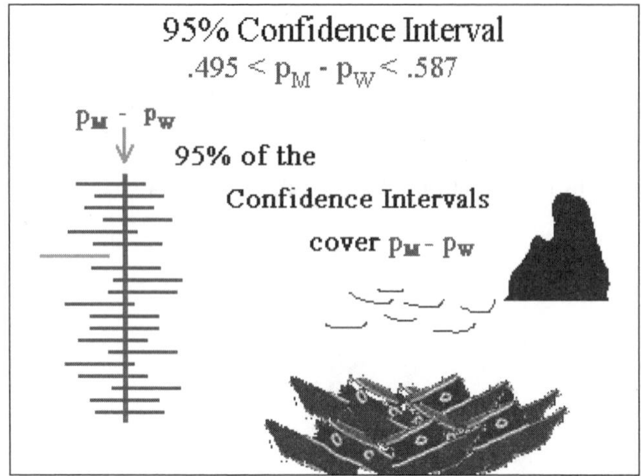

This is equivalent to saying that there is a probability of .95 that the interval between .495 and .587 includes the true population difference in proportions.

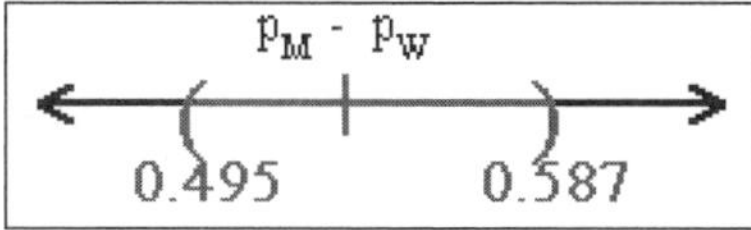

The true difference might actually lie outside this interval, but there is only a 5% chance of this happening.

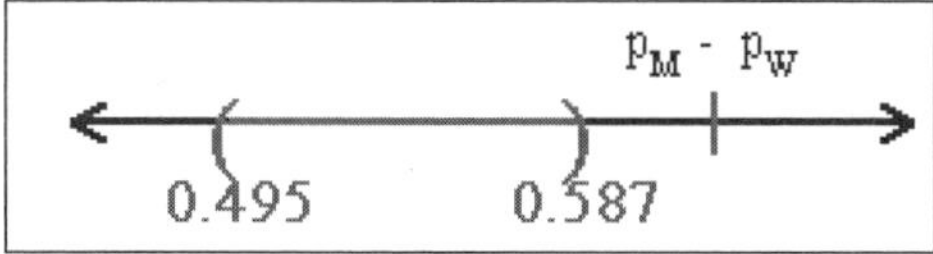

The probability statement that describes the confidence interval, which has the population parameter, $P_M - P_W$, without any hats, at its center, suggests that this parameter is a random variable. This is not so. The parameter $P_M - P_W$ does not vary at all; it has a single fixed population value. The random elements of the interval are the limits 0.495 and .587, which are computed from the sample data and will vary from sample to sample.

$$Pr\ (0.495 < p_M - p_W < 0.587) = 0.95$$

Random Fixed Random

In general, a confidence interval is a measure of the precision of an estimate of some parameter of interest, which for our example, is the difference between two population proportions. The narrower the width of the confidence interval, the more precise the estimate.

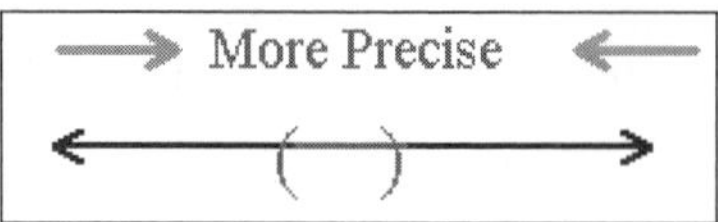

In contrast, the wider the width is, the less precise the estimate will be.

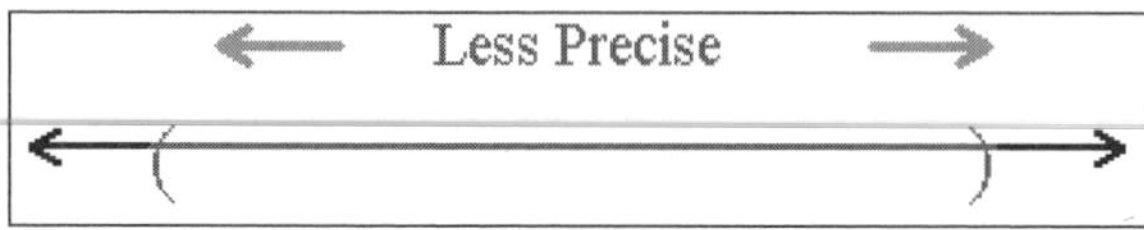

The extreme case of no precision at all would occur for difference measures (e.g., risk difference and incidence rate difference) where the confidence interval goes from minus infinity to infinity; for ratio measures (e.g., odds ratio, risk ratio, and incidence density ratio), it would be a confidence interval from zero to infinity; and for proportions, it would be a confidence interval from zero to 1.

Study Questions (Q12.6)

1. For a confidence interval that goes from minus infinity to infinity, how much confidence do we have that the true parameter is being covered by the interval?
2. If 90%, 95%, and 99% confidence intervals were obtained for the difference between two proportions based on the same data, which confidence interval would be the widest, and which would be the narrowest?
3. Suppose two different datasets yielded 95% confidence intervals for the difference between two proportions. Which dataset (A or B below) gives the more precise estimate?
   Dataset A: $.49 < p1 - p2 < .58$
   Dataset B: $.40 < p1 - p2 < .52$

Summary

- ❖ A 95% CI can be interpreted using the probability statement P(L < the parameter < U) = .95
- ❖ If a CI is computed for several repeats of the same study, we would expect about 95% of the CI's to cover the true population parameter.
- ❖ The random elements of a confidence interval are the limits L and U.
- ❖ It is incorrect to assume that the parameter in the middle of a confidence interval statement is a random variable.
- ❖ The larger the confidence level chosen, the wider will be the confidence interval.

A Debate: Does the Titanic data represent a population or a sample?

Our example, as previously indicated, describes the survival data for all men and women passengers on the Titanic, the "unsinkable" ocean liner that struck an iceberg and sank in 1912. Since these data consider all men and women passengers, it can be argued that the proportions being compared are actually population proportions, so that it is not appropriate to carry out either a statistical test of significance or to compute a confidence interval with these data. Nevertheless, a counter-argument is that the 1667 men and 425 women passengers represent a sample of men and women who were eligible to be chosen for the Titanic's journey, whereas the population difference in proportions refers to the proportions of all those eligible for the trip.

These two arguments are debatable, and from our point of view, there is no clear-cut reason to conclude that either argument is correct. In fact, similar debates often occur when analyzing data from an epidemiologic outbreak investigation. For example, when seeking the source of an outbreak of diarrhea from a picnic lunch, statistical tests are often carried out on data that represent everyone who attended the picnic. Such tests are justifiable only if the data being analyzed is considered a sample rather than a population.

Quiz (Q12.7)

Fill in the blanks.

1. A large-sample 95% confidence interval for the difference in two proportions adds and subtracts from the estimated difference in the two proportions 1.96 times the **???** of the estimated difference.
2. The confidence interval example shown below does not contain the null value for the **???** of the two proportions.
3. The **???** within a confidence interval has a single fixed value and does vary at all.

Choices

confidence level	difference	estimated mean	estimated standard error
estimated variance	population parameter	ratio	

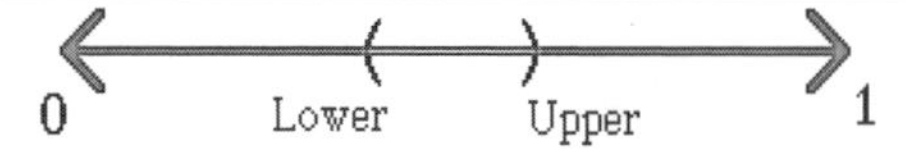

0	Lower	Upper	1

Example: $L = .495 < P_M - P_W < U = .587$

4. The **???** of a confidence interval may vary from sample to sample.
5. For a **???** confidence interval, the probability is 0.95 that the interval between the upper and lower bounds includes the true population parameter.
6. The true population parameter might actually lie outside this interval, but there is only a **???** chance of this happening.

Choices
5% **95%** **97.5%** **confidence level** **population parameter** **upper limit**

12-4 Simple Analyses (continued)

COHORT STUDIES INVOLVING RISK RATIOS

Hypothesis Testing for Simple Analysis in Cohort Studies

We return to the data from a cohort study to assess whether quitting smoking after a heart attack will reduce one's risk for dying. The effect measure in this study was a risk ratio and its estimate was 2.1. What can we say about the population risk ratio based on the estimated risk ratio obtained from the sample?

Heart Attack Patients	Smoke	Quit	Total
Death	27	14	41
Survival	48	67	115
Total	75	81	156

COHORT
Smoking → Mortality
$\hat{RR} = 2.1 \xrightarrow{?} RR$

We wish to know if we have evidence from the sample that the risk ratio is statistically different from the null value. That is, we wish to perform a **test of hypothesis** to see if the risk ratio is significantly different from 1. The null hypothesis being tested is that the population risk ratio is 1. The logical alternative hypothesis here is that the risk ratio is >1, since prior to looking at the data the investigators were interested in whether continuing smoking was more likely than quitting smoking to affect mortality.

TEST OF HYPOTHESIS $H_0: RR = 1$
$H_A: RR > 1$

Because the risk ratio is the ratio of cumulative incidences for the exposed group (CI_1), divided by cumulative incidences for the unexposed group (CI_0), we can equivalently state the null hypothesis in terms of the difference in population cumulative incidences as shown here:

$$H_0: RR = \frac{CI_1}{CI_0} = 1$$

$$H_0: p_1 - p_0 = 0$$

$$H_0: ROR = 1$$

Equivalent

Because the risk ratio equals one if and only if the risk odds ratio equals one, we can also equivalently state the null hypothesis in terms of the risk odds ratio.

Because cumulative incidence is a proportion, the cumulative incidence version of the null hypothesis implies that our test about the risk ratio is equivalent to testing a hypothesis about the difference between two proportions: H_0: $p_1 - p_0 = 0$.

The test statistic is the difference in the two estimated cumulative incidences divided by the estimated standard error of this difference, under the null hypothesis that the risk ratio is one. Because the sample sizes in both groups are reasonably large, this test statistic has approximately a standard normal distribution under the null hypothesis.

$$T = \frac{\hat{p}_1 - \hat{p}_0}{\sqrt{\hat{p}(1-\hat{p})\left(\frac{1}{n_1} + \frac{1}{n_0}\right)}} \quad \text{where } p_1 = CI_1,\ p_0 = CI_0$$
$$p = \text{pooled CI}$$

The computed value of the test statistic is obtained by substituting the estimated cumulative incidences and corresponding sample sizes into the test statistic formula as shown here. The resulting value is 2.65.

$$T^* = \frac{(27/75) - (14/81)}{\sqrt{\frac{41}{156}\left(1 - \frac{41}{156}\right)\left(\frac{1}{75} + \frac{1}{81}\right)}} = 2.65$$

The P-value for this test is then obtained by finding the area in the right tail of the standard normal distribution above the computed value of 2.65. The exact P-value turns out to be .0040. Because the P-value of .0040 is well below the conventional significance level of .05, we reject the null hypothesis and conclude that the risk ratio is significantly greater than the null value of one. In other words, we have found that among heart attack patients who smoke, continuing smokers have a significantly higher risk for dying than smokers who quit after their heart attack.

Summary

- When testing the hypothesis about a risk ratio (RR) in a cumulative-incidence cohort study, the null hypothesis can be equivalently stated as either RR = 1, $CI_1 - CI_2 = 0$, or ROR = 1, where CI_1 and CI_2 are the cumulative incidences for the exposed and unexposed groups, respectively.
- The alternative hypothesis can be stated in terms of the RR either as RR $\neq$ 1, RR > 1, or RR < 1 depending on whether the alternative is two-sided, upper one-sided, or lower one-sided, respectively.
- To test the null hypothesis that RR = 1, the test statistic is the same as that used to compare the difference between two proportions.
- Assuming large samples, the test statistic has approximately the N(0,1) distribution under H_0.

Chi Square Version of the Large-sample Test

The large-sample test for a risk ratio can be carried out using either the **normal distribution** or the **chi square distribution**. The reason for this equivalence is that if a standard normal variable Z is squared, then Z square has a chi square distribution on 1 degree of freedom.

More specifically, for our mortality study of heart attack patients, here is the test statistic that we previously described, it follows a standard normal distribution under the null hypothesis that the risk ratio equals 1. The square of this statistic is shown next to its corresponding chi square distribution:

$$T = \frac{\hat{p}_1 - \hat{p}_0}{\sqrt{\hat{p}(1-\hat{p})\left(\frac{1}{n_1} + \frac{1}{n_0}\right)}}$$

$$T^2 = \frac{(\hat{p}_1 - \hat{p}_0)^2}{\hat{p}(1-\hat{p})\left(\frac{1}{n_1} + \frac{1}{n_0}\right)}$$

0 Z

0 $\chi^2_{1\ df}$

With a little algebra, we can rewrite the statistic in terms of the cell frequencies a, b, c and d of the general 2 by 2 table that summarizes the exposure disease information in a cohort study that estimates cumulative incidence.

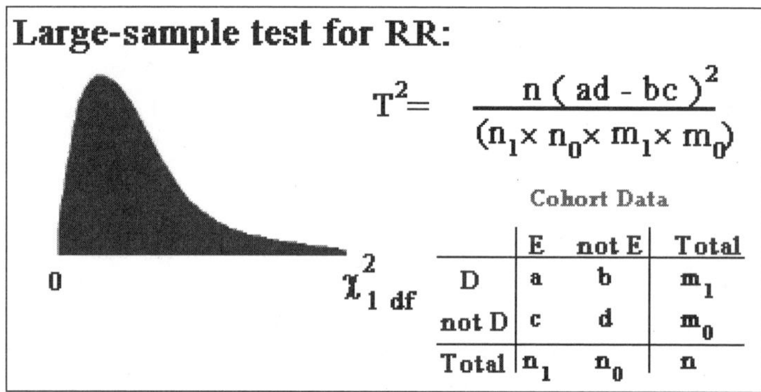

Large-sample test for RR:
$T^2 = \frac{n\,(ad - bc)^2}{(n_1 \times n_0 \times m_1 \times m_0)}$
0
$\chi^2_{1\ df}$
Cohort Data
E
not E
Total
D
a
b
m_1
not D
c
d
m_0
Total
n_1
n_0
n

For the mortality study of heart attack patients, the values of the cell frequencies are shown following this paragraph. Substituting these values into the chi square statistic formula, we obtain the value 7.04. This value is the square of the computed test statistic we found earlier ($2.65^2 \approx 7.04$)

$$\frac{156\,[(27)(67)-(48)(14)]^2}{(75)\,(81)\,(41)\,(115)} = 7.04$$

Mortality Study

	E	not E	Total
D	27	14	41
not D	48	67	115
Total	75	81	156

Is the chi square version of our test significant? The normal distribution version was significant at the .01 significance level, so the chi square version had better be significant.

But in comparing these two versions of the test, we need to address a small problem. A chi-square statistic, being the square of a Z statistic, can never be negative, but a standard normal statistic can either be positive or negative. Large values of a chi square statistic that might indicate a significant result could occur from either large positive values or large negative values of the normal statistic. In other words, if we use a chi square statistic to determine significance, we are automatically performing a test of a **two-sided** alternative hypothesis even though we are only looking for large values in the right tail of the distribution.

Study Questions (Q12.8)

The .99 and .995 percentage points of the chi square distribution with 1 df are given by the values 6.635 and 7.789, respectively.

1. Does the computed chi square value of 7.04 fall in the upper 1 percent of the chi square distribution?
2. Does the computed chi square value of 7.04 fall in the upper .5 percent of the chi square distribution?
3. Would a test of a two-sided alternative for RR be significant at the .01 significance level? (Note: The computed test statistic is 7.04.)
4. Would a test of a two-sided alternative for RR be significant at the .005 significance level?

So, if our chi square statistic allows us to assess a two-sided alternative hypothesis about the risk ratio, how can we assess a one-sided alternative? One way is simply to carry out the normal distribution version of the test as previously illustrated. The other way is to divide the area in the right tail of the chi square curve in half.

Study Questions (Q12.8) continued

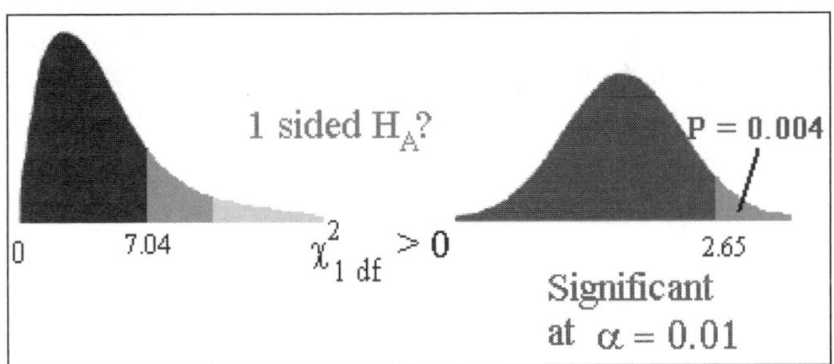

5. Using the above results, would a test of an upper one-sided alternative RR > 1 be significant at the .01 significance level?
6. Would a test of an upper one-sided alternative RR > 1 be significant at the .005 significance level?

Summary

- An alternative but equivalent way to test for the significance of a risk ratio is to use a chi square test.
- The square of a standard normal variable has the chi square distribution with one degree of freedom.
- We can directly compute the chi square statistic using a formula involving the cell frequencies of a 2 x 2 table for cohort data that allows for estimation of risk.
- When we use a chi square statistic to determine significance, we are actually performing a test of a two-sided alternative hypothesis.
- We can assess a one-sided alternative either using the normal distribution version of the test or by using the chi square distribution to compute a P-value.

The P-value for a One-Sided Chi Square Test

We now describe how to compute a P-value when using a chi square test involving a one sided alternative hypothesis about a risk ratio. We will illustrate this computation once again using data from the mortality study on smoking behavior of heart attack patients. To determine the P-value, we must first compute the area shaded as pink (lighter shaded in the right tail of the distribution) under the chi-square distribution above the value of the computed chi-square statistic 7.04. This area actually gives the P-value for a 2-sided alternative hypothesis. In fact, it is equivalent to the combined area in the two tails of the corresponding normal distribution defined by the computed value of 2.65.

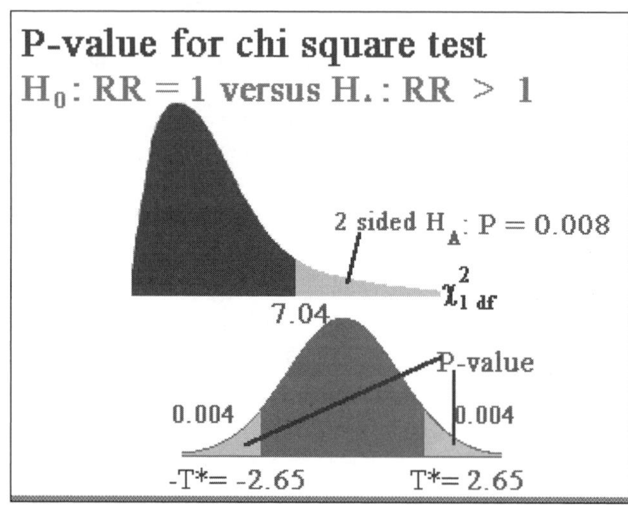

For the upper one-sided alternative that the population risk ratio is greater than 1, we previously found the P-value to be the area only in the right tail of the normal curve above 2.65. This right tail area is but one-half of the corresponding area in the right tail of the chi-square curve above 7.04. Consequently, the area in the right tail of the chi square curve gives twice the P-value for a one-tailed alternative. Therefore, this area must be divided by 2 in order to get a one-sided P-value.

$$P = \frac{.008}{2} = .004$$

Study Questions (Q12.9)

1. Suppose the alternative hypothesis in our example above was a **lower** one-sided alternative and the computed value of the chi square statistic is 7.04, corresponding to a computed Z statistic of +2.65. What is the corresponding P-value?
2. Based on your answer to the previous question, what would you conclude about the null hypothesis that the RR equals 1?
3. Suppose the alternative hypothesis was a **lower** one-sided alternative and the computed value of the chi square statistic is 7.04, but this time corresponding to a computer Z statistic of –2.65. What is the corresponding P-value?
4. Based on your answer to the previous question, what would you conclude about the null hypothesis that the RR equals 1?

Summary

- When using a chi square test of a two-sided alternative hypothesis about a risk ratio, the P-value is obtained as the area under the chi square curve to the right of the computed test statistic.
- If the alternative hypothesis is one-sided, e.g., RR > 1, then the P-value is one-half the area in the right tail of the chi square curve.

Testing When Sample Sizes Are Small

This table displays hypothetical results from a five-month randomized clinical trial comparing a new anti-viral drug for shingles with the standard anti-viral drug "Valtrex". Only 13 patients were involved in the trial, which was a pilot study for a larger clinical trial. The estimated risk ratio from these data is 3.2, which indicates that, in the sample of 13 patients, the new drug was 3.2 times more successful than the standard drug.

Pilot Clinical Trial for Shingles Patients

	Drug: New	Drug: Standard	Total
Success	4	2	6
Failure	1	6	7
Total	5	8	13

$$\begin{array}{ll} \text{new drug:} & 4/5 = 0.80 \\ \text{standard drug:} & 2/8 = 0.25 \end{array} \Rightarrow \hat{RR} = 3.2$$

Is this risk ratio significantly different from the null value of one? We are considering a small sample size here, so a large-sample test of hypothesis is not appropriate. (Note: How large is large is somewhat debatable here, but it is typically required that the sample size for a proportion must be large enough (e.g., >25) for the sample to be normally distributed.) However, there is a statistical test for sparse data, called **Fisher's Exact Test**, that we can use here. Fisher's Exact Test is appropriate for a 2x2 table relating a dichotomous exposure variable and a dichotomous disease variable. To use Fisher's exact test, we must assume that the values on the margins of the tables are fixed values prior to the start of the study. If we make this 'fixed marginals assumption', we can see that once we identify the number in anyone cell in the table, say the exposed cases or **a** cell, the numbers in the other three cells can be determined by using only the frequencies on the marginals.

Study Questions (Q12.10)

1. Determine the formulas for calculating the values for **b**, **c**, and **d** in terms of **a** and the fixed marginal values.

General Layout	E	not E	Total
D	4		6
not D			7
Total	5	8	13

2. Calculate the values in the other cells of the table above knowing only the marginal values and the value in one cell.

To test our hypothesis about the risk ratio, we need only consider the outcome for one of the four cells of the table. For example, we would like to know what values we might get in the **a** cell that would be unlikely to occur under the null hypothesis that the risk ratio, or for that matter the risk odds ratio, equals 1. In particular, we would like to determine whether the value obtained in the **a** cell for our study, which turned out to be 4, is a rare enough event under the null hypothesis for us to reject the null hypothesis and conclude that the observed risk ratio is statistically significant. We can answer this question by computing the P-value for our test.

Study Questions (Q12.10) continued

General Layout	E	not E	Total
D	5		6
not D			7
Total	5	8	13

3. Suppose the **a** cell value was 5 instead of 4, assuming the marginals are fixed. What would be the corresponding revised values for **b**, **c**, and **d**?
4. What would be the risk ratio for the revised table?
5. Is the revised risk ratio further away from the null risk ratio than the risk ratio of 3.2 actually observed?
6. Are there any other possible values for the **a** cell that would also be further away from the null risk ratio than 3.2?
7. Based on the previous questions, why would we want to compute the probability of getting an **a** cell value of 4 or 5 under the null hypothesis?

To compute the P-value for Fisher's Exact test, we need to determine the probability distribution of the **a** cell frequency under the null hypothesis. Assuming fixed-marginals, this distribution is called the **hypergeometric distribution**. The formulae for the hypergeometric distribution and the corresponding P-value for Fisher's Exact Test are described in the box that follows this activity.

Study Questions (Q12.10) continued

Using the hypergeometric distribution, the (Fisher's Exact Test) one-sided P-value for the study described in this activity is P(a =4 or 5|Rr=1) = P(a=4|RR=1) + P(a=5|RR=1) = .0816 + .0046 = .0863.

8. What are your conclusions about the null hypothesis that RR = 1?

Summary

- Fisher's Exact Test provides a test of significance for a risk ratio or an odds ratio when the data are sparse.
- To use Fisher's exact test, we must assume that the values on the margins of a 2x2 table are fixed values prior to the start of the study.
- The marginal frequencies of a 2x2 table provide no information concerning the strength of the association.
- To compute the P-value for Fisher's Exact test, we need to determine the probability distribution of the **a** cell frequency under the null hypothesis.
- Assuming fixed-marginals, the **a** cell has the **hypergeometric distribution**.
- Computer programs, including a DataDesk program for ActivEpi, are available to calculate the P-value for this test.

How to carry out Fisher's Exact Test using the hypergeometric distribution

To compute the P-value for Fisher's Exact test, we need to use the probability distribution of the "a" cell frequency (i.e., the number of exposed cases) under the null hypothesis of no effect of exposure on disease.

Data layout with fixed marginal totals and with cell "a" = "j"

	E	Not E	
D	j	$m_1 - j$	m_1
Not D	$n_1 - j$	$n_0 - m_1 + j$	m_0
	n_1	n_0	

Assuming fixed-marginals, this distribution is called the hypergeometric distribution, which is given by the following probability formula for a random variable A:

$$pr(A = j \mid H_0) = \frac{C_j^{n_1} C_{m_1 - j}^{n_0}}{C_{m_1}^{n_1 + n_0}} \quad \text{where} \quad C_s^r = \frac{r!}{s!(r-s)!}$$

denotes the number of combinations of *r* items taken *s* at a time, and

$r! = r(r\text{-}1)(r\text{-}2)...(3)(2)(1)$ denotes *r* factorial,

$s!$ and $(r\text{-}s)!$ denote *s* factorial and $(r\text{-}s)$ factorial, respectively.

It then follows that the exact P-value for a test of H_0 versus H_A using the above hypergeometric formula is given by:

$$P = pr(A \geq a \mid H_0) = \sum_{j=a}^{\min(n_1, m_1)} pr(A = j \mid H_0)$$

We illustrate the calculation of this P-value using the data from the pilot clinical trial for a new treatment for patients with shingles. Here again is the data:

Clinical trial data on shingles patients.

	Drug: New	Drug: Standard	
Success	4	2	6
Failure	1	6	7
	5	8	13

The P-value is thus calculated as the probability that the **a** cell is 4 or 5 under the null hypothesis that RR=l:

$$P = pr(A \geq 4 \mid H_0) = \sum_{j=4}^{5} pr(A = j \mid H_0)$$

$$= \frac{C_4^5 C_2^8}{C_6^{13}} + \frac{C_5^5 C_1^8}{C_6^{13}} = .0816 + .0047 = .0863$$

Since p is greater than .05 we would fail to reject the null hypothesis that the RR = 1 and conclude that there is not a significant effect of the new drug at the .05 level of significance.

Compute a test of Hypothesis using Fisher's Exact test in DataDesk

An exercise is provided to compute a Fisher's exact test using the DataDesk program included on the CD ROM.

12-5 Simple Analyses (continued)

COHORT STUDIES INVOLVING RISK RATIOS (continued)

Large-sample version of Fisher's Exact Test - The Mantel-Haenszel Test

Let's consider again the mortality study data on smoking behavior of heart attack patients. For these data, we have previously described a large-sample chi square statistic for testing the null hypothesis that the risk ratio is 1.

Heart Attack Patients	Smoke	Quit	Total
Death	27	14	41
Survival	48	67	115
Total	75	81	156

$$\chi^2 = \frac{n(ad - bc)^2}{(n_1 \times n_0 \times m_1 \times m_0)}$$

H_0: RR = 1

Because the mortality data involves a large-sample, we do not need to use Fisher's Exact Test for these data. Nevertheless, we could still compute the Fisher's Exact test statistic. To compute Fisher's Exact test, we assume that the frequencies on the margins of the table are fixed and then we compute the probability under the null hypothesis of getting an **a** cell value at least as large as the value of 27 that was actually obtained. We would therefore compute and sum 15 probability values, from a= 27 to a= 41 to obtain the P-value for Fisher's Exact Test.

Fisher's Exact Test

compute $P(a \geq 27 | H_0)$

P(27|RR=1) + P(28|RR=1) + P(29|RR=1) +
P(30|RR=1) + P(31|RR=1) + P(32|RR=1) +
P(33|RR=1) + P(34|RR=1) + P(35|RR=1) +
P(36|RR=1) + P(37|RR=1) + P(38|RR=1) +
P(39|RR=1) + P(40|RR=1) + P(41|RR=1)

Although such a calculation can be accomplished with an appropriate computer program, a more convenient large-sample approximation to Fisher's Exact Test is often used instead when the cell frequencies in the two by two table are moderately large. This large-sample approach is called the **Mantel-Haenszel** (**MH**) test for simple analysis. The Mantel-Haenszel statistic is shown below. This statistic has an approximate chi square distribution with one degree of freedom under the null hypothesis that the risk ratio is 1. Consequently, the P-value for a one-sided alternative using this statistic is obtained in the usual way by finding the area under the chi-square distribution above the computed test statistic and then dividing this area by 2.

Mantel-Haenszel test (MH)

$$\chi^2(MH) = \frac{(n-1)(ad - bc)^2}{(n_1 \times n_0 \times m_1 \times m_0)}$$

approx. $\chi^2_{1\ df}$

under H_0: RR=1

If we compare the large-sample chi square statistic for Fisher's exact test with the large-sample chi square statistic previously described for comparing two proportions, we see that these two test statistics are remarkably similar. In fact, they differ only in that the statistic for approximating Fisher Exact Test contains n-1 in the numerator but the earlier large-sample version contains n in the numerator. When n is large, using either n or n-l in the numerator will have little effect on the computed chi square statistic.

Mantel-Haenszel test (MH)

$$\chi^2(MH) = \frac{(n-1)(ad - bc)^2}{(n_1 \times n_0 \times m_1 \times m_0)}$$

$$\chi^2 = \frac{n(ad - bc)^2}{(n_1 \times n_0 \times m_1 \times m_0)}$$

In our mortality data example, for instance, the value of the chi square approximation to Fisher's exact test is equal to 6.99. The large-sample chi square statistic for comparing two proportions was previously shown to be 7.04. Clearly, the two chi square statistics are very close, although not exactly equal. The corresponding one-sided P-values are .0040 and .0041, respectively, essentially equal.

Heart Attack Patients	Smoke	Quit	Total
Death	27	14	41
Survival	48	67	115
Total	75	81	156

Mantel-Haenszel test (MH)

$$\chi^2(MH) = \frac{(n-1)(ad - bc)^2}{(n_1 \times n_0 \times m_1 \times m_0)} = \frac{(156-1)(27x67-14x48)^2}{(75x81x41x115)}$$

$$= 6.99 \ (P = .0040)$$

$$\chi^2 = \frac{n(ad - bc)^2}{(n_1 \times n_0 \times m_1 \times m_0)} = \frac{(156)(27x67-14x48)^2}{(75x81x41x115)}$$

$$= 7.04 \ (P = .0041)$$

*Note: in the box above one-sided p-values are provided.

This example illustrates that in large-samples, these two chi square versions are essentially equivalent and will lead to the same conclusions about significance.

Study Questions (Q12.11)

	Drug		
	New	Standard	Total
Success	4	2	6
Failure	1	6	7
Total	5	8	13

For the above data, we previously showed that the one-sided P-value for Fisher's Exact test is .0863. The MH statistic computed from these data turns out to be 3.46.

1. The P-value for a two-sided MH test is .0630. What is the P-value for a one-sided Mantel-Haenszel test?
2. Why is the P-value for the Mantel-Haenszel test different from the P-value for Fisher's Exact test?
3. The computed value for the large-sample chi square statistic for comparing two proportions is 3.75. Why is this latter value different from the computed Mantel-Haenszel test statistic of 3.46?

Summary

- A large-sample approximation to Fisher's Exact test is given by the Mantel-Haenszel (MH) chi-square statistic for simple analysis.
- The MH statistic is given by the formula:

$$\chi^2 = \frac{(n-1)(ad-bc)^2}{n_1 n_0 m_1 m_0}$$

- The MH chi square statistic contains n-1 in the numerator whereas the large-sample chi square version for comparing two proportions contains n in the numerator.
- In large-samples, either of these two chi square versions are essentially equivalent and will lead to the same conclusions about significance.

Large-Sample Confidence Interval for a Risk Ratio

We once again use the data from the mortality study on smoking behaviors of heart attack patients, this time to describe how to obtain a large-sample confidence interval for a risk ratio.

Heart Attack Patients	Smoke	Quit	Total
Death	27	14	41
Survival	48	67	115
Total	75	81	156

A risk ratio is a ratio of two proportions, each of which is a measure of cumulative incidence. If we were interested in the **difference** rather than the **ratio** between two cumulative incidences, the large sample confidence interval would be given by the commonly used confidence interval formula for two proportions shown here:

Large-sample confidence interval for $p_1 - p_2$

$$(\hat{p}_1 - \hat{p}_2) \pm 1.96\sqrt{\frac{\hat{p}_1(1-\hat{p}_1)}{n_1} + \frac{\hat{p}_2(1-\hat{p}_2)}{n_2}}$$

This formula says that we must add and subtract from the difference in the two estimated proportions 1.96 times the estimated standard error of this difference. The corresponding 95 percent confidence interval formula for a risk ratio is slightly more complicated. In contrast to the risk difference formula, the risk ratio formula looks like this:

Large sample confidence interval for RR $\left(= \frac{p_1}{p_2}\right)$?

$$\hat{RR}\exp\left[\pm 1.96\sqrt{\frac{1-\hat{CI}_1}{n_1\hat{CI}_1} + \frac{1-\hat{CI}_0}{n_0\hat{CI}_0}}\right]$$

Note: $\exp[\,a\,] = e^a$

From this formula, the lower and upper limits are then given by the following expressions:

$$L = \hat{RR}\exp\left[-1.96\sqrt{\frac{1-\hat{CI}_1}{n_1\hat{CI}_1}+\frac{1-\hat{CI}_0}{n_0\hat{CI}_0}}\right]$$

$$U = \hat{RR}\exp\left[+1.96\sqrt{\frac{1-\hat{CI}_1}{n_1\hat{CI}_1}+\frac{1-\hat{CI}_0}{n_0\hat{CI}_0}}\right]$$

An equivalent version of the risk ratio formula, which helps explain where this formula comes from, is shown here:

$$e^{\ln(\hat{RR}) \pm 1.96\sqrt{\hat{Var}\{\ln(\hat{RR})\}}}$$

The confidence interval for a risk ratio is obtained by exponentiating a large sample confidence interval for the natural log of the risk ratio. The formula used for the estimated variance of the log of the risk ratio is actually an approximate, not an exact formula.

$$\frac{1-\hat{CI}_1}{n_1\hat{CI}_1}+\frac{1-\hat{CI}_0}{n_0\hat{CI}_0} \approx \hat{Var}\{\ln(\hat{RR})\}$$

There are two reasons why the formula for the risk ratio is more complicated than for the risk difference. First, the estimated difference in two proportions is approximately normally distributed, but the estimated ratio of two proportions is highly skewed. In contrast, the log of the risk ratio is more closely normally distributed.

Second, the variance of a ratio of two proportions is complicated mathematically and is not equal to the ratio of the variances of each proportion. However, since the log of a ratio is the difference in logs, approximating the variance of a difference is much easier.

We now apply the risk ratio formula to the mortality study data set. Substituting the values for the estimated risk ratio, the cumulative incidences in each group, and the sample sizes, we obtain the lower and upper confidence limits shown here:

Heart Attack Patients	Smoke	Quit	Total
Death	27	14	41
Survival	48	67	115
Total	75	81	156

$$2.08\exp\left[\pm 1.96\sqrt{\frac{1-27/75}{75*27/75}+\frac{1-14/81}{81*14/81}}\right]$$

$$1.185 < RR < 3.661$$

Study Question (Q12.12)

1. Interpret the above results.

Summary

- The confidence interval formula for a ratio of two proportions is more mathematically complicated than the formula for a difference in two proportions.
- The 95% risk ratio formula multiplies the estimated risk ratio by the exponential of plus or minus the quantity 1.96 times the square root of the variance of the log of the estimated risk ratio.
- This risk ratio formula is obtained by exponentiating a large sample confidence interval for the natural log of the risk ratio.
- The estimated ratio of two proportions is highly skewed whereas the log of the risk ratio is more closely normally distributed.
- The variance of the log of a risk ratio is mathematically easier to derive than is the variance of the risk ratio of itself, although an approximation is still required.

Large-Sample Approximation Formula for a Confidence Interval for the Risk Ratio

To obtain a 95% confidence interval for a risk ratio (**RR**), we might be inclined initially to use the interval:

$$\hat{RR} \pm 1.96\sqrt{\hat{Var}(\hat{RR})}$$

However, this interval is not recommended because there is asymmetry in the distribution of the estimated RR, i.e., $\hat{RR}$ is not normally distributed. The recommended approach involves working with the estimate of **lnRR** rather than **RR** itself, since the natural log transformation tends to change and skewed distribution to an approximately normal distribution. A two-step procedure is used: First, a 95% large-sample confidence interval for **lnRR** is obtained of the form:

$$\ln\hat{RR} \pm 1.96\sqrt{\hat{Var}(\ln\hat{RR})}$$

Then, since **RR = exp(lnRR)**, the desired upper and lower values of the 95% confidence interval for **RR** are found by taking the anti logarithms of lower and upper values for the confidence interval for **lnRR**. The resulting formula takes the general form:

$$\hat{RR}\exp\left[\pm 1.96\sqrt{\hat{Var}(\ln\hat{RR})}\right]$$

To complete the calculation, however, we need to find an expression for $\hat{Var}(\ln \hat{RR})$. Although an exact mathematical expression for this variance estimate is not readily available, a good approximation to this variance can be obtained by using what is called a **Taylor Series approximation** (for further details, see Kleinbaum, Kupper, and Morgenstern, Epidemiologic Research, p.298-299, John Wiley and Sons, 1982). Using what is called a **first-order Taylor-series approximation**, the approximate formula for this variance is given by:

$$\hat{Var}(\ln\hat{RR}) \approx \frac{(1-\hat{CI}_1)}{n_1\hat{CI}_1} + \frac{(1-\hat{CI}_0)}{n_0\hat{CI}_0}$$

where $\hat{CI}_1$ and $\hat{CI}_0$ denote the cumulative incidence for exposed persons and unexposed persons, respectively. Thus, the general 95% large-sample approximation confidence interval formula can be written as:

$$\hat{RR}\exp\left[\pm 1.96\sqrt{\frac{(1-\hat{CI}_1)}{n_1\hat{CI}_1} + \frac{(1-\hat{CI}_0)}{n_0\hat{CI}_0}}\right]$$

Compute Large-Sample Tests of Hypothesis and Confidence Intervals for Data from Cohort Studies using DataDesk

An example of calculating large sample tests of hypothesis and confidence intervals for cohort studies is demonstrated using the DataDesk program.

Quiz (12.13)

For the cohort study data shown below, the estimated probability of death among continuing smokers (CS) is 0.303, and the estimated probability of death among smokers who quit (OS) is 0.256.

Heart Attack

Patients	Smoke	Quit	Total
Death	277	243	520
Survival	638	705	1343
Total	915	948	1863

$$T = \frac{\hat{p}_{CS} - \hat{p}_{QS}}{\sqrt{\hat{p}(1-\hat{p})\left(\frac{1}{n_{CS}} + \frac{1}{n_{QS}}\right)}}$$

1. What is the estimated probability of death among the entire sample: **???**.

2. Based on the computed probability values, use a calculator or computer to compute the value of the test statistic T for testing the null hypothesis that the RR equals 1. Your answer is T* = **???**.

Choices
0.279 **0.475** **0.491** **0.533** **0.721** **2.20** **2.23** **2.27** **2.31**
2.36

The normal density tool that is described in an activity at the bottom of lesson book page 12-2 can be used to determine P-values. The actual normal density tool is located on Lesson book page 16-2 in the Reference section of the ActivEpi CD-ROM.

3. Use the normal density tool to find the P-value for based on the computed T statistic (2.23) and assuming an upper-one-sided alternative hypothesis. Your answer is: P-value = **???**.

4. Based on the P-value, you should reject the null hypothesis that RR=1 at the **???** significance level but not at the **???** significance level.

Choices
0.003 **0.013** **0.021** **0.120** **1%** **5%**

The computed Z statistic for testing RR = 1 for these data is 2.23.

5. Using the computed Z statistic, what is the value of the Mantel-Haenszel chi-square statistic for these data? MH CHISQ = **???**.

6. What is the P-value for a two-sided test of the null hypothesis that RR=1? (You may wish to use the chi square distribution tool located on lesson page 16-2 of the ActivEpi CD-ROM.) P-value =**???**.

7. Fisher's exact test is not necessary here because the sample size for this study is **???**.

Choices
0.006 **0.013** **0.026** **0.120** **0.240** **4.970** **4.973** **large** **small**

Calculating Sample Size for Clinical Trials and Cohort Studies

When a research proposal for testing an etiologic hypothesis is submitted for funding, it is typically required that the proposal demonstrate that the number of subjects to be studied is "large enough" to reject the null hypothesis if the null hypothesis is not true. This issue concerns the **sample size** advocated for the proposed study. The proposer typically will prepare a section of his/her proposal that describes the sample size calculations and resulting decisions about sample size requirements.

Since most epidemiologic studies, even if considering a single dichotomous exposure variable, involve accounting for several (**control**) variables in the analysis, the methodology for determining sample size can be very complicated. As a result, many software programs have been developed to incorporate such multivariate complexities, e.g., **Egret SIZ**, **PASS**, and **Power and Precision** (a web-based package). The use of such programs is likely the most mathematically rigorous and computationally accurate way to carry out the necessary sample size deliberations.

Nevertheless, there are basic principles from which all sophisticated software derive, and such principles are conveniently portrayed in the context of a 2x2 table that considers the simple (i.e., crude) analysis of the primary exposure-disease relationship under study. Moreover, the use of sample size formulae for a simple analysis is often a convenient and non-black-box approach for providing a reasonable as well as understandable argument about sample size requirements for a given study. A description of such formulae now follows.

All formulae for sample size requirements for hypothesis testing consider the two types of error that can be made from a statistical test of hypothesis. A **Type I error** occurs if the statistical test (incorrectly) rejects a true null hypothesis, and a **Type II error** occurs if the test (incorrectly) does not reject a false null hypothesis. The probability of making a Type I error is usually called **α**, the **significance level of the test**. The probability of making a Type II error is usually called **β**, and **1-β** is called the **power of the test**. All sample size formulae that concern hypothesis testing are aimed at determining that sample size for a given study that will achieve desired (small) values of **α** and **β** and that will detect a specific departure from the null hypothesis, often denoted as **Δ**. Consequently, the investigator needs to specify values for **α**, **β**, and **Δ** into an appropriate formula to determine the required sample size.

For clinical trials and cohort studies, the sample size formula for detecting a risk ratio (RR) that differs from the null value of 1 by at least **Δ**, i.e. (**Δ = RR –1**) is given by the formula:

$$n = \frac{(Z_{1-\alpha/2} + Z_{1-\beta})^2 \bar{p}\bar{q}(r+1)}{[p_2(RR-1)]^2 r}$$

where

$Z_{1-\alpha/2}$ = the 100(1 - α/2) percent point of the N(0,1) distribution

$Z_{1-\beta}$ = the 100(1 – β) percent point of the N(0,1) distribution

p_2 = expected risk for unexposed subjects

$\bar{p} = p_2(RR + 1) / 2$

$\bar{q} = 1 - \bar{p}$

r = ratio of unexposed to exposed subjects

(Note: if the sample sizes are to be equal in the exposed and unexposed groups, then **r** = 1. When **r** does not equal 1, the above formula provides the sample size for the exposed group; to get the sample size for the unexposed group, use **n x r**.)

To illustrate the calculation of **n**, suppose **α** = .05, **β** = .20, **RR** = 2, $\mathbf{p_2}$ = .04, and **r** = 3. Then:

$\bar{p} = (.04)(2 + 1) / 2 = .06$

and substituting these values into the formula for **n** yields:

$$n = \frac{(1.96 + 0.8416)^2 (.06)(.94)(3+1)}{[(.04)(2-1)]^2 3} = 368.9$$

Thus, the **sample size (n)** needed to detect a risk ratio (**RR**) of 2 at an **α** of .05 and a **β** of .20, when the expected risk for exposed ($\mathbf{p_2}$) is .04 and the ratio of unexposed to exposed subjects (**r**) is 3, is 369 exposed subjects and 368.9 x 3 = 1,107 unexposed subjects. The above sample size formula can also be used to determine the sample size for estimating a prevalence ratio (i.e., **PR**) in a cross-sectional study; simply substitute **PR** for **RR** in the formula.

12-6 Simple Analyses (continued)

CASE-CONTROL STUDIES

Large-sample (Z or Chi Square) Test

The table shown below has been described in previous lessons illustrating data from a case-control study to determine the source of an outbreak of diarrhea at a resort in Haiti. The odds ratio estimate here is larger than the null value of one, and therefore indicates that there is a moderately large effect of the exposure. What can we conclude about the **population exposure odds ratio** based on the **estimated odds ratio** obtained from the sample?

Resort Study	Raw Hamburger Ate	Did not eat	Total
Cases	17	20	37
Controls	7	26	33
Total	24	46	70

CASE-CONTROL

Source → Diarrhea

$\hat{OR} = 3.2 \xrightarrow{?} EOR$

We can answer this question by performing a test of hypothesis that the population exposure odds ratio is 1. The logical alternative hypothesis here is that the odds ratio is > 1, since prior to looking at the data the investigators were interested in whether eating raw hamburger was more likely than not eating raw hamburger to affect whether a person would develop a diarrheal illness. The null hypothesis that the odds ratio equals 1 (H_0: EOR = 1) can equivalently be expressed as the difference in two proportions being equal to zero. The proportions in this case are the proportion exposed among cases and the proportion exposed among controls. These **exposure proportions are not risks** since the study is a case-control study. Nevertheless, we can still use a **large-sample Z statistic** to test for the difference between these two proportions.

$$H_0: p_1 - p_0 = 0 \qquad H_0: EOR = 1 \quad H_A: EOR > 1$$

$$\hat{p}_1 = \hat{p}(\text{exposed among cases}) = 17/37 = 0.46$$

$$\hat{p}_0 = \hat{p}(\text{exposed among controls}) = 7/33 = 0.21$$

Below is the test statistic, which gives the difference in the two estimated exposure proportions divided by the estimated standard error of this difference under the null hypothesis. If the sample sizes in both case and control groups are reasonably large, which we will assume is the case here, this test statistic has approximately a standard normal distribution under the null hypothesis.

$$T = \frac{\hat{p}_1 - \hat{p}_0}{S_{\hat{p}_1 - \hat{p}_0}}$$

standard error under H_0 → $S_{\hat{p}_1 - \hat{p}_0}$

If n_{cases}, $n_{controls}$ large

approx. N(0,1) under H_0

Because the large-sample statistic for testing the difference between two proportions is normally distributed, the square of test statistic has approximately a chi square distribution with 1 degree of freedom. (Note: We will describe only the chi square version of this test here. However, you may wish to either click on the asterisk on the lesson page next to the icon for this activity or see the box following this activity in this companion textbook to see the compute value of the T statistic and its corresponding P-value based on the normal distribution.)

$$T^2 = \frac{(\hat{p}_1 - \hat{p}_0)^2}{(S_{\hat{p}_1 - \hat{p}_0})^2}$$

If n_{cases}, $n_{controls}$ large

approx. N(0,1) under H_0

The general data layout for this chi square test from case-control data is essentially the same as previously described for cohort data. The only difference is that for case-control data, the row margins are fixed prior to the start of the study, whereas for cohort data, the column margins are fixed in advance.

Case-Control

	E	not E	Total
D	a	b	m_1
not D	c	d	m_0
Total	n_1	n_0	n

Cohort

	E	not E	Total
D	a	b	m_1
not D	c	d	m_0
Total	n_1	n_0	n

Consequently the exact same chi square formula previously described for cohort data can be used for case-control data.

$$X^2 = \frac{(\hat{p}_1 - \hat{p}_0)^2}{\left(S_{\hat{p}_1 - \hat{p}_0}\right)^2} = \frac{n(ad - bc)^2}{(n_1 n_0 m_1 m_0)}$$

Substituting the, cell frequencies from our case-control study data into the chi square statistic formula we obtain the value 4.74. If the Mantel-Haenszel chi square formula is used instead, so that the numerator is replaced by n - 1, the chi square statistic changes only slightly to 4.67.

$$X^2 = \frac{70(17*26-20*7)^2}{(24*46*37*33)} = 4.74$$

$$X^2_{MH} = \frac{(70-1)(17*26-20*7)^2}{(24*46*37*33)} = 4.67$$

Study Question (Q12.14)

1. Using the Mantel Haenszel chi square value of 4.67, describe how you would determine the P-value for the one-sided alternative that the exposure odds ratio is greater than 1? (This question asks for a process, not a numerical value.)
2. The 95 percent and 97.5 percent points of the chi square distribution with 1 df are 3.84 and 5.02, respectively. Based on this information, what can you say about the P-value for the Mantel-Haenszel test? (You are being asked for a number or range of numbers for the P-value.)
3. Is the estimated exposure odds ratio from the case-control study statistically significant at the 0.05 level of significance? At the 0.01 level of significance?

Summary

- When testing the hypothesis about an odds ratio (OR) in a case-control study, the null hypothesis can be equivalently stated as either EOR = 1 **or** $p_1 - p_2 = 0$, where p_1 and p_2 are estimated exposure probabilities for cases and non-cases.
- The alternative hypothesis can be stated in terms of the EOR either as EOR $\neq$ 1, EOR > 1, or EOR < 1, depending on whether the alternative is two-sided, upper one-sided, or lower one-sided, respectively.
- One version of the test statistic is a large-sample N(0,1) statistic used to compare two proportions.
- An alternative version is a large-sample chi square statistic, which is the square of the N(0,1) statistic.
- A Mantel-Haenszel large-sample chi square statistic can alternatively be used for the chi square test.

The Z Statistic for a Large-Sample Test of an Odds Ratio

The large-sample Z statistic to test for the difference between two exposure probabilities from a case-control study is given by the formula:

$$T = \frac{(\hat{p}_{CA} - \hat{p}_{CO})}{\sqrt{\hat{p}(1-\hat{p})(\frac{1}{n_{CA}} + \frac{1}{n_{CO}})}}$$

where $\hat{p}_{CA} = \hat{P}r(E \mid D)$, $\hat{p}_{CO} = \hat{P}r(E \mid \text{not } D)$, and $\hat{p} = \hat{P}r(E)$

This statistic gives the difference in the two estimated exposure proportions divided by the estimated standard error of this difference under the null hypothesis that the odds ratio is one. If the sample sizes in both case and control groups are reasonably large, which we will assume is the case here, this test statistic has approximately the standard normal distribution under the null hypothesis.

Using the data from the outbreak of diarrhea at a Haitian resort, the computed value of the test statistic, which we call **T***, is obtained by substituting the estimated cumulative incidences and corresponding sample sizes into the test statistic formula. These values are:

$$\hat{p}_{CA} = \frac{17}{37} = .459, \ \hat{p}_{CO} = \frac{7}{33}, \text{ and } \hat{p} = \frac{24}{70} = .343$$

where $n_{CA} = 37$ and $n_{CO} = 33$

The resulting value of the **Z** statistic is **T* = 2.18**. The **P-value** for this test is then obtained by finding the area in the right tail of the standard normal distribution above the computed value of **2.18**. This area is **.0146**. This P-value is below the conventional significance level of .05 but above .01, we would therefore reject the null hypothesis at the 5 percent level but not at the 1 percent level. In other words, we have found that the odds ratio for the association between eating raw hamburger and the development of diarrheal illness is of borderline statistical significance.

Testing When Sample Sizes Are Small

This table displays hypothetical case-control results from the diarrheal outbreak study based on a sample of 19 persons from the resort in Haiti. The total sample size here is less than one-third the sample size previously found for the actual outbreak data described in the previous activity. The estimated odds ratio is 3.5.

Hypothetical Diarrheal Outbreak Data

Resort Study	Raw Hamburger Ate	Did not eat	Total
Cases	5	5	10
Controls	2	7	9
Total	7	12	19

$\hat{OR} = 3.5$

Is this odds ratio significantly different from the null value of one? We are considering a small sample size here, so a large-sample test of hypothesis is not appropriate. We can, however, use Fisher's Exact Test for these data. We again consider the general data layout for a 2 by 2 table with a dichotomous exposure and dichotomous disease result. To use Fisher's exact test, we must assume that the values on the margins of the table are fixed prior to the start of the study.

General Layout	E	not E	Total	
D	a	b	m_1	assume
not D	c	d	m_0	fixed
Total	n_1	n_0	n	margins

Assuming "fixed margins", we only need to consider the outcome for one of the four cells of the table, say the **a** cell. We therefore would like to know what values we might get in the **a** cell that would be unlikely to occur under the null hypothesis. In particular, we would like to determine whether the value obtained in the **a** cell for our case-control study, which turned out to be 5, is a rare enough event under the null hypothesis for us to reject the null hypothesis and conclude that the observed odds ratio is statistically significant. We can answer this question by computing the P-value for our test.

Study Questions (Q12.15)

Resort Study	Raw Hamburger Ate	Did not eat	Total
Cases	4	b	10
Controls	c	d	9
Total	7	12	19

Suppose the **a** cell value was 4 instead of 5, assuming the marginals are fixed.

1. What would be the corresponding revised values for **b**, **c**, and **d**?
2. What would be the odds ratio for the revised table?
3. Is the revised odds ratio further away from the null odds ratio than the odds ratio of 3.5 actually observed?
4. Are there any other possible values for the **a** cell that would be further away from the null odds ratio than 3.5?
5. Based on previous questions, why would we want to compute the probability of getting an **a** cell value of 5, 6, or 7 under the null hypothesis?

To compute the P-value for Fisher's Exact test, we need to determine the probability distribution of the **a** cell frequency under the null hypothesis. Assuming fixed-marginals, this distribution is the **hypergeometric distribution**. The formula for this distribution and the corresponding P-value for Fisher's Exact Test are given in the asterisk on the ActivEpi CD-ROM and in the box following the activity in this Companion Textbook.

Study Questions (Q12.15) continued

Using the hypergeometric distribution, this P-value is:

$$\Pr(a = 5 \text{ or } 6 \text{ or } 7|OR = 1) = \Pr(a = 5|OR = 1) + \Pr(a = 6|OR = 1) + \Pr(a = 7|OR = 1)$$
$$= .1800 + .0375 + .0024 = .2199$$

6. What do you conclude about the null hypothesis?

Summary

- Fisher's exact test provides a test of significance for an odds ratio as well as a risk ratio with the data are sparse.
- To use Fisher's exact test, we must assume that the values on the margins of a 2x2 table are fixed values prior to the start of the study.
- To compute the P-value for Fisher's exact test, we need to determine the probability distribution of the **a** cell frequency under the null hypothesis.
- Assuming fixed-marginals, the **a** cell has the hypergeometric distribution.
- Computer programs, including DataDesk for ActivEpi, are available to calculate the P-value for this test.

How to carry out Fisher's Exact Test using the hypergeometric distribution in a Case-Control Study

Fisher's Exact Test procedure is the same regardless of whether the study design is case-control, cohort or cross-sectional, as long as the assumption of **fixed marginals** is assumed. The derivation of the procedure is now described (as previously done in an earlier asterisk/box when discussing small samples in cohort studies or clinical trials). An example of the use of Fisher's Exact test with case-control data is then provided. To compute the **P-value** for Fisher's Exact test, we need to use the probability distribution of the a-cell frequency (i.e., the number of exposed cases) under the null hypothesis of no effect of exposure on disease.

Data layout with fixed marginal totals and with cell "a" = "j"

	E	Not E	
D	j	$m_1 - j$	m_1
Not D	$n_1 - j$	$n_0 - m_1 + j$	m_0
	n_1	n_0	

Assuming fixed-marginals, this distribution is called the **hypergeometric distribution**, which is given by the following probability formula for a random variable A:

$$pr(A = j \mid H_0) = \frac{C_j^{n_1} C_{m_1 - j}^{n_0}}{C_{m_1}^{n_1 + n_0}} \qquad \text{where} \qquad C_s^r = \frac{r!}{s!(r-s)!}$$

denotes the number of combinations of r items taken s at a time, and
$r! = r(r-1)(r-2)...(3)(2)(1)$ denotes r factorial,
$s!$ and $(r-s)!$ denote s factorial and $(r-s)$ factorial, respectively.

It then follows that the **exact P-value** for a test of H_0:OR = 1 versus H_A: OR > 1 using the above hypergeometric formula is given by

$$P = pr(A \geq a \mid H_0) = \sum_{j=a}^{\min(n_1, m_1)} pr(A = j \mid H_0)$$

We illustrate the calculation of this P-value using the data from a hypothetical case-control study of an outbreak of a diarrheal disease at a Haitian resort involving 19 subjects (instead of the actual 70 subjects in the real outbreak). Here again is the data:

Hypothetical case-control study of an outbreak at a Haitian resort.

	Raw Hamburger		
	Ate	**Did not Eat**	
Case	5	5	10
Control	2	7	9
	7	12	19

The P-value is thus calculated as the probability that the a cell is > 5 (but smaller than the minimal marginal values of 7 and 10) under the null hypothesis that OR=1:

Continued on next page

How to carry out Fisher's Exact Test using the hypergeometric distribution in a Case-Control Study (continued)

$$P = pr(A \geq 5 \mid H_0) = \sum_{j=5}^{7} pr(A = j \mid H_0)$$

$$= \frac{C_5^7 C_{10-5}^{12}}{C_{10}^{19}} + \frac{C_6^7 C_{10-6}^{12}}{C_{10}^{19}} + \frac{C_7^7 C_{10-7}^{12}}{C_9^{19}} = .1800 + .0375 + .0024 = .2199$$

Since this P-value is much larger than .05, we conclude that the null hypothesis could not be rejected, i.e., there is not sufficient evidence from these 19 subjects that eating raw hamburger had a significant effect on whether or not a subject became ill.

Large-Sample Confidence Interval

We once again use the case-control data from an outbreak of diarrheal disease at a resort in Haiti, this time, to describe how to obtain a large-sample confidence interval for an odds ratio. The estimated odds ratio that describes the association of eating raw hamburger with diarrheal illness was 3.2.

Resort Study	Raw Hamburger Ate	Did not eat	Total
Cases	17	20	37
Controls	7	26	33
Total	24	46	70

$\hat{OR} = 3.2$

Below is the general formula for the 95% confidence interval for any odds ratio based on the general 2x2 data layout for a case-control study. The lower and upper limits are also provided.

$$\hat{OR}e^{\pm 1.96\sqrt{\frac{1}{a}+\frac{1}{b}+\frac{1}{c}+\frac{1}{d}}}$$

Lower Limit

$$\hat{OR}e^{-1.96\sqrt{\frac{1}{a}+\frac{1}{b}+\frac{1}{c}+\frac{1}{d}}}$$

Upper Limit

$$\hat{OR}e^{+1.96\sqrt{\frac{1}{a}+\frac{1}{b}+\frac{1}{c}+\frac{1}{d}}}$$

General Layout	E	not E	Total
D	a	b	m_1
not D	c	d	m_0
Total	n_1	n_0	n

An equivalent version of an odds ratio formula, which helps explain where this formula comes from, is shown here:

$$e^{\ln\hat{OR} \pm 1.96\sqrt{\hat{Var}(\ln\hat{OR})}}$$

This expression says that the confidence interval for an odds ratio is obtained by exponentiating a large sample confidence interval for the natural log of the odds ratio. The formula used for the estimated variance of the log of the odds ratio is actually an approximate, not an exact, formula.

We now apply the odds ratio formula to the case-control outbreak data set. Substituting into the formula the values for the estimated odds ratio and the four cell frequencies, we obtain the lower and upper confidence limits shown here.

$$3.16\, e^{\pm 1.96\sqrt{\frac{1}{17}+\frac{1}{20}+\frac{1}{7}+\frac{1}{26}}}$$

$$1.099 < OR < 9.074$$

Study Question (12.16)

1. What interpretation can you give to this confidence interval?

Summary

- ❖ The formula for a 95% confidence interval for an odds ratio formula multiplies the estimated odds ratio by the exponential of plus or minus the quantity 1.96 times the square root of the variance of the log of the estimated odds ratio.
- ❖ The odds ratio formula is obtained by exponentiating a large sample confidence interval for the natural log of the odds ratio.
- ❖ The variance of the log of an odds ratio is approximately equal to the sum of the inverses of the four cell frequencies in the 2x2 table layout.

Deriving the Large-Sample Approximation Confidence Interval Formula for the Odds Ratio

As with the risk ratio (RR), to obtain a 95% confidence interval for an odds ratio (**OR**), we might be inclined initially to use the interval:

$$\hat{OR} \pm 1.96\sqrt{\hat{Var}(\hat{OR})}$$

However, this interval is **not** recommended because there is asymmetry in the distribution of the estimated OR, i.e., $\hat{OR}$, is not normally distributed. The recommended approach involves working with the estimate of **lnOR** rather than **OR** itself, since the natural log transformation tends to change and skewed distribution to an approximately normal distribution. A two-step procedure is used: First, a 95% large-sample confidence interval for **lnOR** is obtained of the form:

$$\ln\hat{OR} \pm 1.96\sqrt{\hat{Var}(\ln\hat{OR})}$$

Then, since **OR = exp(lnOR)**, the desired upper and lower values of the 95% confidence interval for **OR** are found by taking the anti logarithms of lower and upper values for the confidence interval for **lnOR**. The resulting formula takes the general form:

$$\hat{OR}\exp\left[\pm 1.96\sqrt{\hat{Var}(\ln\hat{OR})}\right]$$

To complete the calculation, however, we need to find an expression for $\hat{Var}(\ln\hat{OR})$. Although an exact mathematical expression for this variance estimate is not readily available, a good approximation to this variance can be obtained by using what is called a **Taylor Series approximation** (for further details, see Kleinbaum, Kupper, and Morgenstern, Epidemiologic Research, p.298-299, John Wiley and Sons, 1982). Using what is called a **first-order Taylor-series approximation**, the approximate formula for this variance is given by:

Continued on next page

Deriving the Large-Sample Approximation Confidence Interval Formula for the Odds Ratio (continued)

$$\hat{Var}(\ln\hat{OR}) \approx \frac{1}{a}+\frac{1}{b}+\frac{1}{c}+\frac{1}{d}$$

where **a**, **b**, **c**, **d** denote the cell frequencies in the 2x2 data layout for a case-control study. Thus, the general 95% large-sample approximation confidence interval formula can be written as:

$$\hat{O}R\exp\left[\pm 1.96\sqrt{\frac{1}{a}+\frac{1}{b}+\frac{1}{c}+\frac{1}{d}}\right]$$

Compute Large-Sample Tests of Hypotheses and Confidence Intervals for Case-Control Studies using DataDesk

An example of calculating large sample tests of hypothesis and confidence intervals for case-control studies is demonstrated using the DataDesk program.

Quiz (12.17)

1. In the general data layout for a chi-square test from case-control data, the **???** are fixed prior to the start of the study.
2. The general data layout for a chi-square test from cohort data has fixed **???**.
3. The large-sample chi square formula for case-control data is the **???** the large-sample chi square formula for cumulative incidence cohort data.
4. The computing formula for Fisher's exact test for case-control data is **???** the corresponding formula for cumulative incidence cohort data.

Choices
cell frequencies   chi-square distributions   column margins   different from   normal distributions
row margins   same as

Fill in the table's blanks.

Resort Study	Raw Hamburger Ate	Did not eat	Total
Cases	6	B	11
Controls	C	D	11
Total	10	12	22

5. B = **???**
6. C = **???**
7. D = **???**

Choices
3   4   5   6   7   8   9

Quiz continued on next page

confidence interval exponentiating confidence interval

8. The expression below says that the **???** for an odds ratio is obtained by **???** a large sample **???** for the natural log of the **???**. odds ratio

$$e^{\ln\hat{OR} \pm 1.96\sqrt{\hat{Var}(\ln\hat{OR})}}$$

where $\hat{Var}(\ln\hat{OR}) \approx 1/a + 1/b + 1/c + 1/d$

Choices
confidence interval **exponentiating** **hypothesis test** **natural log** **odds ratio** **risk ratio**

Calculating Sample Size for Case-Control and Cross-Sectional Studies

In an asterisk/box on the previous lesson page (12-5), we described a formula for determining the sample size (**n**) for estimating a **risk ratio** (**RR**) in clinical trials and cohort studies. This formula also applies to sample size calculations for the **prevalence ratio** (**PR**) in cross-sectional studies (i.e., replace **RR** with **PR** in the formula). We now describe and illustrate a variation of the above mentioned formula that can be used to determine the sample size for estimation an **odds ratio** (**OR**) in case-control or cross-sectional studies.

As with sample size formula for **RR**, when considering the **OR** instead, the investigator must specify values for the significance level **α**, the probability of a Type II error **β**, and the extent of departure of the study effect from the null effect, i.e., **Δ**, into an appropriate formula to determine the required sample size. For case-control and cross-sectional studies, the sample size formula for detecting an odds ratio (**OR**) that differs from the null value of 1 by at least **Δ**, i.e. (**Δ = OR –1**) is given by the formula:

$$n = \frac{(Z_{1-\alpha/2} + Z_{1-\beta})^2 \bar{p}\bar{q}(r+1)}{[p_1 - p_2]^2 r}$$

where

$Z_{1-\alpha/2}$ = the 100(1 - α/2) percent point of the N(0,1) distribution

$Z_{1-\beta}$ = the 100(1 – β) percent point of the N(0,1) distribution

p_1 = expected proportion of cases with exposure
p_2 = expected proportion of controls with exposure
$\bar{p} = (p_1 + p_2) / 2$
$\bar{q} = 1 - \bar{p}$
r = ratio of the number of controls to cases

(Note: if the sample sizes are to be equal in the case and control groups, then **r = 1**. When **r** does not equal 1, the above formula provides the sample size for the **case** group; to get the sample size for the control group, use **n x r**.).

To use the above formula, one typically specifies (a guess for) **p2** and the **OR** to be detected, and then solves the **OR** formula for p1 in terms of OR and p2:

$$p_1 = \frac{p_2 OR}{1 - p_2 + p_2 OR}$$

To illustrate the calculation of **n**, suppose **α** = .05, **β** = .20, **OR** = 2, $\mathbf{p_2}$ = .040, and **r** = 3. Then:

Continued on next page

Calculating Sample Size for Case-Control and Cross-Sectional Studies (continued)

$$p_1 = \frac{(.04)(2)}{1-.04+(.04)(2)} = .0769$$

$$\bar{p} = (.04 + .0769) / 2 = .0585$$

and finally solve for **n** to yield:

$$n = \frac{(1.96+0.8416)^2(.0585)(.9415)(3+1)}{[.0769-.04]^2 3} = 423.3$$

Thus, the **sample size** (**n**) needed to detect an odds ratio (**OR**) of 2 at an **α** of .05 and a **β** of .20, when the expected proportion of exposed among controls (p_2) is .04 and the ratio of controls to cases (**r**) is 3, is **424 cases** and 423.3 x 3 = **1,270 controls**.

12-7 Simple Analyses (continued)

COHORT STUDIES INVOLVING RATE RATIOS

Testing for Rate Ratios

This table has been described in previous lessons to illustrate data from a **person-time cohort study** to assess the relationship between serum cholesterol level and mortality. The effect measure in this study is a **rate ratio**, also called an **incidence density ratio** (**IDR**), and its point estimate is 3.46.

PERSON-TIME COHORT

Cholesterol Study	Cholesterol Level: Borderline high	Normal
Deaths	26	14
P -Yrs	36,581	68,239

$\hat{IDR} = 3.46$

What can we conclude about the population rate ratio based on the estimated rate ratio obtained from the sample? We can perform a test of hypothesis that the population rate ratio is 1. This null hypothesis can equivalently be expressed as the difference in two incidence rates being equal to zero.

H_0: IDR = 1

or H_0 : $IR_1 - IR_0 = 0$

The rates in this case are the mortality rate for persons with borderline or high cholesterol ($\hat{IR}_1$)and the mortality rate for persons with normal cholesterol ($\hat{IR}_0$).

$$\hat{IR}_1 = \hat{Rate}\text{ (borderline or high)} = \frac{26}{36581}$$

$$= 7.11 \text{ per 10,000 py}$$

$$\hat{IR}_0 = \hat{Rate}\text{ (normal)} = \frac{14}{68239}$$

$$= 2.05 \text{ per 10,000 py}$$

The logical alternative hypothesis here is that the rate ratio is > 1, since prior to looking at the data the investigators were interested in whether persons with borderline high cholesterol have a higher mortality rate than persons with normal cholesterol. Because we are dealing with rates, and not risk estimates, we cannot use a large sample Z test that compares two proportions or its equivalent chi square test. But, we can use a different large-sample Z statistic or its equivalent chi square statistic for comparing two rates. Here is the general layout for computing a rate ratio.

H_0: IDR = 1 H_A: IDR > 1

2 x 2 Layout for Person-Time Cohort Data

	Exposed	Unexposed	Total
New Cases	I_1	I_0	I
Person Time	PT_1	PT_0	PT

I_1 and I_0 denote the number of new cases in the exposed and unexposed groups, and PT_1 and PT_0 denote the corresponding person time accumulation for these two groups. Here is the test statistic, **T**, where p_0 denotes the person-time for the exposed group divided by the total person-time for exposed and unexposed combined:

$$T = \frac{I_1 - I p_0}{\sqrt{I p_0 (1 - p_0)}} \quad \text{approx. } N(0, 1) \text{ under } H_0$$

$$p_0 = \frac{PT_1}{PT}$$

This test statistic will have approximately a normal distribution under the null hypothesis provided the total number of new cases, I, is sufficiently large and p_0 is not close to one or zero. We will refer to this statistic as a Z statistic. If we square this Z statistic, we obtain an approximate one degree of freedom (1 df) chi square statistic. A simplified formula for this chi square statistic is shown here:

$$Z^2 = \frac{(I_1 PT_0 - I_0 PT_1)^2}{I\, PT_1 PT_0}$$

We now compute Z^2 for our cholesterol mortality dataset. The computed chi square value is 15.952.

$$Z^2 = \frac{(26 \times 68{,}239 - 14 \times 36{,}581)^2}{40 \times 36{,}581 \times 68{,}239} = 15.952$$

approx. chi square with 1 d.f. under H_0

Study Question (12.18)

1. Using the large-sample chi square value of 15.592, describe how you would determine the P-value for this one-sided alternative that the rate ratio (i.e., IDR) is greater than 1? (You are being asked to describe a process, not to obtain a numerical value.)
2. The 99.95 percent point of the chi square distribution with 1 df is 12.116. Based on this information, what can you say about the P-value for the large sample chi square test? (You are now being asked to give a number or range of numbers.) $\hat{IR}_1$
3. Based on the above P-value, what do you conclude from this test of hypothesis?
4. What is the computed value of the Z statistic that provides a test that is equivalent to the above chi square test?

Summary

- When testing the hypothesis about a rate ratio in a person-time cohort study, the null hypothesis can be equivalently stated as either IDR = 1 *or* $IR_1 - IR_0 = 0$.
- IDR denotes the rate ratio (i.e., incidence density ratio) and IR_1 and IR_0 are the incidence rates for the exposed and unexposed groups.
- The alternative hypothesis can be stated in terms of the IDR ≠ 1, IDR > 1, or IDR < 1 depending on whether the alternative is two-sided, upper one-sided, or lower one-sided, respectively.
- One version of the test statistic is a large sample N(0,1) or Z statistic used to compare two incidence rates.
- An alternative version is a large-sample chi square statistic, which is the square of the Z statistic.

The rationale behind the Z-test for a rate ratio

Here, we describe how to test **H0: IDR = 1**, or, equivalently, **H0: IR_1 - IR_0 = 0**, where **IDR** denotes the **rate ratio** (i.e., incidence density ratio) in a simple analysis of a person-time cohort study, and $\mathbf{IR_1}$ and $\mathbf{IR_0}$ are the incidence rates for exposed and unexposed groups. The data layout is given as follows:

Data layout for person-time cohort study

	Exposed	Not Exposed	Total
New Cases (D)	I_1	I_0	I
Person-Time	PT_1	PT_0	PT

Since the estimates of $\hat{IR}_1$ and $\hat{IR}_0$ are not binomial proportions, a standard test that compares two binomial proportions is not appropriate here. However, a testing approach is possible if we assume that each of the **I** cases represents an independent **Bernoulli** trial, with "success" and "failure" defined as being in the exposed and unexposed categories, respectively.

Under this assumption, we can assume that the number of exposed cases, I_1, has a binomial distribution for I trials and probability of success $\mathbf{p_0}$ under the null hypothesis, where $\mathbf{p_0 = PT_1/PT}$, the proportion of total person-time associated with the exposed group.

Thus, if **A** denotes the random variable for the number of cases out of **I** total cases that are exposed, it follows from the binomial distribution that the **exact P-value** for a test of **H0: IDR = 1** versus **HA: IDR > 1** using the above binomial formula is given by

$$\Pr(A \geq I_1 \mid H_0) = \sum_{j=I_1}^{I} C_j^I p_0^j (1-p_0)^{I-j} \qquad \text{where} \qquad C_s^r = \frac{r!}{s!(r-s)!}$$

denotes the number of combinations of **r** items taken **s** at a time, and
r! = r(r-l)(r-2)...(3)(2)(1) denotes **r factorial**,
s! and (r-s)! denote **s factorial** and **(r-s) factorial**, respectively.

We illustrate the calculation of this **P-value** using the data from the person-time study of that compares the mortality rates for persons with borderline high cholesterol with persons with normal cholesterol. Here again is the data:

Person-time cholesterol cohort study

	Exposed	Not Exposed	Total
Deaths (D)	26	14	40
Person-Years	36,581	68,239	104,820

The P-value is thus calculated as the probability that the number of exposed cases is 26 or higher (up to 40 total cases) under the null hypothesis that IDR=1:

$$Pr(A \geq 26 \mid H_0) = \sum_{j=26}^{40} C_j^{40} p_0^j (1-p_0)^{40-j}$$

Continued on next page

The rationale behind the Z-test for a rate ratio (continued)

This expression involves the sum of 15 terms and is difficult to calculate without a computer program. However, a good approximation to this **exact P-value** can be obtained from a large-sample **Z** test. The **Z** test is therefore recommended unless the total number of cases is small, in which case one should use the exact P-value formula given above. The large-sample Z-statistic is given by the following formula:

$$Z = \frac{I_1 - Ip_0}{\sqrt{Ip_0(1-p_0)}}$$

For the cholesterol-mortality data given above, the computed Z is calculated as follows:

$$Z = \frac{26 - 40(36{,}581/104{,}820)}{\sqrt{40(36{,}581/104{,}820)\{1-(36{,}581/104{,}820)\}}} = 3.994$$

The computed **Z** statistic is quite large; in particular, the **P-value** for a test of $\mathbf{H_0}$**: IDR = 1** versus $\mathbf{H_A}$**: IDR > 1** is extremely small, in fact, smaller than .0001. Consequently, the null hypothesis should be rejected and it can be concluded that persons with borderline high cholesterol have a significantly higher mortality rate than persons with normal cholesterol.

Large-Sample Confidence Interval (CI) for a Rate Ratio

To describe the general formula for a 95% confidence interval for a rate ratio, we again consider the data layout for a person-time cohort study. This confidence interval formula looks as follows:

Large-Sample 95% Confidence Interval for IDR

2 x 2 Layout for Person-Time Cohort Data

	Exposed	Unexposed	Total
New Cases	I_1	I_0	I
Person Time	PT_1	PT_0	PT

$$\widehat{IDR} = \frac{I_1/PT_1}{I_0/PT_0}$$

$$\widehat{IDR}\,e^{\pm 1.96\sqrt{\frac{1}{I_1}+\frac{1}{I_0}}}$$

The lower and upper limits are then given by the following expressions:

$$L = \widehat{IDR}\,e^{-1.96\sqrt{\frac{1}{I_1}+\frac{1}{I_0}}} \quad U = \widehat{IDR}\,e^{+1.96\sqrt{\frac{1}{I_1}+\frac{1}{I_0}}}$$

An equivalent version of the rate ratio formula looks like this:

$$e^{\ln(\widehat{IDR}) \pm 1.96\sqrt{\widehat{Var}\{\ln(\widehat{IDR})\}}}$$

The formula used above for the estimated variance of the log of the rate ratio is an approximate, not an exact formula for the variance.

Let's apply the rate ratio formula to the mortality study data set. Substituting into the formula the values for the estimated rate ratio and the cell frequencies for exposed and unexposed groups, we obtain the lower and upper confidence limits: $1.81 < IDR < 6.63$

$$3.46e^{\pm 1.96\sqrt{\frac{1}{26}+\frac{1}{14}}}$$

$$=1.81 < IDR < 6.63$$

Study Question (12.19)

1. Interpret these results above. What do they mean?

Summary

- ❖ The 95% CI formula for a rate ratio multiplies the estimate rate ratio by the exponential of plus or minus the quantity 1.96 times the square root of the variance of the log of the estimated rate ratio.
- ❖ This rate ratio formula is obtained by exponentiating a large sample confidence interval for the natural log of the rate ratio.
- ❖ The variance of the log of a rate ratio is approximately equal to the sum of the inverses of the exposed and unexposed cases.

Deriving the Large-Sample Approximation Confidence Interval Formula for the Rate Ratio

As with the risk ratio (**RR**), to obtain a 95% confidence interval for rate ratio (**IDR**), we might be inclined initially to use the interval:

$$\hat{IDR} \pm 1.96\sqrt{\hat{Var}(\hat{IDR})}$$

However, this interval is **not** recommended because the distribution of the estimated IDR is not normally distributed. The recommended approach involves working with the estimate of **lnIDR** rather than **IDR** itself, since the natural log transformation tends to change and skewed distribution to an approximately normal distribution. A two-step procedure is used: First, a 95% large-sample confidence interval for **lnIDR** is obtained of the form:

$$\ln\hat{IDR} \pm 1.96\sqrt{\hat{Var}(\ln(\hat{IDR})}$$

Then, since **IDR = exp(lnIDR)**, the desired upper and lower values of the 95% confidence interval for **IDR** are found by taking the anti logarithms of lower and upper values for the confidence interval for **lnIDR**. The resulting formula takes the general form:

$$\hat{IDR}\exp\left[\pm 1.96\sqrt{\hat{Var}(\ln\hat{IDR})}\right]$$

To complete the calculation, however, we need to find an expression for $\hat{Var}(\ln\hat{IDR})$. Using what is called a first-order **Taylor Series approximation,** an approximate formula for this variance is given by:

$$\hat{Var}(\ln\hat{IDR}) \approx \frac{1}{I_1} + \frac{1}{I_0}$$

where $\mathbf{I_1}$ and $\mathbf{I_0}$ denote the number of exposed and unexposed cases, respectively. Thus, the general 95% large-sample approximation confidence interval formula can be written as:

Continued on next page

Deriving the Large-Sample Approximation Confidence Interval Formula for the Rate Ratio (continued)

$$\hat{IDR}\exp\left[\pm 1.96\sqrt{\frac{1}{I_1}+\frac{1}{I_0}}\right]$$

A Large-Sample Approximation CI Formula for the Rate Difference

To obtain a 95% confidence interval for a rate difference (i.e., **IDD = IR$_1$ - IR$_0$**), where **IR$_1$** and **IR$_0$** denote the incidence rates for exposed and unexposed persons, respectively, the following formula may be used:

$$\hat{IDD} \pm 1.96\sqrt{\hat{Var}(\hat{IDD})}$$

where $\hat{IDD} = \hat{IR}_1 - \hat{IR}_0$ and an approximate large-sample formula for $\hat{Var}(\hat{IDD})$ is given by:

$$\hat{Var}(\hat{IDD}) \approx \frac{I_1}{PT_1^2} + \frac{I_0}{PT_0^2}$$

where **I$_1$, I$_0$, PT$_1$, PT$_0$** are new cases and person-time information in the following data layout for a person-time cohort study:

Data layout for person-time cohort study

	Exposed	**Not Exposed**	Total
New Cases (D)	I_1	I_0	I
Person-Time	PT_1	PT_0	PT

The above variance approximation formula derives from assuming that the number of exposed and unexposed cases, i.e., **I$_1$** and **I$_0$**, each have a **Poisson distribution** from which the variances for each group can be estimated by **I$_1$/(PT$_1$)^2** and **I$_0$/(PT$_0$)^2** for exposed and unexposed groups, respectively.

We illustrate the calculation of the large-sample CI for the IDD using the data from the person-time cohort study that compares persons with borderline high cholesterol with persons with normal cholesterol:

Person-time cholesterol cohort study

	Exposed	**Not Exposed**	Total
Deaths (D)	26	14	40
Person-Years	36,581	68,239	104,820

Substituting the values in this table for **I$_1$, I$_0$, PT$_1$, PT$_0$** in the confidence interval formula yields the following 95% confidence interval for the rate difference **IDD**:

$$\frac{26}{36581} - \frac{14}{68239} \pm 1.96\sqrt{\frac{26}{36581^2}+\frac{14}{68239^2}} = \frac{7.1075}{10^4} - \frac{2.0516}{10^4} \pm 1.96\sqrt{\frac{1.9430}{10^8}+\frac{0.30065}{10^8}}$$

$$= \frac{5.0559}{10^4} \pm 1.96\sqrt{\frac{2.2437}{10^8}} = \frac{5.0559 \pm 2.9359}{10^4}$$

Thus, the lower and upper limits of the CI for the rate difference is given by **L = 2.1200** per 10,000 person-years, and **U = 7.9918** per 10,000 person-years, where $\hat{IDD}$ =5.0559 per 10,000 person-years.

Compute Large-Sample Tests of Hypotheses and Confidence Intervals for Rate data from Cohort Studies

An example of calculating large sample tests of hypothesis and confidence intervals for rate data from cohort studies is demonstrated using the DataDesk program.

Quiz (12.20)

1. The expression below says that the confidence interval for a **???** is obtained by **???** a large sample confidence interval for the **???** of the rate ratio.
2. The formula used for the **???** variance of the log of the rate ratio is an **???**, not an **???**, formula for this variance.

Choices

approximate **estimated** **exact** **exponentiating** **exponentiation** **natural log**
odds ratio **rate ratio** **testing**

$$e^{\ln(\hat{IDR}) \pm 1.96\sqrt{\hat{Var}\{\ln(\hat{IDR})\}}}$$

$$\hat{Var}\{\ln(\hat{IDR})\} \approx 1/I_1 + 1/I_0$$

Nomenclature

Table setup for cohort, case-control, and prevalence studies:

	Exposed	Not Exposed	Total
Disease/cases	a	b	n_1
No Disease/controls	c	d	n_0
Total	m_1	m_0	n

Table setup for cohort data with person-time:

	Exposed	Not Exposed	Total
Disease (New cases)	I_1	I_0	I
No Disease	-	-	-
Total disease-free person-time	PT_1	PT_0	PT

χ^2	Chi square
χ^2_{MH}	Mantel-Haenszel chi square
CI	Confidence interval
$\hat{CI}_0$	Cumulative incidence or "risk" in the nonexposed (b/m_0)
$\hat{CI}_1$	Cumulative incidence or "risk" in the exposed (a/m_1)
CID	Cumulative incidence difference or risk difference, $CI_1 - CI_0$; same as risk difference (RD)
e	Exponentiation or "antilog"; $e \approx 2.71828$
H_0	Null hypothesis
H_A	Alternative hypothesis
I	Number of new cases
I_0	Number of new cases in nonexposed
I_1	Number of new cases in exposed
$\hat{ID}$	Incidence density (or "rate") in the population (I/PT)
$\hat{ID}_0$	Incidence density (or "rate") in the not exposed (I_0/PT_0)
$\hat{ID}_1$	Incidence density (or "rate") in the exposed (I_1/PT_1)
IDD	Incidence density difference or rate difference, $ID_1 - ID_0$
IDR	Incidence density ratio or rate ratio: ID_1 / ID_0; same as Incidence Rate Ratio (IRR)
IRR	Incidence rate ratio or rate ratio: ID_1 / ID_0; same as Incidence Density Ratio (IDR)
L	Lower limit of the confidence interval
ln	Natural log
n	Size of population under study
MH	Mantel-Haenszel
OR	Odds ratio: ad/bc
p	Population proportion
$\hat{p}$	Sample proportion *or* the weighted average of two proportions
P or Pr	Probability
P_1	Proportion with outcome in exposed
P_0	Proportion with outcome in unexposed
PT	Disease-free person-time
PT_0	Disease-free person-time in nonexposed
PT_1	Disease-free person-time in exposed
RD	Risk difference: risk in exposed minus risk in unexposed; same as cumulative incidence difference (CID)
RR	Risk ratio: risk in exposed divided by risk in unexposed; same as cumulative incidence ratio (CIR)
ROR	Risk odds ratio
S	Standard error
T	T test statistic
T*	Computed T test statistic
U	Upper confidence limit
Var	Variance
Z	Z test statistic

Formulae

Statistical Tests

T statistic for the difference between two sample proportions

$$T = \frac{\hat{p}_1 - \hat{p}_0}{\sqrt{\hat{p}(1-\hat{p})\left(\frac{1}{n_1} + \frac{1}{n_0}\right)}} \quad \text{where} \quad \hat{p} = \frac{n_1 \hat{p}_1 + n_0 \hat{p}_0}{n_1 + n_0}$$

Chi square statistic for a 2x2 table

$$\chi^2 = \frac{n(ad - bc)^2}{n_1 n_0 m_1 m_0}$$

Mantel-Haenszel chi square statistic for a 2x2 table

$$\chi^2_{MH} = \frac{(n-1)(ad - bc)^2}{n_1 n_0 m_1 m_0}$$

T statistic for the difference between two sample rates

$$T = \frac{I_1 - Ip_0}{\sqrt{Ip_0(1-p_0)}} \quad \text{where} \quad p_0 = \frac{PT_1}{PT}$$

Chi square statistic for comparing two rates

$$\chi^2 = \frac{(I_1 PT_0 - I_0 PT_1)^2}{IPT_1 PT_0}$$

Confidence intervals

Large sample 95% confidence interval for the difference between two proportions (cumulative or risk difference)

$$\hat{p}_1 - \hat{p}_0 \pm 1.96\sqrt{\frac{\hat{p}_1(1-\hat{p}_1)}{n_1} + \frac{\hat{p}_0(1-\hat{p}_0)}{n_0}}$$

Large sample 95% confidence interval for the risk ratio (ratio of two proportions)

$$\hat{RR}\ \exp\left[\pm 1.96\sqrt{\frac{1-\hat{CI}_1}{n_1\hat{CI}_1} + \frac{1-\hat{CI}_0}{n_0\hat{CI}_0}}\right]$$

Large sample 95% confidence interval for the odds ratio

$$\hat{OR}\ \exp\left[\pm 1.96\sqrt{\frac{1}{a} + \frac{1}{b} + \frac{1}{c} + \frac{1}{d}}\right]$$

Large sample 95% confidence interval for the incidence density ratio

$$\hat{IDR}\ \exp\left[\pm 1.96\sqrt{\frac{1}{I_1}+\frac{1}{I_0}}\right]$$

Large sample 95% confidence interval for the incidence density difference

$$\hat{IDD} \pm 1.96\sqrt{\frac{I_1}{PT_1^2}+\frac{I_0}{PT_0^2}}$$

References

Basic Statistics References

There are a great many beginning texts that cover the fundamental concepts and methods of statistics and/or biostatistics. (Note: biostatistics concerns applications of statistics to the biological and health sciences.) Consequently, it is not our intention here to provide an exhaustive list of all such texts, but rather to suggest a few references that this author has used and/or recommends. Among these, we first suggest that you consider Velleman's ActivStats CD ROM text, which has the same format has ActivEpi. Suggested references include:

Kleinbaum DG, Kupper LL, Muller KA, Nizam A Applied Regression Analysis and Other Multivariable Methods, 3rd Edition. Duxbury Press, 1998. (Chapter 3 provides a compact review of basic statistics concepts and methods.)

Moore D. The Active Practice of Statistics. WH Freeman Publishers, 1997 (This book is designed specifically to go with ActivStats and matches it closely.)

Remington RD and Schork MA. Statistics with Applications to the Biological and Health Sciences. Prentice Hall Publishers, 1970.

Velleman P. ActivStats - A Multimedia Statistics Resource (CDROM), Addison-Wesley Publishers, 1998.

Weiss N. Introductory Statistics, 6th Edition. Addison-Wesley Publishing Company, Boston, 2002.

References on Simple Analysis

Fleiss JL. Confidence Intervals for the odds ratio in case-control studies: the state of the art. J Chronic Dis 1979;32(1-2):69-77.

Goodman SN. Toward evidence-based medical statistics. 1: The P value fallacy. Ann Intern Med 1999;130(12):995-1004.

Goodman SN. Toward evidence-based medical statistics. 2: The Bayes factor. Ann Intern Med 1999;130(2):1005-13.

Kleinbaum DG, Kupper LL, Morgenstern H. Epidemiologic Research: Principles and Quantitative Methods, Chapter 15 , John Wiley and Sons, 1982.

Miettinen O. Estimability and estimation in case-referent studies. Am J Epidemiol 1976;103(2):226-235.

Mantel N, Haenszel W. Statistical aspects of the analysis of data from retrospective studies of disease. J Natl Cancer Inst 1959;22(4):719-48.

Thomas DG. Exact confidence limits for an odds ratio in a 2x2 table. Appl Stat 1971;20:105-10.

Homework

ACE-1. Exposure Odds Ratio

The following data are from a case-control study designed to investigate the theory that a certain study factor (E) is a determinant of some rare disease (D). A representative group of incident cases of the disease arising in a given population over a five-year period was identified. These cases were then compared to a random sample of an equal number of noncases from the same population.

	Exposed	**Not Exposed**	Total
Case	27	23	50
Control	18	32	50
Total	45	55	100

a. Estimate the exposure odds ratio (EOR) for the data. Can we say (using this point estimate alone) that there is statistical evidence of an association between the study factor and the disease? Explain.
b. Assess the statistical significance of the observed EOR value, using a Mantel-Haenszel chi-square test. Determine the one-sided p-value and interpret the result.
c. Calculate a 92% (!) large-sample confidence interval for the EOR. (You may wish to use the **Data-desk** template **Crude OR/RR.ise** to help with the calculations.)
d. What do you conclude about the E-D relationship based on your findings in parts a, b, and c?

ACE-2. Difference Measure vs. Ratio Measure

The following data are from a study of breast cancer among women with tuberculosis. Women who were exposed repeatedly to multiple x-ray fluoroscopies were compared to women not so exposed:

	Radiation Exposure		
	Yes	No	Total
Breast Cancer	41	15	56
Person-years	28,010	19,017	47,027

a. Calculate an appropriate difference measure of association for these data and interpret your result.
b. Calculate a 90% confidence interval for the difference measure in part a. What do you conclude based on this result? (You may wish to use the **Datadesk** template **Crude IDR.ise** to help with the calculations.)
c. Calculate an appropriate ratio measure of association for these data and interpret your result.
d. Calculate a 90% confidence interval for the ratio measure in part c. (You may wish to use the **Datadesk** template **Crude IDR.ise** to help with the calculations.)
e. Is your conclusion in part d regarding the association between the exposure and the outcome the same as previously (i.e. part b)?

ACE-3. Test of Hypothesis

Helicobacter pylori (HP) is a bacterium that infects the cells that line the stomach, causing acute and chronic inflammation. The organism is considered a causal factor in peptic ulcer disease and has been linked epidemiologically to the development of gastric adenocarcinoma. A group of investigators was interested in assessing the relationship between alcohol consumption and HP infection. They identified 300 subjects whose blood contained antibodies to HP, 63 of whom consumed alcohol. Of 267 subjects who were antibody-negative for HP, 91 consumed alcohol.

a. Use this information to fill in the 2 x 2 table below:

	Alcohol	No Alcohol
HP		
no HP		

b. Calculate and interpret the appropriate ratio measure of association for these data.
c. Use these data to carry out a hypothesis test of the association between alcohol consumption and HP infection. Be sure to state the null (Ho) and alternative (HA) hypotheses and provide an appropriate p-value. What is your conclusion?
d. Calculate a 95% confidence interval for the measure in part b above.

ACE-4. Fisher's Exact Test

The following table contains data from a case-control study relating an exposure (E) to a disease (D).

	Exposed	**Unexposed**	
Diseased	5	6	11
Non-diseased	3	2	5
	8	8	16

If the significance of the association between exposure and disease were to be assessed using a Fisher's Exact test, how many terms would be summed in order to calculate the p-value?

i. 11
ii. 8
iii. 4
iv. 7

a. Use the **Contingency Tables** command from the **Calc** menu in **Datadesk** to carry out the computation of Fisher's exact test. What do you conclude from this test?

ACE-5. Incidence Density Ratio

Suppose you wish to analyze the person-time cohort data given in the following table:

	Exposure		
	Yes	No	Total
Disease	50	150	200
Person-years	10,000	40,000	50,000

a. What is the incidence density ratio?
b. Carry out a large-sample test of the hypothesis (H0: IDR=1).
c. Provide a formula for computing a 95% confidence interval for the IDR. In stating this formula, put numerical values in place of the symbols that describe the general confidence interval formula for this situation.

d. Calculate the 95% confidence interval described in part c. (You may wish to use the Datadesk template **Crude IDR.ise** to carry out the calculations.
e. Based on your answers to the previous question, what do you conclude about the E-D relationship?

ACE-6. Sample Size: Clinical Trial

A clinical trial is being planned to evaluate the effectiveness of a new therapy for osteoporosis. The new therapy is to be compared to the standard therapy, which is known to produce a positive response (increased bone density) in 20% of patients. Preliminary data suggest that the new therapy may produce a positive response in twice the proportion of patients, compared to the standard. If this is true, how many subjects will need to participate in the clinical trial in order to show a statistically significant response? Assume that $\alpha = 0.05$ (2-sided) and $\beta = 0.10$ (1-sided). Show your calculations. (Hint: use the sample size formula:

$$\mathbf{n} = \frac{(\mathbf{Z}_{1-\alpha/2} + \mathbf{Z}_{1-\beta})^2\, \overline{\mathbf{p}}\overline{\mathbf{q}}(\mathbf{r}+1)}{[\mathbf{p}_2(\mathbf{RR}-1)]^2\mathbf{r}}$$

ACE-7. Sample Size: Case-Control Study

Data from the Tricontinental Seroconverter Study were used for a case-control analysis of the potential association between HIV status and substance use. Recent HIV seroconverters were compared to subjects who tested negative for HIV; all subjects were asked about their substance use in the year prior to study enrollment. The following table summarizes the data on amphetamine use:

	Amphetamine Use		
	Yes	**No**	**Total**
HIV +	78	267	345
HIV -	35	310	345

a. Calculate the exposure odds ratio (EOR) for these data.
b. How many cases and controls would the investigators have needed in order to detect the same EOR if they had decided to select **r**= 3 controls per case? Let α **= 0.05** (2-sided) and **Power = 90%.**

Hint: You may use the following sample size formula:

$$\mathbf{n} = \frac{(\mathbf{Z}_{1-\alpha/2} + \mathbf{Z}_{1-\beta})^2\, \overline{\mathbf{p}}\overline{\mathbf{q}}(\mathbf{r}+1)}{[\mathbf{p}_1 - \mathbf{p}_2)]^2\mathbf{r}}$$

ACE-8. Factors Affecting Sample Size

An epidemiologist is planning a study to evaluate the association between an exposure (E) and a disease (D). For each of the following situations, indicate whether the required sample size would increase, decrease, or remain unchanged.

a. An increase in α from 0.05 to 0.10.
b. A decrease in the expected impact of a new therapy (compared to standard).
c. An increased willingness to risk a type I error.

ACE-9. Sample Size and Power

The authors of an epidemiologic study were concerned that, due to small sample size, their study would have low power. Describe briefly the basis for their concern, i.e. what is the potential problem associated with low power in an epidemiologic study?

Answers to Study Questions and Quizzes

Q12.1

1. RR, the ratio or two proportions, or OR, the ratio of two odds, each of the form p/(1-p).
2. The usual null hypothesis for a risk ratio is RR = 1, where 1 is the null value of the risk ratio.
3. The usual null hypothesis for an odds ratio is OR = 1, where 1 is the null value of the odds ratio.

Q12.2

1. The pooled estimate of the common proportion is given by {(.7972 x 1667) + (.2565 x 425)} / {1667 + 425} = .6874.

Q12.3

1. The 2-sided alternative A1, in the lower right corner.
2. Because the normal curve is symmetric, the left and right tails have equal areas. Thus, compute one tail's area and then multiply by two.
3. For an upper one-sided alternative, the P-value is the area under the normal curve to the right of T*=12.0. From the table, we find P<.0001. Thus, if the null hypothesis were true and our alternative hypothesis had been one-tailed, our results had less than a .01% chance of occurring.
4. The P-value is twice the area beyond the value of 12.0 under the curve, so P < 0.0002. This is because a computed T* less than –12.0 in the left tail of the normal distribution would also represent a worse value under the null hypothesis than the T* = 12.0 that was actually observed. Thus, if the null hypothesis were true and our alternative hypothesis had been two-tailed, our results had less than a .02% chance of occurring.
5. Yes, the P-value, which represents the chance that our results would occur if the null hypothesis were true, is extremely small.
6. Conclude that the test is significant, i.e., reject the null hypothesis and conclude that the proportions for men and women are significantly different.
7. a) P > .01: Do not reject H_0. b) P = .023: Reject H_0.
8. c) P < .001: Reject H_0. d) P = .54: Do not reject H_0. e) P = .0002: Reject H_0.
9. a) P > .01: Do not reject H_0. b) P = .023: Do no reject H_0.
10. c) P < .001: Reject H_0. d) .01 < P < .05: Do not reject H_0. e) P = .0002: Reject H_0.

Q12.4

1. hypothesis testing
2. interval estimation
3. hypothesis testing
4. interval estimation
5. significance level
6. alternative hypothesis
7. rejection region, null hypothesis, statistically
8. P-value, null hypothesis
9. not reject
10. .01, .001

Q12.5

1. The standard error used here does not assume that the null hypothesis is true. Thus, the variance for each proportion must be computed separately using its sample proportion value.

Q12.6

1. 100% confidence.
2. The 99% confidence interval would be the widest and the 90% confidence interval would be the narrowest.
3. Dataset A gives the more precise estimate because it's confidence interval is narrower than that for Dataset B.

Q12.7

1. estimated standard error
2. difference
3. population parameter
4. upper limit
5. 95%
6. 5%

Q12.8

1. Yes, because 7.04 is larger than 6.635, which is the .99 percent point of the chi square distribution with 1 df.
2. No, because 7.04 is less than 7.879, which is the .995 percent point of the chi square distribution with 1 df.
3. Yes, because the P-value for a two-sided test is the area above 7.04, which is less than the area above 6.635, which is .01.

4. No, because the P-value is the area above 7.04, which is greater than the area above 7.879, which is .005.
5. Yes, because the upper one-sided test using the normal curve was significant at the 1 percent level.
6. Yes, the one-tailed P-value using the normal distribution was .0040, which is less than .005.

Q12.9

1. The correct P-value is the area under the normal curve below +2.65, which is 1 minus the area above 2.65, which is calculated to be 1 - .004 = .996. The value of .996 can equivalently be obtained from the chi square curve by taking one half of .008 and subtracting from 1, i.e., 1 – (.008 / 2) = .996. It would be incorrect, therefore, to take one-half of the area of .008 under the chi-square curve nor would it be correct to take one-half of 1 minus .008.
2. Do not reject the null hypothesis because the P-value is very high.
3. The correct P-value is the area under the normal curve below –2.65, which is .004. This can also be obtained by taking half of the area above the chi square value of 7.04, which is .008/2 = .004.
4. Reject the null hypothesis because the P-value is very small, and much smaller than .05 and .01.

Q12.10

1. $b = m_1 - a$; $c = n_1 - a$; $d = m_0 - n_1 + a$
2. b = 2; c = 1; d = 6
3. b = 1; c = 0; d = 7
4. RR = (5/5) / (1/8) = 8
5. Yes
6. No. An **a** cell value greater than 5 is not possible because the assumed fixed column marginal of 5 would then be exceeded.
7. P(a=4 or 5|RR=1) = P(a=4|RR=1) + P(a=5|RR=1). This tells us how rare our observed results are under the null hypothesis. It is the P-value for testing this hypothesis.
8. Since P is greater than .05, we fail to reject at the .05 significance level for the null hypothesis that the RR = 1.

Q12.11

1. The one-sided p-value is .0630 / 2 = .0315
2. The large-sample assumption does not hold.
3. The sample size n = 13 is not large, so that the difference between n = 13 and n – 1 = 12 has a stronger effect on the calculation of each test statistic.

Q12.12

1. This confidence interval has a 9% probability of covering the true risk ratio that compares *continuing smokers* to *smokers who quit.* Even though the confidence interval contains the null value of 1, it is wide enough to suggest that the true risk ratio might be either close to one or as large as 3.6. In other words, the point estimate of 2.1 is somewhat imprecise, and the true effect of quitting smoking after a heart attack may be either very weak or very strong.

Q12.13

1. 520/1863 = 0.279
2. 2.23: T* = (.303 - .256) / [sqrt{.279 * (1-.279) * [(1/915) + (1/948)]}] = 2.23
3. .013
4. 5%, 1%
5. 4.970: MH chi square = (1862 * 2.23^2)/1863 = 4.970
6. 0.026
7. large

Q12.14

1. Determine the area under the chi-square distribution with 1 df above the value 4.67. The P-value is one-half this area.
2. The area under the chi square curve above 4.67 lies between .025 and .05. Consequently, after dividing each limit by 2, the P-value for this test lies between .0125 and .025.
3. At the 0.05 level, yes (significant). At the 0.01 level, no.

Q12.15

1. b=6, c=3, d=6
2. OR = (4*6)/(6*3) = 1.3
3. No. An OR of 1.3 is closer to 1 than is an OR of 3.5.
4. Yes, **a** cell values of either 6 or 7 would give odds ratios larger than 3.5.
5. Because Pr(a=5, 6, or 7|OR=1) = Pr(a=5|OR=1) + Pr(a=6|OR=1) + Pr(a=7|OR=1) tells us how rare is our observed result under the null hypothesis.
6. Since P is much greater than .05, we would not reject the null hypothesis that the OR = 1 at the .05 level of significance. In other words, there is not statistically significant evidence in this hypothetical dataset to say that eating raw hamburger was the source of the outbreak.

Q12.16

1. The confidence interval, ranging between 1.1 and 9.1, is very wide, indicating a lack of precision in the estimated odds ratio of 3.2. So, even though the estimated odds ratio is statistically significant at the .05 level, the true odds ratio might be either very weak (i.e., close to 1) or very strong (i.e., just over 9).

Q12.17

1. row margins
2. column margins
3. same as
4. same as
5. 5
6. 4
7. 7
8. confidence interval, exponentiation, confidence interval, odds ratio

Q12.18

1. Determine the area under the chi square distribution with 1 df above the value 15.952. The P-value is one-half this area.
2. The area under the chi square curve above 15.952 is less than .0005. Consequently, after dividing by 2, the P-value for this test is less than .00025, which is extremely small.
3. Persons with borderline high cholesterol have a significantly higher mortality rate than persons with normal cholesterol.
4. Z-squared = 15.952, so Z = sqrt(125.952) = 3.994. This computed Z must be a positive number because persons with borderline high cholesterol have a higher estimated mortality rate (7.11 per 10,000 py) than persons with normal cholesterol (2.05 per 10,000 py).

Q12.19

1. The confidence limits, which range from 1.81 to 6.63, are quite wide, indicating imprecision in the rate ratio estimate of 3.46. The true rate ratio might be either close to 2 or higher than 6. Nevertheless, we are 95% confident that the true rate ratio is much larger than 1.

Q12.20

1. rate ratio, exponentiating, natural log
2. estimated, approximate, exact

LESSON 13

CONTROL OF EXTRANEOUS FACTORS

In previous lessons, we have discussed and illustrated several important concepts concerning the ***control*** *of additional (**extraneous**) variables when assessing a relationship between an exposure variable and a health-outcome variable. In this lesson, we briefly review these concepts and then provide an overview of several options for the process of control that are available at both the design and analysis stages of a study.*

13-1 Control of Extraneous Factors

What do we Mean by Control?

Suppose we are studying whether there is a link between exposure to a toxic chemical and the development of lung cancer in a chemical industry. To answer this question properly, we would want to isolate the effect of the chemical from the possible influence of other variables, particularly age and smoking status, two known risk factors for lung cancer. That is, our goal is to determine whether or not exposure to the chemical contributes anything over and above the effects of age and smoking to the development of lung cancer.

Variables such as age and smoking in this example are often referred to as **control variables**. When we assess the influence of such control variables on the **E→D** relationship, we say we are **controlling for extraneous variables**. By extraneous, we simply mean that we are considering variables other than **E** and **D** that are not of primary interest but nevertheless could influence our conclusions about the **E→D** relationship.

In general, we typically carry out a simple analysis of an exposure-disease relationship as the starting point for more complicated analyses that we will likely have to undertake. A simple analysis allows us to see the **crude association** between exposure and disease and therefore allows us to make some preliminary insights about the exposure-disease relationship. Unfortunately, a simple analysis by definition ignores the influence that variables other than the exposure may have on the disease. If there are other variables already known to predict the disease, then the conclusions suggested by a simple analysis may have to be altered when such risk factors are taken into account.

Consequently, when we control for extraneous variables, we assess the effect of the exposure **E** on the disease **D** at different combinations of values of the variables we are controlling. When appropriate, we evaluate the overall **E→D** relationship by combining the information over the various combinations of control values.

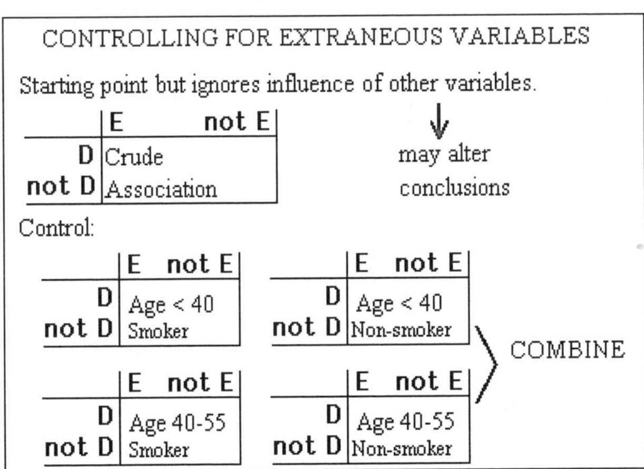

Study Questions (Q13.1)

Consider a case-control study to assess whether a certain toxic chemical (**E**) is associated with the development of lung cancer (**D**) in a chemical industry. Suppose we wish to consider the control of age and smoking status. Assume that we categorize age into three groups: below 40, 40-55, and over 55. We also categorize smoking as "ever smoked" versus "never smoked".

1. How many combinations are there of the categories of age and smoking?

Two kinds of pooled analyses with these data are:

- Pool 2x2 tables of these combinations into one overall "pooled" table and compute an odds ratio for this pooled table, and
- Compute an odds ratio for each 2x2 table corresponding to each combination and then average these separate odds ratios in some way.

2. Which of these two analyses controls for age and smoking?

Several questions arise when considering the control of extraneous variables. Why do we want to control in the first place? That is, what do we accomplish by control? What are the different options that are available for carrying out control? Which option for control should we choose in our study? Which of the variables being considered should actually be controlled? What should we do if we have so many variables to control that we run out of data? These questions will be considered in the activities to follow.

Questions about Control:

Why control?

What are the options for control?

Which option should we choose?

Which variables should actually be controlled?

What if we run out of data?

Summary

- ❖ When assessing an E→D relationship, we determine whether **E** contributes anything over and above the effects of other known predictors (i.e., **control variables**) of **D**.
- ❖ When we assess the influence of control variables, we say we are **controlling for extraneous variables**.
- ❖ A simple analysis ignores the control of extraneous variables.
- ❖ Controlling assesses the **E→D** relationship at combinations of values of the control variables.
- ❖ When appropriate, controlling assesses the **overall E→D** relationship after taking into account control variables.

Reasons for Control

The typical epidemiologic research question assesses the relationship between one or more health outcome variables, **D**, and one or more exposure variables, **E**, taking into account the effects of other variables, **C**, already known to predict the outcome.

D = health outcome variables
E = exposure variables
C = control variables

When there is only one **D** and one **E**, and there are several control variables, the typical research question can be expressed as shown here, where the arrow indicates that the variable **E** and the **C** variables on the left are to be evaluated as predictors of the outcome **D**, on the right.

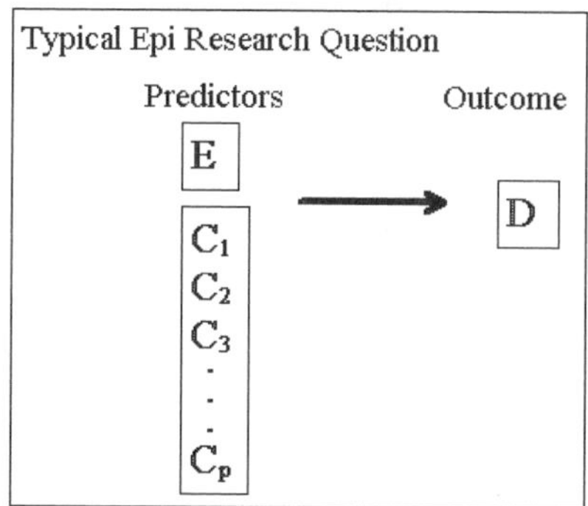

Why are the **C** variables here? That is, what are the reasons why we want to control for the **C**'s? One reason for control is to ensure that whatever effect we may find of the exposure variable cannot be explained away by variables already known to have an effect on the health outcome. In other words, we want to make sure we have accounted for the possible confounding of the **E→D** relationship due to the influence of known risk factors for the health outcome.

A second reason for control is to ensure that we remove any variability in the estimate of the E→D effect contributed by other known predictors. We might gain **precision** in our effect estimate, for example, a narrow confidence interval, as a result of controlling. In some situations there may be a loss of precision when controlling for confounders.

A third reason for control is to allow us to assess whether the effect of the exposure may vary depending on the characteristics of other predictors. For example, there may be a strong effect of exposure for smokers but no effect of exposures for non-smokers. This issue concerns **interaction**, or **effect modification**.

These are the primary three reasons for controlling: 1) to control for confounding; 2) to increase precision; and 3) to account for the possibility of **effect modification**.

1. **Confounding**
2. **Precision**
3. **Effect Modification**

All three reasons are important, but there is nevertheless an ordering of when they should be considered in the course of an analysis.

Ordering for Analysis ?

1. **Effect Modification**
2. **Confounding** (validity)
3. **Precision** (random error)

The possibility of **effect modification** should be considered first, because if there is an effect modification, then a single adjusted estimate that controls for confounding may mask the fact that the **E→D** relationship differs for different categories of a control variable.

Once effect modification is addressed or found to be absent, **confounding** should be considered, particularly in terms of those control variables **not** found to be effect modifiers. Confounding should be assessed prior to precision because confounding concerns the **validity** of an estimate. **Precision** only concerns **random error**. We would rather have a valid estimate than a narrow confidence interval around a biased estimate.

Study Questions (Q13.2)

Consider again a case-control study to assess whether a certain toxic chemical (**E**) is associated with the development of lung cancer (**D**) in a chemical industry, where we wish to control for age and smoking status. Also, assume that we categorize age into three groups: below 40, 40-55, and over 55 years of age; and we categorize smoking as "ever smoked" versus "never smoked."

1. True of False. We can assess confounding of either age or smoking by determining whether the **E→D** relationship differs within different categories of either age or smoking or both combined.
2. True or False. A more precise estimate of the odds ratio (OR) for the **E→D** association will be obtained if we control for both age and smoking status.

Suppose that when controlling for both age and smoking status, the OR for the **E→D** association is 3.5, but when ignoring both age and smoking status, the corresponding crude OR is 1.3.

3. Does this indicate confounding, precision, or interaction?

Suppose that when controlling for age and smoking status, the adjusted OR is 3.5, as above, with a 95% confidence interval ranging from 2.7 to 4.5, but that the crude OR of 1.3 has a 95% confidence interval from 1.1 to 1.5.

4. Which of these to OR's is more appropriate?

Suppose in addition to the above information, you learned that the estimated OR relating **E** to **D** is 5.7 for smokers but only 1.4 for non-smokers?

5. Would you want to control for confounding of both age and smoking?

Summary

- The typical epi research question assesses the relationship of one or more **E** variables to one or more **D** variables controlling for several **C** variables.
- The three reasons to control are confounding, precision, and effect modification.
- The possibility of effect modification should be considered first, followed by confounding, and then precision.

Options for Control

Design Options

Suppose you wish to assess the possible association of personality type and coronary heart disease. You decide to carry out a cohort study to compare the CHD risk for a group of subjects with Type A personality pattern with the corresponding risk for Type B subjects. You plan to follow both groups for the same duration, say 5 years.

Your exposure variable, **E**, is therefore dichotomous. You recognize that age, gender, ethnicity, blood pressure, smoking status, and cholesterol level are important CHD risk factors that you need to observe or measure for control in your study. You also recognize that there are other factors such as genetic factors, daily stress level, physical activity level, social class, religious beliefs, that you might also like to consider but you don't have the resources to measure.

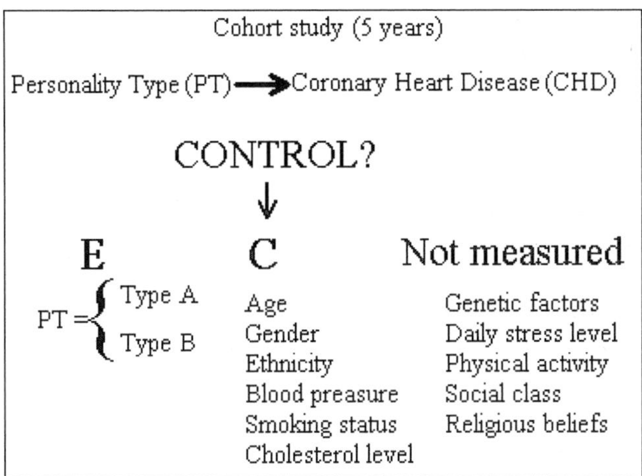

How do you carry out your study to control for any or all of the variables we have just mentioned? That is, what are your options for control? Some of your options need to be carried out at the study design stage **prior** to data collection. Other options are carried out **during the analysis stage** after the data has been obtained. It is possible to choose more than one option in the same study.

Design Options- prior to data collection

Analysis Options- after data collection

This activity focuses only on the **design options**. One design option is **randomization**. You might wish to randomly assign an initial disease-free individual to either Type A or Type B personality type. If you could randomize, then the variables you want to control for, even those you don't actually measure, might be distributed similarly for both exposure groups. However, this option is unavailable here. You can't force people to be one personality type or another; they are what they are.

A second design option is called **restriction**. This means specifying a narrow range of values for one or more of the control variables. For example, you might decide to restrict the study to African-Americans, to women, to persons older than 50, or to all three, but not to restrict any of the other variables on your list.

Study Questions (Q13.3)

1. What is a drawback to limiting the study to women only?
2. Other than generalizing to other age groups, what is another drawback to limiting the age range to persons over 50?

A third design option is **matching**. Matching imposes a "partial restriction" on the control variable being matched. For example, if we match on smoking status using what is called **pair matching**, our cohort would consist of pairs of subjects, where each pair would have a Type A subject and Type B subject who are either both smokers or both non-smokers. Smoking status would therefore not be restricted to either all smokers or all non-smokers. What would be restricted, however, is the smoking status distribution, which would be the same for both Type A and Type B groups. Matching is often not practical in cohort studies such as the study described here, so it is rarely used in cohort studies. Rather it is most often used in case-control studies or in clinical trials.

Study Questions (Q13.3) continued

Suppose you carry out a case-control study to compare CHD risks for Type A with Type B subjects. You decide to pair-match on both smoking status (ever versus never) and gender. Your cases are CHD patients identified from a cardiovascular disease registry and your controls are disease-free non-CHD subjects that are community-based.

3. What is the smoking status and gender of a control subject who is matched with a male non-smoking case?
4. If there are 110 cases in the study, what is the total number of study subjects? (Assume pair matching, as described above.)
5. Is there any restriction on the smoking status or gender of the cases?
6. Is there any restriction on the smoking status or gender of the controls?

Summary

Three design options for controlling extraneous variables are:

- **Randomization** – randomly allocating subjects to comparison groups
- **Restriction** – specifying a narrow range of possible values of a control variable.
- **Matching** – a partial restriction of the distribution of the comparison group.

Analysis Options

What control options are available at the analysis stage of a study? We'll continue the illustration of a 5-year cohort study to assess the possible association of personality type and coronary heart disease. Once the data are collected, the most direct and logical analysis option is a **stratified analysis**.

In our cohort study, if we have not used either restriction or matching at the design stage, **stratification** can be done by categorizing all control variables and forming combinations of categories called **strata**. For example, we might categorize age in three groups, say under 50, 50-60, and 60 and over; ethnicity into non-whites verses whites; diastolic blood pressure into below 95 and 95 or higher; and HDL level at or below 35 versus above 35. Because there are 6 variables being categorized into 3 categories for age and 2 categories for the other 5 variables, the total number of category combinations is 3 times 2^5, or 96 strata.

Control: AGE, GENDER, ETHNIC, DBP, SMK, HDL

Categorize all control variables	Combine categories
AGE - Under 50, 50 - 60, Over 60	6 Variables
ETHNIC - Non-white, White	$3 \times 2^5 = 96$ (Strata)
DBP - Below 95, 95 and higher	
HDL - At or below 35, Above 35	
GENDER - Male, Female	
SMK - Yes, No	

An example of a stratum is subjects over 60, female, non-white, with a diastolic blood pressure greater than 95, a smoker, and with HDL level below 35. For each stratum we form the 2x2 table that relates exposure, here personality type, to the disease, here, CHD status.

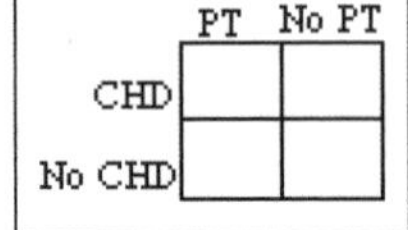

A stratified analysis is then carried out by making decisions about the **E-D** relationship for individual strata and if appropriate, combining the information over all strata to provide an overall adjusted estimate that controls for all variables together.

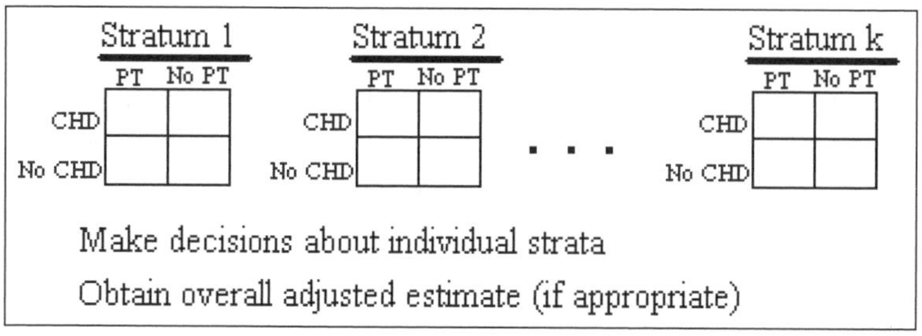

	Stratum 1			Stratum 2			Stratum k	
	PT	No PT		PT	No PT		PT	No PT
CHD			CHD			. . . CHD		
No CHD			No CHD			No CHD		

Make decisions about individual strata

Obtain overall adjusted estimate (if appropriate)

Study Questions (Q13.4)

1. When would it not be appropriate to compute an overall adjusted estimate that combines the information over all strata?
2. How would you assess the E→D relationship within any given stratum?
3. For the cohort study illustrated here, what is the biggest obstacle in carrying out stratum-specific analyses?
4. How might you carry out stratified analyses that avoid dealing with a large number of strata containing zero cells?
5. What would be an advantage and a disadvantage of doing several stratified analyses one variable at a time?

A major problem with doing stratified analysis when there are many variables to control is that you quickly run out of subjects. **An alternative option that gets around this problem is to use a mathematical *model*.** A mathematical model is a mathematical expression or formula that describes how an outcome variable, like CHD status in our example, can be predicted from other variables, which in our example are the exposure variable and the control variables we have measured or observed. In other words, we have a variable to be **predicted**, often referred to as the **dependent variable** and typically denoted **Y**, and **predictors**, often called **independent variables** and typically denoted with **X**'s.

Mathematical Model

Formula that describes how an outcome variable can be predicted from other variables

$$\mathbf{Y} = f(\mathbf{X}_1, \mathbf{X}_2, \mathbf{X}_3, \ldots, \mathbf{X}_p)$$

(dependent variable) (independent variables)

f denotes "formula"

When modeling is used, we do not have to split up the data into strata. Instead, we obtain a formula to predict the dependent variable from the independent variables. We can also use the formula to obtain estimates of effect measures such as risk ratios or odds ratios. But modeling has difficulties of its own. These include the choice of the model form to use, the variables to be included in the initial and final model, and the assumptions required for making statistical inferences.

Difficulties:

Choice of model form to use

Variables to be included

Assumptions for inferences

For dichotomous dependent variables like CHD in our example, the most popular mathematical model is called the **logistic model**.

Study Questions (Q13.4) continued

Suppose we want to use mathematical modeling to assess the relationship between personality type (**E**) and CHD status (**D**) controlling for age, gender, ethnicity, diastolic blood pressure (DBP), smoking (SMK), and high-density lipoprotein (HDL).

6. True or False. The only possible choices for the independent variables in this model are E and the above 6 control variables.
7. True or False. In a mathematical model, continuous variables must be categorized.
8. True or False. In a case-control study, the dependent variable is exposure status.

Suppose f(X_1, X_2, X_3, X_4, X_5, X_6, X_7) represents a mathematical formula that provides good prediction of CHD status, where X_1 through X_7 denote E and the 6 control variables described above.

9. True or False. If we substitute a person's specific values for X_1 through X_7 into the formula, we will determine that person's correct CHD status.

Summary

At the analysis stage, there are two options for control:

- **Stratified analysis** - categorize the control variables and form combinations of categories or strata.
- **Mathematical modeling** – use a mathematical expression for predicting the outcome from the exposure and the variables being controlled.
- Stratified analysis has the drawback of running out of numbers when the number of strata is large.
- Mathematical modeling has its drawbacks, including the choice of model and the variables to be included in the initial and final model.

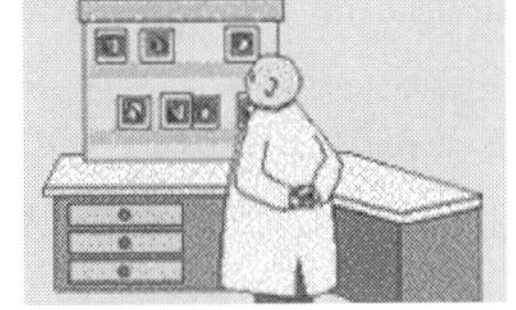

Quiz (Q13.5)

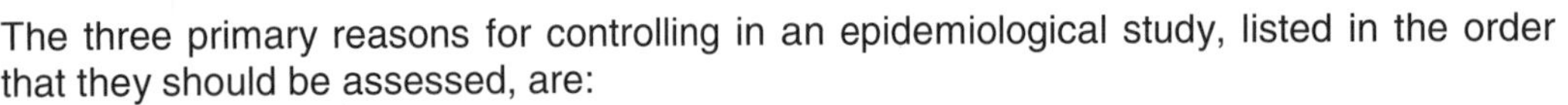

The three primary reasons for controlling in an epidemiological study, listed in the order that they should be assessed, are:

1. ???
2. ???
3. ???

Choices
Confounding Effect Modification Matching Mathematical Modeling Precision
Randomization Restriction Stratification

There are three options for control that can be implemented at the design stage.

4. ??? is a technique for balancing how unmeasured variables are distributed among exposure groups.
5. ??? limits the subjects in the study to a narrow range or values for one or more of the control variables.
6. In a case-control study, ??? ensures that the some or possibly all control variables have the same or similar distribution among case and control groups.

Choices
Matching Mathematical Modeling Optimization Randomization Restriction Stratification

Quiz continued on next page

7. Stratification is an analysis technique that starts by dividing the subjects into different <u>???</u> based on categories of the <u>???</u> variables.
8. A major problem with stratified analyses is having too many <u>???</u> variables, which can result in <u>???</u> data in some strata.

Choices
<u>control</u> **<u>disease</u>** **<u>exposure</u>** **<u>large numbers</u>** **<u>random samples</u>** **<u>sparse</u>** **<u>strata</u>** **<u>treatment</u>**

9. In mathematical modeling we use a formula to predict a <u>???</u> variable from one or more <u>???</u> variables.
10. The problems with using a mathematical model include the choice of the <u>???</u> of the model, deciding what ??? to include in the model, and the <u>???</u> required for making statistical inferences from the model.

Choices
<u>assumptions</u> **<u>cases</u>** **<u>complexity</u>** **<u>control</u>** **<u>dependent</u>** **<u>form</u>** **<u>independent</u>**
<u>subjects</u> **<u>treatments</u>** **<u>variables</u>**

Randomization

Randomization allocates subjects to exposure groups at random. In epidemiologic research, randomization is used only in **experimental studies** such as **clinical** or **community trials**, and is never used in observational studies.

What does randomization have to do with the control of extraneous variables? The goal of randomization is **comparability**. Randomization tends to make the comparison groups similar on demographic, behavioral, genetic, and other characteristics **except** for exposure status. The investigator hopes, therefore, that if the study finds any difference in health outcome between the comparison groups, that difference can only be attributable to their difference in exposure status.

Randomization

- Allocates subjects to exposure groups at random
- Only in experimental studies
- Never in observational studies
- Comparison groups have similar characteristics except for exposure status
- Unmeasured variables are likely to be similarly distributed among exposure groups

For example, if subjects are randomly allocated to either a new drug or a standard drug for the treatment of hypertension, then it is hoped that other factors, such as age and sex, might have approximately the same distribution for subjects receiving the new drug as for subjects receiving the standard drug. Actually, there is no guarantee even with randomization that the distribution of age, for example, will be the same for the two treatment groups. The investigator can always check the data to see what has happened regarding any such characteristic, providing the characteristic is measured or observed in the study. If, for example, the age distribution is found to be different between the two treatment groups, the investigator can take this into account in the analysis by stratifying on age.

An important advantage of randomization is what it offers for those variables **not** measured in the study. Variables that are not measured obviously cannot be taken into account in the analysis. Randomization offers insurance, though no guarantee, that such unmeasured variables are similarly distributed among the different exposure groups. In observational studies, on the other hand, the investigator can account for only those variables that are measured, allowing more possibility for spurious conclusions because of unknown effects of important unmeasured variables.

Study Questions (Q13.6)

Suppose you plan to do a case-control study to assess whether personality type is a risk factor for colon cancer.

1. Can you randomly assign your study subjects to different exposure groups?

Suppose you plan a clinical trial to compare two anti-hypertensive drugs. You wish to control for age, race, and gender, but you also wish to account for possible genetic factors that you cannot measure.

2. Can you control for specific genetic factors in your analysis?
3. Will the two drug groups have the same distributions of age, race, and gender?
4. What do you hope randomization will accomplish regarding the genetic factors you have not measured?

Summary

- **Experimental studies** use **randomization** whereas observational studies do **not** use randomization.
- The goal or randomization is comparability.
- Randomization tends to make comparison groups similar on other factors to be controlled.
- An important advantage of randomization is that it tends to make variable not measured similarly distributed among comparison groups.
- There is not guarantee that randomization will automatically make comparison groups similar on other factors.

13-2 Control of Extraneous Factors (continued)

Restriction

Restriction is another design option for control in which the eligibility of potential study subjects is narrowed by restricting the categories of one or more control variables. Restriction can be applied to both continuous and categorical variables. For a categorical variable, like gender, restriction simply means that the study is limited to one or more of the categories. For a continuous variable, restriction requires limiting the range of values, such as using a narrow age range, say from 40 to 50 years of age.

Restriction typically provides complete control of a variable. It is convenient, inexpensive, and it requires a simple analysis to achieve control. For example, if a study is restricted to females only, the analysis does not require obtaining an adjusted effect that averages over both genders. The main disadvantage of restriction is that we cannot generalize our findings beyond the restricted category. For continuous variables, another disadvantage is that the range of values being restricted may not be sufficiently narrow, so there may still be confounding to be controlled within the chosen range.

Restriction (Design Option)

eligibility narrowed

Continuous Limiting the range of values

Categorical Limited to one or more of the categories

Advantages	**Disadvantages**
Complete control	Cannot generalize
Convenient	Not sufficiently narrow (continuous variables)
Inexpensive	
Simple analysis	

Given a list of several control variables to measure, we typically use restriction on a small number of variables. This allows hypotheses to be assessed over several categories of most control variables, thereby allowing for more generalizability of the findings.

Control Variables

C_1 restrict
C_2 restrict
C_3
C_4 } Allows generalizability
C_5
C_6

For example, if we want to control for age, gender, ethnicity, diastolic blood pressure, smoking, and HDL level, we would likely use restriction for no more than two or three of these variables, say age and gender.

Restriction may be used at the analysis stage even if not used at the design stage. For example, even though the study sample may include several ethnic groups, we may decide to only analyze the data for one of these groups, particularly if other ethnic groups have relatively few subjects. However, it's more advantageous to choose restriction at the design stage to gain precision in the estimated effect or to reduce study costs. For example, for a fixed study size or fixed study cost, restricting ethnicity to African-Americans at the design stage will provide more African-American subjects, and therefore more precision in effect measures for African-Americans, than would be obtained if the design allowed several ethnic groups to be eligible.

Summary

- Restriction is a design option that narrows the eligibility of potential study subjects by restricting the categories of one or more control variables.
- Restriction can be applied to both categorical and continuous variables.
- Restriction typically provides complete control, is convenient, inexpensive, and requires a simple analysis to achieve control.
- The main disadvantage of restriction is not being able to generalize findings beyond the restricted category.
- For continuous variables, another disadvantage is the possibility of residual confounding within the range of restricted values.
- Restriction may be used in the analysis stage, but if used at the design stage, precision may be gained and/or study costs may be reduced.

Quiz (Q13.7)

Label each of the following statements as **True** or **False**.

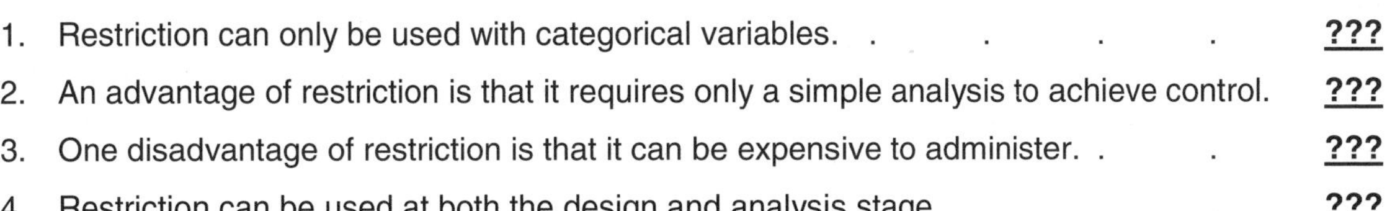

1. Restriction can only be used with categorical variables. . . . . . ???
2. An advantage of restriction is that it requires only a simple analysis to achieve control. ???
3. One disadvantage of restriction is that it can be expensive to administer. . . ???
4. Restriction can be used at both the design and analysis stage. . . . ???

Matching

Matching is a design option that can be used in **experimental studies** and in **observational cohort studies**, but is most widely used in **case-control studies**. A general definition of matching that allows other designs is given in Lesson 15.

There are generally two types of matching: **individual matching** and **frequency matching**. When individual matching is used in a case-control study, one or more controls are chosen for each case so that the controls have the same or similar characteristics on each of the variables involved in the matching. For example, if we match on age, race, and sex, and a given case is, say, 40 years old, black, and male, then the one or more controls matched to this case must also be close to or exactly 40 years old, black, and male. For continuous variables, like age, the categories used for matching must be specified prior to the matching process. For age, say, if the matching categories are specified as 10-year age bands that include the age range 35-45, then the control match for a 40-year-old case must come from the 35-45 year old age range.

Here, we are restricting the distribution of age, race, and gender in the control group to be the same as in the case group. But we are **not** restricting either the values or the distribution of age, race, and sex for the **cases**. That's why we say that matching imposes a **partial restriction** on the control variables being matched.

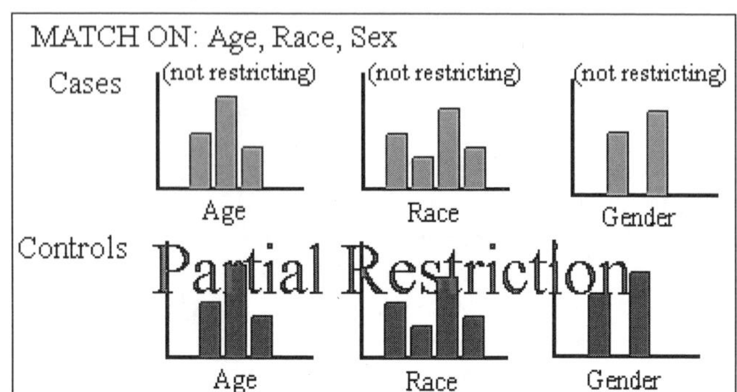

If **frequency matching** is used in a case-control study, then the matching is done on a group rather than individual basis. For example, suppose we wish to frequency match on race and gender in a case-control study, where the cases have the race-by-gender breakdown shown here:

	White Male	**White Female**	**Black Male**	**Black Female**
Cases	100 (33%)	100 (33%)	40 (13%)	60 (20%)

The controls then must be chosen as a group to have the same distribution as the cases over the four races by gender strata. If we want to have twice as many controls as cases, then the race by gender breakdown for controls will follow the same distribution pattern as the cases.

Several issues need to be considered when matching.

- ✓ First, what are the **advantages** and **disadvantages** of matching? A major reason for matching is to gain efficiency or precision in the estimate of effects, say, the odds ratio.
- ✓ **Should we match** at all, or should we choose the cases and controls without matching? There is no simple answer to this question, but a rough guideline is to only match on variables that you think will be strong confounders in your data.
- ✓ **How many controls** should we choose for each case? A rough guideline here is that usually no more than four controls per case will be necessary in order to gain precision.
- ✓ How do we **analyze matched data**? The answer here depends in part on whether or not there are other variables to be controlled in the analysis besides the matching variables. In particular, if the only variables being controlled are involved in the matching, than the appropriate analysis is a special kind of **stratified analysis**. But if in addition to the matching variables, there are other variables to be controlled, then the appropriate analysis involves **mathematical modeling**, usually using **logistic regression** methods.

Matching - Issues

What are the Advantages/Disavantages?
Gain precision (efficiency)

Should we match at all?
Match on variables that are strong confounders

How many controls for each case?
4 controls per case

How do we analyze matched data?
No other variables: Stratified analysis
Other variables: Mathematical modeling

Summary

- Matching can be used in both experimental and observational studies, and is most often used in case-control studies.
- There are two types of matching: **individual matching** and **frequency matching**.
- A major reason for matching is to gain efficiency or precision in the estimate of effect.
- Usually no more than four controls per case will be necessary in order to gain precision.
- If the only variables being controlled are involved in the matching, then use stratified analysis.
- If there are other variables to be controlled, then use mathematical modeling.

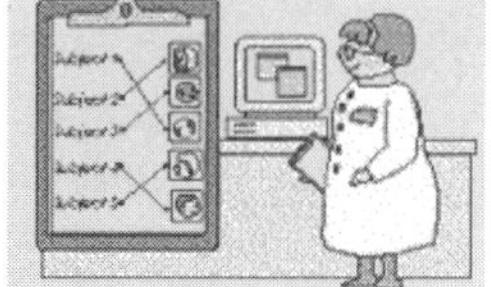

Quiz (Q13.8)

Label each of the following statements as **True** or **False**.

1. Matching is used mostly with observational cohort studies. . . . . **???**
2. For continuous variables, the ranges used for creating matching categories must be specified prior to the matching process. . . . . **???**
3. For frequency matching, there can be more controls than cases. . . . . **???**
4. Stratified Analysis is typically used for matched data to control for variables other than those involved in the matching. . . . . **???**

Stratified Analysis

Stratified analysis is an analysis option for control that involves categorizing all study variables, and forming combinations of categories called **strata**. If both the exposure and the disease variables are dichotomous, then the strata are in the form of several two by two tables. The number of strata will depend on how many variables are to be controlled and how many categories are defined for each variable.

	E	not E			E	not E			E	not E
D				D			. . .	D		
not D				not D				not D		

Study Questions (Q13.9)

1. If three variables are to be controlled and each variable is dichotomized, how many strata are obtained?
2. If the control variables are age, race, and gender, give an example of one of the strata.

Once the strata are defined, a stratified analysis is carried out by making **stratum-specific** simple analyses and, if appropriate, by making an overall summary assessment of the **E→D** relationship that accounts for all control variables simultaneously. Both the stratum-specific analyses and the overall summary analyses will typically involve computing and interpreting a point estimate of the effect, say a risk ratio, a confidence interval for the point estimate, and a test of hypothesis for the significance of the point estimate.

For an overall **summary assessment**, the point estimate is an adjusted estimate that is some form of weighted average of stratum-specific estimates. The confidence interval is an interval estimate around this weighted average, and the test of hypothesis is a generalization of the Mantel-Haenszel chi-square that now considers several strata.

Summary

Stratified analysis involves the following steps:

- Categorize all variables
- From combinations of categories (i.e., strata)
- Carry out stratum-specific analyses
- Carry out an overall E→D assessment, if appropriate
- Both stratum-specific analyses and overall assessment require point and interval estimates, and a test of hypothesis.

For overall assessment:

- The point estimate is an adjusted estimate that is a weighted average of stratum-specific estimates.
- The confidence interval is an interval estimate around the adjusted (weighted) estimate.
- The test of hypothesis is a generalization of the Mantel-Haenszel chi square test.

Quiz (Q13.10)

Label each of the following statements as **True** or **False**.

1. Stratified analysis is an analysis option for control that involves categorizing all study variables, and forming combinations of categories called strata. . . . . . ???
2. Tests of hypothesis are not appropriate for stratified analyses. . . . . ???
3. When carrying out stratum-specific analyses, the point estimate is typically computed as a weighted average. . . . . ???
4. When carrying out stratum-specific analyses, an appropriate test statistic for large samples is a Mantel-Haenszel chi square statistic. . . . . ???
5. If it is appropriate to carry out overall assessment over all strata, a recommended test statistic for large samples is a Mantel-Haenszel chi square statistic. . . . . . ???
6. When carrying out overall assessment over all strata, a Mantel-Haenszel chi square statistic is always appropriate for large samples. . . . . ???

13-3 Control of Extraneous Factors (continued)

Mathematical Modeling

Introduction to Mathematical Modeling

A **mathematical model** is a mathematical expression or formula that describes how an **outcome variable** can be predicted from **explanatory variables** that affect the outcome. Below is a general expression for a mathematical model. In this formula, **Y** denotes the **outcome variable**, often called the **dependent variable**. The **X**s in the formula denote the **predictors**, often called the **independent variables**. **f** denotes a mathematical formula or function that involves the **X**s in some way. Because we can rarely perfectly predict **Y** from the **X**s, **e** denotes the error that represents what is left over after we predict **Y** from the **X**s using the formula **f**.

Math Model:

$$Y = f(X_1, X_2, \ldots, X_p) + e$$

dependent variable	independent variables	error

In epidemiology, regardless of the mathematical form of the function, the **Y** variable is typically the **health outcome** or **disease variable**, which we have generally called **D**. At least one of the **X** variables is an **exposure variable**, which we have called **E**. Other **X** variables can be control variables, which we've denoted by the letter **C**. Still other **X** variables can be functions of the **C** variables, like age^2 if age is a continuous **C** variable, or **product terms** like **E x age**. Product terms involving the exposure variable and one or more control variables are often considered to evaluate **interaction**. The product term **E x age**, for example, considers the interaction of exposure with age, where age is viewed as a possible effect modifier of exposure.

Epidemiology: Y health outcome (D)
X_1 = Exposure (E)
X_2, X_3 control variables (C)
$X_4 = AGE^2$
X_5 = E x AGE (interaction)

Study Questions (Q13.11)

Suppose we want to use mathematical modeling to assess the relationship between personality type (PT) and CHD status (Y) controlling for gender, race, and SMK, all (0, 1) variables.

1. Give a general expression (i.e., statement involving **f**) for a mathematical model containing these variables.

The function **f** can take a variety of forms. One of the most popular functions is called a **linear function**, and the corresponding model is called a **linear model**. As an example, suppose the **X**s in the model are the 0, 1 variables personality type, denoted PT, gender, race, and smoking status, denoted as SMK. Then, the function **f** for a linear model involving these variables might look as shown here:

Example: (0,1) variables
X_1= PT X_3= RACE
X_2= GENDER X_4= SMK
Linear f (PT, GENDER, RACE, SMK)
$= b_0 + b_1(PT) + b_2(GENDER) + b_3(RACE) + b_4(SMK)$

In this function, the quantities b_0 through b_4 are unknown parameters called **regression coefficients** that need to be estimated from the study data. A more general form for a linear function is shown here, where we have replaced the four variables in the above example by several **X**s.

General Linear Function:
$f(X_1, X_2, ..., X_p) = b_0 + b_1X_1 + b_2X_2 + ... + b_pX_p$

The **X**s in a linear model can be **categorical** or **quantitative** factors. The model is still considered to be a linear model because it is a sum involving the regression coefficients.

Study Questions (Q13.11) continued

X_1= PT X_3= RACE
X_2= GENDER X_4= SMK
General Linear Function:
$f(X_1, X_2, ..., X_p) = b_0 + b_1X_1 + b_2X_2 + ... + b_pX_p$

Suppose the **b**'s in the linear model involving PT, gender, race, and SMK are, respectively:

$b_0 = 0.10$, $b_1 = 0.20$, $b_2 = 0.25$, $b_3 = 0.10$, and $b_4 = 0.15$

2. What is the predicted value for CHD (i.e., Y) for a person with variable values PT = 1, gender = 1, race = 1, and SMK = 1?
3. What is the predicted value for CHD for a person with variable valued PT = 0, gender = 1, race = 1, and SMK = 1?

Suppose that in addition to the variables PT, gender, race, and SMK, the following product terms are added to the model:

PT x gender, PT x race, PT x SMK

4. State the form of the linear model that includes these new variables in addition to the original four variables.

Suppose in the new model, the **b**'s are:

b_0= 0.08, b_1= 0.18, b_2= 0.35, b_3= 0.07, b_4= 0.12, b_5= 0.03, b_6= 0.02, and b_7= 0.03

5. What is the predicted value for CHD (i.e., Y) for a person with variable values PT = 1, gender = 1, race = 1, and SMK = 1?
6. What is the predicted value for CHD for a person with variable values PT = 0, gender = 1, race = 1, and SMK = 1?

Summary

- A mathematical model is a mathematical expression or formula that describes how an outcome variable can be predicted from explanatory variables.
- A general expression for a mathematical model is where **Y** is the **dependent variable**, the **X**'s are the **independent variables, f** is a mathematical function, and **e** is the error term.
  $$\mathbf{Y = f(X_1, X_2, \ldots, X_p) + e}$$
- In epidemiology, **Y** is usually the **health outcome**, and the **X**'s may include exposure variables, control variables, and more complicated variables like product terms.
- A very popular function is the linear function, which is generally of the form $\mathbf{f(X_1, X_2, \ldots, X_p) = b_0 + b_1X_1 + b_2X_2 + \ldots + b_pX_p}$.
- A linear function is linear in the regression coefficients, even though it may include product terms as **X**'s.

Expected Value

Another way to write a mathematical model in terms of the **expected value** of **Y** given values for the **X**'s. This expected value is the average value of **Y** over all persons in our study population with the same **X** values for each **X** in the model:

$$Exp(Y \mid X_1, X_2, \ldots, X_p) = f(X_1, X_2, \ldots, X_p)$$
expected value

(Note: the expected value may be presented as **Exp(Y|...)** or **E(Y|...)** in this Lesson.)

For example, if the **X**'s are age, race, and gender, then the expected value of **Y** for a person who is 30-years-old, black, and female is the average **Y** for all persons in the study population who are 30, black, and female. Similarly, the expected value for a 50-year-old white male is the average **Y** for all persons who are 50, white, and male.

The expected value of **Y** given the **X**s equals the function value **f** (without the error term) because the expected value of error term always equals zero. That is, the average error over all subjects in the study population is assumed to be zero.

When we estimate a mathematical model using data from our study, we write the estimated model as shown here:

Estimated Model (fitted model) $\hat{Y} = \hat{f}(X_1, X_2, \ldots, X_p)$ predicted Y

The $\hat{Y}$ on the left side of this formula denotes the predicted value of **Y** that results from specifying values for the **X**s in the estimated model for a particular individual. The estimated model is denoted as $\hat{f}$. We often refer to this model as the **fitted model** since this is the model we obtain when we fit the model to our study data.

We can predict a **Y** using a linear model involving the **X**'s. The formulae for the expected value and the fitted model will look like this:

Linear Model

$$\text{Exp}(Y|X_1, X_2, \ldots, X_p) = b_0 + b_1(X_1) + b_2(X_2) + \ldots + b_p(X_p)$$

$$\hat{Y} = \hat{b}_0 + \hat{b}_1(X_1) + \hat{b}_2(X_2) + \ldots + \hat{b}_p(X_p)$$

We have placed 'hats' for the regression coefficients in the fitted model, since the coefficients here represent estimates obtained from **fitting the model** to the study data.

Study Questions (Q13.12)

Suppose we wish to use a linear model to predict a person's systolic blood pressure (SBP) from his or her age (as a continuous variable) and smoking status (SMK=1 if ever smoked and SMK = 0 if never smoked).

1. State a simple linear model involving the above predictors using the expected value form of the model.

Suppose you fit a linear model to your study data using the variables age, SMK, and age x SMK as your predictors.

2. State the form of the predicted (i.e., estimated) model involving these variables.

Suppose you fit the following linear model, Exp(SBP|age, SMK) = $b_0 + b_1(\text{age}) + b_2(\text{SMK})$ and your estimated regression coefficients are $\hat{b}_0 = 50$, $\hat{b}_1 = 2$, $\hat{b}_2 = 10$.

3. What is the predicted value for SBP for a 30-year-old smoker?
4. What is the predicted value for SBP for a 30-year-old non-smoker?

Summary

- An alternative way to write a mathematical model is in terms of the **expected value** of **Y** given values for the **X**'s.
- The expected value of **Y** given the **X**'s can be written as:

$$\mathbf{Exp(Y|X_1, X_2,\ldots, X_p) = f(Y|X_1, X_2,\ldots, X_p)}$$

- The expected value of the error term is always zero.
- The predicted value of **Y** is given by the expression $\hat{Y} = \hat{f}(X_1, X_2,\ldots, X_p)$.
- The expected value for a linear model can be written as:

$$\mathbf{Exp(Y|X_1, X_2,\ldots,X_p) = b_0 + b_1X_1 + b_2X_2 + \ldots + b_pX_p}$$

- The predicted value of **Y** for a linear model is given by $\hat{Y} = \hat{b}_0 + \hat{b}_1X_1 + \hat{b}_2X_2 + \ldots + \hat{b}_pX_p$, where the $\hat{b}$'s denote estimated regression coefficients obtained by fitting the model to the study data.

The Logistic Model

In epidemiology, many studies consider health outcome variables that are dichotomous. We may seek to determine whether or not a person develops a given disease, lives or dies, has high or normal blood pressure, gets ill or is not ill from an outbreak, and so on. The **dependent variable**, **Y**, analyzed in such studies therefore takes on one of two possible values, often conveniently defined as either 1 or 0.

When the dependent variable is dichotomous, the most popular mathematical model is a non-linear model called the **logistic model**. Here we show the general mathematical formulation of both a **linear model** and a **logistic model**:

Linear Model

$$E(Y|X_1,X_2,\ldots,X_p)=b_0+b_1X_1+b_2X_2+\ldots+b_pX_p$$

Logistic Model - non-linear

$$E(Y|X_1,X_2,\ldots,X_p)=\frac{1}{1+e^{-(b_0+b_1X_1+b_2X_2+\ldots+b_pX_p)}}$$

These two models look quite different. The logistic model, in particular, is a non-linear model because it is not written as a linear sum of the regression coefficients. There is, nevertheless, a linear component within the denominator of the logistic model, which gives the model special meaning to epidemiologists, as we will explain. The **e** in this model is the **base for natural logarithms**.

In general, given specific values for the **b**'s and the **X**'s, these two models will yield different predicted values.

Study Questions (Q13.13)

Suppose Y = CHD (1=yes, 0=no), X_1 = gender (1 = female, 0 = male), and X_2 = SMK (1 = smoker, 0 = non-smoker). Suppose also that $b_0 = 0.10$, $b_1 = 0.25$, and $b_2 = 0.15$.

1. Compute and compare the predicted values using both the linear model and the logistic model for a female smoker. (You should use a calculator here, but it will help to know that exp(-.50) = .6065.)

The logistic model has several features that are important to epidemiologists.

Logistic Model

- When Y equals one or zero, the expected value is a probability
- The logistic model will always give a value between 0 and 1
- The linear component can be expressed with a simple transformation

First, when the dependent variable **Y** takes on the values 1 or 0 depending on whether the health outcome occurs or not, the expected value expression simplifies to a probability statement. This means that using a logistic model describes the probability or risk that a person develops the health outcome as a function of predictor variables of interest. The predictors can include an exposure variable, control variables, and product terms of exposure with control variables. The logistic model thus conveniently allows for risk to be modeled from epidemiologic data.

Another feature of the logistic model is that the non-linear function that defines the model has the property that, no matter what numerical values are specified for either the regression coefficients or the **X**'s, the logistic function will always yield a value in the range between 0 and 1. This property ensures that the predicted value obtained for any individual is constrained to the range of values for a probability (i.e., $0 \leq Pr \leq 1$). In contrast, a linear model may yield predicted values either below zero or above one for specified values of the predictors.

Thirdly, the linear component in the logistic model can be expressed through a simple transformation of the probability expression for the model. In particular, the natural log of the probability divided by 1 minus the probability (i.e., **ln[P/(1 – P)]**) turns out from algebra to yield this linear component.

$$\ln \frac{P}{1-P} = b_0 + b_1X_1 + b_2X_2 + \ldots + b_pX_p$$

$$P = \Pr(Y=1|X_1, X_2, \ldots, X_p)$$

<u>**Study Questions (Q13.13) continued**</u>

Lets do a quick introduction and review of exponentials and natural logs (i.e., the latter is usually denoted by ln).
Here are some rules:

e = 2.7183
ln(e) = 1.0000
exp(z) = e raised to the power z (i.e., e^z)
ln[exp(z)] = z
exp(z + w) = e^{z+w} = exp(z)exp(w)
ln(z * w) = ln(z) + ln(w)
ln(z/w) = ln(z) – ln(w)

2. Express ln(P/(1-P)] as a difference between two logs.
3. Solve ln(A) = Z (in terms of exp)
4. Solve ln[P/(1-P)] = Z for P/(1 – P)
5. Solve Ln[P/(1 – P)] = Z for P (this one is tricky)
6. If ln(OR) = 1, what is the OR?
7. If ln(OR) = 0, what is the OR?

This transformation (i.e., **ln[P/(1 – P)]**) is called the logit transformation, and it is meaningful to epidemiologists because the quantity **P/(1-P)** denotes an odds. Consequently, the logit function allows us to express the log odds:

ln[P/(1 – P)] = ln(Odds)

and therefore the odds for an individual with specific values for the **X**'s as a linear function of the **X**'s. If we compare the odds for one individual divided by the odds for another individual, we obtain an odds ratio, which is the fundamental measure of effect that can be estimated from the logistic model.

$$\frac{\frac{P_1}{1 - P_1}}{\frac{P_2}{1 - P_2}} = \text{Odds Ratio}$$

Study Questions (Q13.13) continued

Logit Transformation

$$\ln \frac{P}{1-P} = b_0 + b_1X_1 + b_2X_2 + \ldots + b_pX_p$$

Consider a logistic model for which Y = CHD (1 = yes, 0 = no), X_1 = gender (1 = female, 0 = male), and X_2 = SMK (1 = smoke, 0 = non-smoker). Suppose that for this model, the regression coefficients are $b_0 = 0.10$, $b_1 = 0.25$, and $b_2 = 0.15$.

8. What is the odds for a female smoker?
9. What is the odds for a male smoker?
10. What is the odds ratio that compares a female smoker to male smoker?
11. What is the odds ratio that compares a female non-smoker to a male non-smoker?

Summary

- When **Y** is dichotomous, the most popular mathematical model is a non-linear model called the **logistic model**.
- The expected value of the logistic model is:

$$\mathbf{E(Y|X_1,\ldots,X_p)} = \frac{1}{1+e^{-(b_0+b_1X_1+b_2X_2+\ldots+b_pX_p)}}$$

- For Y(0, 1), the expected value simplifies to $\mathbf{Pr(Y=1|X_1,\ldots,X_p)}$, which describes the risk for developing the health outcome.
- The logistic function will always yield predicted values that lie between 0 and 1.
- Using the logit transformation, **ln[P/(1-P)]**, the logistic model can estimate the odds for an individual and the odds ratio that compares two individuals.

Risk Function – An Example

Patients are often concerned about their prognosis. For example, a man who has been diagnosed with diabetes mellitus will ask his physician to give an indication of the risk of having a heart attack in the next ten years. The physician will base her risk prediction not only on the fact that this man is a diabetic, but also on the presence or absence of several other cardiovascular risk factors.

Study Questions (Q13.14)

1. What specific (risk factor) information will be important for the doctor to give an accurate risk prediction of the patient's risk for having a heart attack in the next 10 years?

The physician must integrate all of this information to predict the risk of cardiovascular disease for this particular patient. One way to make this prediction involves use of a **risk function**. Risk functions can be computed using various mathematical models. One of these models is the **logistic model**. Here again is the general formula for this model.

$$\text{Risk of Disease} = \frac{1}{1 + e^{-(b_0 + b_1x_1 + b_2x_2 + \ldots + b_px_p)}}$$

The **b**'s in the formula are estimated from the data. The **x**'s are the values of various cardiovascular risk factors.

In the following example, the 10-year risk of dying can be computed using the values of the **b**'s of various risk factors that are given in the table shown here:

Variable Name:	b_0 = -9.194
Age years	b_1 = 0.104
Gender; men = 1; woman = 0	b_2 = 0.658
Body mass index kg/m^2	b_3 = -0.041
Systolic blood pressure mmHg	b_4 = 0.007
Serum cholesterol mg/100ml	b_5 = 0.001
Diabetes mellitus yes = 1; no = 0	b_6 = 1.251
Smoking cigarettes/day	b_7 = 0.023
Heart rate beats/min	b_8 = 0.0053

Study Questions (Q13.14) continued

Using a calculator with the formula: $\dfrac{1}{1 + e^{-(b_0 + b_1x_1 + b_2x_2 + \ldots + b_px_p)}}$

2. Compute the 10-year risk of dying for a 62-year-old non-diabetic man who smokes 20 cigarettes/day, had a body mass index of 25 kg/m^2, a blood pressure of 140 mmHg, a cholesterol level of 245 mg/100mL, and a heart rate of 80 beats/min?
3. Compute the 10-year risk for a 62-year-old non-diabetic man who does not smoke and has a body mass index of 25 kg/m^2, a blood pressure of 140 mmHg, a cholesterol level of 245 mg/100ml, and a heart rate of 80 beats/min.

The answers to the previous two questions were 26.8% for the risk for a male non-diabetic who smokes 20 cigarettes per day and 18.8% for a male non-diabetic non-smoker, both of whom have identical values on 5 other risk factors.

4. What is the risk ratio that compares these two subjects? What does this say about the effect of smoking on the risk for a heart attack?
5. What do the above risk estimates say about the risk ratio that compares a diabetic male to a non-diabetic male, controlling for the other risk factors on the list?
6. How might you use the fitted model to compare a diabetic subject with a non-diabetic subject?

Summary

- ❖ The risk function can be used to determine an individual's risk for a particular disease
- ❖ Risk functions can be computed using various mathematical models, such as the logistic regression model. The formula for this model is:

$$\text{Risk of disease} = \frac{1}{1 + e^{-(b_0 + b_1X_1 + b_2X_2 + \ldots + b_pX_p)}}$$

Basic Features of the Logistic Model

The following is a list of some important properties of the **logistic model**:

1. **General formula**:

$$\Pr(D=1 \mid X_1,\ldots,X_p) = \frac{1}{1+e^{-(b_0+b_1X_1+b_2X_2+\ldots+b_pX_p)}}$$

2. An equivalent way to write this model is the **logit** form, which simplifies to the following linear function:

$$\text{logit } P(X) = \ln\frac{P(X)}{1-P(X)} = b_0 + b_1X_1 + b_2X_2 + \ldots + b_pX_p$$

where the parameters $b_0, b_1, b_2, \ldots b_p$ in this model represent unknown regression coefficients that need to be estimated. (Note: Often Greek symbols rather than Latin symbols are used to denote regression coefficients, e.g., β_i instead of $\mathbf{b_i}$, but we will stay with **b**'s in the description here.)

3. The usual estimation method used is called **Maximum Likelihood (ML) Estimation**, and the estimators of these regression coefficients are called **maximum likelihood estimators** (**MLE's**) and are typically written with "hats" over the parameters as follows: $\hat{b}_0, \hat{b}_1, \hat{b}_2, \ldots \hat{b}_p$

4. If $\mathbf{X_1}$ is a (0,1) variable and <u>none</u> of the other **X**'s are **product terms** of the form $\mathbf{X_1 \times W_j}$, then the **adjusted odds ratio** for the effect of $\mathbf{X_1}$, controlling for the other **X**'s is given by the formula:

$$\hat{OR}_{adj} = \exp(\hat{b}_1)$$

where $\mathbf{b_1}$ is the coefficient of $\mathbf{X_1}$ in the model.

5. If $\mathbf{X_1}$ is a (0,1) variable and <u>some</u> of the other **X**'s are product terms of the form $\mathbf{X_1 \times W_j}$, then the **adjusted odds ratio** for the effect of $\mathbf{X_1}$, controlling for the other X's is given by the formula:

$$\hat{OR}_{adj} = e^{[\hat{b}_1 + \sum \hat{b}_j W_j]}$$

where $\mathbf{b_1}$ is the coefficient of $\mathbf{X_1 \times W_j}$ in the model.

6. If $\mathbf{X_1}$ is a (0,1) variable and none of the other **X**'s are **product terms** of the form $\mathbf{X_1 \times W_j}$, then a **95% confidence interval** for the **adjusted odds ratio** for the effect of $\mathbf{X_1}$, controlling for the other **X**'s is given by the formula:

$$\exp(\hat{b}_1 \pm 1.96 \times s_{\hat{b}_1})$$

where $\mathbf{b_1}$ is the coefficient of $\mathbf{X_1}$ and

$$s_{\hat{b}_1} = \sqrt{\hat{Var}(\hat{b}_1)}$$

is the estimated **standard error** of the estimate of $\mathbf{b_1}$.

Continued on next page

Basic Features of the Logistic Model (continued)

7. The **Wald test** is one form of test of the null hypothesis that an adjusted odds ratio based on a logistic model is equal to the null value of 1. If we again assume that our exposure variable of interest is the (0, I) variable $\mathbf{X_1}$ and that there are no product terms in the model, two alternative ways to state the null hypothesis are:

$$H_0 : b_1 = 0 \text{ or } H_0 : OR = e^{b_1} = 1$$

where $\mathbf{b_1}$ is the coefficient of $\mathbf{X_1}$. The Wald test statistic is a chi-square statistic given by the formula:

$$Z^2 = \left(\frac{\hat{b}_1}{s_{\hat{b}_1}} \right)^2 = \frac{\hat{b}_1^{\,2}}{\hat{Var}(\hat{b}_1)}$$

This statistic has the **chi square distribution** with 1 d.f. under the null hypothesis.

8. An alternative test procedure, which is considered to have better statistical properties than the Wald test) is called the **Likelihood Ratio** (**LR**) test. Although both these tests may yield different computational results, they are both large-sample tests and will more often than not give similar conclusions. When in doubt, however, use the **LR** test.

Example:

We now illustrate several of these features (except for the test procedures) of the logistic model for a study of the effect of smoking history ($\mathbf{X_l}$ = **SMK**, I if ever, 0 if never) on the development of coronary heart disease (**D = CHD**, 1 if yes, 0 if no) in a cohort of 609 men from Evans County, Georgia followed from 1960-1967. Two known risk factors, $\mathbf{X_2}$ = **AGE** (continuous) and $\mathbf{X_3}$ = **GENDER** (1 if female, 0 if male) are to be controlled. A no-interaction logistic model for these data is given as follows:

$$\text{logit } P(X) = b_0 + b_1 SMK + b_2 AGE + b_3 GENDER$$

The above model is called a **no-interaction model** because it does not contain any predictors of the form $\mathbf{X_1 \times W_j}$, where $\mathbf{W_j}$ is a potential effect modifier. The formula for the estimated **adjusted odds ratio** for the effect of **SMK**, adjusted for **AGE** and **GENDER**, is then equal to:

$$\hat{OR}_{adj} = \exp(\hat{b}_1)$$

where $\mathbf{b_1}$ is the coefficient of the exposure variable $\mathbf{X_l}$ = **SMK**. (Note: the coefficients of the $\mathbf{X_2}$ and $\mathbf{X_3}$, which do not involve the exposure variable $\mathbf{X_l}$ = **SMK**, do not appear in the **OR** formula).

An alternative logistic model for these data that allows for the possibility of **interaction** of the exposure variable **SMK** with both control variables **AGE** and **GENDER** is given by the following model:

$$\text{logit } P(X) = b_0 + b_1 SMK + b_2 AGE + b_3 GENDER + b_4 SMK \times AGE + b_5 SMK \times GENDER$$

where

$$X_1 = SMK, X_2 = AGE, X_3 = GENDER, X_4 = SMK \times AGE, X_5 = SMK \times GENDER$$

For this interaction model, there are two **W**'s, i.e.,$\mathbf{W_1}$= AGE and $\mathbf{W_2}$ = GENDER. Thus, the formula for the adjusted odds ratio is given by:

Continued on next page

Basic Features of the Logistic Model (continued)

$$\hat{OR}_{adj} = \exp(\hat{b}_1 + \hat{b}_4 AGE + \hat{b}_5 GENDER$$

so that the value of the "interaction" odds ratio will change (i.e., is modified) depending on the values of the potential effect modifiers **AGE** and **GENDER**.

As a (hypothetical) numerical example, suppose that the **MLE**'s for $\mathbf{b_1}$, $\mathbf{b_4}$, and $\mathbf{b_5}$ are given by:

$$\hat{b}_1 = .25, \hat{b}_4 = .01, \hat{b}_5 = .03$$

Then the adjusted odds ratio estimate takes the following computational form:

$$\hat{OR}_{adj} = \exp(.25 + .01AGE + .03GENDER)$$

Therefore, if **AGE = 40** and **GENDER =1** (female), the estimated odds ratio is given by:

$$\hat{OR}_{adj} = \exp[.25 + .01(40) + .03(1)] = 1.97$$

whereas if **AGE = 60** and **GENDER=1**, the estimated odds ratio is given by (a different value):

$$\hat{OR}_{adj} = \exp[.25 + .01(60) + .03(1)] = 2.41$$

Reference: Kleinbaum DG and Klein M, Logistic Regression- A Self-learning Text, 2nd edition, Springer-Verlag Publishers, 2002.

The EVW Logistic Model

An important special case of the logistic model that has special relevance for the analysis of epidemiologic data is the **EVW** version of the model, which is given by the following formula:

$$\Pr(D=1 \mid X_1, \ldots, X_p) = P(X) = \frac{1}{1 + e^{-(\alpha + \beta E + \sum_{i=1}^{p1} \gamma_i V_i + \sum_{j=1}^{p2} \delta_j EW_j)}}$$

An equivalent way to write this model is the logit form, defined as **logit P(X)**, which is obtained by taking the natural log of **P(X)/[1 - P(X)]**. This logit form, from algebra, simplifies to the following linear function:

$$\text{logit } P(\mathbf{X}) = \alpha + \beta E + \sum_{i=1}^{p_1} \gamma_i V_i + \sum_{j=1}^{p_2} \delta_j EW_j$$

The $\boldsymbol{\alpha}$, $\boldsymbol{\gamma}$'s, $\boldsymbol{\beta}$'s, and $\boldsymbol{\delta}$'s in this model (previously denoted as **b**'s in the activity on logistic modeling corresponding to this asterisk/box) represent unknown regression coefficients that need to be estimated. The usual estimation method used is called **maximum likelihood (ML) estimation**, and the estimators of these regression coefficients are called **maximum likelihood estimators (MLE's)** and are typically written with "hats" over the parameters as follows: $\hat{\alpha}, \hat{\beta}, \hat{\gamma}$, and $\tilde{\delta}$.

The **E**, **V**'s and **EW**'s represent the predictor variables (i.e., the X's) in this model. More specifically, the predictor variable **E** denotes a single exposure variable of interest, the **V**'s denote control variables (i.e., **potential confounders**), and the **W**'s denote potential **effect modifiers** that go into the model as product terms with the exposure variable **E**. (Note: we refer to the **V**'s and **W**'s as **potential** confounders and **potential** effect modifiers because any of these variables may eventually be dropped from the model when the study data is analyzed.)

Continued on next page

The EVW Logistic Model (continued)

The general formula for the estimated **adjusted odds ratio** for the effect of a **dichotomous** (0,1) exposure variable **E**, adjusted for the **V** and **W** variables being controlled as potential confounders and potential effect modifiers, respectively, is given by the following expression:

$$\hat{OR}_{adj} = \exp(\hat{\beta} + \sum_{j=1}^{p_2} \hat{\delta}_j W_j)$$

where $\hat{\beta}$ and the $\hat{\delta}$'s are the **MLE** estimates of the coefficients of only those variables in the model that involve the exposure variable **E** (note: the coefficients of the **V** variables, which do not involve **E**, do not appear in the above **OR** formula).

If the logistic model being considered contains no potential effect modifiers (i.e., no W's in the model), then the above adjusted odds ratio expression (again assuming a 0,1 exposure variable) simplifies to:

$$\hat{OR}_{adj} = \exp(\hat{\beta})$$

which only involves **β**, the coefficient of the exposure variable **E**.

We now illustrate the **EVW** logistic model for a study of the effect of smoking history (**E =SMK**, 1 if ever, 0 if never) on the development of coronary heart disease (**D =CHD**, 1 if yes, 0 if no) in a cohort of 609 men from Evans County, Georgia followed from 1960-1967. Two known risk factors, **AGE** (continuous) and **GENDER** (1 if female, 0 if male) are to be controlled.

A **no-interaction** logistic model for these data based on the above general **EVW** formula is given as follows:

$$\text{logit } P(X) = \alpha + \beta E + \gamma_1 V_1 + \gamma_2 V_2$$

In this example, **E = SMK**, $\mathbf{V_1}$ **= AGE**, and $\mathbf{V_2}$ **= GENDER**. The model is called a **no-interaction model** because it does not contain any predictors of the form **EW**, where **W** is a potential effect modifier. The formula for the estimated adjusted odds ratio for the effect of **SMK**, adjusted for **AGE** and **GENDER**, is then equal to:

$$\hat{OR}_{adj} = \exp(\hat{\beta})$$

where **β** is the coefficient of the exposure variable **E = SMK**. (Again note: the coefficients of the **V** variables, which do not involve **E**, do not appear in the **OR** formula).

An alternative logistic model for these data that allows for the possibility of interaction of the exposure variable **SMK** with both control variables **AGE** and **GENDER** is given by the following model:

$$\text{logit } P(X) = \alpha + \beta E + \gamma_1 V_1 + \gamma_2 V_2 + \delta_1 EW_1 + \delta_2 EW_2$$

where **E = SMK**, $\mathbf{V_1}$ **= AGE =** $\mathbf{W_1}$, and $\mathbf{V_2}$ **= GENDER =** $\mathbf{W_2}$. For this interaction model, the formula for the estimated adjusted odds ratio is given by:

$$\hat{OR}_{adj} = \exp(\hat{\beta} + \hat{\delta}_1 AGE + \hat{\delta}_2 GENDER)$$

so that the value of the "interaction" odds ratio will change (i.e., is modified) depending on the values of the potential effect modifiers **AGE** and **GENDER**.

As a (hypothetical) numerical example, suppose that the **MLE's** for β, δ_1, and δ_2 are given by:

$$\hat{\beta}_1 = .25, \hat{\delta}_1 = .01, \hat{\delta}_2 = .03$$

Then the adjusted odds ratio estimate takes the following computational form:

Continued on next page

The EVW Logistic Model (continued)

$$\hat{OR}_{adj} = \exp(.25 + .01AGE + .03GENDER)$$

Therefore, if **AGE = 40** and **GENDER =1** (female), the estimated odds ratio is given by:

$$\hat{OR}_{adj} = \exp[.25 + .01(40) + .03(1)] = 1.97$$

whereas if **AGE = 60** and **GENDER=1**, the estimated odds ratio is given by (a different value):

$$\hat{OR}_{adj} = \exp[.25 + .01(60) + .03(1)] = 2.41$$

Reference: Kleinbaum DG and Klein M, Logistic Regression- A Self-learning Text, 2nd edition, Springer-Verlag Publishers, 2002.

Quiz (Q13.15)

True or **False**, each of the following can be included as an X variable in a mathematical model:

1. Control variables. . . . **???**
2. Disease variable. . . . **???**
3. Dependent variables. . . . **???**
4. Product terms of two or more X variables. **???**
5. Functions of control variables. . . **???**

Fill in the Blanks

6. For dichotomous dependent variables, the most popular form of a mathematical model is a **???** model.
7. A mathematical model includes **???**, which are estimated from the data in order to compute a predicted value for the y variable.

Choices
curved **linear** **logistic** **probabilities** **regression coefficients**

A linear model has been built from a set of study data. The formula is given below:

$$Y = 0.15 + 0.10 \times PT + 0.2 \times Gender + 0.15 \times Race + 0.25 \times SMK$$

8. What is the predicted value for CHD for the following X variable values? PT =1, Gender=O, Race=O, and SMK=O. **???**
9. What is the predicted value for CHD with these X variable values? PT =1, Gender=1, Race=O, and SMK=O? **???**
10. If a logistic model involving the same data and same variables was used instead of a linear model, the answers to parts a. and b. would likely be **???**.

Choices
.15 **.25** **.35** **.45** **.50** **different** **identical**

11. The expected value form of a mathematical model gives the **???** value of Y over all persons in the study population who have the same values for corresponding X's in the model.
12. The expected value model does not include an error term because the average error over all the subjects in the study population is assumed to be **???**

Choices

1	average	largest	smallest	zero

13. The logistic model predicts the probability that a person will develop the **???** of interest based on that person's values for the **???** and control variables used in the model.
14. The **???** transformation allows the logistic model to provide information about an odds ratio.

Choices

arcsin	dependent	expected value	exposure	health outcome	linear
logit	square root				

Nomenclature

b_x	In mathematical models, the regression coefficients, usually numbered b_o-b_p
C	Control variable or covariate
D	Disease or outcome variable
E	Exposure variable, or in mathematical models, the expected value
X_x	In mathematical models, the independent or predictor variables, usually numbered X_1-X_p
Y	In mathematical models, the dependent or outcome variable

General Linear Model

$$E(Y \mid X_1, X_2, ..., X_p) = b_o + b_1X_1 + b_2X_2 + ... + b_pX_p$$

General Logistic Model (non-linear)

$$E(Y \mid X_1, X_2, ..., X_p) = \frac{1}{1 + e^{-(b_o + b_1X_1 + b_2X_2 + ... + b_pX_p)}}$$

Logit Transformation

$$\ln \frac{P}{1-P} = b_o + b_1X_1 + b_2X_2 + ... + b_pX_p$$

References

Feinstein AR. Clinical Biostatistics. Mosby Publishers, St. Louis, 1977.

Hoes AW, Grobbee DE, Valkenburg HA, Lubsen J, Hofman A. Cardiovascular risk and all-cause mortality: a 12 year follow-up study in the Netherlands. Eur J Epidemiol 1993;9(3):285-92.

Kleinbaum DG, Kupper LL, Morgenstern H. Epidemiologic Research: Principles and Quantitative Methods. John Wiley and Sons Publishers, New York, 1982.

Kleinbaum DG, Klein M. Logistic Regression: A Self-Learning Text, 2nd Ed. Springer Verlag Publishers, 2002.

Miettinen OS. Matching and design efficiency in retrospective studies. Am J Epidemiol 1970;91(2):111-8.

Homework

ACE-1. Control of Extraneous Variables

a. What does it mean to say that "we are controlling for extraneous variables?"
b. Describe three reasons why "we control for extraneous variables" and explain the order in which these three reasons should be considered when analyzing one's data?
c. State at least one advantage and one disadvantage of the following options for control:

Randomization
Restriction
Matching
Stratified Analysis
Mathematical Modeling

d. What is the main reason why mathematical modeling is used instead of or in addition to stratified analysis?
e. Suppose Y=CHD, X1=AGE, X2=SMK and X3=Age*SMK State the logistic model in terms of these variables. What part of this model is a "linear component"
f. State the logit form of the logistic model described in part e.
g. Why is the logit form of the model of particular importance to epidemiologists?

ACE-2. Logistic Model: Calculate an Odds Ratio

An important rule involving the logistic model is:

"If the independent variables in the model are E, $C_1, \ldots, C_p$, where E is a (0,1) exposure variable and **none** of the C's are product terms of the form E x C (e.g., E x Age), then the adjusted odds ratio for the effect of E on the outcome D, controlling for $C_1, \ldots, C_p$ is given by the expression

$$OR(E, D \mid C_1, \ldots, C_p) = \mathbf{exp}(b)$$

where b is the coefficient of E in the model."

a. Consider the logistic model given by the expression:

$$P(CHD=1 \mid SMK, WGT, HPT) = \frac{1}{1 + \exp[-(b_0 + b_1 SMK + b_2 WGT + b_3 HPT)]}$$

where
SMK=1 if ever-smoked and SMK=0 if never-smoked
WGT=1 if heavy and WGT=0 if normal weight
HPT=1 if hypertensive and HPT=0 if normotensive.

For this model, what is the **formula** for the odds ratio for the effect of SMK on CHD, controlling for WGT and HPT?

b. Suppose when fitting the above model to study data, the estimated regression coefficients are given by

$$\hat{b}_0 = -6.80\,,\ \hat{b}_1 = 0.84\,,\ \hat{b}_2 = 0.37 \quad \hat{b}_3 = .44$$

What is your **estimate** of the odds ratio for the effect of SMK on CHD, controlling for WGT and HPT?

c. Using the same set of estimates given in part b, what is your estimate of the odds ratio for the effect of HPT on CHD controlling for SMK and WGT?

d. How would you revise the logistic model defined in part a if you wanted to estimate the odds ratio for the effect of SMK on CHD controlling only for HPT? What would be the formula for adjusted odds ratio in this revised model? Would you expect to get the same answer as you obtained in part b?

ACE-3. Logistic Model: Calculate a Confidence Interval

Suppose as in question 2 above that the independent variables in the model are E, C_1, …, C_p, where E is a (0,1) exposure variable and **none** of the C's are product terms of the form E x C (e.g., E x Age), Then, a large-sample **95% confidence interval** for the adjusted odds ratio for the effect of E on the outcome D, controlling for C_1, …, C_p is given by the expression

$$\mathbf{exp}\ [\hat{b} \pm 1.96\ s_{\hat{b}}]$$

where $\hat{b}$ is the coefficient of E in the model and $s_{\hat{b}}$ is the estimated standard error of $\hat{b}$.

Consider, as in question 2, the logistic model involving (0,1) predictor variables given by the expression:

$$P(CHD=1 \mid SMK, WGT, HPT) = \frac{1}{1 + \exp[-(b_0 + b_1 SMK + b_2 WGT + b_3 HPT)]}$$

Here is the computer output from fitting this model:

Variable	Coefficient	Std. Error	Wald Statistic	P-value
Constant	- 6.7727	1.1402	35.3	.0000
SMK	0.8347	0.3052	7.5	.0062
WGT	0.3695	0.2936	1.6	.2083
HPT	0.4393	0.2908	2.3	.1309

Calculate a 95% confidence interval for OR(SMK, CHD | WGT, HPT).

ACE-4. Wald Statistic

The Wald Statistic is a large-sample test of the null hypothesis that the coefficient of a given variable in a logistic model equal to zero, i.e., H_0: b = 0. Since **exp**(b) gives an odds ratio, it follows that an equivalent null hypothesis is H_0: OR = 1. The formula for the Wald Statistic is given by:

$$\mathbf{Z}^2 = \left(\frac{\hat{b}}{s_{\hat{b}}}\right)^2$$

and this statistic has the chi-square distribution under the null hypothesis.

Using the computer output provided in question 3 above, carry out **separate** Wald tests for the significance of SMK, WGT, and HPT, respectively. (In answering this question state the null hypothesis in terms of an adjusted odds ratio, state the value of the Wald statistic, its P-value, and draw a conclusion about statistical significance.)

ACE-5. Logistic Model: Interaction

Another important rule involving the logistic model is:

"If the independent variables in the model are E, C_1, …, C_p, where E is a (0,1) exposure variable and **some** of the C's are product terms of the form E x W (e.g., E x Age), then the adjusted odds ratio for the effect of E on the outcome D, controlling for C_1, …, C_p is given by the expression

$$OR(E, D \mid C_1, \ldots, C_p) = \mathbf{exp}(b + d_1W_1 + d_2W_2 + \ldots + d_qW_q)$$

where b is the coefficient of E in the model , W_1, W_2, … , W_q are q variables that are in the model in the form C = E x W, and the d_i are the coefficients of the product terms $C_i = E_i$ x W_i, respectively

a. Consider the logistic model given by the expression:

$$P(CHD=1 \mid SMK, WGT, HPT) = \frac{1}{1 + \exp[-(b_0 + b_1SMK + b_2WGT + b_3HPT + d_1(SMK \times WGT) + d_2(SMK \times HPT)]}$$

where SMK, WGT, and HPT are (0,1) variables as defined in question 2.

For this model, what is the **formula** for the odds ratio for the effect of SMK on CHD, controlling for WGT and HPT? In order to get a numerical value for this OR, what must you specify in addition to the values of certain regression coefficients? Which formula makes sense with regard to the meaning of interaction/effect modification?

Answers to Study Questions and Quizzes

Q13.1

1. 3 x 2 = 6 combinations
2. Approach 2 controls for age and smoking, since it considers what happens when variables are controlled. Approach 1 ignores control of age and smoking.

Q13.2

1. False. The statement addresses the question of interaction/ effect modification, not confounding.
2. False. The precision obtained will depend on the data; there is no guarantee that precision is always gained by controlling for extraneous variables.
3. Confounding, since the statement concerns what happens when we compare a crude estimate of effect with an adjusted estimate of effect.
4. If age and smoking are risk factors for lung cancer (they are), then the OR of 3.5 is more appropriate because it controls for confounding, even though it is less precise than the crude estimate; i.e., validity is more important than precision.
5. No, estimated OR's of 5.7 for smokers and 1.4 for non-smokers indicate strong interaction due to smoking (provided the observed interaction is statistically significant). An assessment of confounding would require comparing a crude estimate to an adjusted estimate, but use of the latter would not be appropriate because it would mask the presence of strong interaction.

Q13.3

1. You cannot generalize your results to men; that is, generalizing to men is an "external validity" problem.
2. The age group of persons over 50 is not necessarily narrow enough to completely control for age. In particular, there may still be "residual" confounding due to age within the age group over 50.
3. The control will have the same smoking status and gender as the case, i.e., the control will be a non-smoking male.
4. 220, since there will be 110 cases and 110 matched controls.
5. No, cases can be either male or female and either smokers or non-smokers, and they can have any distribution possible of each of these variables.
6. Yes, the controls are restricted to have the same distribution of both smoking status and gender as the cases.

Q13.4

1. It would not be appropriate to compute an overall adjusted estimate of there is strong evidence of interaction, e.g., if the estimated risk ratios in two or more strata are both statistically and meaningfully different.
2. Carry out a simple analysis for the given stratum by computing a point estimate of effect (e.g., a risk ratio), a confidence interval around the point estimate, and a test of hypothesis about the significance of this estimate.
3. There are 96 strata in all, so that it is highly likely that the entire dataset will be greatly thinned out upon stratification, including many strata containing one or more zero cells.
4. Do a stratified analysis one variable or two variables at a time, rather than all the variables being controlled simultaneously.
5. An advantage is that you can make some preliminary insights about confounding and interaction for every control variable. A disadvantage is that you will not be controlling for all variables simultaneously.
6. False. If age is continuous, then age^2 might also be used. Similarly for other continuous variables.

Also, product terms like E x age, age x gender might be used.

7. False. Continuous variables can be treated as either continuous or categorical, depending on the investigator's judgment. However, for a stratified analysis, continuous variables must be categorical.
8. False. Even though cases and controls are selected first and previous exposure then determined, case-control status is the dependent variable in a mathematical model because it represents the health outcome variable being predicted.
9. False. Mathematical models rarely, if ever, perfectly predict the outcome variable. There is always some amount of error that represents the difference between a predicted value and the observed value.

Q13.5

1. Effect modification
2. Confounding
3. Precision
4. Randomization
5. Restriction
6. Matching
7. Strata, Control
8. Control, Sparse
9. Dependent, Independent
10. Form, Variables, Assumptions

Q13.6

1. No. In a case-control study, exposure status is determined only after cases and controls are selected. Therefore, randomization to exposure groups (i.e., personality types) is not possible in case-control studies.
2. No. You cannot control for factors in your analysis that you have not measured.
3. Not necessarily. Randomization would tend to make the distribution of age, race, and gender similar in the two drug groups, but there is no guarantee that they will be the same.
4. You hope that the distribution of genetic factors is similarly distributed within the two drug groups, even though these factors have not been measured. Moreover, you hope that, for any other unmeasured factors, randomization will distribute such factors similarly over the groups being considered.

Q13.7

1. False – Restriction can also be used to limit the range of values of a continuous variable.
2. True – An advantage of restriction is that it is inexpensive to administer.
3. False
4. True

Q13.8

1. False – Matching is used mostly with case-control studies.
2. True
3. True
4. False – If in addition to the matching variables, there are other variables to be controlled, then the appropriate analysis involves mathematical modeling using logistic regression methods.

Q13.9

1. 2 x 2 x 2 = 8 strata
2. One stratum would contain all study subjects who are in a categorized age group and have the same race and gender (e.g., white females 30-40 years old).

Q13.10

1. True
2. False – both the stratum-specific and the overall summary statistics will typically involve computing a point estimate, a confidence interval, and a test of hypothesis.
3. False – when carrying out stratum-specific analyses, the point estimate is a simple point estimate calculated for a specific stratum.
4. True
5. True
6. False – a Mantel-Haenszel test is not appropriate if there is significant and meaningful interaction over the strata.

Q13.11

1. CHD = f(PT, gender, race, SMK) + e
2. 0.10 + 0.20(1) + 0.25(1) + 0.10(1) + 0.15(1) = 0.80
3. 0.10 + 0.20(0) + 0.25(1) + 0.10(1) + 0.15(1) = 0.60
4. f(PT, gender, race, SMK) = b_0 + b_1(PT) + b_2(gender) + b_3(race) + b_4(SMK) + b_5(PT x gender) + b_6(PT x race) + b_7(PT x SMK)
5. 0.08 + 0.18(1) + 0.35(1) + 0.07(1) + 0.12(1) + 0.03(1x1) + 0.02(1x1) + 0.03(1x1) = 0.88
6. 0.08 + 0.18(0) + 0.35(1) + 0.07(1) + 0.12(1) + 0.03(1x0) + 0.02(1x0) + 0.03(1x0) = 0.62

Q13.12

1. Exp(SBP|age, SMK) = b0 + b_1(age) + b_2(SMK). Other possible models include:

Exp(SBP|age, SMK) = b0 + b_1(age) + b_2(SMK) + b_3(age x SMK)
Exp(SBP|age, SMK) = b0 + b_1(age) + b_2(SMK) + b_3(age x SMK) + b_4(age^2)

2. $\hat{SBP} = \hat{b}_0 + \hat{b}_1(\text{age}) + \hat{b}_2(\text{SMK}) + \hat{b}_3(\text{age} \times \text{SMK})$
3. 50 + 2(30) + 10(1) = 120
4. 50 + 2(30) + 10(0) = 110

Q13.13

1. Linear model: $\hat{Y} = .10 + .24(1) + .15(1) = 0.50$
   Logistic model:
   $$\hat{Y} = \frac{1}{1+e^{-(.10+.25(1)+.15(1))}} .6225$$
   The two predicted values are quite different.
2. ln[P/(1-P)] = ln(P) – ln(1 – P)
3. A= exp(Z)
4. P/(1 - P) = exp(Z)
5. P = 1/[1 + exp{-Z}], a logistic function. If Z = b_0 + b_1Z_1 + ... + b_pX_p, then P = a logistic model.
6. OR = 2.7183 = e
7. OR = 1
8. ln[P/(1–P)] = 0.10 + 0.25(1) + 0.15(1) = .50. So the odds = exp(.50) = 1.65.
9. ln[P/(1–P)] = 0.10 + 0.25(0) + 0.15(1) = .25. So the odds = exp(.25) = 1.28.
10. Odds ratio (female smoker to male smoker) = 1.65/1.28 = 1.29.
11. Odds ratio (female non-smoker to male non-smoker) = exp(.35)/exp(.10) = 1.29.

Q13.14

1. To give an accurate prediction of the man's risk, the doctor will collect information on smoking behavior, body mass index, systolic blood pressure, cholesterol levels, medication use, and other possible cardiovascular risk factors.
2. The 10-year risk of dying = 1/{1 + exp –(9.194 + .62*0.104 + 1*0.658 + 25*(-0.041) + 140*0.007 + 245*0.001 + 0*1.251 + 20*0.023 + 80*0.0053)} = 26.8%.
3. The 10-year risk of dying = 1/{1 + exp –(9.194 + 1*0.658 + 62*0.104 + 25*(-0.041) + 140*0.007 + 245*0.001 + 0*1.251 + 0*0.023 + 80*0.0053)} = 18.8%.
4. RR = 26.8 / 18.8 = 14.4. Without considering statistical significance, the risk for a non-diabetic male who smokes 20 cigarettes/day is about 1.4 times the risk for a non-diabetic male non-smoker.
5. The two risk estimates provide no information that compares a diabetic with a non-diabetic subject.
6. Compare two computed risk function estimates, one in which the diabetes variable equals 1 and the other in which the diabetes variable equals 0, keeping corresponding values of all other variables at the same value.

Q13.15

1. True
2. False – In a typical epidemiologic study, we try to determine predictors of a health outcome. Therefore the dependent variable is the health outcome of interest.
3. False – The dependent variable by definition is the Y variable.
4. True
5. True
6. Logistic
7. Regression coefficient
8. .25
9. .45
10. different
11. average
12. zero
13. health outcome, exposure
14. logit

LESSON 14

STRATIFIED ANALYSIS

This is an analysis option for the control of extraneous variables that involves the following steps:

1. *Categorize all variables.*
2. *Form combinations of categories (i.e., strata).*
3. *Perform stratum-specific analyses.*
4. *Perform overall E-D assessment if appropriate.*

*Both stratum-specific analyses and overall assessment require a **point estimate**, an **interval estimate**, and a **test of hypothesis**. In this lesson, we focus on **overall assessment**, which is the most conceptually and mathematically complicated of the four steps. For overall assessment, the point estimate is an **adjusted estimate** that is typically in the form of a **weighted average** of stratum-specific estimates. The **confidence interval** is typically a large-sample interval estimate around the adjusted (weighted) estimate. The test of hypothesis is a generalization of the **Mantel-Haenszel chi square test**.*

14-1 Stratified Analysis

An Example – 1 Control Variable

We illustrate the four steps of a **stratified analysis** with an example:

STRATIFIED ANALYSIS
1. Categorize all variables
2. Form combinations of categories (strata)
3. Perform stratum-specific analyses
4. Perform overall E → D assessment

The tables below show data from a hypothetical retrospective cohort study to determine the effect of exposure to a suspected toxic chemical called TCX on the development of lung cancer. Suppose here that the only control variable of interest is smoking. First, we categorize this variable into two groups, smokers and non-smokers. Second, we form two-way tables for each stratum. Third, we perform **stratum specific analyses** as shown here. These data illustrate confounding. The crude data that ignores the control of smoking yields a moderately strong risk ratio estimate of 2.1. This is meaningfully different from the two estimates obtained when smoking is controlled, both of which indicate no association.

Retrospective Cohort Study TCX → Lung Cancer

Crude Data	TCX	no TCX	Total
LC	27	14	41
No LC	48	67	115
Total	75	81	156

Control variable: Smoking

$c\hat{RR} = 2.1$

1. Categorize all variables
2. Form combinations of categories (strata)
3. Perform stratum-specific analyses

Non-smokers	TCX	no TCX	Total
LC	1	2	3
No LC	24	48	72
Total	25	50	75

$\hat{RR} = 1.0$

Smokers	TCX	no TCX	Total
LC	26	12	38
No LC	24	19	43
Total	50	31	81

$\hat{RR} = 1.3$

Step 3 also involves computing **interval estimates** and a **P-value** for each stratum and then interpreting the results separately for each stratum as well as for the crude data. Each **stratum-specific** analysis is essentially a simple analysis for a two-way table. Here are the computed results:

TCX → Lung Cancer

Crude Data $c\hat{RR} = 2.1$

95% CI: (1.2,3.7)

P = 0.02

Non-smokers $\hat{RR} = 1.0$

95% CI: (0.1, 10.6)

P = 1.00

Smokers $\hat{RR} = 1.3$

95% CI: (0.8, 2.3)

P = 0.49

Study Questions (Q14.1)

1. What is your interpretation of the stratum-specific results?
2. Does there appear to be interaction due to smoking?
3. Does there appear to be an overall effect of TCX exposure after controlling for smoking status?

Step 4, the overall **E→D** assessment, should only be performed when appropriate. When evidence of **confounding** is present, this assessment should be conducted. However, when there is sufficient evidence of **interaction** or **effect modification**, this step is considered inappropriate. In our example, the risk ratio estimates for both smoking groups are essentially the same, which indicates that it is reasonable to go ahead with a summary or overall assessment.

To perform this step, we must do three things: compute an **overall adjusted estimate** of the exposure-disease effect over all the strata, carry out a **test of hypothesis** of whether or not there is an overall effect controlling for the stratification, and compute and interpret an **interval estimate** around the adjusted point estimate.

4. Perform overall E-D assessment

a. Adjusted estimate : weighted average

b. Test procedure : Mantel-Haenszel test for stratified analysis

c. Interval estimate : large sample confidence interval

The adjusted estimate typically is some form of **weighted average** of stratum-specific estimates. The test procedure is the **Mantel-Haenszel** test for stratified analysis. The interval estimate is typically computed as a large sample confidence interval based on percentage points of the normal distribution. These three components of overall assessment will be described further in the activities to follow.

Study Questions (Q14.1) continued

A precision-based adjusted risk ratio estimate of the TCX to lung cancer relationship is computer to be 1.25. A 95% confidence interval around this estimate turns out to be (.78, 2.00). The Mantel-Haenszel test statistic has a P-value of .28.

4. What do you conclude from these results about the overall assessment of the E-D relationship in this study?

Summary

- The simplest form of stratification occurs when there is a single dichotomous variable to be controlled.
- In this case, only one variable is categorized (step 1) and two strata are obtained (step 2).
- Step 3 typically involves computing a point estimate, an interval estimate, and a P-value for each stratum.
- Overall assessment (step 4) may not be appropriate if there is interaction/effect modification.
- Step 4 involves computing an overall adjusted estimate of effect, a large-sample confidence interval for the adjusted effect, and a test of significance (the Mantel-Haenszel test).

Overall Assessment?

Because the risk ratio estimates for both smoking groups are essentially the same, we have concluded that it is reasonable to go ahead with an overall assessment using an **adjusted estimate**, a **confidence interval** around this adjusted estimate, and a **Mantel-Haenszel** test for the stratified data. The results are presented below. They clearly indicate that there is no meaningful or significant effect of TCX on the development of lung cancer when controlling for smoking.

TCX ⟶✗ Lung Cancer when controlling for smoking

	Non-smokers	Smokers	Crude
☑	$\hat{RR}=1.0$	$\hat{RR}=1.3$	$c\hat{RR}=2.1$

$a\hat{RR}=1.25$,
95% CI {.78,2.00},
$\chi^2=1.18$ (P=.28)

But, what if we obtained a different set of stratum specific estimates, for example, the results shown below (examples 2 and 3)? Would we still want to compute an adjusted estimate, obtain a confidence interval around it and compute a Mantel-Haenszel test?

	Non-smokers	Smokers	Crude
☑	$\hat{RR}=1.0$	$\hat{RR}=1.3$	$c\hat{RR}=2.1$
☐	$\hat{RR}=.52$	$\hat{RR}=3.5$	$c\hat{RR}=1.2$
☐	$\hat{RR}=1.1$	$\hat{RR}=4.2$	$c\hat{RR}=2.7$

Note: The rows of risk ratio results are, from top to bottom, examples 1, 2, and 3, respectively.

These two examples show a very strong interaction due to smoking. And, the type of interaction in example 2 is quite different from the interaction in example 3. The stratum-specific risk ratio estimates of 0.52 and 3.5 in example 2 are on the opposite side of the null value of 1. In contrast, the stratum-specific risk ratio estimates of 1.1 and 4.2 from example 3 are on the same side of the null value, although they are also quite different.

When stratum specific effects are on opposite sides of 1, as in example 2, it is possible that they can cancel each other in the computing of an adjusted effect. Consequently, in this situation, the use of such an adjusted estimate, corresponding confidence interval, and Mantel-Haenszel test is **not** recommended. The important results in this case are given by the contrasting stratum-specific effects, and these are likely to be masked by carrying out overall assessment.

When stratum specific effects are all in the same direction, as in example 3, a spurious appearance of no association cannot arise from cancellation of opposite effects. It may therefore be worthwhile, despite the interaction, to perform overall assessment, depending on the investigator's judgment of how large the difference between stratum-specific effects is or how stable these estimates are.

Summary

- Overall assessment (step 4) may not be appropriate if there is interaction/effect modification.
- The most compelling case for not carrying out an overall assessment is when significant stratum-specific effects are on opposite sides of the null value.
- When all stratum-specific effects are on the same side of the null value, overall assessment may be appropriate even if there is interaction.
- The most appropriate situation for performing overall assessment is when stratum-specific effects are all approximately equal, indicating no interaction over the strata.

The Breslow-Day (BD) Test for Interaction

This is a test for interaction that can be used for stratified data obtained from a cumulative-incidence cohort study, case-control study or cross-sectional study, and is based on the following data layout for stratum *i*:

Data Layout for Stratum *i*

	E	**not E**	
D	a_i	b_i	m_{1i}
not D	c_i	d_i	m_{0i}
	n_{1i}	n_{0i}	n_i

The **null hypothesis** being tested is that there is **no interaction** over the strata, or equivalently, that the effects estimates over the strata are uniform (i.e., the same). The test statistic has the following form:

$$\chi^2_{BD} = \sum_{i=1}^{G} \frac{(a_i - a'_i)^2}{\hat{Var}_0(a_i)}$$

In this formula, the quantity a'_i denotes the **expected value in the *a* cell** of the table for stratum ***i*** assuming a common adjusted effect $\hat{\theta}$ (e.g., **aOR**, **aRR**, **mOR**, or **mRR**). This expected value is obtained by solving the following equation for the **a** term separately each stratum.

$$\frac{a'_i(m_{0i} - n_{1i} - a'_i)}{(m_{1i} - a'_i)(n_{1i} - a'_i)} = \hat{\theta}$$

The variance term in the **BD** formula is given by the following expression:

Continued on next page

The Breslow-Day (BD) Test for Interaction (continued)

$$\hat{Var}_0(a_i) = \frac{1}{\frac{1}{a'_i} + \frac{1}{m_{1i} - a'_i} + \frac{1}{n_{1i} - a'_i} + \frac{1}{m_{0i} - n_{1i} + a'_i}}$$

We illustrate the calculation of the BD test using hypothetical data given by the following two tables:

Stratum 1

	E	**not E**	
D	15	5	20
not D	85	95	180
	100	100	200

$\hat{OR} = 3.35$

Stratum 2

	E	**not E**	
D	5	15	20
not D	95	85	180
	100	100	200

$\hat{OR} = .30$

Based on these data, and letting θ = **aOR**, the terms in the **BD** formula are calculated to give the following results:

$$a\hat{OR} = 1, a'_1 = a'_2 = 10, \hat{Var}_0(a_1) = \hat{Var}_0(a_2) = 4.5$$

Substituting these terms in the B-D formula yields the following results:

$$\chi^2_{BD} = \frac{(a_1 - a'_1)^2}{\hat{Var}_0(a_1)} + \frac{(a_2 - a'_2)^2}{\hat{Var}_0(a_2)} = \frac{(15-10)^2}{4.5} + \frac{(5-10)^2}{4.5} = 11.11$$

Since there are **G = 2** strata, the degrees of freedom for the chi square is **G** - 1 = 1. The **P-value** for this test is .0008, which is very small. Consequently, we would conclude that there is significant interaction over the two strata, which supports the observed interaction seen from comparing the point estimates in the two tables.

An Example – Several Explanatory Variables

In recent years, antibiotic resistance has become a major problem in the treatment of bacterial infections. Many antibiotics that used to provide effective treatment against certain bacteria, particularly Staphylococcus aureus, or Staph, no longer work because newer strains of Staph aureus are resistant to antimicrobial drugs. When someone is diagnosed with infection due to Staph aureus, the first line of treatment typically involves methicillin-based antimicrobial drugs. However, strains of Staph aureus resistant to those drugs are now are considered a major problem for patients seen in emergency rooms. Resistant bacteria of this type are called methicillin-resistant Staph infections or MRSA.

We may wonder what are the characteristics or risk factors associated with having an MRSA infection? To study this question, a cross-sectional study was carried out at Grady Hospital in Atlanta, Georgia involving 297 adult patients seen in an emergency department whose blood cultures taken within 24 hours of admission were found to have Staph aureus infection. Information was obtained on several variables, some of which were previously described risk factors for methicillin resistance:

Analysis on Several Variables

- MRSA status: (1=yes, 0=No)
- Previous Hospitalization: (1=yes, 0=No)
- Age: (continuous) ⟶ 1=age>55, 0=age ≤55
- Gender: (1=male, 0=female)
- Antimicrobial Drugs: (1=yes, 0=no)

We use this information to illustrate a stratified analysis to assess whether previous hospitalization is associated with methicillin resistance, controlling for age, gender, and prior use of antimicrobial drugs. Age is continuous so we will categorize age into two groups (1=age greater than 55 years; 0=age less than or equal to 55 years).

We first consider the crude data relating previous hospitalization to MRSA status:

CRUDE DATA

Crude Data		Prev Hosp Yes	Prev Hosp No	Total
MRSA	Yes	103	12	115
	No	75	102	177
Total		178	114	292

Prev Hosp ⟶ MRSA Status

$c\hat{OR} = 11.67\ (5.99, 22.77)$

$\chi^2_{MH} = 65.01\ (p<0.001)$

Study Questions (Q14.2)

1. Looking at the crude data only, is previous hospitalization associated with methicillin resistance?
2. What reservations should you have about your answer to question 1?
3. Should you automatically stratify on age, sex, and prior antimicrobial use since they were measured in the study?
4. Since there were 297 persons in the study, why does the overall total equal 292?

Now let's see what happens when we stratify separately on age, sex, and prior antimicrobial drug use? Each stratified table is depicted separately in the following.

Relation between MRSA and Previous Hospitalization Stratified on Age

age ≤ 55

Crude Data		Prev Hosp Yes	Prev Hosp No	Total
MRSA	Yes	51	6	57
	No	52	72	124
Total		103	78	181

$\hat{OR} = 11.77$

age > 55

Crude Data		Prev Hosp Yes	Prev Hosp No	Total
MRSA	Yes	52	6	58
	No	23	30	53
Total		75	36	111

$\hat{OR} = 11.30$

$\chi^2_{MHS} = 62.49\ (P < .0001)$

$a\hat{OR} = 11.56\ (5.87, 22.76)$

B-D Test for Interaction (P=.95)

5. Focusing on stratifying age only, does there appear to be interaction/effect modification due to age based on the stratum-specific results?
6. The Breslow Day (BD) Test for Interaction provides a P-value for testing the null hypothesis that there is no interaction over the strata. Based on this test with stratifying on age, is there evidence of interaction?
7. Based on your answers to the above questions, is an overall assessment of the E-D relationship appropriate when stratifying on age?
8. Is there confounding due to age? (Hint: $c\hat{OR} = 11.67$.)
9. Does there appear to be a significant effect of previous hospitalization on MRSA when controlling for age?
10. What does the confidence interval for the adjusted estimate say about this estimate?

Relation between MRSA and Previous Hospitalization Stratified on Sex

Gender = 1 (Male)

Crude Data		Prev Hosp Yes	Prev Hosp No	Total
MRSA	Yes	69	9	78
	No	44	70	114
	Total	113	79	192

Gender = 0 (Female)

Crude Data		Prev Hosp Yes	Prev Hosp No	Total
MRSA	Yes	34	3	37
	No	31	32	63
	Total	65	35	100

$\widehat{OR}=12.20$ $\chi^2_{MHS}= 65.66$ $(P<.0001)$ $\widehat{OR}=11.70$

$a\widehat{OR}= 12.06$ (6.15, 23.62)

B-D Test for Interaction (P=.96)

11. Is an overall assessment of the E-D relationship appropriate when stratifying only on gender?
12. Is there confounding due to gender? (Hint: $c\widehat{OR} = 11.67$.)
13. Does there appear to be a significant effect of previous hospitalization on MRSA status when controlling for gender?

Relation between MRSA and Previous Hospitalization Stratified on Prior Antimicrobial Drug use ("PAMDU")

PAMDU = 1

Crude Data		Prev Hosp Yes	Prev Hosp No	Total
MRSA	Yes	92	3	95
	No	47	13	60
	Total	139	16	155

PAMDU = 0

Crude Data		Prev Hosp Yes	Prev Hosp No	Total
MRSA	Yes	10	9	19
	No	27	89	116
	Total	37	98	135

$\widehat{OR} = 8.48$ $\chi^2_{MHS}= 20.08$ $(P< .0001)$ $\widehat{OR} = 3.66$

$a\widehat{OR}= 5.00$ (2.26,11.04)

B-D Test for Interaction (P=.31)

14. Is an overall assessment of the E-D relationship appropriate when stratifying only on PAMDU?
15. Is there confounding due to PAMDU? (Hint: $c\widehat{OR} = 11.67$.)
16. Does there appear to be a significant effect of previous hospitalization on MRSA status when controlling for PAMDU?

Summary

- When several variables are being controlled using stratified analysis, the typical first step in the analysis is to analyze and interpret the crude data.
- The next step typically is to stratify separately on each control variable including carrying out an overall assessment of the E-D relationship, if appropriate.
- One approach to determine whether overall assessment is appropriate is to assess whether stratum-specific effects are more or less the same.
- Another approach is to carry out a Breslow Day test of the null hypothesis that there is no interaction/effect modification due to the variable(s) being stratified.

An Example – Several Explanatory Variables (Continued)

Here is a summary table that results from stratifying on each control variable separately.

STRATIFIED ANALYSIS FOR ASSOCIATION OF PREV. HOSP. WITH MRSA STATUS

Control Variable	aOR (95% CI)	MH P-Value	B-D P-Value
None	11.67 (5.99, 22.77)	P<.0001	–
Age	11.56 (5.87, 22.76)	P<.0001	.95
Gender	12.06 (6.15, 23.62)	P<.0001	.96
PAMDU	5.00 (2.26, 11.04)	P<.0001	.31

Crude Odds Ratio = 11.67 (5.99, 22.77)

Study Questions (Q14.3)

1. Based on the information in the above table, which, if any, of the variables age, gender, and previous antimicrobial drug use needs to be controlled?
2. Is there a gain in precision from the control of any of the variables age, gender, and previous antimicrobial drug use?

We now add to the summary table the results from controlling for two and three variables at a time.

STRATIFIED ANALYSIS FOR ASSOCIATION OF PREV. HOSP. WITH MRSA STATUS

Control Variable	aOR (95% CI)	MH P-Value	B-D P-Value
None	11.67 (5.99, 22.77)	P<.0001	–
Age	11.56 (5.87, 22.76)	P<.0001	.95
Gender	12.06 (6.15, 23.62)	P<.0001	.96
PAMDU	5.00 (2.26, 11.04)	P<.0001	.31
Age, Gender	11.59 (5.91, 22.76)*	P<.0001	.90
Age, PAMDU	4.63 (2.08, 10.29)	P<.0001	.59
Gender, PAMDU	5.04 (2.31, 11.03)*	P<.0001	.60
Age, Gender, PAMDU	4.66 (2.14, 10.14)*	P<.0001	.74

Crude Odds Ratio = 11.67 (5.99, 22.77)

* These estimates use a correction of .5 in every cell of those strata that contain a zero frequency

3. Does controlling for age, gender, or both have an affect on the results after already controlling for previous antimicrobial drug use (PAMDU
4. Using the BD P-value, is there any evidence that there is interaction when stratifying on any or all of these three variables being controlled?
5. Why do you think it is necessary to use a correction factor such as .5 in strata that contain a zero frequency?
6. Based on all the information in the table, what is the most appropriate estimate of the odds ratio of interest? (You may choose two alternatives here.)
7. Is there evidence that previous hospitalization has an effect on whether or not a person is methicillin resistant to Staph aureus?

The stratum-specific results when simultaneously controlling for age, gender, and previous antimicrobial drug use are shown in the box at the end of this Activity. There are 8 strata, because three variables are being controlled and each variable has two categories.

Study Questions (Q14.3) continued
(Note: there is no question 8)

9. What is the most obvious characteristic that describes the stratified results just shown?
10. What does your answer to the previous question indicate about stratum-specific analyses with these strata?
11. Based on comparing stratum-specific odds ratio estimates, does there appear to be interaction within the stratified data?
12. Give three reasons that justify doing an overall Mantel-Haenszel test using these data?

Summary

- ❖ When several variables are being controlled simultaneously using stratified analysis, not all of these variables may need to be controlled depending on whether a variable contributes to confounding or precision.
- ❖ The simultaneous control of several variables typically leads to strata with small numbers and often zero cell frequencies.
- ❖ When there are small numbers in some strata, stratum-specific conclusions may be unreliable.
- ❖ There are three things to consider when assessing interaction in stratified data:
    - o Are stratum-specific estimates essentially the same?
    - o Is the Breslow-Day test for interaction significant?
    - o Are stratum-specific estimates unreliable because of small numbers?

Stratum Specific Results

Here are the stratum specific results when simultaneously controlling for age, gender, and previous antimicrobial drug use.

1. Age ≤ 55, Male, PAMDU=Yes

		Prev. Hosp. Yes	Prev. Hosp. No	
MRSA	Yes	37	2	39
	No	22	7	29
		59	9	68

$\hat{OR}$ =5.89

2. Age ≤ 55, Male, PAMDU = No

		Prev. Hosp. Yes	Prev. Hosp. No	
MRSA	Yes	5	4	9
	No	13	49	62
		18	53	71

$\hat{OR}$ =4.71

3. Age ≤ 55, Female, PAMDU=Yes

		Prev. Hosp. Yes	Prev. Hosp. No	
MRSA	Yes	9	0	9
	No	14	3	17
		23	3	26

$\hat{OR}$ =4.59 with .5 adjustment

4. Age ≤ 55, Female, PAMDU = No

		Prev. Hosp. Yes	Prev. Hosp. No	
MRSA	Yes	0	0	0
	No	2	13	15
		2	13	15

$\hat{OR}$ =5.4 with .5 adjustment

5. Age > 55, Male, PAMDU=Yes

		Prev. Hosp. Yes	Prev. Hosp. No	
MRSA	Yes	24	1	25
	No	2	2	4
		26	3	29

$\hat{OR}$ =24.00

6. Age > 55, Male, PAMDU = No

		Prev. Hosp. Yes	Prev. Hosp. No	
MRSA	Yes	2	2	4
	No	7	12	19
		9	14	23

$\hat{OR}$ =1.71

7. Age > 55, Female, PAMDU=Yes

		Prev. Hosp. Yes	Prev. Hosp. No	
MRSA	Yes	22	0	22
	No	9	1	10
		31	1	68

$\hat{OR}$ =7.11 with .5 adjustment

8. Age > 55, Female, PAMDU = No

		Prev. Hosp. Yes	Prev. Hosp. No	
MRSA	Yes	3	3	6
	No	5	15	20
		8	18	26

$\hat{OR}$ =3.00

Quiz (Q14.4)

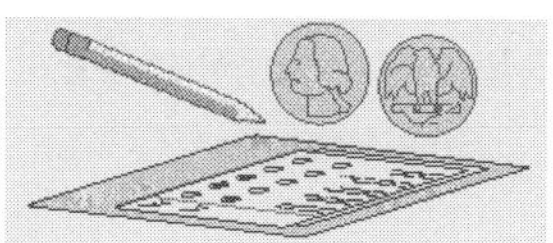

Label each of the following statements as **True** or **False**.

1. Stratification only involves categorizing variables into two groups and conducting separate analysis for each group. . . . . ???
2. One of the four steps of a stratified analysis is to compute an overall summary E-D assessment when appropriate. . . . . ???
3. The calculation of an overall summary estimate may be considered inappropriate if there is considerable evidence of statistical interaction. . . . . ???
4. The calculation of an overall summary estimate may be considered inappropriate if there is considerable evidence of confounding. . . . . ???
5. When considering the appropriateness of computing overall summary results, the investigator must exercise some judgment regarding the clinical importance of the observed differences among stratum-specific estimates as well as to the stability of these estimates. . . . . ???

6. Compare the stratum specific RR estimates for each of the three situations below. Fill in the blank with **yes**, **no** or **maybe** regarding the appropriate use of a summary estimate.

Situation:	RR: Stratum 1	RR: Stratum 2	Overall Est.
Opposite direction	0.7	3.5	???
Same direction	1.5	4.8	???
Uniform effect	2.3	2.9	???

Choices
Maybe **No** **Yes**

14-2 Stratified Analysis (continued)

Testing for Overall Association – The Mantel-Haenszel Test

General Purpose and Characteristics

The **Mantel-Haenszel test** is the most widely used and recommended procedure for testing an overall association in a stratified analysis. The **null hypothesis** being tested is that there is no association over all the strata, or equivalently, that the adjusted effect measure over all the strata equals the null value. The typical alternative hypothesis is usually one-sided, since the investigator often wants to determine whether the adjusted effect is larger than the null value; however, a two-sided alternative or a lower one-sided alternative may also be considered. The test statistic has approximately the chi square distribution with one degree of freedom under the null hypothesis.

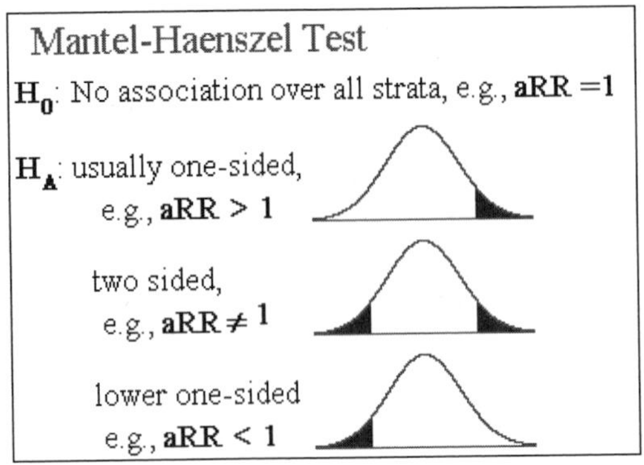

For dichotomous disease and exposure variables, the data layout in a given stratum can be described by one of two formats as shown here. The format depends on the study design:

Cohort (RR) Case-Control or X-sectional Studies

	E	Not E	Total
D	a_i	b_i	m_{1i}
not D	c_i	d_i	m_{0i}
Total	n_{1i}	n_{0i}	n_i

Person-Time Cohort (IDR) Studies

	E	Not E	Total
D	I_{1i}	I_{0i}	I_i
Person-Time	PT_{1i}	PT_{0i}	PT_i

But regardless of the study design, the test statistic has the basic computational structure shown here:

$$\chi^2_{MHS} = \frac{[a - E_0(a)]^2}{Var_0(a)}$$

a = total # of exposed cases

$E_0(a)$= expected total # of exposed cases under H_0

$Var_0(a)$= variance of total # of exposed cases under H_0

In this formula, **a** denotes the total number of exposed cases; that is, **a** is the sum of the **a** cell frequencies over all the strata. The term $\mathbf{E_0(a)}$ in the numerator gives the **expected** total number of exposed cases under the null hypothesis of no association between exposure and disease in any of the strata. The term $\mathbf{Var_0(a)}$ describes the variance of the total number of exposed cases under the same null hypothesis. The specific computational form for the expected value and the variance term depend on the study design. For the first of the two data layouts, these terms are derived assuming the hypergeometric distribution, which we introduced in Lesson 12. For a person-time cohort study, the expected value and variance terms are derived using the binomial distribution. (Note: see the two boxes at then end of this activity for a description of how the expected values and variance terms are derived based on either the hypergeometric distribution or the binomial distribution.)

For cumulative incidence cohort studies, case-control studies and cross-sectional studies the computing formula simplifies to the expression shown here:

Cohort (RR),
Case-Control or
X-sectional Studies

	E	Not E	Total
D	a_i	b_i	m_{1i}
not D	c_i	d_i	m_{0i}
Total	n_{1i}	n_{0i}	n_i

$$\chi^2_{MHS} = \frac{[a - E_0(a)]^2}{Var_0(a)}$$

$$\chi^2_{MHS} = \frac{\left(\sum_{i=1}^{G} \frac{a_i d_i - b_i c_i}{n_i}\right)^2}{\sum_{i=1}^{G} \frac{n_{1i} n_{0i} m_{1i} m_{0i}}{(n_i - 1)n_i^2}}$$

This formula is written in terms of the cell frequencies and total frequencies given in the stratum layout. When there are only two strata, so that **G** equals two, the formula simplifies further:

$$\chi^2_{MHS} = \frac{\left(\frac{a_1 d_1 - b_1 c_1}{n_1} + \frac{a_2 d_2 - b_2 c_2}{n_2}\right)^2}{\frac{n_{11} n_{01} m_{11} m_{01}}{(n_1 - 1)n_1^2} + \frac{n_{12} n_{02} m_{12} m_{02}}{(n_2 - 1)n_2^2}}$$

We now apply this formula to the retrospective cohort study data relating TCX to lung cancer. Substituting the data from the two strata into the Mantel-Haenszel formula, the computed value turns out to be 1.18. For a two-sided alternative, the P-value is the area under the chi square distribution with 1 degree of freedom above the value 1.18. This turns out to be 0.28.

Retrospective Cohort Study TCX → Lung Cancer

Non-smokers	TCX	no TCX	Total
LC	1	2	3
No LC	24	48	72
Total	25	50	75

$\hat{RR} = 1.0$

Smokers	TCX	no TCX	Total
LC	26	12	38
No LC	24	19	43
Total	50	31	81

$\hat{RR} = 1.3$

$$\chi^2_{MHS} = \frac{\left(\frac{1*48 - 2*24}{75} + \frac{26*19 - 12*24}{81}\right)^2}{\frac{25*50*3*72}{(75-1)75^2} + \frac{50*31*38*43}{(81-1)81^2}} = 1.18$$

<u>**Study Questions (Q14.5)**</u>

1. Since the P-value for a two-sided Mantel-Haenszel test is .28, what is the P-value for a one-sided alternative?
2. Based on your answer to the previous questions, what do you conclude about the overall effect of TCX on lung cancer controlling for smoking?

(To be discussed in the next activity.) In the MH formula for two strata, notice that the numerator sums two terms, one from each stratum, before squaring this sum.

3. What might happen if, instead of the results previously described, the risk ratio for nonsmokers was on the opposite side of the null value than the risk ratio for smokers?

Summary

- The Mantel-Haenszel (MH) test is the most widely used and recommended procedure for testing an overall association in a stratified analysis.
- The null hypothesis is that there is no overall association over all the strata, or equivalently, that the adjusted effect measure equals the null value.
- The typical alternative hypothesis is one-sided, but may instead be two-sided.
- The MH test statistic is approximately chi square with 1 df under the null hypothesis.
- For cumulative incidence cohort, case-control, and cross-sectional studies, the MH statistic is derived assuming the hypergeometric distribution.
- For person-time cohort studies, the MH statistic is derived using the binomial distribution.

Derivation of the Mantel-Haenszel Test for Cumulative-Incidence Cohort Studies, Case-Control Studies, and Cross-Sectional Studies

Here we describe how the **Mantel-Haenszel (MH) Test** is derived for **Cumulative-Incidence Cohort Studies**, **Case-Control Studies**, and **Cross-Sectional Studies** where the data layout for the i-th stratum is given as follows:

Data Layout for Stratum *i*

	E	**not E**	
D	a_i	b_i	m_{1i}
not D	c_i	d_i	m_{0i}
	n_{1i}	n_{0i}	n_i

For each of the above mentioned study designs, the Mantel-Haenszel test assumes that the marginal frequencies $\mathbf{n_{1i}}$, $\mathbf{n_{0i}}$, $\mathbf{m_{1i}}$, $\mathbf{m_{0i}}$ are **fixed** within each stratum. It follows, as in a simple analysis with fixed margins, that the distribution of the number of exposed cases (in the **a** cell) has the **hypergeometric distribution** under the null **hypothesis of no overall association controlling for the stratified variables**. From the hypergeometric distribution, the expected value and null variance of the number of exposed cases within each stratum are then given by the following formulae:

$$E_0(a_i) = \frac{n_{1i}m_{1i}}{n_i}$$

and

$$Var_0(a_i) = \frac{n_{1i}n_{0i}m_{1i}m_{0i}}{(n_i - 1)n_i^2}$$

so that, summing over all strata, we get the expected mean total and variance mean total exposed cases to be:

$$E_0(a_i) = \sum_{i=1}^{G} E_0(a_i) = \sum_{i=1}^{G} \frac{n_{1i}m_{1i}}{n_i}$$

and

$$Var_0(a_i) = \sum_{i=1}^{G} Var_0(a_i) = \sum_{i=1}^{G} \frac{n_{1i}n_{0i}m_{1i}m_{0i}}{(n_i - 1)n_i^2}$$

Substituting the expressions for the expected mean and variance of total exposed cases into the general Mantel Haenszel (MH) test statistic formula:

Continued on next page

Derivation of the Mantel-Haenszel Test for Cumulative-Incidence Cohort Studies, Case-Control Studies, and Cross-Sectional Studies (continued)

$$\chi^2_{MH} = \frac{[a - E_0(a)]^2}{Var_0(a)}$$

the test statistic simplifies (from algebra) into the following computing formula:

$$\chi^2_{MH} = \frac{\left[\sum_{i=1}^{G} \frac{a_i d_i - b_i c_i}{n_i}\right]^2}{\sum_{i=1}^{G} \frac{n_{1i} n_{0i} m_{1i} m_{0i}}{(n_i - 1) n_i^2}}$$

Under the null hypothesis of no overall association, this MH statistic is approximately chi square with 1 degree of freedom.

Derivation of the Mantel-Haenszel Test for Person-Time Studies

Here we describe how the **Mantel-Haenszel (MH) Test** is derived for **Person-Time Cohort Studies** where the data layout for the i-th stratum is given as follows:

Person-Time Cohort Study
Data Layout for Stratum i

	E	not E	
D	I_{1i}	I_{0i}	I_i
PT	PT_{1i}	PT_{0i}	PT_i

Under the **null hypothesis of no overall association controlling for the stratified variables**, this version of the MH test assumes that the distribution of the number of exposed cases (in the 'a' cell) within each stratum has the **binomial distribution**. In particular, the number of exposed cases in stratum i, is assumed to be a binomial random variable with probability of success given by:

$$p_{0i} = \frac{PT_{1i}}{PT_i}$$

and with number of trials equal to the total number of cases in the stratum, i.e.,

$$N_i = I_{1i} + I_{0i} = I_i$$

so that the variance of the binomial probability is given by:

$$Var_0(p_{0i}) = N_i p_{0i}(1 - p_{0i}) = \frac{I_i PT_{1i} PT_{0i}}{PT_i^2}$$

It follows, then, that the null mean and variance for the a-cell in the i-th stratum are given by the formulae:

Continued on next page

Derivation of the Mantel-Haenszel Test for Person-Time Studies (continued)

$$E_0(a_i) = \frac{I_i PT_{1i}}{PT_i}$$

and

$$Var_0(a_i) = \frac{I_i PT_{1i} PT_{0i}}{PT_i^2}$$

so that, summing over all strata, we get the expected mean total and variance mean total exposed cases to be:

$$E_0(a) = \sum_{i=1}^{G} E_0(a_i) = \sum_{i=1}^{G} \frac{I_i PT_{1i}}{PT_i}$$

and

$$Var_0(a) = \sum_{i=1}^{G} Var_0(a_i) = \sum_{i=1}^{G} \frac{I_i PT_{1i} PT_{0i}}{PT_i^2}$$

The specific form that the MH test for person-time data is then obtained by substituting the expressions for the expected mean and variance of total exposed cases into the general Mantel Haenszel (MH) test statistic formula:

$$\chi^2_{MH} = \frac{[a - E_0(a)]^2}{Var_0(a)}$$

where *a* denotes the total number of observed exposed cases over all strata, i.e.,

$$a = \sum_{i=1}^{G} I_{1i}$$

Under the null hypothesis of no overall association, this MH statistic is approximately chi square with 1 degree of freedom.

When Not to Use the MH Test

Let's consider an example in which the **Mantel-Haenszel test** would **not** be appropriate. Suppose the results from the retrospective cohort study relating TCX to lung cancer looked like this:

Results of Retrospective Cohort Study TCX ⟶ LC

Crude Data	TCX	no TCX	Total
LC	20	20	40
No LC	180	180	360
Total	200	200	400

$\hat{RR} = 1.0$ (P=1.00)

Non-smokers	TCX	no TCX	Total
LC	15	5	20
No LC	85	95	180
Total	100	100	200

$\hat{RR} = 3.0$ (P=.01)

Smokers	TCX	no TCX	Total
LC	5	15	20
No LC	95	85	180
Total	100	100	200

$\hat{RR} = .33$ (P=.01)

These data show, for non-smokers, a moderately large positive risk ratio that is highly significant, and for smokers, a risk ratio that describes a moderately large negative association but is also highly significant. In contrast, the crude data show a non-significant risk ratio exactly equal to one. The two stratum specific effects are both strong and significant, but indicate **opposite direction interaction**.

Let's see what happens if we inappropriately conducted a Mantel-Haenszel test anyway. Here again is the Mantel-Haenszel formula when there are two strata:

$$\chi^2_{MHS} = \frac{\left(\frac{a_1 d_1 - b_1 c_1}{n_1} + \frac{a_2 d_2 - b_2 c_2}{n_2}\right)^2}{\frac{n_{11} n_{01} m_{11} m_{01}}{(n_1 - 1) n_1^2} + \frac{n_{12} n_{02} m_{12} m_{02}}{(n_2 - 1) n_2^2}}$$

Substituting the data from the two strata into the Mantel-Haenszel formula, the computed value turns out to be exactly zero. In particular, notice that the two terms in the numerator of the formula are the negative of each other, so when they are summed together, their sum is zero.

$$\chi^2_{MHS} = \frac{\left(\frac{15*95 - 5*85}{200} + \frac{5*85 - 15*95}{200}\right)^2}{\left(\frac{100*100*20*180}{(200-1)200^2} + \frac{100*100*20*180}{(200-1)200^2}\right)} = 0.0$$

When there is opposite direction interaction, use of the Mantel-Haenszel test is often inappropriate because it may mask a strong interaction effect that reflects the true exposure disease relationship.

<u>Study Questions (Q14.6)</u>

1. If there are four or more strata involved in the computation of the MH statistic, is it still possible that the test can give a misleading result?
2. What can you say about the MH test result if there is strong interaction but on the same side of the null value?

Summary

- ❖ The MH test may give misleading results and is often not appropriate when stratum-specific estimates are on opposite sides of the null value.
- ❖ If there are strong and significant effects on the opposite sides of the null value, the MH test might be non-significant, thereby giving a misleading impression of no exposure-disease effect.
- ❖ The numerator in the MH formula is the square of a sum rather than a sum of squared terms.
- ❖ Consequently, positive and negative terms in the sum may cancel each other out, thereby leading to a non-significant MH test.

The MH test for Person Time Cohort Data

Let's consider data from a **person-time cohort study** of the possible association between obesity and all-cause mortality among women ages 60 to 75 from a northeastern US urban population. We consider 3 strata defined by categorizing age into 3 groups.

Person-Time Cohort Study Obesity ➔ Mortality

AGE 60-64	OB	not OB	Total
DTH	7	9	16
P-years	234.5	544.5	779

$\widehat{IDR} = 1.81$ (N.S.)

AGE 65-69	OB	not OB	Total
DTH	11	11	22
P-years	264.5	444.5	709

$\widehat{IDR} = 1.68$ (N.S.)

AGE 70-74	OB	not OB	Total
DTH	12	16	28
P-years	200	410	610

$\widehat{IDR} = 1.54$ (N.S.)

Crude	not OB	not OB	Total
DTH	30	36	66
P-years	699	1399	2098

$\widehat{IDR} = 1.67$ (P = .02)

In looking at these data, the incidence density ratios for the 3 age strata are all about the same size and are also relatively close to the crude incidence density ratio of 1.67. So there is little if any evidence of interaction. An overall adjusted estimate appears justified, as does a corresponding confidence interval around this estimate and a Mantel-Haenszel test for overall association. Here, we focus only on the Mantel Haenszel test.

The general data layout for the i-th stratum is shown here for person-time cohort studies.

Person-Time Cohort (IDR) Studies

	E	Not E	Total
D	I_{1i}	I_{0i}	I_i
Person-time	PT_{1i}	PT_{0i}	PT_i

Notice that the **a** cell in the upper table, denoted as $\mathbf{I_{1i}}$ and the **b** cell as $\mathbf{I_{0i}}$ instead of $\mathbf{a_i}$ and $\mathbf{b_i}$, respectively. Here again is the general form of the Mantel-Haenszel statistic:

$$\chi^2_{MHS} = \frac{[a - E_0(a)]^2}{Var_0(a)}$$

a = Total # of exposed cases $= \sum_{i=1}^{G} I_{1i}$

$E_0(a)$ = Expected total # of exposed cases under H_0: no overall association

$Var_0(a)$ = Variance of the total # of exposed cases under same H_0

Here, **a** denotes the total number of exposed cases. $\mathbf{E_0(a)}$ in the numerator denotes the expected number of exposed cases under the null hypothesis of no overall association. $\mathbf{Var_0(a)}$ denotes the variance of the total number of exposed cases under the same null hypothesis.

The expected value and variance terms can be derived using a binomial distribution for which the probability of success is the proportion of total **PT** in the *i*-th strata that come from exposed persons, and the number of trials is the total number of cases in the *i*-th stratum.

Binomial distribution (n,p)

$$p = \frac{PT_{1i}}{PT_i}$$

$$n = I_{1i} + I_{0i} = I_i$$

We now apply this formula to the person-time cohort data relating obesity to mortality. Substituting the data from the three strata into the Mantel-Haenszel formula, the computed chi-square value turns out to be 4.14. The area under the chi square distribution with 1 degree of freedom above the value 4.14 is .04.

$$\chi^2_{MHS} = \frac{[30 - E_0(a)]^2}{Var_0(a)} = 4.14 \quad a = 30$$

<u>**Study Questions (Q14.7)**</u>

1. What is the P-value for a one-sided alternative?
2. Based on your answer to the previous question, what do you conclude about the overall effect of obesity on mortality controlling for age?

<u>**Summary**</u>

❖ For person-time cohort studies as well as cumulative incidence cohort, case-control, and cross-sectional studies, the MH statistic is approximately chi square with one df under the null hypothesis of no overall association.

❖ The particular form that the MH formula takes for person-time data is derived by assuming that the number of exposed cases in a given stratum has the binomial distribution.

❖ For the above binomial distribution, the probability of success is the proportion of total PT in the g-th stratum that come from the exposed persons, and the number of trials is the total number of cases in the g-th stratum.

Quiz (Q14.8)

Fill in the Blanks

A cohort of physical activity (PA) and incidence of diabetes was conducted over a six-year period among Japanese-American men in Honolulu. Data from that study are summarized below, stratified on body mass index (BMI):

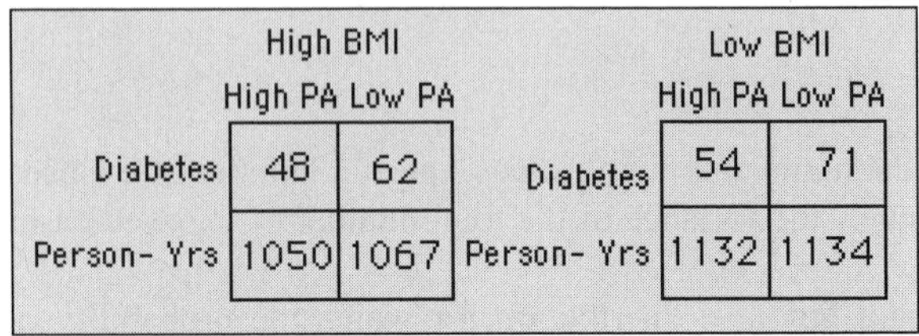

	High BMI			Low BMI	
	High PA	Low PA		High PA	Low PA
Diabetes	48	62	Diabetes	54	71
Person- Yrs	1050	1067	Person- Yrs	1132	1134

1. What is the rate ratio (IDR) for the High BMI group? . . . . **???**
2. What is the rate ratio (IDR) for the Low BMI group? . . . . **???**
3. What is the crude IDR for these data? (Hint: combine data over both strata.) . **???**
4. Is there evidence of confounding in these data? . . . . . . **???**
5. Is there evidence of interaction in these data? . . . . . . **???**

Choices
0.76 **0.77** **0.79** **1.27** **1.30** **1.32** **No** **Yes**

6. What would be the advantage of conducting an overall assessment for these data? **???**
7. The Mantel-Haenszel chi-square turns out to be 3.83, which corresponds to a 2-sided p-value of 0.051. Assuming a 5% significance level, do you reject or fail to reject the null hypothesis? **???**
8. Suppose this had been a one-tailed test, what would be your conclusion? . **???**

Choices
decreased precision **fail to reject the null hypothesis** **increased precision**
reject the null hypothesis

14-3 Stratified Analysis (continued)

Overall Assessment using Adjusted Estimates

How do we obtain overall adjusted estimates of effect in a stratified analysis? The answer depends on: 1) the type of **mathematical approach** considered; 2) the type of **effect measure** appropriate for the study design and study purpose; and 3) the **sparseness** of the data within the strata. The mathematical form may be either a **weighted average** or a **maximum likelihood** estimate. The study design and purpose may call for either a **ratio measure**, such as a risk ratio, odds ratio or rate ratio, or **difference effect measure**, such as a risk difference or a rate difference. And if some of the strata contain zero cells, the adjusted estimate must accommodate stratum-specific estimates that are undefined.

How to Obtain Adjusted Estimates?

Mathematical approach

Weighted Average
Maximum Likelihood Estimate (MLE)

Study design and purpose

Ratio measure : RR, OR, IDR
Difference measure : RD, IDD

Sparseness of data

Zero cells: stratum-specific estimates are undefined

Weighted averages, which are easy to understand, are more often used than **maximum likelihood** estimates in summarizing stratified analyses, so we will focus only on weighted average estimates here. Nevertheless, maximum likelihood estimates are conveniently obtained by fitting mathematical models such as the logistic model, and such estimates are typically used whenever modeling is considered appropriate. Fortunately, in many instances, the results obtained from these two mathematical approaches are either very close or identical, particularly for matched data.

Adjusted estimates that are weighted averages come in two forms shown following this paragraph in the simple case where there are only two strata. We denote the effect measure as **theta** (**θ**). So, depending on the study design and purpose, **θ** may be either a ratio measure such as a risk ratio or a difference measure, such as a risk difference.

How to Obtain Adjusted Estimates?

Weighted Average — θ: ratio measure (e.g., RR) or difference measure (e.g., RD)

Linear Weighting:

$$\theta_{adj} = \frac{(w_1 \times \theta_1) + (w_2 \times \theta_2)}{w_1 + w_2}$$

Difference effect measures
Ratio effect measures when there are zero cells

Log-linear Weighting:

$$\theta_{adj} = \exp\left[\frac{(w_1 \times \ln\theta_1) + (w_2 \times \ln\theta_2)}{w_1 + w_2}\right]$$

Ratio effects when the weights are chosen to reflect the precision of a stratum specific estimate

Linear weighting is always used for difference effect measures like the risk difference or the rate difference, and it is also used in certain kinds of ratio effect measures, particularly when there are zero cells in some of the strata. In contrast, log-linear weighting is used to obtain adjusted estimates for ratio effects whenever the weights are chosen to reflect the precision of a stratum specific estimate. We call such adjusted estimates **precision-based**.

When there are zero cells or even sparse data without zero cells, a special form of weighted average called a **Mantel-Haenszel adjusted estimate** is typically used. Mantel-Haenszel estimates are often used instead of precision-based estimates even when stratum-specific data are not sparse. This is because the Mantel Haenszel estimates have been shown to have desirable statistical properties, including being identical to maximum likelihood estimates for case-control data. The specific form that precision-based and Mantel-Haenszel estimates take for ratio effect measures will be described and illustrated in the activities that follow.

Summary

- ❖ The mathematical form of an adjusted estimate may be either a weighted average or a maximum likelihood estimate.
- ❖ The study design and purpose may call for either a ratio measure, such as a risk ratio, odds ratio, or rate ratio, or difference effect measure, such as a risk difference or rate difference.
- ❖ If some of the strata contain zero cells, the adjusted estimate must accommodate stratum-specific estimates that are undefined.
- ❖ Weighted average may involve either linear weighting or log-linear weighting, where the latter is used for ratio measures when the weights are precision-based.
- ❖ Mantel-Haenszel estimates are used when there is sparse (i.e., zero cell) data, but also have good statistical properties even when there are no zero cells.

Adjusted Estimates Using Weighting by Precision

One common way to find an **adjusted estimate of effect**, say an adjusted risk ratio, is to **weight the stratum-specific estimates** according to their **precision**. This approach is available in most computer programs that compute adjusted estimates. Mathematically, we define the precision of an estimate as one over the estimated variance of the estimate. We can also think of precision as a measure of how small the confidence interval around an estimate is. The smaller the confidence interval, the more precise the estimate.

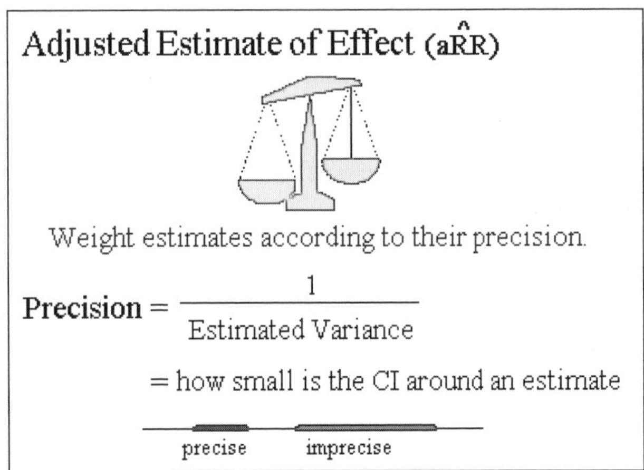

Adjusted Estimate of Effect ($a\hat{RR}$)
Weight estimates according to their precision.
Precision = $\frac{1}{\text{Estimated Variance}}$
= how small is the CI around an estimate
precise
imprecise

For two strata, the formula for the precision-based adjusted relative risk is a weighted average of the natural logs of the relative risks for each stratum, rather than a weighted average of the relative risk estimates themselves. Given suitable weights $\mathbf{w_1}$ and $\mathbf{w_2}$, this expression is computed by first obtaining a weighted average of the log of the stratum-specific risk ratios and then exponentiating this weighted average to get a value that is on a risk ratio scale.

Precision-based ($a\hat{RR}$) (2 strata)

$$a\hat{RR} = \exp\left[\frac{w_1 \times \ln(\hat{RR}_1) + w_2 \times \ln(\hat{RR}_2)}{w_1 + w_2}\right]$$

We take logs to simplify some of the underlying mathematics. The weights are precision-based and are obtained by computing the inverse of the estimated variance of the natural log of the stratum-specific risk ratio estimates. Using the two-way data layout for stratum **i**, the actual formula for the weights is shown below in terms of cell frequencies **a**, **b**, **c** and **d**, and the exposure totals $\mathbf{n_1}$ and $\mathbf{n_0}$ for stratum **i**.

$$w_i = \frac{1}{\widehat{Var}(\ln \widehat{RR}_i)} = \frac{1}{\left(\frac{c_i}{n_{1i}a_i}\right) + \left(\frac{d_i}{n_{0i}b_i}\right)}$$

stratum i	E	Not E	Total
D	a_i	b_i	m_{1i}
not D	c_i	d_i	m_{0i}
Total	n_{1i}	n_{0i}	n_i

Because the weights as well as the formula are complicated to compute, particularly when there are more than 2 strata, we recommend that you use a computer to carry out the calculations. For the retrospective cohort example involving two smoking history strata, the formula is shown here.

$$a\widehat{RR} = \exp\left[\frac{0.69 \times \ln(1.0) + 14.38 \times \ln(1.3)}{0.69 + 14.38}\right]$$

Retrospective cohort study of chemical workers

Stratum 1: non-smokers

	TCX	No TCX	Total
LC	1	2	3
no LC	24	48	72
Total	25	50	75

$\widehat{RR}_1 = 1.0$
$w_1 = 0.69$

Stratum 2: smokers

	TCX	No TCX	Total
LC	26	12	38
no LC	24	19	43
Total	50	31	81

$\widehat{RR}_2 = 1.3$
$w_2 = 14.38$

Our retrospective cohort example has wide discrepancies among the four cell frequencies of the table. In particular, for non-smokers one of the four cells has only one subject and another cell has only two subjects, but the other two cells have fairly large frequencies. We say that such a table is **unbalanced**. Using a computer program to do the computations, we find these following 95% confidence intervals for the estimated risk ratio in each table. The first table shows a much wider, less precise confidence interval than the second table. So more weight should be given to the second table than the first table in calculating the precision-based adjusted estimate.

$\widehat{RR}_1 = 1.0$	$\widehat{RR}_2 = 1.3$
$w_1 = 0.69$	$w_2 = 14.38$
95% CI:	95% CI:
(.095 , 10.506)	(.801 , 2.252)
less precise	more precise

$a\widehat{RR} = 1.3$

Using a computer program to carry out the calculations, we find that the adjusted estimate of the risk ratio rounds to 1.3. This value is the same as obtained for the risk ratio estimate for smokers. Thus, the adjusted risk ratio gives almost all the weight to smokers, whose risk ratio estimate is much more precise than the estimate obtained for the non-smokers.

When weighting by precision, the risk ratios from **balanced tables** are given more weight than risk ratios from **unbalanced tables** of the same total sample size. When calculating precision-based adjusted estimates, the weight used for each stratum is influenced not only by the total sample size, but also by the amount of balance in the four cells of the table.

<u>Summary</u>

- ❖ When computing adjusted estimates, the typical weights are based on precision.
- ❖ The smaller the variance, the higher the precision. Also, the narrower the confidence interval, the higher the precision.
- ❖ The precision-based formula for an aRR is obtained by exponentiating a weighted average of the log of the RR estimates.
- ❖ Each weight is the inverse of the variance of the natural log of a stratum-specific estimate.
- ❖ When weighting by precision, the risk ratios in balanced tables are given more weight than in unbalanced tables of the same total sample size.

<u>Quiz (Q14.9)</u>

Fill in the Blanks

1. When weighting by precision, which of the following is NOT a factor in determining the precision-based stratum specific weights? **???**

   A. The sample size
   B. The magnitude of the risk ratio
   C. The table balance
   D. The variance of the estimate

<u>True</u> or <u>False</u>

2. When weighting by precision, the larger the confidence interval is around the estimate, the larger the weight will be. . . . . **???**
3. When weighting by precision, if the stratum specific risk ratios are the same, then the strata weights will be the same. . . . . **???**
4. Unbalanced data sets typically have smaller weights than balanced data sets. . . **???**
5. The value of the adjusted risk ratio will lie between the two stratum-specific risk ratio estimates. **???**

General Formula for a Precision-Based Adjusted Risk Ratio

We can expand the **precision-based formula** for the adjusted risk ratio for two strata to a general precision-based formula for **G** strata. The general formula contains **G** weights. Each weight gives the estimated precision of the log of the estimated risk ratio for its stratum. We can rewrite this formula using standard summation notation. We calculate the weight times the natural log of the risk ratio for each stratum then sum these over all strata. Then we divide by the sum of the weights over all the strata and exponentiate the result:

General Precision-Based Formula

$$\hat{aRR} = \exp\left[\frac{\sum w_i \ln(RR_i)}{\sum w_i}\right]$$

Consider data with four strata, below are the four risk ratio estimates, their corresponding precision-based weights the table sample sizes, the crude risk ratio estimate and the corresponding sample size. Don't worry how to find these numbers, particularly the weights, a computer can do this for you easily. Notice that the ordering of the weights does not correspond directly to the ordering of the stratum-specific sample sizes:

Example with 4 strata

Stratum	1	2	3	4	crude
$\hat{RR}$	2.01	1.49	1.88	1.54	2.45
weight	1.075	2.4974	6.9473	4.4856	----
n	282	76	161	90	609

Study Question (Q14.10)

1. In the summary table, the largest weight is given to stratum 3, which has the second highest sample size, whereas the smallest weight is given to stratum 1, which has the highest sample size. How can this happen?

We use these weighted average formulas to compute the adjusted risk ratio by substituting the values for the **w's** and stratum-specific risk ratio estimates from the table into the formula. The resulting adjusted risk ratio is 1.71. This value is somewhat lower than the crude risk ratio estimate of 2.45. Thus, there is some confounding due to age. The adjusted risk ratio should be used instead of the crude risk ratio to summarize the effect of exposure on disease.

$$\hat{aRR} = \exp\left[\frac{w_1 \times \ln(\hat{RR}_1) + w_2 \times \ln(\hat{RR}_2) + w_3 \times \ln(\hat{RR}_3) + w_4 \times \ln(\hat{RR}_4)}{w_1 + w_2 + w_3 + w_4}\right]$$

$$= \exp\left[\frac{1.075 \times \ln(2.01) + 2.4974 \times \ln(1.49) + 6.9473 \times \ln(1.88) + 4.4856 \times \ln(1.54)}{1.075 + 2.4974 + 6.9473 + 4.4856}\right]$$

$$= 1.71$$

Summary

- When there are **G** strata, the precision-based formula for an adjusted risk ratio is a weighted average of the natural logs of the risk ratio estimates of the **G** strata.
- The weights are estimates of the precision of the logs of the stratum-specific risk ratio estimates.
- The formula should be calculated using a computer program.

The Precision-Based Adjusted Rate Ratio

For stratified person-time data, the measure of effect of typical interest is the rate ratio or IDR (i.e., incidence density ratio), The following formula is used to obtain a precision-based adjusted rate ratio (i.e., **aIDR**)

$$a\hat{IDR} = \frac{\sum_{i=1}^{k} w_i \ln(\hat{IDR}_i)}{\sum_{i=1}^{k} w_i}$$

The weights in this formula are approximate estimates of the inverse variance of the rate ratio in each stratum, and are given by the expression:

$$w_i = \frac{I_{1i} I_{0i}}{I_i}$$

where $\mathbf{I_{1i}}$, $\mathbf{I_{0i}}$, and $\mathbf{I_i}$ for $\mathbf{i} = 1, 2, \ldots, \mathbf{k}$ denote the disease frequencies among exposed, unexposed, total number of subjects, respectively, in the study.

Compute a Precision-Based Adjusted Risk Ratio (aRR) in DataDesk

An exercise is provided to demonstrate how to compute a precision-based adjusted risk ratio (aRR) using the DataDesk program.

Compute a Precision-Based Adjusted Rate Ratio (aIDR) in Data Desk

An exercise is provided to demonstrate how to compute a precision-based rate ratio (aIDR) using the DataDesk program.

14-4 Stratified Analysis (continued)

Precision-Based Adjusted Odds Ratio

Thus far we have considered stratified data only from cohort studies where the adjusted measure of effect is a **precision-based risk ratio** (**aRR**). We now describe how to compute a **precision-based adjusted odds ratio** (**aOR**) for stratified data from either case-control, cross-sectional, or cohort studies in which the odds ratio is the effect measure of interest. Consider a general two-way data layout for one of several strata. The precision-based adjusted odds ratio is the exponential of a weighted average of the natural log of the stratum-specific odds ratios. This formula applies whether the study design used calls for the **risk odds ratio**, the **exposure odds ratio**, or the **prevalence odds ratio**:

stratum i	Exposure Yes	Exposure No	Total
D	a_i	b_i	m_{1i}
not D	c_i	d_i	m_{0i}
Total	n_{1i}	n_{0i}	n_i

Precision-based **aOR**

$$\widehat{\mathbf{a}OR} = \exp\left[\frac{\sum w_i \times \ln(\widehat{OR}_i)}{\sum w_i}\right]$$

Below is the formula for the weights, which is different for the adjusted odds ratio than for the adjusted risk ratio, because precision is computed differently for different effect measures.

$$w_i = \frac{1}{\frac{1}{a_i} + \frac{1}{b_i} + \frac{1}{c_i} + \frac{1}{d_i}}$$

Let's focus on the weights for the adjusted odds ratio. If all the cell frequencies for a given stratum are reasonably large, then the denominator will be small, so its reciprocal, which gives the precision, will be relatively large.

stratum i	Exposure Yes	Exposure No	Total
D	100	100	m_{1i}
not D	100	100	m_{0i}
Total	n_{1i}	n_{0i}	400

Precision-based **aOR**

$$\widehat{\mathbf{a}OR} = \exp\left[\frac{\sum w_i \times \ln(\widehat{OR}_i)}{\sum w_i}\right]$$

$$w_i = \frac{1}{\frac{1}{100} + \frac{1}{100} + \frac{1}{100} + \frac{1}{100}} = 25$$

On the other hand, if one of the cell frequencies is very small, the precision will tend to be relatively small. This explains, at least for the odds ratio, why an unbalanced stratum with at least one small cell frequency is likely to yield a relatively small weight even if the total stratum size is large.

stratum i	Exposure Yes	No	Total
D	1	199	m_{1i}
not D	100	100	m_{0i}
Total	n_{1i}	n_{0i}	400

Precision-based **aOR**

$$a\hat{OR} = \exp\left[\frac{\sum w_i \times \ln(\hat{OR}_i)}{\sum w_i}\right]$$

$$w_i = \frac{1}{\frac{1}{1} + \frac{1}{199} + \frac{1}{100} + \frac{1}{100}} = 0.98$$

Summary

- ❖ The measure of association used to assess confounding will depend on the study design.
- ❖ The formula for the adjusted risk ratio applies when the study design used calls for the prevalence ratio
- ❖ The formula for the adjusted odds ratio applies whether the study design used calls for the risk odds ratio, the exposure odds ratio, or prevalence odds ratio.
- ❖ The weights are computed differently for the adjusted odds ratio than for the adjusted risk ratio.
- ❖ An unbalanced stratum with at least one small cell frequency is likely to yield a relatively small weight even if the total stratum size is large.

Computing the Adjusted OR – An Example

These tables show results from a case-control study to assess the relationship of alcohol consumption to oral cancer. The tables give the crude data when age is ignored and the stratified data when age has been categorized into three groups.

An Example: Case-control Study

Alcohol consumption → oral cancer

Crude:

	Alc	no Alc	
OCa	27	47	
no OCa	90	443	n = 607

Stratified:

	Ages 40-49 Alc	no Alc	Ages 50-59 Alc	no Alc	Ages 60+ Alc	no Alc
OCa	4	25	12	10	11	12
no OCa	22	309	37	67	31	67
	n = 360		n = 126		n = 131	

Below is the formula for the adjusted odds ratio for these data. Now, given that the weights for each strata are as follows:

$\hat{OR}$ = 2.8 (crude)

$\hat{OR}$ = 2.2, 2.2, 2.0 (stratum specific)

w_1=2.95
w_2=4.44
w_3=4.52

$$a\hat{OR} = \exp\left[\frac{w_1 \ln(\hat{OR}_1) + w_2 \ln(\hat{OR}_2) + w_3 \ln(\hat{OR}_3)}{w_1 + w_2 + w_3}\right]$$

1. What is the estimated adjusted odds ratio for these data? (Hint: Try to answer without calculations.)

2.1	2.3	1.9

The estimated odds ratio is 2.1.

2. Are the crude and adjusted estimates meaningfully different?

yes	no

Yes, the crude ratio of 2.8 indicates there is almost a three-fold excess risk but the adjusted estimate of 2.1 indicates approximately a two-fold excess risk.

3. Which estimate is more appropriate, the crude or the adjusted estimate?

crude	adjusted

The adjusted estimate is more appropriate because it is meaningfully different from the crude estimate and controls for the confounding due to age.

Quiz (Q14.11)

The data below are from a cross-sectional seroprevalence survey of HIV among prostitutes in relation to IV drug use. The crude prevalence odds ratio is 3.59. (You may wish to use a calculator to answer the following questions.)

	Black or Hispanic			White	
	IV Drug Use	No IV Drug Use		IV Drug Use	No IV Drug Use
HIV +	31	12	HIV +	16	3
HIV -	93	144	HIV -	141	124

1. What is the estimated POR among the Black or Hispanic group? . . . . . . ???
2. What is the estimated POR among the Whites? . . . . . . ???
3. Which table do you think is more balanced and thus will yield the highest precision-based weight? . ???

Choices
3.25 3.59 4.00 4.31 4.69 Black or Hispanic White

In the study described in the previous question, the estimated POR for the Black or Hispanic group was 4.00 and the estimated POR for the Whites was 4.69. The precision-based weight for the Black or Hispanic group is calculated to be 7.503 and the weight for the whites is 2.433.

4. Using the formula below, calculate the adjusted POR for this study. . . . . . ???

5. Recall that the crude POR was 3.59. Does this provide evidence of confounding? . . . **???**

$$a\hat{POR} = \exp\left[\frac{\sum w_i \times \ln(\hat{OR}_i)}{\sum w_i}\right]$$

Choices

1.00 **4.16** **4.35** **4.97** **debatable** **no** **yes**

14-5 Stratified Analysis (continued)

Mantel-Haenszel Adjusted Estimates

The Zero-Cell Problem

Let's consider the following set of results that might have occurred in a case control study relating an exposure variable **E** to a disease D:

Hypothetical Results: Case-Control Study

E → D

Crude Data	E	not E	Total
D	10	11	21
Not D	1	14	15
Total	11	25	36

$\hat{OR} = 12.7$

Stratum 1	E	not E	Total
D	5	6	11
Not D	0	5	5
Total	5	11	16

$\hat{OR}$ =undefined

Stratum 2	E	not E	Total
D	5	5	10
Not D	1	9	10
Total	6	14	20

$\hat{OR} = 9.0$

The odds ratio for stratum 1 is undefined because of the zero cell frequency in this stratum. So we cannot say whether there is either confounding or interaction within these data. One approach sometimes taken to resolve such a problem is to add a small number, usually .5, to each cell of any table with a zero cell. If we add .5 here, the odds ratio for this modified table is 9.3.

Stratum 1	E	not E	Total
D	5.5	6.5	12.0
Not D	.5	5.5	6.0
Total	6.0	12.0	18.0

Stratum 2	E	not E	Total
D	5	5	10
Not D	1	9	10
Total	6	14	20

$\hat{OR}$ =undefined

$\hat{OR} = 9.3$ ← **Interaction? NO** → $\hat{OR} = 9.0$

Now it appears that there is little evidence of interaction since the two stratum-specific odds ratios are very close. But, there is evidence of some confounding because the crude odds ratio is somewhat higher than either stratum specific odds ratios. Although this approach to the zero-cell problem is reasonable, and is often used, we might be concerned that the choice of .5 is arbitrary.

Study Questions (Q14.12)

Stratum 1	E	not E	Total
D	5	6	11
Not D	0	5	5
Total	5	11	16

1. Given the values a = 5, b = 6, c = 0, and d = 5 for stratum 1, what is the estimated odds ratio if 0.1 is added to each cell frequency?
2. Given the values a = 5, b = 6, c = 0, and d = 5 for stratum 1, what is the estimated odds ratio if 1.0 is added to each cell frequency?
3. What do your answers to the above questions say about the use of adding a small number to all cells in a stratum containing a zero cell frequency?

We might also be concerned about computing a precision-based adjusted odds ratio that involves a questionably modified stratum-specific odds ratio.

Study Questions (Q14.12) continued

The stratum-specific odds ratios obtained with .5 is added to stratum 1 are 9.31 for stratum 1 and 9.00 for stratum 2.

4. If a precision-based adjusted odds ratio is computed by adding .5 to each cell frequency in stratum 1 the weights are .3972 for stratum 1 and .6618 for stratum 2. What value do you obtain for the estimated aOR?

The stratum-specific odds ratios obtained with .1 is added to stratum 1 are 41.64 for stratum 1 and 9.00 for stratum 2.

5. If a precision-based adjusted odds ratio is computed by adding .1 to each cell frequency in stratum 1 the weights are .0947 for stratum 1 and .6618 for stratum 2. What value do you obtain for the estimated aOR?
6. How do your answers to the previous two questions compare?
7. What do your answers to the previous questions say about the use of a precision-based adjusted odds ratio?

Fortunately, there is an alternative form of adjusted estimate to deal with the zero-cell problem called the **Mantel-Haenszel odds ratio**, which we describe in the next activity.

Summary

- When there are sparse data in some strata, particularly zero cells, stratum-specific odds ratios become unreliable and possibly undefined.
- One approach when there are zero cell frequencies in a stratum is to add a small number, typically 0.5, to each cell frequency in the stratum.
- A drawback to the latter approach is that the resulting modified stratum-specific effect estimate may radically change depending on the small number (e.g., .5, .1) that is added.
- The use of a precision-based adjusted estimate in such a situation then becomes problematic.
- Fortunately, there is an alternative approach to the zero-cell problem, which involves using what is called a Mantel-Haenszel adjusted estimate.

The Mantel-Haenszel Odds Ratio

The **Mantel-Haenszel adjusted estimate** used in case-control studies is called the **Mantel-Haenszel odds ratio**, or the **mOR**. Here is its formula:

Mantel-Haenszel odds ratio (mOR)

$$mOR = \frac{\sum_{i=1}^{G} \frac{a_i d_i}{n_i}}{\sum_{i=1}^{G} \frac{b_i c_i}{n_i}}$$

G = # of strata

This formula can also be used to compute an adjusted odds ratio in cross-sectional and cohort studies. A key feature of the **mOR** is that it can be used without having to modify any stratum that contains a zero-cell frequency. For example, if there is a zero-cell frequency as shown below for the **c**-cell, then the computation simply includes a zero in a sum in the denominator of the **mOR**, but the total sum will not necessarily be zero.

Mantel-Haenszel odds ratio (mOR)

$$mOR = \frac{\sum_{i=1}^{G} \frac{a_i d_i}{n_i}}{\sum_{i=1}^{G-1} \frac{b_i c_i}{n_i} + 0}$$

G = # of strata

	E	$\bar{E}$	
D	a_G	b_G	m_{1G}
$\bar{D}$	$c_G = 0$	d_G	m_{0G}
	n_{1G}	n_{0G}	n_G

Study Questions (Q14.13)

1. Suppose there are G=5 strata and that either the *b*-cell or *c*-cell is zero in each and every strata. What will happen if you compute the mOR for these data?
2. Suppose there are G=5 strata and that either the *a*-cell or *d*-cell is zero in each and every strata. What will happen if you compute the mOR for these data?
3. What do your answers to the above questions say about using the mOR when there are zero cell frequencies?

Another nice feature of the **mOR** is that, even though it doesn't look it, the **mOR** can be written as a weighted average of stratum-specific odds ratios provided there are no zero cells in any strata. So the **mOR** will give a value somewhere between the minimum and maximum-specific odds ratios over all strata, as will any weighted average.

$$mOR = \frac{\sum_{i=1}^{G} \frac{a_i d_i}{n_i}}{\sum_{i=1}^{G} \frac{b_i c_i}{n_i}} = \frac{\sum_{i=1}^{G} \frac{b_i c_i}{n_i} \hat{OR}_i}{\sum_{i=1}^{G} \frac{b_i c_i}{n_i}}$$

* The mOR can be equivalently written as a weighted average, provided NO ZERO CELLS.

Still another feature of the **mOR** is that it equals 1 only when the Mantel-Haenszel chi square statistic equals zero. It is possible for the precision-based aOR to be different from 1 even if the Mantel-Haenszel chi square statistic is exactly equal to zero. In general, the **mOR** has been shown to have good statistical properties, particularly when used with matched case-control data. So it is often used instead of the precision-based **aOR** even when there are no zero-cells or the data are not sparse.

We now apply the mOR formula to the stratified data example we have previously considered. Substituting the cell-frequencies from each stratum into the **mOR** formula, the estimated **mOR** turns out to be 15.25:

Case-Control Data

Crude Data	E	not E	Total
D	10	11	21
Not D	1	14	15
Total	11	25	36

$\hat{OR} = 12.7$

Stratum 1	E	not E	Total
D	5	6	11
Not D	0	5	5
Total	5	11	16

$\hat{OR} = \text{undefined}$

Stratum 2	E	not E	Total
D	5	5	10
Not D	1	9	10
Total	6	14	20

$\hat{OR} = 9.0$

$$\hat{mOR} = \frac{\frac{5\times 5}{16} + \frac{5\times 9}{20}}{\frac{6\times 0}{16} + \frac{5\times 1}{20}} = 15.25$$

This adjusted estimate is somewhat higher than the crude odds ratio of 12.7 and much higher than the odds ratio of 9.0 in stratum 2. Because stratum 1 has an undefined odds ratio, we cannot say whether there is evidence of interaction. However, because the crude and adjusted odds ratios differ, there is evidence of confounding.

Study Questions (Q14.13) continued

If a precision-based adjusted odds ratio is computed by adding .5 to each cell in stratum 1, the aOR that is obtained is 9.11. If, instead, .1 is added to each cell frequency in stratum 1, the aOR is 10.65.

4. Compare the aOR results above with the previously obtained mOR of 15.25. Which estimate do you prefer?

Summary

- For case-control studies as well as other studies involving the odds ratio, an alternative to a precision-based adjusted odds ratio is the Mantel-Haenszel odds ratio (mOR)
- Corresponding to the mOR, the mRR or mIDR can be used in cohort studies.
- A key feature of the mOR is that it may be used without modification when there are zero cells in some of the strata.
- The mOR can also be written as a weighted average of stratum-specific odds ratios **provided** there are no zero cells in any strata.
- The mOR equals 1 only when the Mantel-Haenszel chi square statistic equals zero.
- The mOR has been shown to have good statistical properties, particularly when used with matched case-control data.

Formulae for Mantel-Haenszel Risk Ratio and Rate Ratio

The Mantel-Haenszel risk ratio (i.e., mRR) is given by the following formula:

$$m\hat{R}R = \frac{\sum_{i=1}^{k} \frac{a_i n_{0i}}{n_i}}{\sum_{i=1}^{k} \frac{b_i n_{1i}}{n_i}}$$

where the quantities in the formula come from the cell frequencies in the 2x2 table of the *i*-th stratum for a cumulative incidence cohort study:

Data Layout for Stratum i

	E	not E	
D	a_i	b_i	m_{1i}
not D	c_i	d_i	m_{0i}
	n_{1i}	n_{0i}	n_i

The Mantel-Haenszel rate ratio (i.e., mIDR) is given by the following formula:

$$mI\hat{D}R = \frac{\sum_{i=1}^{k} \frac{I_{1i} PT_{0i}}{PT_i}}{\sum_{i=1}^{k} \frac{I_{0i} PT_{1i}}{PT_i}}$$

where the quantities in the formula come from the cell frequencies in the 2x2 table of the *i*-th stratum for a person-time study:

Person-Time Cohort Study
Data Layout for Stratum i

	E	not E	
D	I_{1i}	I_{0i}	I_i
PT	PT_{1i}	PT_{0i}	PT_i

Compute Adjusted Odds Ratios (aOR and mOR) in DataDesk

An exercise is provided to demonstrate how to compute adjusted odds ratios using the DataDesk program.

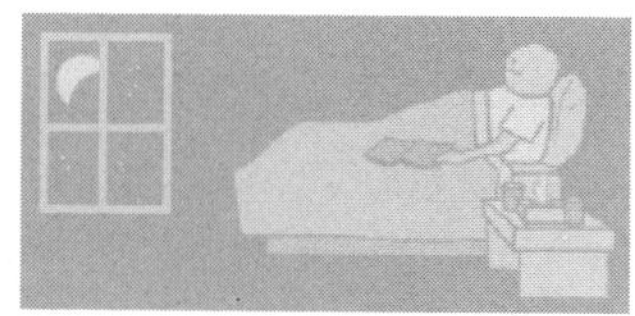

Quiz (Q14.14)

True or **False**

1. The practice of adding a small number to each of the cells of a two by two table to eliminate a zero cell is arbitrary and should be used with caution. . . . . . . ???
2. It is the adjusted estimate that is most affected by adding a small number to each cell rather than the stratum specific estimates. . . . . ???
3. When there are zero cell frequencies, the use of Mantel-Haenszel adjusted estimates should be preferred since they can usually be calculated without adjustment. . . . . . ???

Interval Estimation

Interval Estimation - Introduction

We now describe how to obtain a large-sample confidence interval around an adjusted estimate obtained in a stratified analysis. This interval estimate can take one of the two forms shown here:

Large-sample Confidence Interval for an Adjusted Estimate

Difference measures: $\hat{\theta} \pm Z_p\sqrt{\hat{Var}(\hat{\theta})}$

Ratio measures: $\hat{\theta}\exp\left[\pm Z_p\sqrt{\hat{Var}(\ln\hat{\theta})}\right]$

The **Z** in each expression denotes a percentage point of the standard normal distribution. The **θ** ("theta") in each expression denotes the effect measure of interest. It can be either a difference measure, such as risk difference, or a ratio measure, such as a risk ratio. Typically, **θ** is a weighted average of stratum specific effects. In particular, for **risk difference measures**, **θ** will have a linear weighting as shown here:

Linear-weighting:

$$a\hat{RD} = \frac{\sum w_i \hat{RD}_i}{\sum w_i}$$

where $\hat{RD}_i$ is the estimated risk difference in stratum i

Mantel-Haenszel adjusted estimates for ratio effect measures also have linear weighting.

Linear-weighting:

$$m\hat{OR} = \frac{\sum w_i \hat{OR}_i}{\sum w_i}$$

provided $\hat{OR}_i$ is defined for each stratum

For **precision-based ratio estimates**, **θ** will have log-linear weighting.

Log Linear-weighting:

$$\theta = \exp\left[\frac{\sum w_i \ln(\theta_i)}{\sum w_i}\right] \quad \text{e.g.,} \quad a\hat{OR} = \exp\left[\frac{\sum w_i \ln(\hat{OR}_i)}{\sum w_i}\right]$$

The variance component within the confidence interval will take on a specific mathematical form depending on the effect measure. For **precision-based measures**, the variance component conveniently simplifies into expressions that involve the sum of weights. In particular, for **precision-based difference measures**, the confidence interval formula reduces to form shown here:

$$\hat{\theta} \pm \frac{1.96}{\sqrt{\sum w_i}}$$

For **precision-based ratio measures**, the formula is written this way:

$$\hat{\theta}\, e^{\frac{\pm 1.96}{\sqrt{\sum w_i}}}$$

As an example, we again consider cohort data involving four strata. We have four risk ratio estimates, their corresponding precision-based weights and sample sizes, the crude risk ratio estimate, and corresponding sample size.

Example with 4 strata

Stratum	1	2	3	4	crude
$\hat{RR}$	2.01	1.49	1.88	1.54	2.45
w_i	1.0751	2.4974	6.9473	4.4856	----
n	282	76	161	90	609

$$a\hat{RR} = 1.71$$

The adjusted risk ratio turns out to be 1.71. To obtain the 95% precision-based interval estimate for this adjusted risk ratio, we start with the formula shown here:

$$\text{95\% CI: } a\hat{RR}\, e^{\frac{\pm 1.96}{\sqrt{(w_1 + w_2 + w_3 + w_4)}}}$$

We then substitute into the formula 1.71 for $a\hat{RR}$ and the values shown in the table for the four weights:

$$\text{95\% CI: } 1.71\, e^{\frac{\pm 1.96}{\sqrt{(1.0751+2.4974+6.9473+4.4856)}}}$$

The lower and upper limits of the confidence interval then turn out to be 1.03 and 2.84, respectively

$$\text{95\% CI: } 1.71\, e^{\frac{\pm 1.96}{\sqrt{15.0054}}} = (1.03, 2.84)$$

Study Questions (Q14.15)

1. How would you interpret the above confidence interval?

Risk Difference example with 4 strata

Stratum	1	2	3	4	crude
$\hat{RD}$	0.0630	0.0578	0.1078	0.0851	0.1310
w_i	71.9424	96.8891	183.9742	137.4186	----
n	282	76	161	90	609

$\hat{aRD} = 0.0850 \qquad \hat{aRD} \pm \frac{1.96}{\sqrt{\sum w_i}}$

2. Based on the information above, calculate a 95% confidence interval for the adjusted risk difference. (You will need to use a calculator to carry out this computation.)
3. How do you interpret this confidence interval?

Summary

- A large-sample interval estimate around an adjusted estimate can take one of the two forms:

  $\hat{\theta} \pm Z_p \sqrt{\hat{Var}(\hat{\theta})}$ for difference effect measures and

  $\hat{\theta} \exp\left[\pm Z_p \sqrt{\hat{Var}(\ln\hat{\theta})}\right]$ for ratio effect measures

- For risk difference measures and for Mantel-Haenszel estimates, θ will have linear weighting.
- For precision-based ratio estimates, θ will have log-linear weighting.
- For precision-based measures, the variance component involves the sum of weights as follows:

  $\hat{\theta} \pm \frac{Z_p}{\sqrt{\sum w_i}}$ for difference measures

  $\hat{\theta} \exp\left[\pm \frac{Z_p}{\sqrt{\sum w_i}}\right]$ for ratio measures

Interval Estimation for the mOR

Consider again the case-control stratified data involving sparse strata that we previously used to compute a **Mantel-Haenszel odds ratio**. For these data, the estimated Mantel-Haenszel odds ratio is 15.25.

Stratified Data with Sparse Strata (mOR)

Case-Control Data

Crude Data	E	not E	Total
D	10	11	21
not D	1	14	15
Total	11	25	36

$\hat{OR} = 12.7$

Stratum 1	E	not E	Total
D	5	6	11
not D	0	5	5
Total	5	11	16

$\hat{OR} = \text{undefined}$

Stratum 2	E	not E	Total
D	5	5	10
not D	1	9	10
Total	6	14	20

$\hat{OR} = 9.0$

$m\hat{OR} = 15.25$

We find a 95% confidence interval around this estimate with the formula shown here:

95% confidence interval:

$$m\hat{OR} \exp\left[\pm 1.96 \sqrt{\left(\hat{Var}(\ln m\hat{OR})\right)}\right]$$

The variance term in this formula is a complex expression involving the frequencies in each stratum. We present the variance formula here primarily to show you how complex it is. Use a computer program to do the actual calculations, which we will do for you here.

$$\hat{Var}(\ln m\hat{OR}) = A + B + C$$

$$A = \frac{[\Sigma(P_i R_i)]}{2[\Sigma(R_i)]^2} \qquad P_i = \frac{(a_i + d_i)}{n_i}$$

$$B = \frac{[\Sigma(P_i S_i + Q_i R_i)]}{2[\Sigma(R_i)\Sigma(S_i)]} \qquad Q_i = \frac{(b_i + c_i)}{n_i}$$

$$C = \frac{[\Sigma(Q_i S_i)]}{2[\Sigma(S_i)]^2} \qquad R_i = \frac{a_i d_i}{n_i} \qquad S_i = \frac{b_i c_i}{n_i}$$

Substituting the frequencies in each stratum into the formulae for $\mathbf{P_i}$, $\mathbf{Q_i}$, $\mathbf{R_i}$, and $\mathbf{S_i}$, we obtain the values below. The estimated variance is shown here:

$$\hat{Var}(\ln m\hat{OR}) = 0.0878 + 0.7533 + 0.6000 = 1.4411$$

$$A = \frac{[\Sigma(P_i R_i)]}{2[\Sigma(R_i)]^2} = \frac{2.5516}{2(3.8125)^2} = .0878$$

$$B = \frac{[\Sigma(P_i S_i + Q_i R_i)]}{2[\Sigma(R_i)\Sigma(S_i)]} = \frac{1.4359}{2(3.8125)(.2500)} = .7533$$

$$C = \frac{[\Sigma(Q_i S_i)]}{2[\Sigma(S_i)]^2} = \frac{.0750}{2(.2500)^2} = .6000$$

The 95% confidence interval is shown on the next page. Although this interval does not contain the null value of 1, it is nevertheless extremely wide, which should not be surprising given the sparse strata.

95% Confidence Interval $m\hat{O}R = 15.25$

$$= m\hat{O}R \exp\left[\pm 1.96\sqrt{\hat{Var}(\ln m\hat{O}R)}\right]$$

$$= 15.25 \exp\left[\pm 1.96\sqrt{1.4411}\right]$$

$$= 15.25 \exp\left[\pm 2.3529\right]$$

$$= (1.4502, 160.3693)$$

Summary

- A 95% confidence interval around a Mantel-Haenszel odds ratio is given by the formula:

$$m\hat{O}R \exp\left[\pm 1.96\sqrt{\hat{Var}(\ln m\hat{O}R)}\right]$$

- The variance term in this formula is a complex expression involving the frequencies in each stratum.
- You should use a computer to calculate the variance term in the formula.
- You should not be surprised to find a very wide confidence interval if all strata are sparse.

Large-sample Interval Estimation of the Mantel-Haenszel Risk Ratio and Mantel-Haenszel Rate Ratio

A large-sample 95% confidence interval for the Mantel-Haenszel risk ratio (i.e., rnRR) is given by the following formula;

$$m\hat{R}R \exp\left[\pm 1.96\sqrt{\hat{Var}(\ln m\hat{R}R)}\right]$$

where

$$m\hat{R}R = \frac{\sum_{i=1}^{k} \frac{a_i n_{0i}}{n_i}}{\sum_{i=1}^{k} \frac{b_i n_{1i}}{n_i}}$$

and the variance expression in the formula is given by

$$\hat{Var}(\ln m\hat{R}R) = \frac{\sum_{i=1}^{k}\left(\frac{m_{1i} n_{1i} n_{0i}}{n_i^2} - \frac{a_i b_i}{n_i}\right)}{\left(\sum_{i=1}^{k} \frac{a_i n_{0i}}{n_i}\right)\left(\sum_{i=1}^{k} \frac{b_i n_{1i}}{n_i}\right)}$$

The quantities in the variance formula come from the cell frequencies in the 2x2 table of the *i*-th stratum for a cumulative-incidence cohort study:

Continued on next page

Large-sample Interval Estimation of the Mantel-Haenszel Risk Ratio and Mantel-Haenszel Rate Ratio (continued)

Data Layout for Stratum *i*

	E	not E	
D	a_i	b_i	m_{1i}
not D	c_i	d_i	m_{0i}
	n_{1i}	n_{0i}	n_i

A large-sample 95% confidence interval for the Mantel-Haenszel Rate Ratio (i.e., mlDR) is given by the following formula:

$$m\hat{ID}R \exp\left[\pm 1.96\sqrt{\hat{Var}(\ln m\hat{ID}R)}\right]$$

where

$$m\hat{ID}R = \frac{\displaystyle\sum_{i=1}^{k} \frac{I_{1i}PT_{0i}}{PT_i}}{\displaystyle\sum_{i=1}^{k} \frac{I_{0i}PT_{1i}}{PT_i}}$$

and the variance expression in the formula is given by:

$$\hat{Var}(\ln m\hat{ID}R) = \frac{\displaystyle\sum_{i=1}^{k} \frac{I_i PT_{1i} PT_{0i}}{PT_i^2}}{\left(\displaystyle\sum_{i=1}^{k} \frac{I_{1i}PT_{0i}}{PT_i}\right)\left(\displaystyle\sum_{i=1}^{k} \frac{I_{0i}PT_{1i}}{PT_i}\right)}$$

The quantities in the variance formula come from the cell frequencies in the 2x2 table of the *i*-th stratum for a person-time cohort study:

Person-Time Cohort Study
Data Layout for Stratum i

	E	not E	
D	I_{1i}	I_{0i}	I_i
PT	PT_{1i}	PT_{0i}	PT_i

Compute a Mantel-Haenszel Odds Ratios (mOR) and a Large-Sample Confidence Interval for the mOR in DataDesk

An exercise is provided to demonstrate how to compute the Mantel-Haenszel odds ratio with a large sample confidence interval using the DataDesk program.

Quiz (Q14.16)

A case-control study was conducted to determine the relationship between smoking and lung cancer. The data were stratified into 3 age categories. The results were: aOR = 4.51, and the sum of the weights = 3.44.

1. Calculate a precision based 95% confidence interval for these data.. . . . ???
2. Do these results provide significant evidence that smoking is related to lung cancer when controlling for age? . . . . ???

Choices
1.25, 10.78   1.57, 12.98   no   yes

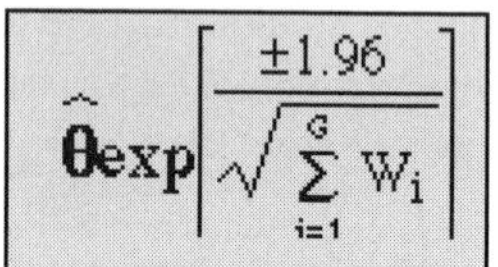

14-6 Stratified Analysis (continued)

Extensions to More Than 2 Exposure Categories

2xC Tables

A natural extension of stratified analysis for 2x2 tables occurs when there are more than two categories of exposure. In such a case, the basic data layout is in the form of a 2xC table where C denotes the number of exposure categories. We now provide an overview of how to analyze stratified 2xC tables.

The tables shown here give the cholesterol level and the coronary heart disease status for 609 white males within two age categories from a 10-year cohort study in Evans County Georgia from 1960 to 1969.

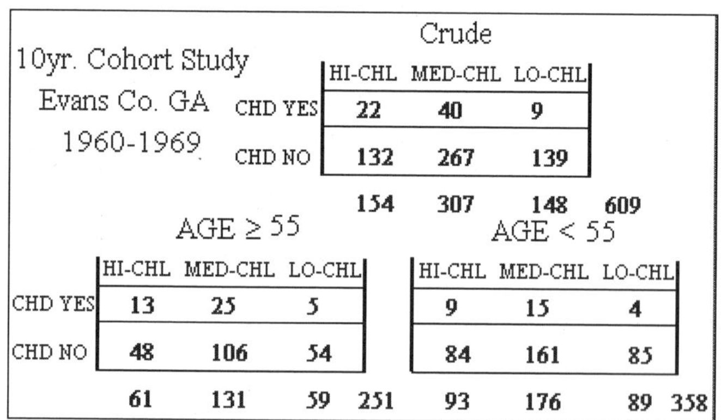

10yr. Cohort Study
Evans Co. GA
1960-1969

Crude

	HI-CHL	MED-CHL	LO-CHL	
CHD YES	22	40	9	
CHD NO	132	267	139	
	154	307	148	609

AGE ≥ 55

	HI-CHL	MED-CHL	LO-CHL	
CHD YES	13	25	5	
CHD NO	48	106	54	
	61	131	59	251

AGE < 55

	HI-CHL	MED-CHL	LO-CHL	
CHD YES	9	15	4	
CHD NO	84	161	85	
	93	176	89	358

These data show three rather than two categories of exposure for each of the two strata. How do we carry out a stratified analysis of such data? We typically carry out stratum-specific analyses and overall assessment, if appropriate, of the exposure-disease relationship over all strata. But because there are three exposure categories, we may wonder how to compute the multiplicity of effect measures possible for each table, how to summarize such information over all strata, and how to modify the hypothesis testing procedure for more than 2 categories.

Typically we compute several effect measures, each of which compares one of the exposure categories to a referent exposure category. In general, if there are **C** categories of exposure, the basic data layout is a **2xC** table. The typical analysis then produces C-1 adjusted estimates, comparing **C-1** exposure categories to the referent category.

	E_1	E_2	E_3	...	E_c
D					
not D					

Adjusted Estimates:

$a\hat{R}R_1\ a\hat{R}R_2\ a\hat{R}R_3 \ldots a\hat{R}R_{c-1}$

In our example, we will designate **low cholesterol** to be the referent category. Because we have cumulative-incidence cohort data, we compute two risk ratios per stratum, one comparing the **High Cholesterol** category to the **Low Cholesterol** Category and the other comparing the **Medium Cholesterol** category to the **Low Cholesterol** category. Below are these estimates for each age group, and precision-based adjusted estimates over both age groups. These results indicate a slight dose-response effect of cholesterol on CHD risk. That is, the effect, as measured by the risk ratio, decreases as the index group changes from high cholesterol to medium cholesterol, when each group is compared to the low cholesterol group.

	AGE ≥ 55				AGE < 55			
	HI-CHL	MED-CHL	LO-CHL		HI-CHL	MED-CHL	LO-CHL	
CHD YES	13	25	5		9	15	4	
CHD NO	48	106	54		84	161	85	
	61	131	59	251	93	176	89	358

AGE ≥ 55: $\hat{RR}$ (HI vs. LO) = 2.51; $\hat{RR}$ (MED vs. LO) = 2.25

AGE < 55: $\hat{RR}$ (HI vs. LO) = 2.15; $\hat{RR}$ (MED vs. LO) = 1.90

$a\hat{RR}$ (HI vs. LO) = 2.29

$a\hat{RR}$ (MED vs. LO) = 2.06

95% confidence intervals for both stratum specific and adjusted odds ratios are shown below. The intervals for adjusted risk ratios are obtained using the previously described formula involving the sum of the weights.

AGE ≥ 55	AGE < 55
$\hat{RR}$ (HI vs. LO) = 2.51 (0.95, 6.60)	$\hat{RR}$ (HI vs. LO) = 2.15 (0.69, 6.73)
$\hat{RR}$ (MED vs. LO) = 2.25 (0.91, 5.59)	$\hat{RR}$ (MED vs. LO) = 1.90 (0.65, 5.56)

$a\hat{RR}$ (HI vs. LO) = 2.29 (1.12, 4.92)

$a\hat{RR}$ (MED vs. LO) = 2.06 (1.04, 4.20)

$c\hat{RR}$ (HI vs. LO) = 2.35 (1.12, 4.94)

$c\hat{RR}$ (MED vs. LO) = 2.14 (1.07, 4.29)

Study Questions (Q14.17)

1. Based on the information provided above, is there evidence that age is an effect modifier of the relationship between cholesterol level and CHD risk?
2. Based on your answer to the previous question, is it appropriate to carry out overall assessment in this stratified analysis?
3. Is there evidence of confounding due to age?

Summary

- ❖ When there are more than two categories of exposure, the basic data layout is in the form of a **2xC** table where **C** denotes the number of exposure categories.
- ❖ As with stratified 2x2 tables, the goal of overall assessment is an overall adjusted estimate, and overall test of hypothesis, and an interval estimate around the adjusted estimate.
- ❖ When there are **C** exposure categories, the typical analysis produces **C=1** adjusted estimates, which compare **C-1** exposure categories to a referent category.

Test for Trend

We now describe how to test hypotheses for stratified data with several exposure categories. We again consider the Evans County data shown here relating cholesterol level to the development of coronary heart disease stratified by age.

Testing Hypotheses- Several Exposure Categories

Crude

	HI-CHL	MED-CHL	LO-CHL	
CHD YES	22	40	9	
CHD NO	132	267	139	
	154	307	148	609

AGE ≥ 55

	HI-CHL	MED-CHL	LO-CHL	
CHD YES	13	25	5	
CHD NO	48	106	54	
	61	131	59	251
Risk	.21	> .19	> .08	

e.g., 13/61 = .21

AGE < 55

	HI-CHL	MED-CHL	LO-CHL	
CHD YES	9	15	4	
CHD NO	84	161	85	
	93	176	89	358
Risk	.10	> .09	> .04	

The exposure variable, cholesterol level, has been categorized into three ordered categories. We can see that, for each stratum, the CHD risk decreases as the cholesterol level decreases.

To see whether these stratum specific results are statistically significant, we must perform a **test for trend**. Such a test allows us to evaluate whether or not there is a significant **dose-response relationship** between the exposure variable and the health outcome. The test for trend can be performed using an **extension of the Mantel-Haenszel test** procedure. This test requires that a numeric value or **score** be assigned to each category of exposure. For example, the three ordered categories of exposure could be assigned scores of 2 when cholesterol is greater than 233, 1 when cholesterol is between 184 and 233, and 0 if cholesterol is below 184.

$S_1 = 2$ if CHL > 233
$S_2 = 1$ if CHL 184 - 233
$S_3 = 0$ if CHL < 184

Alternatively, the scores might be determined by the mean cholesterol value in each of the ordered categories. For these data, then, the scores turn out to be 265.0, 207.8 and 164.4 for the high, medium, and low cholesterol categories, respectively.

$S_1 = 265.0$ if CHL > 233
$S_2 = 207.8$ if CHL 184 - 233
$S_3 = 164.4$ if CHL < 184

The general formula for the test statistic is shown here:

$$\chi^2_{MHT} = \frac{\left[\sum_{i=1}^{G} T_i - E(T_i)\right]^2}{\sum_{i=1}^{G} V(T_i)}$$

G = # of strata

T_i = total score for stratum i

$E(T_i)$ = expected total score for stratum i

$V(T_i)$ = variance of total score for stratum i

The number of strata is **G**. Typically this formula should be calculated using a computer. In our example, there are two age strata, so G=2. The trend test formula then simplifies as shown here (the box at the end of this activity shows how to perform the calculations):

$$\chi^2 = \frac{\left[T_1 - E(T_1) + T_2 - E(T_2)\right]^2}{V(T_1) + V(T_2)}$$

We will use this simplified formula to compute the test for trend where we will use as our scores rounded, mean-cholesterol values for each cholesterol category. Here are the results:

$$\chi^2 = \frac{\left[9465 - 9094.9283 + 6161 - 5932.3240\right]^2}{44088.6783 + 33914.7168} = 4.56$$

(P = 0.0320)

	AGE ≥ 55				AGE < 55			
	HI-CHL	MED-CHL	LO-CHL		HI-CHL	MED-CHL	LO-CHL	
CHD YES	13	25	5		9	15	4	
CHD NO	48	106	54		84	161	85	
	61	131	59	251	93	176	89	358
Risk	.21	.19	.08		.10	.09	.04	

Mean Cholesterol

$S_1 = 265$

$S_2 = 208$

$S_3 = 164$

$T_1 = 9465$ $V(T_1) = 44088.6783$

$T_2 = 6161$ $V(T_2) = 33914.7168$

$E(T_1) = 9094.9283$

$E(T_2) = 5932.3240$

Study Questions (Q14.18)

1. Give two equivalent ways to state the null hypothesis for the trend test in this example.
2. Based on the results for the trend test, what do you conclude?
3. If a different scoring method was used, what do you think would be the results of the trend test?

Let's see what the chi square results would be when we use a different scoring system. Here are the results when we use 2, 1, and 0.

$S_1 = 265, S_2 = 208, S_3 = 164$:

$\chi^2 = 4.56$
(P = 0.0320)

$S_1 = 2, S_2 = 1, S_3 = 0$:

$\chi^2 = 5.04$
(P = .0248)

Study Questions (Q14.18) continued

4. How do the results based on 2, 1, 0 scoring compare to the results based on 265, 208, 164?
5. What might you expect for the trend test results if the scoring was 3, 2, 1 instead of 2, 1, 0?

Summary

- If the exposure variable is ordinal, a one d.f. chi square test for linear trend can be performed using an extension of the Mantel-Haenszel test procedure.
- Such a "trend" test can be used to evaluate whether or not there is a linear dose-response relationship of exposure to disease risk.
- To perform the test for trend, a numeric value or score must be assigned to each category of exposure.
- The null hypothesis is that the risk for the health outcome is the same for each exposure category.
- The alternative hypothesis is that the risk for the health outcome either increases or decreases as exposure level increases.
- The test statistic is best calculated using a computer.

Calculating the Mantel-Haenszel Test for Trend involving Several Exposure Categories

Here we describe how the **Mantel-Haenszel (MH) Test** is extended to test for significant trend over several (ordinal) exposure categories. For this test, the data layout for the *i*-th stratum is given as follows:

	E_1	E_2	E_3	...	E_C	
D	a_{i1}	a_{i2}	a_{i3}		a_{iC}	m_{1i}
not D						m_{0i}
	n_{i1}	n_{i2}	n_{i3}		n_{iC}	n_i

The test statistic is given by the following general formula:

$$\chi^2_{MH=Trend} = \frac{[T_1 - E_0(T_i)]^2}{Var_0(T_i)}, i = 1, 2, ..., G$$

The terms in the above formula are defined as follows:

$$T_i = \text{total score for stratum } i = \sum_{j=1}^{C} a_{ij} S_j$$

S_j = the assigned score for the *j*-th exposure category

Continued on next page

Calculating the Mantel-Haenszel Test for Trend involving Several Exposure Categories (continued)

$$E(T_i) = \text{expected total score for stratum } i = \frac{m_{1i}}{n_i}\left(\sum_{i=1}^{C} n_{ij}S_j\right)$$

$$Var_0(T_i) = \text{variance of total score for stratum } i = \frac{m_{1i}m_{0i}}{n_i}\sigma_i^2$$

$$\sigma_i^2 = \frac{1}{n_i - 1}\left\{\sum_{j=1}^{C} n_{ij}S_j^2 - \left[\sum_{j=1}^{C} n_{ij}S_j\right]^2 \Big/ n_i\right\}$$

Under the **null hypothesis of no significant overall trend** of the effect measure over the exposure categories, the MH test for trend statistic is approximately **chi square** with **1 degree of freedom**. For the special case when there are **G=2**, strata, the MH trend statistic simplifies to:

$$\chi^2_{\text{MH-Trend}} = \frac{[\{T_1 - E(T_1)\} + \{T_2 - E(T_2)\}]^2}{Var_0(T_1) + Var_0(T_2)}$$

where

$$T_1 = \text{total score for stratum } 1 = \sum_{j=1}^{C} a_{1j}S_j$$

$$T_2 = \text{total score for stratum } 2 = \sum_{j=1}^{C} a_{2j}S_j$$

$$E(T_1) = \frac{m_{11}}{n_1}\sum_{i=1}^{C} n_{1j}S_j$$

$$E(T_2) = \frac{m_{12}}{n_2}\sum_{i=1}^{C} n_{2j}S_j$$

$$Var_0(T_1) = \frac{m_{11}m_{01}}{n_1}\sigma_1^2$$

$$Var_0(T_2) = \frac{m_{12}m_{02}}{n_2}\sigma_2^2$$

$$\sigma_1^2 = \frac{1}{n_1 - 1}\left\{\sum_{j=1}^{C} n_{1j}S_j^2 - \left[\sum_{j=1}^{C} n_{1j}S_j\right]^2 \Big/ n_1\right\}$$

$$\sigma_2^2 = \frac{1}{n_2 - 1}\left\{\sum_{j=1}^{C} n_{2j}S_j^2 - \left[\sum_{j=1}^{C} n_{2j}S_j\right]^2 \Big/ n_2\right\}$$

Testing for Overall Association Using Logistic Regression

The **test for trend** that we described in the previous activity can alternatively be carried out using a **logistic regression** model. Here again is the logistic model and its equivalent logit form as defined in an earlier lesson:

Test for Trend - Logistic Regression Alternative

Logistic Model:

$$\Pr(Y=1|X_1,X_2,\ldots,X_p) = \frac{1}{1+e^{-(b_0+b_1X_1+b_2X_2+\ldots+b_pX_p)}}$$

$$\text{logit } P = \ln\frac{P}{1-P} = b_0+b_1X_1+b_2X_2+\ldots+b_pX_p$$

where $P = \Pr(Y=1|X_1,X_2,\ldots,X_p)$

X = independent variable or predictor

Y = dichotomous dependent or outcome variable

The **X**'s in the model are the **independent variables** or **predictors**; **Y** is the **dichotomous dependent** or **outcome** variable, indicating whether or not a person develops the disease. We now describe the specific form of this model for the previously considered Evans County data, relating three categories of cholesterol to the development of CHD within 2 age categories. These data are shown again here:

	AGE > 55				AGE ≥ 55			
	HI-CHL	MED-CHL	LO-CHL		HI-CHL	MED-CHL	LO-CHL	
CHD YES	13	25	5		9	15	4	
CHD NO	48	106	54		84	161	85	
	61	131	59	251	93	176	89	358

HI-CHL: > 233 (Mean = 265)

MED-CHL: 184-233 (Mean = 208)

LO-CHL: <184 (Mean = 164)

The logistic model appropriate for the trend test for these data takes the following logit form:

$$\text{logit } P = \ln\frac{P}{1-P} = b_0+b_1E+b_2A$$

The **E** variable in this model represents cholesterol. The **A** variable represents age. More specifically, **E** is an ordinal variable that assigns scores to the three categories of exposure. The scores can be defined as the mean cholesterol value in each cholesterol category:

$$E = \begin{cases} 265 \text{ if HI-CHL} \\ 203 \text{ if MED-CHL} \\ 164 \text{ if LO-CHL} \end{cases}$$

If the scores are instead, 2, 1, and 0, then we define the **E** variable as shown here:

$$E = \begin{cases} 2 \text{ if CHL} > 233 \\ 1 \text{ if CHL} = 184\text{-}233 \\ 0 \text{ if CHL} < 184 \end{cases}$$

The **A** variable is called a **dummy** or **indicator variable**; it distinguishes the two age strata being considered:

$$A = \begin{cases} 1 \text{ if } \text{age} \geq 55 \\ 0 \text{ if } \text{age} < 55 \end{cases}$$

Study Questions (Q14.19)

In general, if there are **S** strata, then **S** – 1 **dummy variables** are required. For example, if there were 3 strata, then 2 dummy variables are required. One way to define the 2 variables would be:

A1 = 1 if stratum 1, else 0, A2 = 1 if stratum 2, else 0

For such coding:

A1 = 1, A2 = 0 for stratum 1
A1 = 0, A2 = 1 for stratum 2
A1 = 0, A2 = 0 for stratum 3

Using such coding, stratum 3 is called the **referent group**.

Suppose we wanted to stratify by two categories of age (e.g., 1 = Age ≥ 55 vs. 0 = Age < 55) and by gender (1 = females, 0 = males).

1. How many dummy variables would you need to define?
2. How would you define the dummy variables (e.g., D1, D2, etc.) if the referent group involved males under 55 years old?
3. Define the logit form of the logistic model that would incorporate this situation and allow for a trend test the uses mean cholesterol values as the scores.

For the model involving only 2 age strata, the null hypothesis for the test for trend is that the true coefficient of the exposure variable, E, is zero. This is equivalent to saying there is no linear trend in the CHD risk, after controlling for age. The alternative hypothesis is that there is a linear trend in the CHD risk, after controlling for age.

Test for Trend - Logistic Regression Alternative

$$\text{logit } P = \ln \frac{P}{1-P} = b_0 + b_1 E + b_2 A$$

$H_0: b_1 = 0 \Leftrightarrow$ no linear trend in CHD risk (after controlling for age)

$H_A: b_1 \neq 0 \Leftrightarrow$ linear trend exists in CHD risk (after controlling for age)

We use a computer program to fit the logistic model to the Evans County data. When we define the **E** variable from the mean cholesterol in each exposure category, a chi square statistic that tests this null hypothesis has the value 4.56.

$$E = \begin{cases} 265 \text{ if HI-CHL} \\ 203 \text{ if MED-CHL} \\ 164 \text{ if LO-CHL} \end{cases}$$

Likelihood Ratio statistic $\chi^2_{1\,d.f.} = 4.56$

This statistic is called the **Likelihood Ratio** statistic and it is approximately a chi square with 1 d.f. The P-value for this test turns out to be 0.0327. Because the P-value is less than .05, we reject the null hypothesis at the 5% significance level and conclude that there is significant linear trend in these data.

Study Questions (Q14.19) continued

When exposure scores are assigned to be 2, 1, and 0 for HI-CHL, MED-CHL, and LO-CHL, respectively, the corresponding Likelihood Ratio test for the test for trend using logistic modeling yields a chi square value of 5.09 (P = .0240).

4. How do these results compare with the results obtained using mean cholesterol scores? (chi square = 4.56, P = .0327)

   When computing the test for trend in the previous activity (using a summation formula), the results were as follows:

   Mean cholesterol scores: chi square = 4.60 (P = .0320)
   2, 1, 0 scores: chi square = 5.04 (P = .0248)

5. Why do you think these latter chi square results are different from the results obtained from using logistic regression? Should this worry you?

Summary

- Testing hypothesis involving several categories of exposure can be carried out using logistic regression.
- If the exposure is nominal, the logistic model requires dummy variables to distinguish exposure categories.
- If the exposure is ordinal, the logistic model involves a linear term that assigns scores to exposure categories.
- For either nominal or ordinal exposure variables, the test involved a 1 d.f. chi square statistic.
- The null hypothesis is *no overall association between exposure and disease controlling for stratified covariates.*
- Equivalently, the null hypothesis is *the coefficient of the exposure variable in the model is zero.*
- The test can be performed using either a likelihood ratio test or a Wald test, which usually give similar answers, though not always.

Quiz (Q14.20)

The data to the right are from a case-control study conducted to investigate the possible association between cigarette smoking and myocardial infarction (MI). All subjects were white males between the ages of 50 and 54. Current cigarette smoking practice was divided into three categories: nonsmokers (NS), light smokers (LS), who smoke a pack or less each day, and heavy smokers (HS), who smoke more than a pack per day.

	HS	LS	NS
D	38	42	20
not D	25	50	35

1. What is the odds ratio for HS vs. NS? **???**
2. What is the odds ratio for LS vs. NS? **???**

Choices
1.47 1.78 2.66 2.70 3.06

All subjects were categorized as having "high" or "low" social status (SS) according to their occupation, education, and income. The stratum-specific data are shown below.

High SS	HS	LS	NS
D	9	6	5
not D	10	20	20

Low SS	HS	LS	NS
D	29	36	15
not D	15	30	15

Quiz continued on next page

3. Calculate the stratum specific odds ratios:
   a. High SS: HS vs. NS **???**
   b. High SS: LS vs. NS **???**
   c. Low SS: HS vs. NS **???**
   d. Low SS: LS vs. NS **???**

4. Is it appropriate to conduct an overall assessment for these data? **???**

Choices
0.01 **1.20** **1.93** **3.50** **3.60** **5.40** **maybe** **no** **yes**

Consider the following results:
Crude OR, HS vs. NS = 2.66
Crude OR, LS vs. NS = 1.47
Adjusted OR, HS vs. NS = 2.38
Adjusted OR, LS vs. NS = 1.20

5. Do these results provide evidence of trend? **???**

Choices
no **yes**

The Mantel-Haenszel test for trend was performed using scores of 0, 1, 2 for nonsmokers, light smokers, and heavy smokers, respectively. The Mantel-Haenszel Chi-square statistic = 5.1. This corresponds to a one-sided p-value of 0.012.

6. What do you conclude at the 0.05 level of significance? **???**
7. What do you conclude at the 0.01 level of significance? **???**

Choices
fail to reject H_0 **reject H_0**

8. An alternative test for trend for these data can be performed using a logistic model. Define the logit form of the logistic model that would incorporate this situation and allow for a trend test. **???**

Choices
Logit P = $b_0 + b_1$SMK + b_2SES
Logit P = $b_0 + b_1$SMK + b_2SES1 + B_3SES2
Logit P = $b_0 + b_1$SMK + b_2SMK2 + B_3SMK3 + b_4SES

References

Clayton DG. Some odds ratio statistics for the analysis of ordered categorical data. Biometrika 1974;61:525-31.

Kleinbaum DG, Kupper LL, Morgenstern H. Epidemiologic Research: Principles and Quantitative Methods. John Wiley and Sons Publishers, New York, 1982.

Mantel N. Chi-square tests with one degree of freedom: Extensions of the Mantel-Haenszel procedure. J Am Stat Assoc 1963;58:690-700.

Mantel N, Haenszel W. Statistical aspects of the analysis of data from retrospective studies of disease. J Natl Cancer Inst 1959;22(4):719-48.

Rezende NA, Blumberg HM, Metzger BS, Larsen NM, Ray SM, and McGowan JE, Jr. Risk factors for methicillin-resistance among patients with Staphylococcus aureus bacteremia at the time of hospital admission. Am J Med Sci 2002; 323(3):117-23.

Homework

ACE-1. Looking at Stratified Data

A case-control study was performed to assess the relationship between alcohol consumption (ALC) and oral cancer (OCa). The results, stratified on smoking status, are displayed below:

Current Smoker

	ALC	no ALC
OCa	42	7
No OCa	3	4

Former Smoker

	ALC	no ALC
OCa	100	2
no OCa	48	5

Never Smoker

	ALC	no ALC
OCa	158	4
no OCa	125	8

a. What is the crude odds ratio for these data?

b. Calculate the stratum-specific ORs. Is there evidence of effect modification by smoking status? Justify your answer.

c. Which of the following would be appropriate for the analysis of these data? [You may choose MORE than one.]

   i. Calculate an overall summary odds ratio for the relationship between alcohol and oral cancer, adjusted for smoking status.

   ii. Report stratum-specific effects.

   iii. Calculate $\chi 2$ tests of association for the alcohol-oral cancer relationship SEPARATELY for the three strata.

   iv. Compare the crude and adjusted estimated ORs to determine whether there is confounding by smoking status.

   v. Calculate a test of heterogeneity to help determine whether there is effect modification by smoking status.

ACE-2. Overall Assessment

Suppose another case-control study of the same exposure-disease relationship (i.e. alcohol and oral cancer) was performed, but this study was restricted to never smokers. Results of this study, stratified on age, are displayed below:

Age 40-49

	ALC	no ALC
OCa	4	25
No OCa	22	309

Age 50-59

	ALC	no ALC
OCa	12	10
no OCa	37	67

Age 60+

	ALC	no ALC
OCa	11	12
no OCa	31	67

a. What is the crude odds ratio for these data?
b. Calculate the stratum-specific ORs. Is there evidence of effect modification by age? Justify your answer.
c. Is the observed overall association between ALC and OCa, stratified on age, statistically significant? (Answer this question by carrying out an appropriate χ^2 test. Be sure to state the null hypothesis and the p-value for the test.)
d. Calculate a precision-based summary odds ratio. Show your calculations. Which stratum has the smallest variance? Justify your answer.
e. Calculate a Mantel-Haenszel summary odds ratio. Show your calculations. Based on this summary estimate, is there evidence of confounding by age in these data? Justify your answer.

ACE-3. Paternal Radiation Exposure and Birth Defects

A case-control study was conducted to assess whether paternal radiation exposure on the job was associated with birth defects. The investigators were concerned with maternal age as a potential confounder, so they stratified the data as follows:

Maternal age >35

	Radiation	No Radiation
Birth Defect	21	26
Control	17	59

Maternal age $\leq$35

	Radiation	No Radiation
Birth Defect	18	88
Control	7	95

a. Calculate a Mantel-Haenszel summary odds ratio. Show your calculations.
b. Is there evidence of confounding by maternal age in these data? Justify your answer.
c. Is the observed association between paternal radiation exposure and birth defects, controlling for maternal age, statistically significant? (Answer this question by carrying out an appropriate statistical test. Be sure to state the null and alternative hypotheses and provide a p-value.)
d. Calculate a test-based 95 % confidence interval for the summary estimate in part A above.
e. When possible, information on paternal radiation exposure (for the study described above) was taken from employment records rather than from subject interviews. This was done in an effort to MINIMIZE which of the following? [Choose ONE best answer]:

i. Detection bias
ii. Differential misclassification of the outcome
iii. Recall bias
iv. Selection bias

ACE-4. Lead Exposure and Low Birth Weight

The relation between paternal occupational lead exposure and low birth weight was examined in a retrospective cohort study. Men with a blood lead level (BLL) > 50 micrograms per deciliter were considered exposed. Low birth weight (LBW) was defined as a birth weight of less than 2500 grams. Data from the study, stratified on maternal age at child's birth, are provided below:

	Maternal Age ≤20			Maternal Age 21+	
	High BLL	**Low BLL**		**High BLL**	**Low BLL**
LBW	45	30	**LBW**	68	48
No LBW	169	214	**No LBW**	257	354

a. Is maternal age an independent risk factor for LBW in these data? Show any calculations and justify your answer.
b. Use the data-based method to determine whether there is confounding by maternal age in this study. Show any calculations and justify your answer.
c. Is the observed association between paternal blood lead level and low birth weight, controlling for maternal age, statistically significant? Answer this question by carrying out an appropriate statistical test. Be sure to state the null and alternative hypotheses and provide a p-value or p-value range.
d. Calculate an overall summary risk ratio that gives equal weight to the two strata defined above. (i.e. a simple average)
e. Calculate a test-based **93% (!!)** confidence interval for the summary estimate in part 3.D. above.
f. Suppose that, prior to analyzing the data, the investigators were concerned about the possibility of residual confounding. Which of the following would have been a useful method of addressing this concern? [Choose one best answer]:

__Dividing blood lead level (BLL) into additional, narrower categories
__Dividing low birth weight (LBW) into additional, narrower categories
__Dividing maternal age into additional, narrower categories
__All of the above

g. Is there evidence of effect modification (by maternal age) on an additive scale in these data? Show any calculations and justify your answer.
h. Based upon your answer to part 3.G. above, which of the following would be appropriate for the next step in the analysis of these data? [Choose one best answer]:
__Report stratum-specific risk differences
__Calculate an overall adjusted risk difference and associated confidence interval
__Report the crude risk difference
__Perform a statistical test to assess whether there is confounding by maternal age
__None of the above options is appropriate

ACE-5. Stratified Analysis: Effect Modification

When effect modification is present in a stratified analysis, how should the data be presented? (Choose one best answer.)

_____ Provide the crude measure of association

_____ Provide the adjusted measure of association

_____ Provide the stratum-specific measures of association

_____ None of the above

ACE-6. Stratified Analysis: Confounding

When confounding is present in a stratified analysis, how should the data be presented? **[Choose one best answer]**:

_____ Provide the crude measure of association

_____ Provide the adjusted measure of association

_____ Provide the stratum-specific measures of association

_____ None of the above

ACE-7. Alcohol Consumption and Bladder Cancer: Race

A case-control study was conducted to assess the potential relationship between alcohol consumption and bladder cancer. Data from the study are summarized below, stratified on three race categories. In answering some of the questions below you may wish to use the Datadesk template **Stratified OR/RR.ise**.

White

	ALC	no ALC
Case	72	41
Control	106	105

Black

	ALC	no ALC
Case	93	54
Control	113	113

Asian

	ALC	no ALC
Case	68	33
Control	78	142

a. Calculate the stratum-specific ORs. Is there evidence of effect modification by race? Justify your answer. (You might use a statistical test here, i.e., the Breslow-Day test, in addition to comparing the point estimates for the three strata.)
b. Should you do an overall Mantel-Haenszel test for association that controls for race using all three strata? Justify your answer.
c. Considering only the information of black and white subjects, Is the observed overall association between ALC and Bladder Cancer, stratified on race, statistically significant? (Be sure to state the null hypothesis and the p-value for the test.)
d. Should you estimate an overall adjusted odds ratio that controls for race using all three strata? Justify your answer.

e. Considering only the information of black and white subjects, calculate both a precision-based aOR and a mOR. For the aOR, which group, black or white, receives more weight?
f. Calculate and compare 95% CIs for the aOR and mOR.
g. What do you conclude about the ALC, bladder cancer relationship?

ACE-8. Physical Activity and Incidence of Diabetes

A cohort study of physical activity (PA) and incidence of diabetes was conducted over a six-year period among Japanese-American men in Honolulu. Data from that study are summarized below, stratified on body mass index. In answering some of the questions below you may wish to use the Datadesk template **Stratified IDR/IDD.ise**.

	High BMI	
	High PA	**Low PA**
Diabetes	48	62
Person-yrs	1050	1067

	Low BMI	
	High PA	**Low PA**
Diabetes	54	71
Person-yrs	1132	1134

a. Is there evidence of effect modification by BMI? Justify your answer.
b. Calculate and compare a precision-based aIDR and a mIDR.
c. Is there evidence of confounding by BMI in these data?
d. Should you estimate an overall adjusted odds ratio that controls for BMI using all three strata? Justify your answer.
e. Is the observed association between physical activity and diabetes, controlling for BMI, statistically significant? Be sure to state the null and alternative hypotheses being tested, the test statistic, and the P-value.
f. Calculate and compare 95% confidence intervals for the aIDR and mIDR.
g. What do you conclude about the relationship between PA and Diabetes based on these data?

Answers Study Questions and Quizzes

Q14.1

1. For both smokers and non-smokers separately, there appears to be no association between exposure to TXC and the development of lung cancer. Never the less, it may be argued that the RR of 1.3 for smokers indicates a moderate association; however, this estimate is highly non-significant.
2. No, the two stratum-specific risk ratio estimates are essentially equal. Again, the RR of 1.3 for smokers indicates a small effect, but is highly non-significant.
3. No, even though the crude estimate of effects is 2.1, the correct analysis requires that smoking be controlled, from which the data show no effect of TCX exposure. An adjusted estimate over the two strata would provide an appropriate summary statistic that controls for smoking.
4. Since the adjusted point estimate is close to the null value of 1 and the Mantel-Haenszel test statistic is very non-significant, you should conclude that there is no evidence of and E-D relationship from these data

Q14.2

1. Yes, the odds ratio of 11.67 is very high and the MH test is highly significant and, even though the confidence interval is wide, the interval does not include the null value.
2. The association may change when one or more variables are controlled. If this happens and the control variables are risk factors, then an adjusted estimate or estimates would be more appropriate.
3. Not necessarily. If one or more of these variables are not previously known risk factors for MRSA status, then such variables may not be controlled.
4. Some (n=5) study subjects had to having missing information on either MSRA status or on previous hospitalization information. In fact, it was on the latter variable that 5 observations were missing.
5. No, the stratum-specific odds ratios within different age groups are very close (around 11).
6. No, the P-value of .95 is very high, indicating no evidence of interaction due to age.
7. Yes, overall assessment is appropriate because there is no evidence of interaction due to age.
8. No, the crude and adjusted odds ratios are essentially equal.
9. Yes, the Mantel-Haenszel test for stratified data is highly significant (P<.0001).
10. The confidence interval is quite wide, indicating that even though the adjusted estimate is both statistically and meaningfully significant, there is little precision in this estimate.
11. Yes, overall assessment is appropriate because there is no evidence of interaction due to gender.
12. No confounding since, when controlling for gender, the crude and adjusted odds rations are essentially equal.
13. Yes, the Mantel-Haenszel test for stratified data is highly significant (P<.0001).
14. The answer to this question is "maybe." There appears to be interaction because the odds ratio is 8.48 with previous drug use but only 3.66 with no previous drug use. However, both odds ratio estimates are on the same side of 1, so an adjusted estimate will not be the result of opposite effects canceling each other. Moreover, the BD test for interaction is non-significant, which supports doing overall assessment.
15. Yes, when controlling for previous drug use, the crude odds ratio of 11.67 is quite different than the much smaller odds ratio of 5.00.
16. Yes, the Mantel-Haenszel test for stratified data is highly significant (P<.0001), and although the confidence interval is wide, it still does not contain the null value.

Q14.3

1. Previous antimicrobial drug use needs to be controlled because it is a confounder.
2. Yes, precision is gained from controlling for previous antimicrobial drug use, since the width of the confidence interval for the adjusted estimate is much narrower than the width of the corresponding confidence for the crude data.
3. No, neither the adjusted odds ratio nor the confidence interval nor the MH P-value changes either significantly or meaningfully when comparing the results that control for PADMU alone with results that control for additional variables.
4. No, all P-values are quite large, indicating that the null hypothesis of no interaction should not be rejected. However, perhaps a comparison of stratum-specific estimates may suggest interaction when more than one variable is controlled.
5. Because the estimated odds ratio is undefined in a stratum with a zero cell frequency.
6. OR=4.66 is an appropriate choice because it controls for all three variables. being considered for control. Alternatively, OR=5.00 is also appropriate because it results from controlling only for previous antimicrobial drug use, which is the only variable that affects confounding and precision.
7. Yes, the adjusted odds ratio (close to 5.00) indicates a

strong effect that is also statistically significant. The 95% confidence interval indicates a lack of precision, but the results are overall indicative of a strong effect.

8. (Note: there is no question 8)
9. There are small numbers, including a number of zeros in almost all tables.
10. Stratum-specific analyses, even when there are no zero cells, are on the whole unreliable because of small numbers.
11. Yes, the odds ratio estimate in table 5 is 24.00 whereas the odds ratio in table 6 is 1.71 and the odds ratio in table 1 is 5.89, all quite different estimates.
12. The BD test is not significant, all odds ratio estimates, though different, are all on the same side of the null value, and the strata involve very small numbers.

Q14.4

1. F – Stratification also involves performing an overall assessment when appropriate.
2. T
3. T
4. F – An overall summary estimate may be considered inappropriate if there is considerable evidence of interaction.
5. T
6. No, Maybe, Yes

Q14.5

1. For a one-sided alternative, the area in the right tail is twice the P-value, so the one-sided P-value is half of .28, or .14.
2. Do not reject the null hypothesis of no overall effect. There is no evidence that exposure to TCX is associated with the development of lung cancer when controlling for smoking.
3. If there was interaction on opposite sides of the null, then one of the two terms in the sum would be negative and the other would be positive. Consequently, the sum of these terms might be close to zero, yielding a non-significant chi square test, even if the stratum-specific tests were both significant.

Q14.6

1. Most definitely. The numerator is always a sum of quantities that are then squared, in contrast to a sum of quantities that are squared before summing. Consequently, large positive values from some strata may cancel out large negative values of other strata, leading to a non-significant MH test.
2. If the stratum-specific effects are all on the same side of the null value, than all quantities in the numerator of the test statistic have the same sign and therefore cannot cancel each other out.

Q14.7

1. For a one-sided alternative, the area in the right tail is twice the P-value so the one-sided P-value is half of .04, or .02.
2. The null hypothesis of no overall effect would be rejected at the 5 percent level but not at the 1 percent level.

Q14.8

1. 0.79
2. 0.76
3. 0.77
4. No (To compute the crude IDR, you need to combine the data over both strata).
5. No
6. increased precision
7. fail to reject the null hypothesis
8. reject the null hypothesis

Q14.9

1. B
2. F: A larger confidence interval means less precision and hence a smaller weight.
3. F: The magnitude of the risk ratio is not a factor in determining precision-based weights.
4. T
5. T

Q14.10

1. Remember that sample size is not as important as a balanced data set in determining precision. The weights correspond to how balanced the data sets are. The more balance, the higher the weight.

Q14.11

1. 4.00
2. 4.69
3. Black or Hispanic
4. 4.16
5. debatable – The crude POR of 3.59 and the adjusted POR of 4.16 are different but not that far apart, so deciding whether there is a meaningful difference is debatable. Note, however, that if we require a 10% difference for judging confounding, then 3.59 is below a 10% negative change (3.64) in the adjusted POR. Thus, using a 10% change rule, we would conclude that there is confounding.

Q14.12

1. OR(modified by 0.1) = (5.1x5.1) / (0.1x6.1) = 42.6
2. OR(modified by 1.0) = (6x6) / (1x7) = 5.1
3. The modified stratum-specific odds ratio may change radically depending on what value is used to adjust all cell frequencies in a stratum with a zero cell frequency. Consequently, such an approach is quite problematic.
4. $a\hat{O}R = \exp\left[\frac{\{.3972\times\ln(9.31)\}+\{.6618\times\ln(9.00)\}}{.3972+.6618}\right] =$

   $\exp\left[\frac{.8862+1.4541}{1.0590}\right] = 9.11$
5. $a\hat{O}R = \exp\left[\frac{\{.0947\times\ln(42.64)\}+\{.6618\times\ln(9.00)\}}{.0947+.6618}\right] =$

   $\exp\left[\frac{.3354+1.4541}{.7565}\right] = 10.65$
6. They are different (9.11 using a .5 adjustment vs. 10.65 using a .1 adjustment), but not very different.
7. It does not seem that the choice of adjustment factor has a great impact on the resulting aOR, even though it can have a great impact on the value of the odds ratio in the stratum being adjusted.

Q14.13

1. The mOR will be undefined because each term in the sum in the denominator will be zero.
2. The mOR will be undefined because each term in the sum in the numerator will be zero.
3. It is possible, although unlikely, that having zero cell frequencies in every stratum may make the mOR undefined. Nevertheless, if such a situation occurs, the mOR will not work. Instead, you may have to use as an alternative the approach that adds .5 to each cell frequency in any stratum with a zero cell.
4. All three adjusted estimates are different, with the mOR being quite separate from the aOR estimates. A key reason for preferring the mOR is that it avoids using an arbitrary modifying value like .5 or .1. Also, the mOR has good statistical properties as described earlier.

Q14.14

1. True
2. False – It is the stratum-specific estimates that are most affected by adding a small number to each cell.
3. True

Q14.15

1. The null value of 1 is not contained in the confidence interval, but just barely. Although the interval is not very wide, the point estimate of 1.71 is somewhat unreliable since the interval ranges from essentially no association to a moderately strong association.
2. 95% CI for aRD:

   $$.0850 \pm \frac{1.96}{\sqrt{490.2243}} =$$

   $$.0850 \pm .0885 =$$

   $$(-.0035, .1735)$$
3. The null value of 0 is just barely contained in the interval. Although the interval is not very wide, the point estimate of .0850 is somewhat unreliable since the interval ranges from essentially no association to a moderately strong association of .17 for a risk difference.

Q14.16

1. 1.57, 12.98
2. yes

Q14.17

1. The estimated risk ratios comparing HI-CHL with LO-CHL do not differ very much between the two age groups (2.51 vs. 2.15). Similarly, when comparing MED-CHL with LO-CHL, the estimated risk ratios do not differ very much between the two age groups (2.25 vs. 1.90). Overall, these findings indicate no meaningful interaction of age with cholesterol level.
2. Overall assessment is appropriate because there is no evidence of interaction, particularly on opposite sides of the null value.
3. No evidence of confounding, since crude and adjusted estimated are essentially equivalent when comparing HI vs. LO Cholesterol groups and when comparing MED vs. LO Cholesterol groups.

Q14.18

1. (1) There is no significant trend in the CHD risk over the three categories. (2) The risks in each strata are the same.
2. Since the P-value is 0.0320 (two-sided), conclude that there is a significant trend in the CHD risks over the three categories using the scoring method involving mean CHL values within each CHL category.
3. The chi square statistic for trend would likely change

somewhat, but would probably lead to the same conclusion obtained for other scoring methods, although no guarantee that this would always occur.

4. The chi square statistic and corresponding P-values are slightly different, but both statistics would reject the null hypothesis of no linear trend at the .05 significance level.
5. Identical chi square statistics and corresponding P-values. The reason: scores are equally distant. This would also be the case if the scores where 15, 10, and 5.

Q14.19

1. Three dummy variables, because there would be 2x2 = 4 age-by-gender strata.
2. D1 = 1 if female $\geq$ 55 years, else 0; D2 = 1 if female < 55 years, else 0; and D3 = 1 if male $\geq$ 55 years, else 0.
3. logit P = $b_0 + b_1E + b_2D1 + b_3D2 + b_4D3$ where E takes on the three values 265, 208, and 164 for the three cholesterol strata.
4. They are slightly different, as might be expected, but they lead to the same conclusion about the null hypothesis. It is possible, however, that different scoring systems can give different conclusions.
5. The logistic regression approach uses a slightly different approach (called ***maximum likelihood***) for obtaining model estimates and the corresponding tests than the earlier approach (summation formula). Although both approaches can lead to slightly different answers, they are equivalent if the sample sizes are large enough. Nevertheless, the logistic regression approach is preferred by most statisticians because of the properties of maximum likelihood estimates.

Q14.20

1. 2.66
2. 1.47
3. a. 3.60; b.1.20; c. 1.93; d. 1.20
4. maybe: There is some evidence of interaction here, but since it is same side interaction, the investigator must decide whether the difference in 3.6 versus the other estimates is meaningful.
5. yes
6. reject H_0
7. fail to reject H_0
8. Logit P = $b_0 + b_1SMK + b_2SES$

LESSON 15

MATCHING

Matching is an option for control that is available at the study design stage. We previously introduced matching on page 13-2 in Lesson 13. We suggest that you review that activity before proceeding further with this lesson. The primary goal of matching is to gain ***precision*** *in estimating the measure of effect of interest. There are other advantages to matching as well, and there are disadvantages. In this lesson, we define matching in general terms, describe different types of matching, discuss the issue of whether to match or not match, and describe how to analyze matched data.*

15-1 Matching

Definition and Example of Matching

Reye's syndrome is a rare disease affecting the brain and liver that can result in delirium, coma, and death. It usually affects children and typically occurs following a viral illness.

To investigate whether aspirin is a determinant of Reye's syndrome, investigators in a 1982 study carried out a matched case-control study that used a statewide surveillance system to identify all incident cases with Reye's syndrome in Ohio. Population-based matched controls were selected as the comparison group. Potential controls were first identified by statewide sampling of children who had experienced viral illnesses but who had not developed Reye's syndrome. Study controls were then chosen by individually matching to each case one or more children of the same age and with the same viral illness as the case. Parents of both cases and controls were asked about their child's use of medication, including aspirin, during the illness.

Study Questions (Q15.1)

1. Why do you think that **type of viral illness** was considered as one of the matching variables in this study?
2. Why do you think **age** was selected as a matching variable?

This study is a classic example of the use of individual matching in a case-control study. Although the simplest form of such matching is **one-to-one** or **pair matching**, this study allowed for more than one control per case.

Matching typically involves two groups being compared, the index group and the comparison group. In a case-control study, the index group is the collection of cases, for example, children with Reye's syndrome, and the comparison group is the collection of controls.

If the study design was a cohort study or clinical trial, the index group would instead be the collection of exposed persons and the comparison group would be the collection of unexposed persons. Because matching is rarely used in either cohort or clinical trial studies, our focus here will be on case-control studies.

Matching	Case-Control	Cohort / Clin. Trial
INDEX GROUP	Cases	Exposed
versus		
COMPARISON GROUP	Controls	Unexposed

No matter what type of matched design is used, the **key feature** is that the **comparison group is restricted to be similar to the index group** on the matching factors. Thus, in the Reye's Syndrome study, the controls were restricted to have the same distribution as the cases with regard to the variables age and type of viral illness. But we are not restricting the distribution of age or viral illness for the cases. That's why we say that matching imposes a **partial** restriction on the control group in a case-control study.

Summary

- ❖ A 1982 study of the relationship of aspirin to Reye's syndrome in children is a classic example of individually matching in a case-control study.
- ❖ The simplest form of individual matching is one-to-one or pair matching, but can also involve more than one control per case.
- ❖ Typically, matching compares an ***index group*** with a ***comparison group***.
- ❖ In a case-control study, the index group and the comparison group are the cases and controls, respectively.
- ❖ In a cohort study or clinical trial, the index group and the comparison are the exposed and unexposed, respectively.
- ❖ The key feature of matching is that the comparison group is restricted to be similar to the index group with regards to the distribution of the matching factors.

Types of Matching

There are two types of matching, **individual matching** and **frequency matching**. **Individual matching**, say in a case-control study, is carried out one case-at-a-time by sequentially selecting one or more controls for each case so that the controls have the same or similar characteristics as the case on each matching variable. For example, if we match on age, race, and sex, then the controls for a given case are chosen to have the same or similar age, race and sex as the case.

When matching on continuous variables, like age, we need to specify a rule for deciding when the value of the matching variable is "close-enough." The most popular approach for continuous variables is **category matching**. (Note: category matching is one of several ways to carry out individual matching involving a continuous variable. See the box at the end of this activity for a description of other ways to match on a continuous variable.) The categories chosen for this type of matching must be specified prior to the matching process. For example, if the matching categories for age are specified as 10-year age bands then the control match for a 40-year-old case must come from the 36-45 year old age range.

The first step is to categorize each of the matching variables, whether continuous or discrete. Then for each index subject match by choosing one or more comparison subjects who are in the same category as the index subject for every one of the matching variables.

Study Questions (Q15.2)

Consider a case-control study that involves individual category matching on the variables age, gender, smoking status, blood pressure, and body size.

1. What do you need to do first before you can carry out the matching?
2. How do you carry out the matching for a given case?
3. If the case is a 40-year-old male smoker who is obese and has high blood pressure, can its matched control be a 40-year-old male smoker of normal body size with low blood pressure? Explain.

In frequency matching the matching is done on a group rather than individual basis. The controls are chosen as a group to have the same distribution as the cases on the matching variables. For example, we might frequency match on blood pressure and age in a case-control study where the cases have the blood pressure-by-age category breakdown shown below, by insuring that the controls have the same breakdown:

	HiBP, Age>55	HiBP, Age<55	NormBP, Age>55	NormBP, Age<55	Total
Cases	40	60	50	50	200

Study Questions (Q15.2) continued

Suppose you wish to have three times as many total controls as cases. Answer the following questions.

4. What is the BP group by age group breakdown for the number of controls?
5. What is the percentage breakdown by combined BP group and age group for the controls?

How do you decide between individual matching and frequency matching? The choice depends primarily, on which type of matching is more convenient in terms of time, cost, and the type of information available on the matching variables. The choice also depends on how many variables are involved in the matching. The more matching variables there are, the more difficult it is to form matching groups without finding matches individual by individual.

Study Questions (Q15.2) continued

Suppose cases are women with ovarian cancer over 55 years of age in several different hospitals and you wanted to choose controls to be women hospitalized with accidental bone fractures matched on age and hospital.

6. Which would be more convenient, frequency matching or individual matching?

Suppose cases are women with ovarian cancer over 55 years of age in one hospital and controls were women hospitalized with accidental bone fractures and matched on age, race, number of children, age at first sexual intercourse, and age at first menstrual period.

7. Which would be more convenient, frequency matching or individual matching?

Summary

- There are two general types of matching, **individual** versus **frequency matching**.
- **Individual matching** in a case-control study is carried out one case at a time.
- With individual matching, we sequentially select one or more controls for each case so that the controls have the same or similar characteristics as the given case on each matching variable.
- For continuous variables, matching can be carried out using caliper matching, nearest neighbor matching, or category (the most popular) matching.
- **Frequency matching** involves category matching on a group basis, rather than using individual matching.

Other Types of Matching on Continuous Variables

Although category matching is the most popular approach for matching on continuous variables, like age, there are two other approaches that are sometimes used instead. One procedure, called **caliper matching**, involves choosing the control to be within a certain defined interval as the case on the matching factor. For example, when matching on age, this interval, also called a caliper, might require that the control be within plus or minus 5 years of age as the case.

In general, caliper matching is defined as specifying a value c and requiring two subjects to be matched if the comparison subject (e.g., the control) is within plus or minus c units on the matching variable as the index subject (e.g., case).

One problem with caliper matching can occur if the choice of c's very small (i.e., stringent), since it might then be difficult to find matched subjects. Another problem may occur if the choice of c is very large; if so, then the control group is effectively not necessarily restricted to have the same distribution as the cases on the matching factor. If, for example you matched on age using c=30 years, then the age distribution of the controls could be very different than the age distribution of the cases.

A second alternative for matching continuous variables is called **nearest neighbor** matching. This procedure carries out the matching by selecting one or more subjects closest to the case on the matching variable. Nearest neighbor matching be very time-consuming to carry out because it might require a comparison of all potential controls for a given case (provided no potential control has exactly the same value on the matching variable as the case). Also, when there are several matching variables, the nearest neighbor for one variable might not be the nearest neighbor for another variable for the same potential control. Thus, it may be difficult to find the nearest neighbor for several variables simultaneously.

Matching Ratios

An important design issue for a matched study is the ratio of the number of comparison subjects to the number of index subjects in each matched stratum. We call this ratio the **matching ratio** for the matched study. Here is a list of different matching ratios that are possible:

$$\text{Matching ratio} = \frac{\text{\# Comparison Subjects}}{\text{\# Index subjects}}$$

Name	# Cases	# Controls	Type	Why?
1 to 1 (pair-matching)	1	1	Individual	Gain precision (vs. no matching)
R to 1	1	R	Individual	Gain precision (increase n)
R_i to 1	1	R_i	Individual	Try for R, but find < R
R_i to S_i	S_i	R_i	Frequency or Individual	Convenience or 'Pooling'

The smallest and simplest ratio is **1 to 1**, also referred to as **pair matching**. In a case-control study, pair matching matches 1 control to each case and requires individual matching. Why use pair matching? Pair matching can lead to a gain in precision in the estimated effect measure when compared to not matching at all for a study of the same total size. Also, it is easier to find one match than to find several matches per index subject.

R to 1 matching in a case-control study involves choosing **R** controls for each case using individual matching. For example, 3 to 1 matching in a case-control study would require three controls for each case. R to 1 matching is preferable to pair matching because even more precision can be gained from the larger total sample size would be increased. However, from a practical standpoint, it may be difficult to find more than several matched controls for each case.

R_i to 1 matching allows for a varying number of matched subjects for different cases using individual matching. For example, 3 controls may be found for one case, but only 2 for another and perhaps only one control for a third. R_i to 1 matching is often not initially planned but instead results from trying to carry out R to 1 matching and then finding fewer than

R matches for some cases.

R_i by S_i matching allows for one or more controls to be matched as a group to several cases also considered as a group The letter **i** here denotes the **i-th** matched group or stratum containing R_i controls and S_i cases. This matching ratio typically results from frequency matching but can also occur from individual matching when **pooling exchangeable matched strata**.

<u>Study Questions (Q15.3)</u>

A detailed discussion of **pooling** is given in a later activity. Consider an individually matched case-control study involving 2 to 1 matching, where the only matched variable is smoking status (i.e., SMK = 0 for non-smokers and SMK = 1 for smokers). Suppose there are 100 matched sets in which 30 sets involve all smokers and 70 sets involve all non-smokers. Suppose further that we ***pool*** the 30 ("exchangeable") sets involving smokers into one combined stratum and the 70 ("exchangeable") sets involving non-smokers into another combined stratum.

1. How many cases and controls are in the first matched stratum (that combine 30 matched sets)?
2. How many cases and controls are in the second matched stratum (that combines 70 matched sets)?
3. What type of matching ratio scheme is being used involving pooled data, R to 1 or R_i to S_i?

Consider the following table determined by **frequency matching** on race and gender in a case-control study.

Frequency Matching on Race and Gender Case-Control Study

	White Male	White Female	Black Male	Black Female
Cases	100 (33%)	100 (33%)	40 (13%)	60 (20%)
Controls	200 (33%)	200 (33%)	80 (13%)	120 (20%)

4. How many matched strata are there in this frequency-matched study?
5. What type of matching ratio describes this design: R to 1 or R_i to S_i?
6. What are the numbers of controls and cases in each stratum?

<u>Summary</u>

- The **matching ratio** for a matched design is the ratio of the number of comparison subjects to the number of index subjects in each matching stratum.
- Matching ratios may be **1 to 1**, **R to 1**, **R_i to 1**, or **R_i to S_i**.
- The simplest matching ratio is 1 to 1, also called pair matching.
- **R to 1 matching** gives more precision than 1 to 1 matching because of increased sample size, but finding R matches per index subject may be difficult.
- **R_i to 1 matching** typically occurs when trying for R to 1 matching but finding less than R comparison subjects for some index subjects.
- **R_i to S_i matching** typically results from frequency matching but may also result from pooling artificially matched strata from individual matching.

How Many Matched Should You Select?

If **R to 1 matching** is used, how large should R be? The widely accepted answer to this question is that there is little to gain in terms of precision by using an R larger than four. The usual justification is based on the **Pitman Efficiency criterion**, which is approximately the ratio of the variance of an adjusted odds ratio computed from pair matching to the corresponding variance computed from R to 1 matching. Here is the Pittman efficiency formula:

Pitman Efficiency Criterion

approximately

$$\frac{\text{Var}(\hat{aOR}_{1\text{ to }1})}{\text{Var}(\hat{aOR}_{R\text{ to }1})} = \frac{2R}{(R+1)}$$

Computing this criterion for several values of **R** yields the following table:

R	1	2	3	4	5	6	...	∞
2R/(R+1)	1.000	1.333	1.500	1.600	1.667	1.714	...	2
% increase	-	33.3	12.5	6.7	4.2	2.8		-

This table shows diminishing returns once **R** exceeds 4. The Pittman efficiency increases 33.3 percent as **R** goes from 1 to 2, but only 4.2 percent as **R** goes from 4 to 5. The percent increase in efficiency clearly is quite small as **R** gets beyond 4. Moreover, the maximum possible efficiency is 2 and at R = 4 the efficiency is 1.6.

Study Questions (15.4)

1. Why does the Pittman efficiency increase as **R** increases?
2. What does the previous table say about the efficiency of 2 to 1 matching relative to pair-matching?

Summary

- For **R to 1 matching**, there is little to gain in precision from choosing R to be greater than 4.
- A criterion used to assess how large R needs to be is the **Pittman Efficiency criterion**, which compares the precision of R to 1 matching relative to pair matching.
- A table of **Pittman Efficiency** values computed for different values of **R** indicates a diminishing return regarding efficiency once R exceeds 4.

Quiz (15.5)

True or **False**:

1. If we match in a case-control study, the index group is composed of exposed subjects. . ???
2. If we match in a cohort study, the comparison group is composed of non-cases. . ???
3. If we individually category match on age and gender in a case-control study, then the control for a given case must be either in the same age category or have the same gender as the case. . ???
4. When frequency matching on age and race in a case-control study, the age distribution of the controls is restricted to be the same as the age distribution of the cases. . . . ???
5. Five-to-one matching will result in a more precise estimate of effect than obtained from 4-to-one-matching for the same number of cases. . . . . ???
6. Ri-to-1 matching may result when trying to carry out R-to-1 matching. . . . ???
7. Pair matching is a special case of R_i-to-S_i matching. . . . . . ???
8. Not much precision can be gained from choosing more than one control per case. . ???

15-2 Matching (continued)

Reasons for Matching

Why should we use matching to control for extraneous variables when designing an epidemiologic study? The primary advantage of matching is that it can be used to gain **precision** in estimating the effect measure of interest. Matching can allow you to get a **narrower confidence interval** around the effect measure than you could obtain without matching.

Why use matching ?

- Gain precision
- Control for variables difficult to measure
- Practical aspects - convenience, time-saving, cost-saving
- Control for confounding (?)

Another reason for matching is to control for variables that are difficult to measure. For example, matching on neighborhood of residence would provide a way to control for social class, which is difficult to measure as a control variable. Matching on persons from the same family, say brothers or sisters, might be a convenient way of controlling for genetic, social, and environmental factors that would be otherwise difficult to measure.

A third reason for matching is to take advantage of practical aspects of collecting the data, including convenience, timesaving, and cost-saving features. For example, if cases come from different hospitals, it may be practical to choose controls matched on the case's hospital at the same time as you are identifying cases from the hospital's records. In an occupational study involving different companies in the same industry, controls can be conveniently matched to cases from the same company. Such controls will likely have social and environmental characteristics similar to the cases.

Another reason often given for matching is to control for **confounding**. We have placed a question mark after this reason because, even if matching is not used, confounding may be controlled using stratified analysis or mathematical modeling. Also, if you match in a case-control study, you must make sure to do what we later describe as a **matched analysis** in order to properly control for confounding.

Matching is usually limited to a restricted set of control variables. There are typically other variables not involved in the matching that we might want to control. Matching does not preclude controlling for confounding from those risk factors that are measured but not matched.

Summary

The reasons for matching include:

- Gain precision
- Control for variables difficult to measure
- Practical aspects: convenience, time-saving, cost-savings
- Can control confounding for both matched and unmatched variables.

Reasons Against Matching

Why might we decide not to use matching when designing an epidemiologic study? One reason is that matching on a weak or non-risk factor is unlikely to gain precision and might even lose precision relative to not matching. If all potential control variables are at best weak risk factors, the use of matching will not achieve a gain in precision.

Study Questions (Q15.6)

Suppose you match on hair color in a case-control study of occupational exposure for bladder cancer.

1. Do you think hair color is a risk factor for bladder cancer?
2. Based on your answer in the previous question, would you expect to gain precision in your estimate by matching on hair color? Explain.

Another reason not to use matching is the cost of time and labor required to find the appropriate matches, particularly when individual matching is used. To actually carry out the matching, a file that lists potential controls and their values on all matching variables must be prepared and a selection procedure for matching controls to cases must be specified and performed. This takes time and money that would not be required if controls were chosen by random sampling from a source population.

A third reason for not matching is to avoid the possibility of what is called **overmatching**. Overmatching can occur if one or more matching variables are highly correlated with the exposure variable of interest. For example, in an occupational study, if we match on job title, and job title is a surrogate for the exposure being studied, then we will 'match out' the exposure variable. That is, when we overmatch, we are effectively matching on exposure, which would result in finding no exposure-disease effect even if such an effect were present.

Study Questions (Q15.6) continued

3. If the exposure variable is cholesterol level, how might **overmatching** occur from matching on a wide variety of dietary characteristics, including average amount of fast-food products reported in one's diet?

Another drawback of matching is that your study size might be reduced if you were **not** able to find matches for some index subjects. The precision that you hoped to gain from matching could be compromised by such a reduction in the planned study size.

Study Questions (Q15.6) continued

4. What study size problem might occur if you category-match on several variables using very narrow category ranges?

Summary

Reasons for **not** matching:

- ❖ Matching on weak risk factors is unlikely to gain (and may lose) precision.
- ❖ Matching may be costly in terms of time and money required to carry out the matching process.
- ❖ You may inappropriately overmatch and therefore effectively match on exposure.
- ❖ You may have difficulty finding matches and consequently lose sample size and correspondingly the precision you were hoping to gain from matching.

To Match or Not to Match?

How do we decide whether or not we should use matching when planning an epidemiologic study? And if we decide to match, how do we decide which variables to match on? The answer to both of these questions is "**it depends**". Let's consider the list of reasons for and against matching that we described in the previous activities:

Reasons For Matching	Reasons Against Matching
Gain precision	Matching on weak risk factors may lose precision
Control for variables difficult to measure	Matching may be costly in terms of time and money
Practical aspects	You may have difficulty finding matches
Control for confounding of both matched and unmatched variables	You may overmatch
	Can control for confounding even without matching

Your decision whether to match or not to match should be based on a careful review of the items in both columns of this list and on how you weigh these different reasons in the context of the study you are planning.

Study Questions (Q15.7)

Suppose the practical aspects for matching are outweighed by the cost in time and money for carrying out the matching. Also suppose that previously identified risk factors for the health outcome are not known to be very strong predictors of this outcome.

1. Should you match?

Suppose age and smoking are considered very strong risk factors for the health outcome:

2. Should you match or not match on age and smoking?

Suppose you want to control for social and environmental factors.

3. Should you match or not match on such factors?

Although all items listed for or against matching are important, the primary statistical reason for matching is to gain precision. The first items on both lists concern precision and they suggest that whether or not matching will result in a gain in precision depends on the investigator's prior knowledge about the important relationships among the disease, exposure, and potentially confounding variables. If such prior knowledge, is available, for example from the literature, and is used properly, a reasonable decision about matching can be made.

It is widely recommended that, with regards to precision, the safest strategy is to match only on strong risk factors expected to show up as confounders in a study. This recommendation clearly requires subjective judgment about what is likely to happen in one's study regarding the distribution of potential confounders. In practice, a decision to use matching for precision gain applies to only those factors identified in the literature as strong predictors of the health outcome.

Summary

- The answer to the question "to match or not to match?" is "**it depends**".
- Your decision depends on a careful review of the reasons for and against matching and how you weigh these different reasons.
- Whether or not you will gain precision depends on the investigators' prior knowledge about the relationships of the variables being measured.
- Recommendation regarding precision: match only on strong risk factors expected to show up as confounders in one's study.

Quiz (15.8)

True or **False**:

1. One reason for deciding to match in a case-control study is to obtain a valid estimate of the odds ratio of interest. . . . . **???**
2. An advantage of matching over non-matching is that your sample size may be smaller from not matching. . . . . **???**
3. Matching on a weak risk factor may result in a loss of precision when compared to non-matching. **???**

Fill in the Blanks

4. Which of the following choices are reasons against using matching. . . . **???**
   a. You match on a non-risk factor.
   b. You want to control for a variable difficult to measure.
   c. You want to control for both matched and unmatched variables.
   d. Your matching variable is highly correlated with the exposure variable.
   e. It is costly to carry out the matching.

Choices
a only **a, b and c** **a, b and d** **a, d and e** **b only** **b, c and d** **c only** **d only** **e only**

5. If matching is convenient and inexpensive to carry out, it should always be preferred to non-matching. . . . . **???**
6. If your primary reason for considering matching is to gain precision in your estimated odds ratio, should you match or not match in a case-control study? . . . . . **???**

Choices
False **It depends** **True** **Don't match if costly** **Match always** **Match if convenient** **Never match**

15-3 Matching (continued)

Analysis of Matched Data – Options and General Principles

There are two options for analyzing matched data with dichotomous outcomes: **stratified analysis** using **Mantel-Haenszel** methods and **mathematical modeling** using **logistic regression**. Mantel-Haenszel methods are appropriate whenever all the variables being controlled are involved in the matching. Logistic regression methods are appropriate if some variables being controlled have not been matched-on and some variables have been matched-on.

Two options for analyzing matched data:

Stratified Analysis

(using Mantel-Haenszel methods)

- All variables being controlled are involved in matching

Mathematical modeling

(using logistic regression)

- Some variables being controlled have not been matched-on

For example, if we match in case-control study on age, race and sex and these three variables are the only ones being considered for control, then the Mantel-Haenszel methods, for stratified analysis, are appropriate. In contrast, if we match on age, race and sex and we also want to control for other non-matched variables, such as physical activity level, body size, and blood pressure, then it is necessary to use logistic regression methods.

When carrying out a matched analysis, we must consider **four** important principles. **First**, a matched analysis requires that you actually "control" for the matching variables. In particular, if you fail to control for the matching variables in a case-control study, you will not have addressed confounding due to these variables and your estimated odds ratio will be biased towards the null. And, if you don't control for the matching variables in a follow-up study, you are likely not to gain the precision in your estimated risk ratio that you had intended to achieve through matching.

Second, a **matched analysis** is a **stratified analysis**. The strata are the matched sets or pooled matched sets. For example, if you pair match in a case-control study and you have 100 cases, then there are 100 matched sets or strata to analyze. Each matched set would contain two persons, the case and the control:

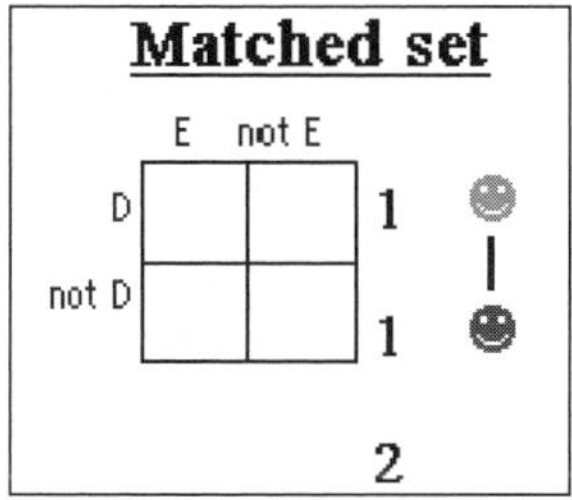

Third, when using logistic regression to do a matched analysis, the strata are defined using dummy or indicator variables. The number of dummy variables will be one less than the number of matching strata. For example, if we pair-match in a case-control study and we have 100 cases, a logistic model for such data will require 99 dummy variables to incorporate the 100 matching strata.

Study Questions (Q15.9)

1. State the logit form of a logistic model that allows for the analysis of 100 case-control matched-pairs to describe the relationship of a dichotomous exposure variable E to a dichotomous outcome D.

Fourth, a key advantage to using logistic modeling with matched data is that you can control for variables involved in the matching as well as variables not involved in the matching. If you use a stratified analysis instead, you will typically have to drop some matching strata from the analysis, and consequently will lose precision in your estimate.

Study Questions (Q15.9) continued

Suppose you match on age, race, and sex in a case-control study involving 100 matched pairs. For your analysis, you wish to control for systolic blood pressure (SBP) and cholesterol level (CHL), neither of which is involved in the matching, as well as controlling for the matching variables.

2. State the logit form of a logistic model for carrying out the analysis described above.
3. What information will be lost if a stratified analysis is carried out to control for the matching as well as for SBP and CHL?

Summary

- Two options for analyzing matched data are stratified analysis using Mantel-Haenszel methods and mathematical modeling using logistic regression.
- A matched analysis requires that you control for the matching variables.
- A matched analysis is a stratified analysis.
- When using logistic regression to do a matched analysis, the strata are defined using dummy (i.e., indicator) variables.
- When using logistic modeling with matched data, you can control for variables involved in the matching as well as variables not involved in the matching.

Does Matching Control for Confounding?

The answer to this question is clearly **no** if we wish to control for variables not matched on in addition to the variables involved in the matching. If, however, we **assume that the only variables being controlled are involved in the matching**, then the answer requires us to consider cohort and case-control studies separately.

In a cohort study, matching automatically controls for confounding without the need to control for the matching variables. Nevertheless, you still need to control for the matching in order to gain the precision that you expected to gain by matching (assuming you made a good decision on which variables to match).

In a case-control study, matching does not automatically control for confounding, so that it is necessary to control for the matching variables in order to control for confounding.

To further explain the above statements, we need to describe **conditions for data-based confounding**. For simplicity in explanation, we will assume that there is only one matching variable, denoted here as **F**. We have previously defined data-based confounding as present if:

$$c\hat{\theta} \neq a\hat{\theta} \quad \text{where} \quad c\hat{\theta} \text{ and } a\hat{\theta}$$

denote the crude and adjusted estimates of the measure of effect θ, respectively. Equivalent conditions for confounding can also be given for a risk ratio (**RR**) and an odds ratio (**OR**) separately in terms of the relationship of the disease variable **D** to the matching variable **F** and the relationship of the exposure variable **E** to the matching variable **F**.

Conditions for confounding of **RR**:

$$\hat{OR}_{DF|\text{unexposed}} \neq 1 \text{ and } \hat{OR}_{EF} \neq 1$$

Conditions for confounding of **OR**:

$$\hat{OR}_{DF|\text{unexposed}} \neq 1 \text{ and } \hat{OR}_{EF|\text{non-diseased}} \neq 1$$

If we apply the conditions for confounding in the **RR** to a matched cohort study, we can see that these conditions are **not satisfied** since

Continued on next page

Does Matching Control for Confounding? (continued)

$$\hat{OR}_{EF} = 1$$

That is, a matched cohort study makes the different exposure groups have the same distribution with respect to the matching variable **F**. Thus, matching in a cohort study automatically controls for confounding in the risk ratio, i.e., the crude risk ratio will always equal the adjusted risk ratio, and so we can ignore controlling for the matching and still get a correct risk ratio.

On the other hand, if we apply the conditions for confounding of the OR to a matched cohort study, we can see that:

$$\hat{OR}_{DF} = 1$$

which is not equivalent to the requirement that:

$$\hat{OR}_{DF|unexposed} = 1$$

That is, even though a matched case-control study makes the different disease groups have the same distribution with respect to the matching variable **F**, it does not automatically follow that the odds ratio relating the **F** to exposure **E**, conditional on disease status (i.e., being non-diseased) equals 1. Matching in a case-control study therefore does not automatically control for confounding of the odds ratio.

Moreover, it can be shown (details omitted here) that the **crude odds ratio for matched case-control data is always biased towards the null value of 1**. Thus, it is necessary to control for the matching variable **F** (using stratified analysis or logistic regression) since ignoring the control of **F** (by using a crude odds ratio estimate) will give a biased answer tending towards concluding an absence of an effect.

What are the Consequences from Not Doing a Matched Analysis (Case-Control Data)?

There are two ways to not carry out a matched analysis:

1. **Ignore the matching**
2. **Break the matching**

1. If we match on one or more variables, but we analyze the resulting study data without controlling for any of the matched variables, then we are **ignoring the matching**. As an example, suppose we have dichotomous **E** and **D** variables, 100 cases, and we pair-match on age and sex. If we ignore the matching and ignore controlling for any other risk factors not matched on, then we are effectively doing a "crude analysis" of the data, i.e., our estimate is a crude odds ratio, **cOR** 'hat'. There are two fundamental criticisms of ignoring the matching:

   a. The estimated **cOR** is always biased towards the null value of 1 in matched (case-control and cohort) studies. Thus, the estimated **cOR** is expected to give a different (biased) odds ratio from the odds ratio (i.e., **mOR**) expected from a matched analysis.
   b. If the matching does its job (i.e., helps precision), the **mOR** estimate is expected to give better precision than the corresponding **cOR** estimate.

2. If we match, but control for the matched variables without doing a matched analysis, then we are **breaking the matching**. As an example, suppose, as above, we have dichotomous **E** and **D** variables, 100 cases, and we pair-match on age and sex. If we break the matching, then we control for age and sex by forming strata from combinations of these 2 variables, and then do a stratified analysis. The number of resulting strata is likely to be considerably less than 100, e.g., if age has 3 categories and sex has 2 categories, then the number of strata is six. If we do not break the matching, we control for age and sex by treating each matched set as a single stratum. Since there are 100 case-control pairs, a matched analysis would then be a stratified analysis involving 100 strata with 2 persons per strata.

Continued on next page

What are the Consequences from Not Doing a Matched Analysis (Case-Control Data)? (Continued)

What is a good reason to break the matching? In the above example, where age has 3 categories and sex has 2 categories, breaking the matching is equivalent to pooling exchangeable matched sets, which is more appropriate than assuming that there are 100 distinct matching strata. [See the activity on "Pooling" on Lesson Page 15-4]. **What's are some of the problems with breaking the matching?**

a. If you break the matching, then it is possible that the precision of the estimated odds ratio might be less than the precision obtained by doing a matched analysis (with or without pooling).
b. The strata resulting from breaking the matching may not be equivalent to the strata that would result from pooling exchangeable matched sets (the correct analysis).
c. As a consequence of b, the estimated odds ratio obtained from a stratified analysis resulting from breaking the matching may be meaningfully different from the estimated odds ratio obtained from a pooled analysis.
d. If you wish to control for variables that have not been matched in addition to the matching variables, then breaking the matching will require you to break up matched sets for those pairs that are in different categories of the unmatched variable(s).

Analysis of Pair-Matched Case-Control Data

*We now illustrate a matched analysis using **pair-matched case-control** data. The data can be analyzed using a stratified analysis to obtain a **mOR**, a **MH test** of hypothesis, and a **confidence interval** around the mOR.*

In the 1970s, several studies were carried out to evaluate whether the use of estrogen as a hormone replacement for menopausal women leads to endometrial cancer. One such study used **individual matching** to carry out a **pair-matched case-control study** involving women living in a Los Angeles retirement community between 1971 and 1975. There were 63 cases. Controls were chosen by individual matching to cases on age, marital status, and date of entry into the retirement community. Each of the 63 matched-pairs represents 63 strata containing 2 persons per stratum. For each stratum, we form the 2 by 2 table that relates exposure, here estrogen use, to disease outcome, here, endometrial cancer status. Each of these strata can take on one of the four forms shown below, depending on the exposure status determined for the case and control persons in a given stratum.

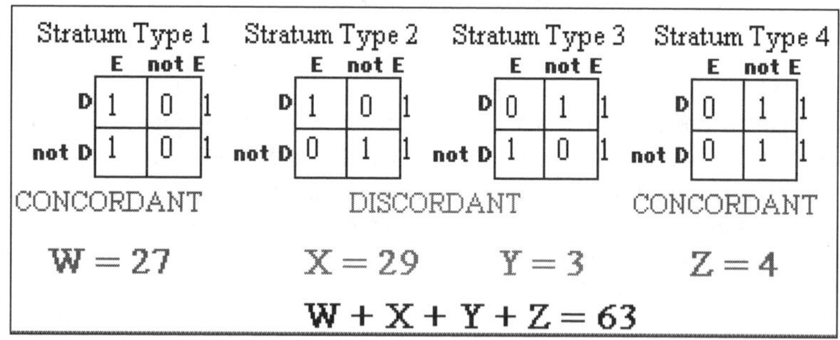

Stratum Type 1 holds any matched pair where **both** the case and the controls are exposed, that is, both used estrogen. A matched pair of this type is called a **concordant** matched pair. We denote the number of concordant matched pairs of this type **W**. The study actually found 27 matched pairs of this type.

Stratum Type 4 holds those matched pairs where **neither** the case nor the control used estrogen. This type of stratum also holds concordant matched pairs since both cases and controls have the same exposure status, this time unexposed. We denote the number of **concordant** matched pairs of this type **Z**. The study actually found 4 matched pairs of this type.

The other two stratum types hold what are called **discordant pairs**. In stratum type 2, the case uses estrogen but the control does not. In stratum type 3, the case did not use estrogen, and the control did. In both these types of strata, the case has a different exposure than its matched control. The numbers of discordant pairs of each type are called **X** and **Y**, respectively. The study found **X** equal to 29 and **Y** equal to 3. Notice that the sum of **W**, **X**, **Y**, and **Z** is 63, the number of matched pairs in the study.

How do we analyze these data? A simple answer to this question is that we use a computer to carry out a stratified analyses of these 63 strata to obtain a **Mantel-Haenszel odds ratio**, a **Mantel-Haenszel test of hypothesis**, and a **95%**

confidence interval around the estimated odds ratio. We will carry out this analysis in the next activity.

Study Question (Q15.10)

1. Why is it necessary to compute an mOR instead of a precision-based adjusted odds ratio (i.e., aOR)?

Summary

- We illustrate a matched analysis using pair-matched case-control data.
- The study involved 63 matched-pairs or strata with 2 persons per stratum.
- There were 4 types of strata, 2 of which involved **concordant** matched pairs and 2 of which involved **discordant** matched pairs.
- **W** = the number of concordant pairs where both the case and control are exposed.
- **X** = the number of discordant pairs where the case is exposed and control unexposed.
- **Y** = the number of discordant pairs where the case is unexposed and the control exposed.
- **Z** = the number of concordant pairs where both the case and control are unexposed.
- The data can be analyzed using a stratified analysis to obtain a mOR, a MH test of hypothesis, and a confidence interval around the mOR.

Analysis of Pair-Matched Case-Control Data (continued)

A stratified analysis of the 63 strata can be carried out using a computer to obtain a **Mantel-Haenszel odds ratio**, a **Mantel-Haenszel test of hypothesis**, and a **95% confidence interval** for the **mOR**. Each of the 63 strata is of one of the four types shown in the table in the previous Activity. A convenient way to carry out this analysis without using a computer is to form the following table using the numbers of **concordant** and **discordant** pairs **W**, **X**, **Y**, and **Z**.

McNemar's Table (pair-matched case-control)		Control E	Control not E
Case	E	W	X
	not E	Y	Z

This table is called **McNemar's table for pair-matched case-control data**. The numbers in this table represent pairs of observations rather than individual observations. Using this table, simple formulas can be written for the mOR, Mantel-Haenszel test statistic, and for a 95 percent confidence interval for the mOR (see table following this paragraph). Notice that all these formulas involve information only on the numbers of **discordant** pairs, **X** and **Y**; the **concordant** pair information is not used.

$$\hat{mOR} = \frac{X}{Y}$$

$$\chi^2_{MHS} = \frac{(X - Y)^2}{X + Y}$$

$$\hat{mOR} \exp\left[\pm 1.96\sqrt{\frac{1}{X} + \frac{1}{Y}}\right]$$

Substituting the values for **X** and **Y** into each of these formulas, we obtain the results shown here:

$$\hat{mOR} = \frac{29}{3} = 9.67$$

$$\chi^2_{MHS} = \frac{(29 - 3)^2}{29 + 3} = 21.13$$

95% CI: **(2.94, 31.73)**

Study Question (Q15.11)

1. How do you interpret the above results in terms of the relationship between estrogen use (E) and endometrial cancer (D) that was addressed by the pair-matched case-control study?

Summary

- ❖ A convenient way to carry out a matched-pairs analysis for case-control data is to use **McNemar's** table containing concordant and discordant matched pairs.
- ❖ Using only the discordant pairs **X** and **Y**, simple formulae can be used to compute the mOR, the MH test of hypothesis, and a 95% CI around the mOR.

The Case-Crossover Design

The **case-crossover design** is a variant of the **matched case-control study** that is intended to be less prone to bias than the standard case-control design because of the way controls are selected. The design incorporates elements of both a matched case-control study and a **nonexperimental retrospective crossover experiment**. (Note: In, a **crossover design**, each subject receives at least two different exposures/treatments at different occasions.) The fundamental aspect of the case-crossover design is that each case serves as its own control. Time-varying exposures are compared between intervals when the outcome occurred (case intervals) and intervals when the outcome did not occur within the same individual.

The case-crossover design was designed to evaluate the effect of brief exposures with transient effects on acute health outcomes when a traditional control group is not readily available. The primary advantage of the case-crossover design lies in its ability to help control confounding. Self-matching subjects against themselves automatically eliminates confounding between subjects and from both measured and unmeasured fixed covariates.

As an example of a case-cross over design, Redlemeier and Tibshirani studied whether the use of a cellular telephone while driving increases the risk of a motor vehicle collision. Their data considered 699 drivers who had cellular telephones and who were involved in motor vehicle collisions resulting in substantial property damage but no personal injury. Each person's cellular-telephone calls on the day of the collision and during the previous week were analyzed through the use of detailed billing records.

Overall, 170 of the 699 subjects had used a cellular telephone during the 10-minute period immediately before collision, 37 subjects had used the telephone during the same period on the day before the collision, and 13 subjects had used the telephone during both periods. This information provided the following McNemar table for analysis:

	Day Before		
Crash Day	**Use**	**Not Use**	
Use	13	157	170
Not Use	24	505	529
	37	662	699

From these data, the following matched analysis results were obtained:

$$m\hat{O}R = \frac{X}{Y} = \frac{157}{24} = 6.5 \quad \text{and} \quad \chi^2 = \frac{(X-Y)^2}{X+Y} = \frac{(157-24)^2}{157+24} = 97.7 \ (P < .001)$$

95% confidence interval for the **mOR**:

$$m\hat{O}R\exp\left[\pm 1.96\sqrt{\frac{1}{X}+\frac{1}{Y}}\right] = 6.5\exp\left[\pm 1.96\sqrt{\frac{1}{157}+\frac{1}{24}}\right] = (4.5, 9.9)$$

The above results indicates a very strong and significant effect that indicates that cell phone use while driving increases the risk for motor vehicle collision. Furthermore, the primary analysis, which adjusted for intermittent driving, yielded an estimated **mOR** of 6.5 with a 95% confidence interval of 4.5 to 9.9.

Analyze Pair-Matched Case-Control Data in DataDesk

An exercise is provided to demonstrate how to analyze pair-matched case-control data using the DataDesk program.

Analysis of R-to-1 Matched Case-Control Data

A 1969 matched case-control study considered the hypothesized relationship between history of induced abortion and tubal pregnancy outcome for women who had at least one earlier pregnancy. This study involved 4 to 1 **individual matching of controls to cases**. There were 18 cases, so that there were 18 matched sets or strata, each containing 5 subjects per stratum. The total number of subjects in the study was therefore 18 times 5, or 90.

Controls were category-matched to each case on order of pregnancy, age, and husband's age. The 2 by 2 data layout for stratum **i** is shown here, where $\mathbf{Y_i}$ equals 1 if the case in stratum **i** had a previous abortion and 0 if not, and where $\mathbf{X_i}$ denotes the number of controls out of 4 in stratum **i** who had a previous abortion.

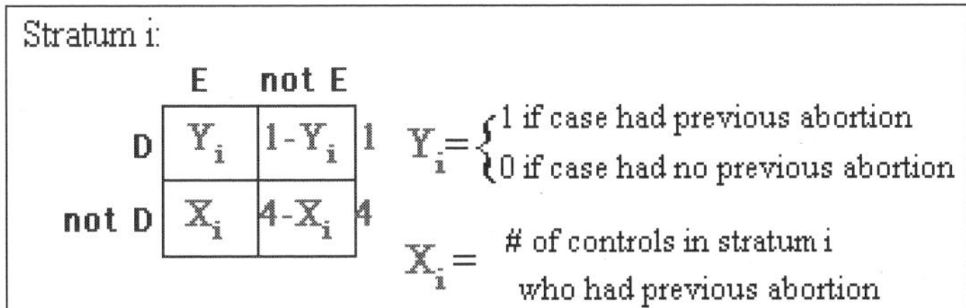

Stratum i:
E
not E
D
Y_i
$1-Y_i$
1
not D
X_i
$4-X_i$
4
$Y_i = \{$ 1 if case had previous abortion
0 if case had no previous abortion
$X_i =$ # of controls in stratum i who had previous abortion

Here is a summary of the study results:

		# of Controls with Previous Abortion				
		$X_i = 4$	$X_i = 3$	$X_i = 2$	$X_i = 1$	$X_i = 0$
Case Abortion History	$Y_i = 1$	1	0	3	5	3
	$Y_i = 0$	0	0	0	1	5

The numbers in the body of the table represent counts of matched sets or strata. If you add up all these numbers, you get 18, which is the total number of matched sets in the study, where each matched set is of the general form shown for stratum **i**. Each number in the body of the table gives the number of matched sets for which $\mathbf{X_i}$ of the controls have had a previous abortion and for which the case either had or did not have a previous abortion. For example, the 1 in the top left cell indicates that 1 of the 18 matched sets had all 4 of the 4 controls with previous abortions and that the case also had a previous abortion. The 5 in the bottom right cell, in contrast, indicates that there were 5 matched sets in which none of the 4 controls nor the case had a previous abortion.

How do we analyze these data? The answer is that we use a computer to carry out a stratified analysis of the 18 strata comprising this dataset to obtain a Mantel-Haenszel odds ratio estimate, a Mantel-Haenszel test of hypothesis, and a 95% confidence interval for the **mOR**. Here are the computed results.

Stratified Analysis

$m\hat{O}R = 33.00$

MH test = 16.0 ($P < .0001$)

95 % CI for mOR: (2.95, 369.17)

Study Questions (Q15.12)

1. Why is the mOR being used here instead of a precision-based aOR?
2. Since there are only 5 subjects in each stratum (i.e., matched set), how can you justify the use of the MH test statistic as a large-sample chi square statistic?
3. What do you conclude from these results?

Summary

- As with pair-matched data, the analysis of **R to 1 matched data** is a stratified analysis, although now there are **R + 1** subjects in each stratum.
- The calculation of the **mOR**, the **MH test**, and a 95% confidence interval for the **mOR** are best carried out using a computer program
- The **i**-th stratum has the data layout shown below where $\mathbf{Y_i}$ denotes the number of cases (either 1 or 0) that are exposed and $\mathbf{X_i}$ denotes the number of controls out of **R** that are exposed in stratum **i**.

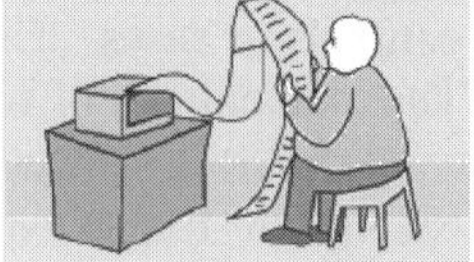

Quiz (Q15.13)

True or **False**:

1. A matched analysis can be carried out using a stratified analysis in which the strata consists of the collection of matched sets or pooled matched sets. . . . . . ???
2. In a pair-matched case-control study, the mOR is computed as: (X-Y)/(X + Y), where X and Y are discordant pair frequencies. . . . . ???
3. Suppose in a pair-matched case-control study, the number of pairs in each of the 4 cells used for McNemar's test is given by W=50, X=30, Y=15, 2=100. Then the computed value of McNemar's test statistic is 2. . . . . ???
4. The Mantel-Haenszel test Statistic is not appropriate when there is R-to-1 matching and R is at least 2. . . . . ???

Consider the McNemar's table shown below from the "Agent Orange" study (Donovan et al., Med. J. Aus., 1984). This is a pair-matched case-control study. Cases are babies born with genetic anomalies and controls are babies born without such anomalies. The exposure factor is status of the father (1 =Vietnam veteran, 0=non-veteran).

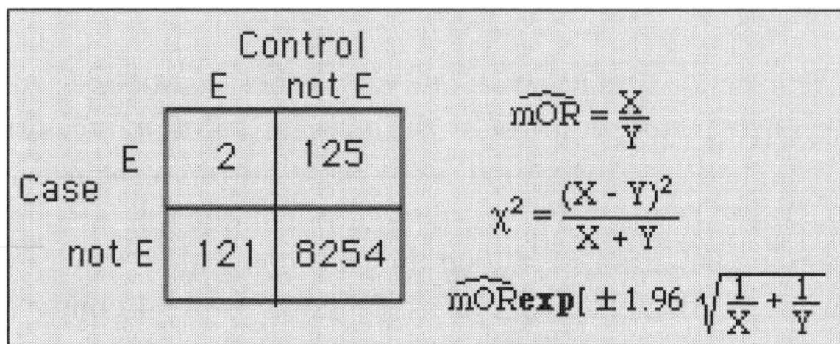

		Control E	Control not E
Case	E	2	125
	not E	121	8254

$$\widehat{mOR} = \frac{X}{Y}$$

$$\chi^2 = \frac{(X - Y)^2}{X + Y}$$

$$\widehat{mOR}\exp[\pm 1.96\sqrt{\frac{1}{X} + \frac{1}{Y}}$$

5. The mOR from this table is . . . . ???
6. The Mantel-Haenszel test statistic is computed to be . . . . . ???
7. A 95 % CI for the mOR is given by the limits . . . . . . ???
8. Based on the data, are Vietnam veterans more likely to have babies with genetic anomalies? ???

Choices

(0.76, 1.25)   (0.80, 1.32)   0.07   0.97   1.03   3.98   no   yes

15-4 Matching (continued)

Pooling Matched Data

Suppose smoking status, defined as **ever** versus **never** smoked, is the only matching variable in a **pair-matched case-control study** involving 100 cases. Suppose further that when the matching is carried out, 60 of the matched pairs contain 2 smokers and 40 of the remaining matched pairs contain 2 non-smokers.

Now let's consider any two of the matched pairs involving smokers, say pair A and pair B. Because the only variable being matched on is smoking, the control in pair A had been eligible to be chosen as the control for the case in pair B before the matching process. Similarly, the control smoker in pair B had been eligible to be the control smoker for the case in pair A.

Even though this did not actually happen after matching took place, the potential **exchangeability** of these two controls suggests that pairs A and B should not be treated as separate strata in a matched analysis. Matched sets such as pairs A and B are called **exchangeable matched sets**. For the entire study involving l00-matched pairs, the 60 matched pairs all of whom are smokers are exchangeable, and the remaining 40 matched pairs of non-smokers are separately exchangeable.

If we ignored exchangeability, the typical analysis of these data would be a stratified analysis that treats all 100 matched pairs as 100 separate strata. The analysis could then be carried out using the discordant pairs information in **McNemar's** table, as we described earlier. But should we actually ignore the exchangeability of matched sets? We say no, primarily because to treat exchangeable strata separately artificially assumes that such strata are unique from each other when in fact they are not.

So how should the analysis be carried out? The answer here is to pool exchangeable matched sets. In our example, pooling would mean that rather than analyzing l00 distinct strata with 2 persons per stratum, the analysis would consider only two pooled strata, one pooling 60 matched sets into a smoker's stratum and the other pooling the other 40 matched sets into a non-smoker's stratum.

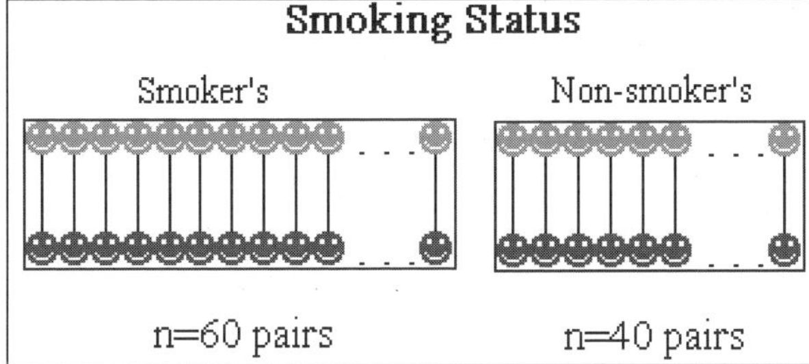

Study Questions (Q15.14)

Consider a 2-to-1 matched case-control study that category-matches on age group (say, below 60 versus 60 or above) and on smoking status (ever versus never).

1. What are the exchangeable matched sets in this study?

Consider a different 2-to-1 matched case-control study that category matches on age group, smoking status, gender, physical activity level, body size, race, and ethnic group.

2. Should the analysis require pooling of matched sets?

Summary

- ❖ Two or more matched sets are **exchangeable** if they have the same combination of matching categories.
- ❖ A **pooled analysis** should be preferred to an **unpooled analysis** whenever there are exchangeable matched sets.
- ❖ A **pooled analysis** involved combining exchangeable matched sets into a single **pooled** stratum and then performing a stratified analysis for the pooled strata.

Summary continued on next page

- ❖ Pooling exchangeable matched sets is appropriate because keeping exchangeable strata separate artificially assumes that such strata are unique from each other when in fact they are not.
- ❖ If several variables are matched, the results from a **pooled analysis** should be negligibly different from the results from an **unpooled analysis**.

Pooling versus Frequency Matching

We address three questions here:

1. Is **pooling of exchangeable matched sets** equivalent to **frequency matching**?
2. If you decide to pool exchangeable matched sets, are the numbers and definitions of the resulting strata identical with the strata that would be defined if you frequency matched instead?
3. Will the results from analyzing pooled matched data always be identical to the results obtained if you had frequency matched instead?

The answer to each of the three questions is **no**. However, under certain assumptions about the way the matching is carried out, the answer to the second question could be yes.

Answer to 1: Pooling requires **individual matching,** which is a different procedure for selecting comparison subjects than is frequency matching. In particular, the number of matching strata resulting from pooling, the way the strata are defined, and the subjects that are selected within each stratum may not be identical to the corresponding items from a frequency matched study (see below).

Answer to 2: If one or more of the matching variables is continuous (e.g., age), then the way exchangeability is defined can be different from the way frequency matched categories might be defined. For example, if individually matching on age, narrower age categories (e.g., 5 years) may be used than chosen and considered exchangeable than the categories chosen if frequency matching (e.g., 15 years) were used instead. In such a situation not only will the number of strata for a pooled analysis be different than for frequency-matched data, but also the strata definitions will be different.

On the other hand, if either a) the exact same matching categories (e.g., 5 year age categories) would have been defined for both individual matching and frequency matching, or b) if the definition of exchangeability is broadened to be equivalent to the category definitions used for frequency matching, then both the number of strata and types of strata will be equivalent for both the pooled and frequency matched analysis.

Answer to 3: The strata resulting from pooling exchangeable matched sets may not be equivalent to the strata that result from frequency matching.

We illustrate with the following example:

Suppose we match on only one variable, say, smoking status (ever versus never), in a case-control study. If we use **individual pair matching** and there are 100 cases, then the original strata will consist of 100 matched pairs, each pair containing one case and one control with the same smoking status.

Since only one variable is involved in the matching, all matched pairs containing ever smokers are exchangeable and all matched pairs containing never smokers are also (separately) exchangeable. If, as recommended, we **pool exchangeable matched sets**, then we wind up with only two strata, one for ever smokers and the other for never smokers.

If. in contrast to individual matching on smoking status, we **frequency-match** on smoking status with the same 100 cases and the same total sample size, we would also wind up with the same two strata as obtained from pooling provided we chose the same control subjects as we would have chosen from frequency matching.

However, even though we will choose the same number (100) of control subjects, there is no guarantee that we will choose the same subjects from the pool of possible controls as we would have obtained using individual matching. Consequently, the two strata obtained from frequency matching may contain different subjects than the two strata obtained from individual pair-matching and the resulting matched analysis are likely to lead to different numerical results (though not necessarily different statistical inference conclusions).

As a second example, suppose we again have 100 cases for a case-control study. This time, we **individually pair-match** on 5 variables, say age, race, sex, smoking status, body size (categorized into obese and not-obese), and social class (say, in three categories). Then, two matched pairs are exchangeable if both case and control are in the same category of age, race, sex, smoking status, body size, and social class. Because 5 variables are involved in the matching here, there are not likely to be very many exchangeable strata, although there may certainly be some exchangeable strata. In fact, it might not even be possible to find a match for certain cases. Moreover, if we **pool exchangeable strata**, the total number of resulting strata is not likely to be reduced greatly from the original number of 100 unpooled strata.

Continued on next page

Pooling versus Frequency Matching (continued)

Now suppose we **frequency match** on the same 100 cases by choosing 100 controls matched on the same 5 variables. As with matching on only one variable, even if the number of strata obtained from frequency matching is the same as that obtained from pooling individual matched strata, the actual subjects may be different depending on the control subjects that actually get selected. Consequently, the resulting analysis from frequency matching and individual matching might yield different results.

Quiz (Q15.15)

ID	STRATUM	CASE	EXP	SMK
1	1	1	1	0
2	1	0	1	0
3	2	1	0	0
4	2	0	1	0
5	3	1	1	1
6	3	0	0	1
7	4	1	1	0
8	4	0	0	0
9	5	1	0	1
10	5	0	0	1

The data to the right are from a hypothetical pair-matched case-control study involving 5 matched pairs, where the only matching variable is smoking (SMK). The disease variable is called CASE and the exposure variable is called EXP. The matched set # is identified by the variable **stratum**.

1. How many concordant pairs are there where both pair members are exposed? . . **???**
2. How many concordant pairs are there where both members are unexposed? . . **???**
3. How many discordant pairs are there where the case is exposed and the control is unexposed? **???**
4. How many discordant pairs are there where case is unexposed and the control is exposed? **???**

Choices
0 **1** **2** **3** **4**

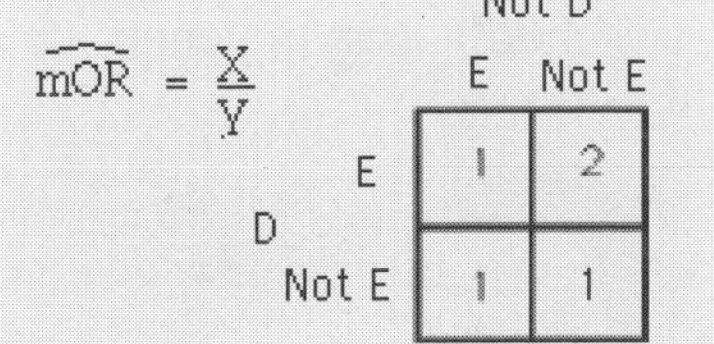

$\widehat{mOR} = \frac{X}{Y}$

		Not D E	Not D Not E
D	E	1	2
D	Not E	1	1

The table to the right summarizes the matched-pairs information described in the previous questions.

5. This table is called **???** table.
6. What is the estimated mOR for these data? . **???**
7. What type of matched analysis is being used with this table? **???**

Choices
0.5 **1** **2** **Berkson's** **Mantel-Haenszel's** **McNemar's** **pooled** **unpooled**

The table to the right groups the matched pairs information described in questions 1-4 into two smoking strata.

$$\frac{\sum \frac{a_i d_i}{n_i}}{\sum \frac{b_i c_i}{n_i}}$$

SMK=1	E	Not E	
D	1	1	2
Not D	0	2	2

SMK=0	E	Not E	
D	2	1	3
Not D	2	1	3

8. What is the estimated mOR from these data? . . **???**
9. What type of matched analysis is being used here? . **???**
10. Which type of analysis should be preferred for these matched data (where smoking status is the only matched variable), pooled or unpooled? . . . **???**

Choices
1 **2** **2.5** **pooled** **undefined** **unpooled**

Quiz continued on next page

The data to the right switches the non-smoker control of stratum 2 with the non-smoker control of stratum 4 from the dataset provided for previous questions 1-4.

ID	STRATUM	CASE	EXP	SMK
1	1	1	1	0
2	1	0	1	0
3	2	1	0	0
4	2	0	0	0
5	3	1	1	1
6	3	0	0	1
7	4	1	1	0
8	4	0	1	0
9	5	1	0	1
10	5	0	0	1

Let:

W = # concordant (E=1 ,E=1) pairs
X = # discordant (E=1 , E=O) pairs
U = # discordant (E=O, E=1) pairs
Z = # concordant (E=O, E=O) pairs for the "switched" data. Then:

11. W= . . **???**
12. X= . **???**
13. Y= . **???**
14. Z= . **???**
15. mOR (unpooled) = **???**
16. mOR (pooled) = **???**

Choices
0 **1** **2** **2.5** **3** **4** **undefined**

For the pair-matched data considered in questions 1-10, mOR(unpooled)=2 whereas mOR(pooled)=2.5. For the ("switched") pair-matched data considered in questions 11-16, mOR(unpooled) was undefined whereas mOR(pooled)=2.5.

PREVIOUS DATA					SWITCHED DATA				
ID	STRATUM	CASE	EXP	SMK	ID	STRATUM	CASE	EXP	SMK
1	1	1	1	0	1	1	1	1	0
2	1	0	1	0	2	1	0	1	0
3	2	1	0	0	3	2	1	0	0
4 *	2	0	1	0	4 *	2	0	0	0
5	3	1	1	1	5	3	1	1	1
6	3	0	0	1	6	3	0	0	1
7	4	1	1	0	7	4	1	1	0
8 k	4	0	0	0	8 *	4	0	1	0
9	5	1	0	1	9	5	1	0	1
10	5	0	0	1	10	5	0	0	1

17. Which of the following helps explain why the pooled mOR estimate should be preferred to the unpooled mOR? . . **???**
   a. The pooled mOR's are equal whereas the unpooled mOR's are different.
   b. The unpooled mOR's assume that exchangeable matched pairs are not unique.
   c. The pooled mOR's assume that exchangeable matched pairs are unique.

Choices
All **None** **a** **b** **c**

Analysis of Frequency Matched Data

The table shown below gives the breakdown of cases and controls for a hypothetical **matched case-control study** that used **frequency matching** on race and gender. The study involves twice as many controls as there are cases. Nevertheless, since numbers of cases and controls differ in each of the four strata, the matching ratio here is of the form $\mathbf{R_i}$ to $\mathbf{S_i}$, where $\mathbf{R_i}$ is the number of controls and $\mathbf{S_i}$ is the number of cases in stratum **i**.

Frequency Matching on Race and Gender

	Stratum 1 White Male	Stratum 2 White Female	Stratum 3 Black Male	Stratum 4 Black Female	
Cases S_i	100 (33%)	100 (33%)	40 (13%)	60 (20%)	300
Controls R_i	200 (33%)	200 (33%)	80 (13%)	120 (20%)	600

Matching Ratio: R_i to S_i

	E	not E	
D	Y_i	$S_i - Y_i$	S_i
not D	X_i	$R_i - X_i$	R_i

$Y_i = \{$# of cases in stratum i who are exposed

$X_i = \{$# of controls in stratum i who are exposed

Following this paragraph are the study data obtained for each of the four strata. How do we analyze these frequency-matched data? The answer is we carryout a **stratified analysis** of the data in the 4 strata to obtain a precision-based odds ratio estimate, a Mantel-Haenszel test of hypothesis, and a 95% confidence interval for the **aOR**. Before computing such summary estimates, however, we must check to see that there is no meaningful or significant interaction over the strata. Here are the computed results:

Frequency Matching on Race and Gender

	Stratum 1 White Male		Stratum 2 White Female		Stratum 3 Black Male		Stratum 4 Black Female		
	E	not E	E	not E	E	not E	E	not E	
Cases S_i	30	70	35	65	11	29	19	41	300
Controls R_i	20	180	22	178	8	72	11	109	600
	$\hat{OR}_1 = 3.86$		$\hat{OR}_2 = 4.36$		$\hat{OR}_2 = 3.41$		$\hat{OR}_2 = 4.59$		

aOR = 4.10

MH test = 64.30 (P < .0001)

95 % CI for aOR: (2.86, 5.88)

B-D test for interaction: P = .96

Now here are some study questions based on these results.

Study Questions (Q15.16)

1. Is there confounding?
2. Based on comparing the stratum-specific odds ratios, is it justifiable to carry out an overall assessment of the exposure-disease relationship?
3. Why is an aOR computed instead of a mOR?
4. Based on the results, what do you conclude about the E→D association?

Summary

- ❖ Frequency matched data may be analyzed using a standard stratified analysis.
- ❖ For frequency-matched data, the strata typically have a varying matching ratio, with $\mathbf{R_i}$ controls and $\mathbf{S_i}$ cases in stratum **i**.
- ❖ Because frequency matched strata typically involve large cell frequencies and zero cell-frequencies are rare, a precision-based aOR is often used instead of a mOR as the overall point estimate.

Quiz (Q15.17)

A case-control study conducted in 1990 in Puerto Rico was aimed at determining risk factors associated with severe measles (D). Controls were frequency matched to cases by region of residence (5 regions). Potential risk factors that were examined included Annual Family Income < $5000 (yes or no), Underlying Illness (yes or no), Mother Without High School Degree (yes or no), Anemia (yes or no).

1. **True** or **False**: When considering each potential risk factor separately as the exposure variable, the appropriate analysis is a stratified analysis involving 5 strata. . . . . ???

In the frequency matched case-control study described previously, 16 cases (i.e., children with severe measles) were compared to selected children with non-severe measles as controls (39 hospitalized and 38 nonhospitalized). An underlying illness was present in 50% of the cases and 16% of the nonhospitalized controls. Stratifying by region, it was found that mOR=5.3 with a 95% CI for mOR = (1.4, 20.2).

2. Which of the following statements **does not** support the use of a mOR for this analysis: ???
   a. There were zero cells in some strata.
   b. Frequency matching was used.
   c. Logistic regression should have been used.
3. **True** or **False**: Since controls were frequency matched to cases on region of residence, it is NOT necessary to control for region of residence in the analysis. . . . ???
4. If we frequency match and choose 3 times as many controls as cases, the matching ratio is . ???
5. Will frequency matching on region of residence result in better precision of the adjusted odds ratio than if individual matching had been used instead? . . . . . ???

Choices
1-to-3 Always False Never Not necessarily R-to-1 R_i-to-S_i True

15-5 Matching (continued)

Analysis of Matched Cohort Data

*We now describe how to analyze pair-matched data from a **cohort study**, where unexposed subjects are paired with exposed subjects on selected matching variables.*

Analysis of 1 to 1 Matched Cohort Data

Thus far we have considered only matched case-control data. We now focus on the analysis of **matched cohort** data. A retrospective cohort study used pair matching to investigate rotating shifts compared to steady shifts were associated with the development of chronic low back pain in a group of factory workers in Ohio from 1950-1975. The study involved 80 **matched pairs** of exposed versus unexposed workers. The exposed workers had been assigned to rotating shifts whereas the unexposed workers had been assigned to steady shifts. Matching involved seven variables: 1) year of birth; 2) year of hire; 3) age at hire; 4) duration of employment; 5) type of job; 6) race; and 7) marital status.

The health outcome, chronic low back pain status, was determined from worker reports obtained during mandatory bi-annual medical exams made through 1975. As with matched case-control data, each of the 80 matched-pairs represent 80 strata containing 2 persons per strata. The table to the right is the 2 by 2 table for stratum **i** that relates exposure, here rotating versus steady shift assignment, to health outcome, here, chronic low back pain. Because this is a matched cohort study, so we are matching on exposure status, the two column totals in stratum **i** are one.

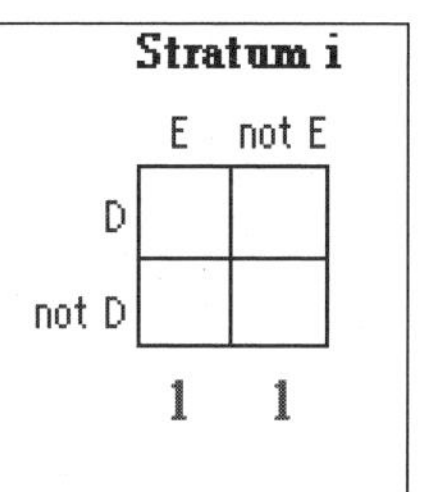

Each of the 80 strata can take on one of the four forms shown below, depending on the health outcome status determined for the exposed and unexposed persons in a given stratum:

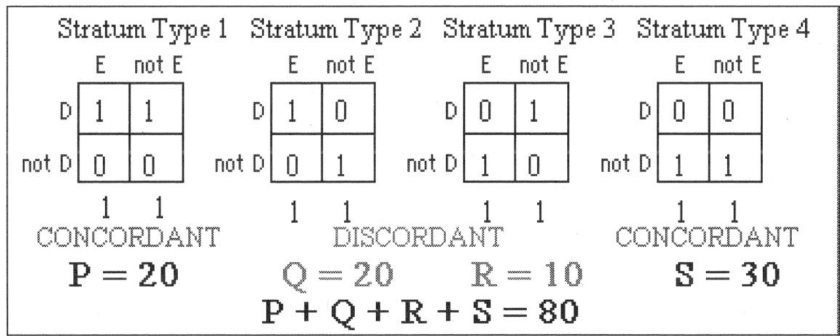

Stratum Type 1 identifies any matched pair where **both** the exposed and the unexposed subjects have the health outcome, that is, both have chronic low back pain. This is called a **concordant matched pair**. We denote the number of concordant matched pairs of this type **P**. The study actually found 20 matched pairs of this type. Stratum Type 4 identifies those matched pairs where **neither** the exposed nor the unexposed subject was determined to have chronic low back pain. These are also **concordant matched pairs** because both exposed and unexposed subjects have the same health outcome, which this time is no chronic back pain. We denote the number of concordant matched pairs of this type **S**. The study actually found 30 matched pairs of this type.

The other two stratum types consider **discordant pairs**. In stratum type 2, the exposed subject develops the health outcome but the unexposed subject does not. In stratum type 3, the exposed subject does not develop the health outcome, but the unexposed subject does. The number of discordant pairs of each type is called **Q** and **R**, respectively. The study found **Q** = 20 and **R** = 10. The sum of **P**, **Q**, **R**, and **S** = 80, the number of matched pairs in the study:

	Stratum Type 1		Stratum Type 2		Stratum Type 3		Stratum Type 4	
	E	not E	E	not E	E	not E	E	not E
D	1	1	1	0	0	1	0	0
not D	0	0	0	1	1	0	1	1
	1	1	1	1	1	1	1	1
	CONCORDANT		DISCORDANT				CONCORDANT	
	P = 20		Q = 20		R = 10		S = 30	

P + Q + R + S = 80

How do we analyze these data? As with any matched analysis, the general answer here is that we use a computer to carry out a stratified analyses of these 80 strata to obtain a Mantel-Haenszel risk ratio (**mRR**), a Mantel-Haenszel test of hypothesis, and a 95 percent confidence interval for the mRR.

Study Question (Q15.18)

1. Why is it necessary to compute an mRR instead of either a mOR or precision-based adjusted risk ratio (i.e., aRR)?

We will carry out this analysis in the next activity.

Summary

- We illustrate a matched analysis using pair-matched cohort data.
- The study involved 80 matched-pairs or strata with two persons per stratum.
- There were four types of strata: two involved **concordant matched pairs** and two **discordant matched pairs**.
- **P** = number of concordant pairs: exposed and unexposed cases
- **Q** = number of discordant pairs: exposed case, unexposed non-case
- **R** = number of discordant pairs: unexposed case, exposed non-case
- **S** = number of concordant pairs: exposed and unexposed non-cases
- The data can be analyzed using a stratified analysis to obtain an mRR, a MH test of hypothesis, and a confidence interval around the mRR.

Analysis of Pair-Matched Cohort Data

We continue to consider the pair-matched retrospective cohort study data shown here involving 80 matched pairs:

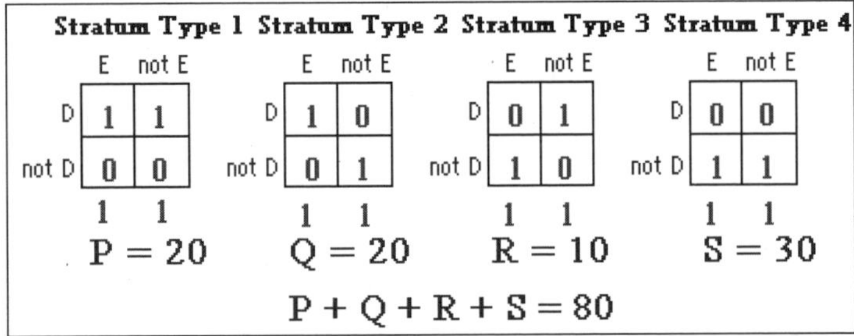

A stratified analysis of the 80 strata can be carried out using a computer to obtain a Mantel-Haenszel risk ratio (mRR), a Mantel-Haenszel test of hypothesis, and a 95% confidence interval for the mRR. Each of the 80 strata will be of one of the four types shown above, involving either concordant or discordant pairs.

A convenient way to carry out this analysis without using a computer is to form the following table using the numbers of concordant and discordant pairs **P**, **Q**, **R**, and **S**. This table is **McNemar's table for pair-matched cohort data**. Each observation in this table represents a pair of observations rather than an individual observation. Using this table we can obtain a formula for the mRR, Mantel-Haenszel test statistic, and a 95 percent confidence interval for the mRR.

McNemar's Table

		Unexposed D	Unexposed not D
Exposed	D	P	Q
	not D	R	S

$$\text{mRR} = \frac{P + Q}{P + R}$$

$$\chi^2_{MHS} = \frac{(Q - R)^2}{Q + R}$$

$$\text{mRR} \exp\left[\pm 1.96\sqrt{\frac{Q + R}{(P + Q)(P + R)}}\right]$$

Notice that the formula for the mRR and its confidence interval involve the number, **P**, of concordant pairs with the health outcome, in addition to the numbers, **Q** and **R** of discordant pairs, although the concordant non-case pair information is not used. Substituting the values for **P**, **Q**, and **R** into each of these formulas we obtain the results shown here:

$$\hat{\text{mRR}} = \frac{20 + 20}{20 + 10} = 1.33$$

$$\chi^2_{MHS} = \frac{(20 - 10)^2}{20 + 10} = 3.33$$

95 % CI for mRR: **(0.98, 1.81)**

Study Question (Q15.19)

1. How do you interpret the above results in terms of the relationship between rotating vs. steady shifts (E) and chronic low back pain (D) that was addressed by the pair-matched cohort study?

Summary

- ❖ A convenient way to carry out a matched-pairs analysis for cohort data is to use **McNemar's table** containing concordant and discordant matched pairs.
- ❖ Simple formulae can be used to compute the mRR, the MH test of hypothesis, and a 95% CI for the mRR.

Quiz (Q15.20)

True or **False**: In a pair-matched cohort study:

1. The total cases in each matched set is always 1. . . . . . . ???
2. The McNemar test of hypothesis involves only the discordant pair information. . . ???

Using the McNemar table below:

		Unexposed D	Unexposed not D
Exposed	D	35	25
	not D	15	24

Formulae

$$\frac{Q}{R} \qquad \frac{P+Q}{P+R} \qquad \frac{(Q-R)^2}{Q+R}$$

$$\widehat{mRR}\exp\left[\pm 1.96\sqrt{\frac{(Q+R)}{(P+Q)(P+R)}}\right]$$

3. The estimated mRR is given by . . . . ???
4. The McNemar test statistic is computed to be . . . . . . ???
5. The 95% CI for the mRR has the following limits . . . . . . ???

Choices
(0.96, 1.50) (1.1, 1.6) 0.25 1.20 1.67 2.5

Analyze Matched Follow-up Data Using McNemar's Table in DataDesk

An exercise is provided to demonstrate how to analyze matched follow-up data using McNemar's table within the DataDesk program.

Logistic Regression – Matched and Unmatched Covariates for Matched Case-Control Studies

Why Stratified Analysis is Inappropriate in Matched Case-Control Studies

We have previously described how to analyze **pair-matched case-control** study data by giving the results for 63 matched pairs by using McNemar's Table to assess the relationship of estrogen to endometrial cancer. A Mantel-Haenszel odds ratio (mOR), Mantel-Haenszel test statistic, and the 95% confidence interval for the mOR are shown here:

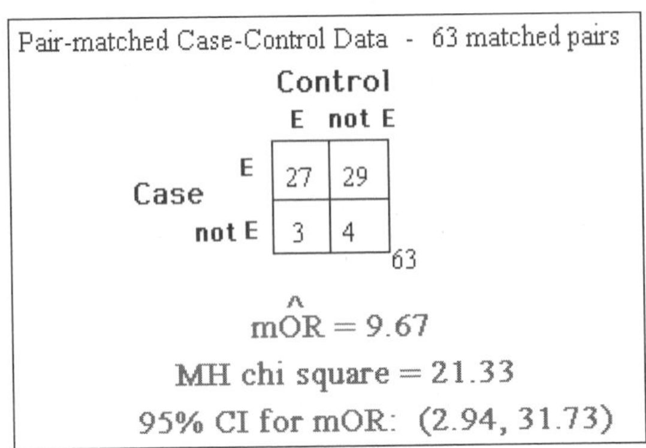

Pair-matched Case-Control Data - 63 matched pairs

		Control E	Control not E
Case	E	27	29
	not E	3	4

63

$\hat{mOR} = 9.67$

MH chi square = 21.33

95% CI for mOR: (2.94, 31.73)

The matching in this study was on the three variables age, marital status, and date of entry into the retirement community, and these were the only variables that were controlled in the analysis. Nevertheless, there were other variables not matched on that were also considered as possible control variables. One of these variables was gall bladder disease status, which the investigators considered to be a risk factor for endometrial cancer that also needed to be controlled. How do we analyze the data to account for not only the matching variables but also for the variable describing gall bladder disease status? Unfortunately, a stratified analysis is not the best approach to use here. To explain this, here are the data on two pairs of subjects in the data set:

Pair	D	E	Gall
A	1	0	1
A	0	1	0
B	1	1	0
B	0	1	1

In match pair A, the case has gall bladder disease but the control does not In pair B, the control has gall bladder disease, but the case does not. If we want to stratify on gall bladder disease status and at the same time retain the matching, we will unfortunately have to drop these two matched pairs from the data set, because both members of each pair are in different categories of the gall bladder status variable. There is a way to carry out the analysis without having to drop any matched pairs. The method to use here is **logistic regression**, which we illustrate for these data in the next activity.

Summary

- Stratified analysis is not the best approach for the analysis of matched data when there are unmatched variables to be controlled in addition to the matching variables.
- When stratifying on unmatched variables as well as the matching variables in a pair-matched study, we must drop from the analysis those matching strata in which the two paired subjects fall in different categories of the unmatched variable.
- Logistic regression can be used without dropping any matched strata to control for both matched and unmatched variables.

Logistic Regression – Matched and Unmatched Covariates in Matched Case-Control Studies

This activity describes how to use **logistic regression** to control for both matched and unmatched variables. We consider again the pair-matched case-control data involving 63 matched pairs to assess the relationship of estrogen use to endometrial cancer. The matching variables were age, marital status and date of entry into the retirement community. The only unmatched control variable is history of gall bladder disease. The logit form of a logistic model for these data is shown here:

$$\text{logit } P = \ln \frac{P}{1-P} = b_0 + \sum_{i=1}^{62} b_i D_i + b_{63} E + b_{64} \text{GALL}$$

D_i = **Dummy variable (i-th matched-pair)**

GALL = **Gall bladder disease (0,1)**

E = **Estrogen use (0,1)**

This model contains 62 dummy variables to distinguish the 63 matched-pairs. The model also contains the unmatched control variable GALL and the exposure variable E which indicates estrogen use status. This model contains 65 parameters, b_0 through b_{64}, which is more than half the total number of subjects. Whenever, as in this example, the number of parameters is large relative to the total number of subjects, the logistic model is fit using the **conditional maximum likelihood estimation** or **CMLE** method. To carry out CMLE, use a computer program.

When such a program is applied to fit this model, the estimated odds ratio that adjusts for both the matching variables and the unmatched variable GALL is given by the expression $e^{\hat{b}_{63}}$, where $\hat{b}$ is the estimated coefficient of the exposure variable. The resulting estimate is $\hat{OR}(adj) = \exp[2.209] = 9.11$

Study Questions (Q15.2

1. Why is the estimated odds ratio obtained from the above logistic model different from the mOR value of 9.67 previously calculated for these data?
2. How would you state the logit form of a logistic model for the effect of estrogen use that controls for the matching but does not account for the variable GALL?
3. The conditional ML estimate of $\hat{b}_{63}$ for the model in question 2 is 2.269. What is the value of the adjusted odds ratio that controls for the matching?

Using the logistic regression approach we can also obtain a chi square statistic to test the null hypothesis that the adjusted odds ratio is 1, that is, that there is no significant effect of estrogen use. An equivalent null hypothesis is that the coefficient of the exposure variable is zero. The resulting chi square statistic, which is called the **Likelihood Ratio statistic** has the value 13.11. This statistic is approximately a chi square with 1 df under the null hypothesis.

Likelihood Ratio

$$\chi^2_{1\ df} = 13.11$$

$$H_0\text{: } OR(adj.) = 1 \Leftrightarrow b_{63} = 0$$

Study Questions (Q15.21) continued

4. What is the P-value for the chi square test? Is the test significant?
5. Why is the chi square statistic different from the McNemar chi square of 21.13 previously computed for these data?

A 95% confidence interval for the adjusted odds ratio can also be obtained from the logistic model output. Here are both the formula for this confidence interval and the resulting confidence limits obtained from the computer output:

95% CI for OR(adj):

$$\exp[\hat{b}_{63} \pm 1.96\ \text{s.e.}(\hat{b}_{63})]$$

$$= (2.76, 30.10)$$

Study Questions (Q15.21) continued

6. How do you interpret this confidence interval?

Summary

- ❖ Logistic regression can be used to control for both matched and unmatched variables.
- ❖ The logistic model here contains **G – 1** dummy variables for **G** matching strata, the exposure variable (**E**), and any unmatched variables also to be controlled.
- ❖ For matched data, the logistic model is estimated using **Conditional Maximum Likelihood Estimation** (**CMLE**).
- ❖ The model is estimated using conditional logistic regression when there are matched data.
- ❖ If **E** is a (0, 1) variable and no predictor variables are product terms, then the adjusted OR for the effect of **E** is given by **exp(b)** where **b** is the coefficient of the **E**.
- ❖ A test of the null hypothesis that the adjusted OR = 1 can be obtained using a **Likelihood Ratio test** based on logistic regression output.
- ❖ A 95% CI for the effect of exposure is given by the formula

$$\exp[b \pm 1.96\ \text{s.e.}(b)]$$

Maximum Likelihood Methods for Analyzing Matched Data Using Logistic Regression

The most popular approach for estimating the parameters in a logistic regression models is **maximum likelihood (ML)** estimation. There are actually two alternative ML approaches: the **unconditional method (UCMLE)** and the **conditional method (CMLE)**. These two methods require different computer programs.

Of these two methods, the **CMLE** approach is typically used to analyze matched data. A survival analysis computer program can be used to carry out the **CLME** method, as provided in the computer software packages SAS and SPSS.

The key distinction between the unconditional and the conditional ML approaches concerns the **number of parameters (p)** in the model being used relative to the **total number of subjects (n)** in the study. If **p** is small relative to **n**, then **UCMLE** is typically used. If **p** is large relative to **n**, then **CMLE** is preferred.

Exactly *how large is large,* however, has never been precisely determined by statisticians. Nevertheless, it is typically argued that the logistic model for matched data usually is in the "large" category.

For example, in a case-control study involving 100 matched pairs with matching on, say, AGE, RACE, and SEX, a logistic model (assuming no pooling and no interaction terms) would have the following form:

$$\text{logit } P(X) = b_0 + \sum_{i=1}^{99} b_i D_i + b_{100} E$$

where the **D**'s denote 99 dummy variables to distinguish the 100 matching strata and **E** denotes a dichotomous exposure variable.

The number of parameters in this model is p = 101 whereas the total number of subjects is n = 100 x 2 = 200. Here, p would usually be considered large relative to n, so the CMLE approach would be recommended. In fact, for pair-matched data, the odds ratio estimate obtained from using CMLE can be mathematically shown to be the square of the odds ratio that would have been obtained if UCLME had been used on the same model instead, i.e.:

$$\hat{OR}_{UCMLE} = (\hat{OR}_{CMLE})^2$$

The odds ratio for the **CMLE** method has been shown to be the correct (i.e., **unbiased**) odds ratio, so this means that the (square) of this odds ratio obtained from the **UCMLE** method is **biased**. In particular, if the correct (i.e., **CMLE**) odds ratio turned out to be 3, then the biased (i.e., **UCMLE**) odds ratio would be 9, which is a very large bias.

In contrast, if matching had not been used in a case-control study involving 200 subjects, a logistic model for assessing the effect of exposure E controlling for, say, AGE, RACE, and SEX would have the following form:

Continued on next page

Maximum Likelihood Methods for Analyzing Matched Data Using Logistic Regression (continued)

$$\text{logit } P(X) = b_0 + b_1\text{AGE} + b_2\text{RACE} + b_3\text{SEX} + b_4\text{E}$$

The number of parameters in this model is **p** = 5 whereas the total number of subjects is **n** = 200. Here, in contrast to the model involving matching, **p** is clearly small relative to **n**, so the **UCMLE** approach would be recommended.

In general, the **CMLE** method always gives the correct (i.e., unbiased) answer (e.g., odds ratio estimate). However, when **p** is small relative to **n**, the **UCMLE** method is preferred because not only will it give essentially the same answer as the **CMLE** method, but the **UCMLE** method will typically give a more precise confidence interval around the point estimate than obtained from the **CMLE** method.

The EVW Logistic Model for Matched Data

A general expression for the logistic model that allows for the analysis of matched data can be written as follows:

$$\text{logit } P(X) = \alpha + \sum_{i=1}^{p_1} \gamma_{1i} V_{1i} + \sum_{i=1}^{p_2} \gamma_{2i} V_{2i} + \beta E + \sum_{j=1}^{p_3} \delta_j E W_j$$

The **α**, **γ**'s, **β**'s, and **δ**'s (previously denoted as **b**'s in the activities on logistic modeling) in this model represent unknown regression coefficients that need to be estimated using (usually) conditional maximum likelihood estimation (i.e., **CMLE**). The **E**, **V**'s and **W**'s represent the predictor variables in this model. Also the **EW**'s represent product terms of **E** with one of the **W**'s. More specifically, the predictor variable **E** denotes a single **exposure variable** of interest, the first set of **V**'s denote **dummy variables** to identify $p_1 + 1$ matching strata, the second set of **V**'s denote control variables (i.e., **potential confounders**) that are not involved in the matching, and the **W**'s denote potential **effect modifiers** that go into the model as product terms with the exposure variable **E**.

We illustrate this model for a study of the relationship of estrogen use (**E** = **EST**, 1 if yes, 0 if no) to endometrial cancer (**D** = **ENC**) involving $p_1 + 1 = 63$ matched pairs of women in a Los Angeles retirement community. The matching variables were age, marital status, and date of entry into the community.

Another variable, not matched, is presence or absence of gall bladder disease (**GALL** = 1 if present, 0 if absent). A no interaction logistic model for these data based on the above general **EVW** formula is given as follows:

$$\text{logit } P(X) = \alpha + \sum_{i=1}^{62} \gamma_{1i} V_{1i} + \gamma_{21}\text{GALL} + \beta\text{EST}$$

The $\mathbf{V_{1i}}$'s In this model denote $\mathbf{p_1} = 62$ dummy variables for the 63 matched pairs. Since there is only one unmatched variable in the model, i.e., **GALL**, there is only one $\mathbf{V_{2i}}$ term in the second sum in the general **EVW** formula, i.e., $\mathbf{p_2} = 1$.

When **CMLE** is used to estimate the parameters in the model, the only parameters actually estimated are the coefficients of EST and GALL, since **α** and the **γ**'s drop out of the conditional likelihood function that is maximized.

The formula for the estimated adjusted odds ratio for the effect of estrogen use, adjusted for the matching variables and for GALL, is then equal to:

$$\hat{OR}(\text{adj}) = \exp(b) \qquad \text{where} \qquad b = \hat{\beta}$$

An alternative logistic model for these data that allows for the possibility of interaction between estrogen use and gall bladder status is given by the following model:

Continued on next page

The EVW Logistic Model for Matched Data (continued)

$$\text{logit } P(X) = \alpha + \sum_{i=1}^{62} \gamma_{1i} V_{1i} + \gamma_{21}\text{GALL} + \beta\text{EST} + \delta_1 \text{EST} \times \text{GALL}$$

For this interaction model, the formula for the estimated adjusted odds ratio is given by.

$$\hat{OR}(\text{adj}) = \exp(b + \hat{\delta}_1 \text{GALL})$$

so that the value of the "interaction" odds ratio will change (i.e., is modified) depending on the value of the potential effect modifier GALL.

Quiz (Q15.22)

Fill in the Blanks

A matched case-control of cervical cancer was conducted in Sydney, Australia (Brock et. al., J. Nat. Cancer Inst., 1988) involving 313 women. The outcome variable was cervical cancer status. The matching variables are age and socioeconomic (SOC) status. Additional variables not matched on were (0, 1) smoking status (SMK), number of lifetime sexual partners (NS), and age at first sexual intercourse (AS).

1. Which of the logistic models shown below is appropriate for analyzing these data? . ???
2. What method of estimation should be used to fit the appropriate model for these data? . ???

Choices
Conditional MLE   Least Squares   Unconditional MLE   a   b   c

a. $\text{logit } P = b_0 + b_1\text{SMK} + b_2\text{NS} + b_3\text{AS}$

b. $\text{logit } P = b_0 + b_1\text{SMK} + b_2\text{NS} + b_3\text{AS} + \sum_{j=1}^{k-1} d_j D_j$

where the D_j's are dummy variables for k strata

c. $\text{logit } P = b_0 + b_1\text{SMK} + b_2\text{NS} + b_3\text{AS} + b_3\text{AGE} + b_3\text{SOC}$

The output shown below was obtained by fitting a logistic regression model to the cervical cancer data. Assuming that the correct model was used, answer the following questions:

Variable	b	S.E.	P-value
SMK	1.4361	0.3167	.0000
NS	0.9598	0.3057	.0017
AS	- 0.6064	0.3341	.0695

3. What is the adjusted odds ratio for the effect of SMK on cervical cancer outcome? . ???
4. Is number of lifetime sexual partners a significant predictor of cervical cancer outcome, controlling for the matching variables and the other non-matching variables? . . . . ???
5. Are the matching variables being controlled? . . . . . . ???
6. A 95% CI for the (adjusted) effect of SMK is given by the expression . . . ???

Choices
1.4361   1.4361±1.96(.3167)   1.4361/.3167   exp(1.4361)   exp[1.4361±.3167)   exp[1.4361±1.96(.3167)]
no   yes

References

References on Matching

Breslow NE, Day NE. Statistical Methods in Cancer Research. Volume 1. The Analysis of Case-Control Studies. International Agency for Research in Cancer, Lyon, 1980.

Brock KE, Berry G, Mock PA, MacLennan R, Truswell AS, Brinton LA. Nutrients in diet and plasma and risk of in situ cervical cancer. J Natl Cancer Inst 1988 Jun 15;80(8):580-5.

Diaz T, Nunez JC, Rullan JW, Markowitz LE, Barker ND, Horan J. Risk factors associated with severe measles in Puerto Rico. Pediatr Infect Dis J 1992;11(10):836-40. (Example of frequency matching)

Donovan JW, MacLennan R, Adena M. Vietnam service and the risk of congenital anomalies. A case-control study. Med J Aust 1984;140(7):394-7.

Halpin TJ, Holtzhauer FJ, Campbell RJ, Hall LJ, Correa-Villasenor A, Lanese R, Rice J, Hurwitz ES. Reye's syndrome and medication use. JAMA 1982;248(6):687-91.

Kleinbaum DG, Klein M. Logistic Regression: A Self-Learning Text, 2nd Ed. Springer Verlag Publishers, 2002.

Kleinbaum DG, Kupper LL, Morgenstern H. Epidemiologic Research: Principles and Quantitative Methods. John Wiley and Sons Publishers, New York, 1982.

Kupper LL, Karon JM, Kleinbaum DG, Morgenstern H, Lewis DK, Matching in epidemiologic studies: validity and efficiency considerations. Biometrics 1981;37(2):271-291.

McNeil D. Epidemiologic Research Methods. John Wiley and Sons Publishers, 1996. (Example of pair matching in a cohort study)

Miettinen OS. Individual matching with multiple controls in the case of all-or-none responses. Biometrics 1969;25(2):339-55.

Miettinen OS. Estimation of relative risk from individually matched series. Biometrics 1970;26(1):75-86.

Ury HK. Efficiency of case-control studies with multiple controls per case: continuous or dichotomous data. Biometrics 1975;31(3):643-9.

Case-Crossover Design

Maclure M, Mittleman MA. Should we use a case-crossover design? Annu Rev Public Health 2000;21:193-221.

Maclure M. The case-crossover design: a method for studying transient effects on the risk of acute events. Am J Epidemiol 1991;133(2):144-53.

Redelmeier DA, Tibshirani RJ. Association between cellular-telephone calls and motor vehicle collisions. N Engl J Med 1997;336(7):453-8.

Homework

ACE-1. Types of Matching

Suppose you are using a case-control study to investigate whether a dichotomous exposure variable E is associated with a dichotomous disease variable D, and you decide to match on age-group (3 categories) and gender. Answer the following questions assuming that all a priori conditions for confounding are satisfied by age-group and gender.

a. If you choose controls by **frequency matching** on age-group and gender, does this mean that you can ignore controlling for age-group and gender in your analysis?
b. If you **individually match** on age-group and gender, does this mean that you can ignore controlling for age-group and gender in the analysis?
c. If you **individually match** on age-group and gender, should you control for age-group and gender by doing a stratified analysis using individually matched sets as your strata. (Assume no other variables are being controlled other than the matching factors.)
d. Is pooling individually matched sets equivalent to frequency matching?
e. If you **frequency match** on age-group and gender, can you consider pooling on one variable (say, age-group) without pooling on the other (say, gender)?
f. If you **frequency match** on age-group and gender, will you obtain better precision (of the adjusted odds ratio) than if we individually match?

g. If you **individually match** on age-group and gender, will you obtain better precision (of the adjusted odds ratio) by pooling (exchangeable) individually matched sets than by using individually matched sets as the strata?
h. If you **individually match** on age-group and gender, will you obtain the same adjusted odds ratio estimate whether or not you pool exchangeable matched sets or use individually matched sets as the strata?
i. Would you expect frequency matching to give roughly the same odds ratio as obtained by pooling individually matched strata?
j. When **individual matching** is used with several matching variables, including continuous variables, will it be advantageous to do a stratified analysis using pooled matched sets as your strata?

ACE-2. Analysis of Matched Data

An investigator plans to conduct a case-control study to examine the relationship between an exposure variable E and a certain disease variable D, controlling for the potential confounding effect of a certain extraneous variable F. The following table describes the total (source) population being studied.

	F yes			F no	
	E	**Not E**		**E**	**Not E**
Cases	120	40	**Cases**	40	40
Controls	1080	1080	**Controls**	360	1080
	1200	1020		400	1020

a. Based on the above data, does the variable F appear to be a "risk factor" for the disease? Explain.
b. Assume that 240 controls are sampled randomly from the source population and that all the 240 cases are to be studied. Compute the "expected frequencies" for the two tables that would result from random sampling
c. Is the variable F a confounder for the (expected) randomly sampled data? Explain.
d. What is the appropriate method of analysis for the (expected) randomly sampled data: stratified or un-stratified? Explain.
e. Assume that all 240 cases are used and that 240 controls are obtained by category matching. Compute the expected frequencies for the two tables that would result from this matched design.
f. What is the appropriate method of analysis for the matched data of part e: stratified or un-stratified? Explain
g. Carry out a stratified (matched) analysis of the expected frequency data. Make sure to compute a point estimate of the effect, a test of significance and a 95% CI.
h. For the given population, which approach would you expect to give more precise results (in the estimation of the adjusted odds ratio), matching or random sampling? Explain.
i. In general, how would you decide whether to match or not match on extraneous variables when doing a case control study?

ACE-3. Matching: Follow-Up vs. Case-Control Studies

This fictitious example illustrates and contrasts a matched follow-up study with a matched case-control study with respect to the control of confounding by a matched factor (age). Consider an underlying cohort of 2,000,000 people in which 50% are young and 50% are old. Suppose 50% of the young are exposed while 10% of the old are exposed. Also suppose that age is a risk factor for the disease. The risk of disease for each exposure-age group, over a given time period, is summarized below:

Group	Number	Risk	Cases
Young Exposed	500,000	0.005	2,500
Young Unexposed	500,000	0.001	500
Old Exposed	100,000	0.030	3,000
Old Unexposed	900,000	0.006	5,400

a. What is the risk ratio for exposed vs. unexposed among the young and among the old ? Is age an effect modifier?

b. Create the crude 2-by-2 table for exposure and disease. What is crude risk ratio for exposure in this cohort? Is age a confounder in this cohort?
c. Suppose you had conducted a matched follow-up study from this cohort in the following manner: One percent of the exposed subjects were enrolled (6,000 subjects). You also enrolled 6,000 unexposed subjects that were matched by age group to the exposed subjects. Create the 2-by-2 table (not stratified) that would result if this study was carried out correctly and there was no random error (i.e., give the expected counts). Calculate the risk ratio. Is the risk ratio confounded?
d. Suppose you conducted a matched case-control study with this cohort in the following manner: One percent of the cases were enrolled (this should be 114 cases). You also enrolled 5 controls for each case yielding 570 controls in which controls were chosen to have the same age distribution as the cases (i.e., controls were matched on age). Create the resultant 2-by-2 table (not stratified) assuming no random error. Choose the controls from the entire cohort (not just from the non-diseased population – the numbers work out easier). Calculate the exposure odds ratio. Is the odds ratio biased?
e. Adjust for the bias from part d (if necessary) by stratifying by age in the case-control study analysis (i.e., calculate stratum specific odds ratios).
f. In part d), controls were chosen from the entire cohort rather from just the non-diseased portion of the cohort. What measure of disease association, does the adjusted exposure odds ratio estimate when controls are chosen in this manner? Why in this study does it make little difference whether the controls were chosen from the entire cohort or just from the non-diseased portion of the cohort?
g. Suppose you conducted a case-control control study in which controls were not matched to cases. As in part d), 114 cases were enrolled, and controls were chosen from the entire cohort. Unlike part d), controls are not matched by age and 600 controls are selected. Create the resultant 2-by-2 table assuming no random error.

Calculate the exposure odds ratio. Is the odds ratio biased?

h. Adjust for the bias from part g) (if necessary) by stratifying by age in the case-control study (i.e., calculate stratum specific odds ratios).

i. Is there any advantage in conducting the matched case-control study (as in parts d and e), as opposed to the unmatched case-control study (as in parts g and h)? Answer this question just in terms of these particular studies and ignore the fact that a slightly different number of controls were chosen for each study (570 vs. 600). The numbers were chosen to make the calculations simpler.
j. Consider a case-control study in which it is desired to control for neighborhood of residence and each case resides in a different neighborhood. Is there any advantage in conducting a matched case control study (matching on neighborhood) as opposed to an unmatched case control study? Explain.
k. Contrast the matched follow-up study conducted in part c) with the matched case-control study conducted in part d) with respect to the control of confounding by age (the matched factor).
l. Below is an excerpt from Kenneth Rothman's text Modern Epidemiology (first edition). The quoted text is the first paragraph of his principles of matching section on page 237. Not withstanding the writing style, there is an important point being made concerning the principles of matching which this entire question attempted to illustrate. Explain the point Rothman is making.

Rothman quote:
The topic of matching in epidemiology is beguiling. What at first seems clear is seductively deceptive. Whereas the clarity of an analysis in which confounding has been securely prevented by perfect matching of the compared series seems indubitable and impossible to misinterpret, the intuitive foundation for this cogency attained by matching is a surprisingly shaky structure that does not always support the conclusions that are apt to be drawn. The difficulty is that our intuition about matching springs from knowledge of experiments or follow-up studies, whereas matching most always applied to case-control studies, which differ enough from follow-up studies to make the implications of matching different and counterintuitive.

Answers to Study Questions and Quizzes

Q15.1

1. Reye's syndrome was associated with only certain types of viruses, so "virus type" was an important risk factor for the outcome.
2. Older children were less likely to develop Reye's syndrome, i.e., age was an important risk factor.

Q15.2

1. You need to categorize the continuous variables age, blood pressure, and body size.
2. Choose one or more controls to be in the same category of age, gender, smoking status, blood pressure, and body size as the case.
3. No. The case and controls have to be in the same category for each of the matching variable. The control choice in the question is not appropriate because both body size and blood pressure categories of the control are different categories than observed on the case.
4. {High BP, age $\geq$ 55} = 120; {High BP, age < 55} = 180; {Normal BP, age $\geq$ 55} = 150; {Normal BP, age < 55} = 150
5. {High BP, age $\geq$ 55} = 20%; {High BP, age < 55} = 30%; {Normal BP, age $\geq$ 55} = 25%; {Normal BP, age < 55} = 25%
6. Frequency matching because it should be logistically easier and less costly to find groups of control subjects, particularly by hospital than to find controls one case at a time.
7. Individual matching because there are many variables to match on, many of which are quite individualized.

Q15.3

1. 30 cases and 60 controls
2. 70 cases and 140 controls
3. R_i to S_i, since $R_1 = 60$, $S_1 = 30$, and $R_2 = 140$, $S_2 = 70$
4. Four
5. R_i to S_i matching ratio. Even though there are twice as many controls overall as there are cases, the numbers of cases and controls vary within each stratum.
6. White Male: R = 200, S = 100; White Female: R = 200, S = 100; Black Male: R = 80, S = 40; Black Female: R = 120, S = 60.

Q15.4

1. For a fixed number of index subjects, total sample size for the study increases as R increases
2. There is a 33% increase in going from R = 1 to R = 2. This indicates that there is considerable precision to be gained by using 2 to 1 matching instead of 1 to 1 matching.

Q15.5

1. False – in a case-control study, the index group is composed of cases.
2. False – in a cohort group, the comparison group is composed of unexposed individuals.
3. False – The control of a given case in this situation would need to be the same regarding both matching factors.
4. True
5. True – Five to one matching will result in an increase in precision of 4.2% compared to four to one matching.
6. True
7. True
8. False – Choosing 2 controls per case versus 1 per case will increase precision by 33%. This increase in precision will continue with each added control per case. However, the table of Pittman Efficiency, values computed for different values of R (number of controls per case) indicates a diminishing return regarding efficiency once R exceeds 4.

Q15.6

1. No, hair color has no known relationship to bladder cancer.
2. Because hair color has no known relationship to bladder cancer, and is not a risk factor needing to be controlled, matching on hair color is unlikely to have any effect on the precision of the estimated exposure-disease effect, i.e., even though matching on hair color will make cases and controls "balanced" with respect to hair color, the estimated effect is unlikely to be more precise than would result from "unbalanced" data obtained from not matching.
3. Fast-food products tend to be high in saturated fats, so if you match on amount of fast-food products in one's diet, you may effectively be matching on cholesterol level.
4. You will have difficulty finding matches for some index subjects, and consequently, are likely to obtain a much smaller sample size than originally planned.

Q15.7

1. There is no strong reason for matching, and the reasons against outweigh the reasons for.
2. You might expect to gain precision from matching, but you also need to weigh the other reasons listed, particularly the cost in time and money to carry out the matching.
3. Again, your decision depends on how you weight all the reasons for and against matching. Matching on neighborhood of residence may be a convenient way to control for social and environmental factors that are difficult to measure. However, if the exposure variable has a behavioral component, you must be careful that you won't overmatch in this situation.

Q15.8

1. False – one reason for deciding to match in a case-control study is to obtain a more precise estimate of the odds ratio of interest.
2. False – a reduction in the sample size due to not finding an appropriate match would be a disadvantage of matching.
3. True
4. a, d, and e
5. False – you still need to be concerning about matching on weak risk factors and overmatching.
6. It depends – it depends on at what cost you gain the precision. It is important to consider other things such as cost, money, time, how strong or weak are the risk factors, overmatching, etc.

Q15.9

1. Logit $P(X) = b_0 + b_1(D1) + b_2(D2) + \ldots + b_{99}(D99) + b_{100}(E)$ where D1 through D99 are 99 dummy variables that distinguish the 100 matched pairs. In particular, the Di may be defined as follows: Di=1 for a subject in the i-th matched pair and Di = 0 for a subject not in the i-th matched pair. Thus, for each of the two subjects in the first matched pair, D1 = 1, D2 = D3 =…= D99 = 0 and for each subject in the 100-th matched pair, D1 = D2 =…D99 = 0.
2. Logit $P(X) = b_0 + b_1(D1) + b_2(D2) + \ldots + b_{99}(D99) + b_{100}(SBP) + b_{101}(CHL) + b_{102}(E)$ where D1 through D99 are 99 dummy variables that distinguish the 100 matched pairs.
3. Any matched pair in which the case is in a different SBP or CHL category than is the corresponding control will have to be dropped from such a stratified analysis. The only matched pairs to be kept for analysis will be those in which both the case and control are in the same SBP and CHL categories.

Q15.10

1. All strata have zero cells, so it is not possible to compute stratum-specific odds ratios.

Q15.11

1. The mOR estimate of 9.67 indicates a very strong relationship between estrogen usage and endometrial cancer, controlling for the matching variables. The MH test has a P-value equal to zero to four decimal places. Therefore, the point estimate is highly significant. The 95% CI is quite wide, so there is considerable imprecision in the point estimate. Overall, however, the results suggest a strong effect of estrogen use on the development of endometrial cancer.

Q15.12

1. All strata have at least one zero cell, since the one case in each stratum is either exposed or unexposed.
2. Even though there are only 5 subjects per stratum, the total study size is 90. The large-sample property of the MH test depends on the total sample size, not on the stratum-specific sample size.
3. There is a very strong and significant effect (mOR = 33, $P < .0001$) of previous abortion and the development of tubal pregnancy for women with at least one previous pregnancy. There is lack of precision in the estimate, however, likely due to the number of cases (18) and total sample size (90) being relatively small.

Q15.13

1. True
2. False – in a pair-matched case-control study, the mOR is computed as X/Y where X and Y are discordant pair frequencies.
3. False – the computed value for McNemar's test statistic is:
   $(X - Y)^2 / (X + Y) = 5$.
4. False- the MH test statistic is appropriate for R-to-1 matching
5. 1.03
6. 0.07
7. (0.80, 1.32)
8. no

Q15.14

1. Four exchangeable sets: smokers below 60,

smokers 60 or above, non-smokers below 60, and non-smokers 60 or above.

2. "It depends!" If there are any exchangeable matched sets, pooling is certainly appropriate. However, there are so many variables being matched on that the study data is likely to contain only a few sets of exchangeable strata. Consequently, the use of a pooled analysis is likely to have a negligible effect on the estimates odds ratio and precision around this estimate when compared to an unpooled matched analysis.

Q15.15

1. 1
2. 1
3. 2
4. 1
5. McNemar's
6. 2
7. unpooled
8. 2.5
9. pooled
10. pooled
11. 2
12. 1
13. 0
14. 2
15. undefined
16. 2.5
17. a

Q15.16

1. There is little evidence of confounding, since the crude (4.09) and adjusted (4.10) odds ratio estimates differ very little.
2. There is very little evidence of interaction, since the stratum-specific odds ratios are quite close to each other (around 4). Consequently, an overall assessment of the E-D relationship appears justified.
3. There are large numbers in each of the four strata and there are no zero cells in any strata. Consequently, a mOR is not necessary (to deal with a zero-cell problem) and an aOR can therefore be computed.
4. There is an approximate fourfold effect of exposure on disease. This effect is significant (based on the MH test) and the estimated confidence interval gives a reasonably precise estimate of the effect.

Q15.17

1. True
2. c – having zero cells in some strata supports the use of an mOR for this or any stratified analysis. Frequency matching does not usually indicate the need to use an mOR, but in this example, the use of frequency matching required an mOR because of at least one zero frequency in stratum 1 and perhaps other zero frequencies in other strata. If logistic regression is used, then an mOR cannot be used to estimate the odds ratio.
3. False – if we ignore control for the matching variable in a case-control study, our estimate is biased towards the null.
4. R_i to S_i – Using frequency matching, the matching ratio of controls to cases will usually be different for different strata.
5. Not necessarily – you should expect to get better precision from frequency matching versus no matching, but there is no guarantee that frequency matching will give better precision than individual matching and vice versa.

Q15.18

1. The study is a cohort study, where we typically use a risk ratio (RR) as the measure of effect. Also, since all strata have zero cells, it is not possible to compute stratum-specific risk or odds ratios, which need to be averaged to obtain an aRR.

Q15.19

1. The mRR estimate of 1.33 indicates a relatively weak relationship between shift type and chronic low back pain controlling for the matching variables. The MH test has a P-value equal to .0680. Consequently, the point estimate is of borderline significance and is not significant at the .05 level. The 95% confidence interval lies between a lower limit (.98) just below 1 and an upper limit of moderate size (1.81) above 1, so there is considerable imprecision in the point estimate. Overall, the results suggest a relatively weak effect of rotating shifts.

Q15.20

1. False – The total number of exposed in each matched set is always 1.
2. True
3. 1.2
4. 2.5
5. (0.96, 1.50)

Q15.21

1. The mOR value controls only for the matching variable, i.e., it does not control for GALL.

2. $$\text{logit } P(X) = b_0 + \sum_{i=1}^{62} b_i D_i + b_{63} E$$
3. The adjusted odds ratio is exp(2.269) = 9.67, which is the same value as the mOR.
4. P=.0001, which is extremely small, so we reject the null hypothesis at the .05 and .05 levels and conclude that estrogen use is a significant predictor of endometrial cancer status control for the matching variables and for the unmatched variable GALL.
5. The McNemar chi square statistic ignores the control of GALL, whereas the chi square statistic here controls for GALL.
6. The confidence interval is quite wide, indicating that there is considerable imprecision in the estimated odds ratio.

Q15.22

1. b
2. Conditional MLE
3. exp(1.436)
4. yes
5. yes
6. exp[1.4361 ± 1.96(.3167)]

LESSON 16

ActivEpi Reference

16-1 Data Desk Reference

We hope that you will find the ActivEpi resources a valuable reference. This page on the ActivEpi CD ROM is designed to make it easy for you to find and use the reference materials in ActivEpi. In addition to the icons on this page, you may want to take advantage of some special web services found by clicking the WEB icon in the control bar.

A number of icons are presented on the ActivEpi CD-ROM for the following Data Desk Activities:

- Launch Data Desk
- See the Data Desk documentation
- Learn how to dichotomize a variable in Data Desk
- See Data Desk Interactive examples
- Perform a sample analysis in Data Desk.

16-2 Tables Reference

This page on the ActivEpi CD ROM provides access to tables found in many statistics textbooks. For each of these distributions, the user can click on the distribution graphic or move around with the tables. The distributions included are the Z, t, chi square, and F distributions. A screen image for the Z (Standard normal) is presented below; the other distributions have similar types of presentation.

Selected Statistical Tables

Standard Normal Distribution Table

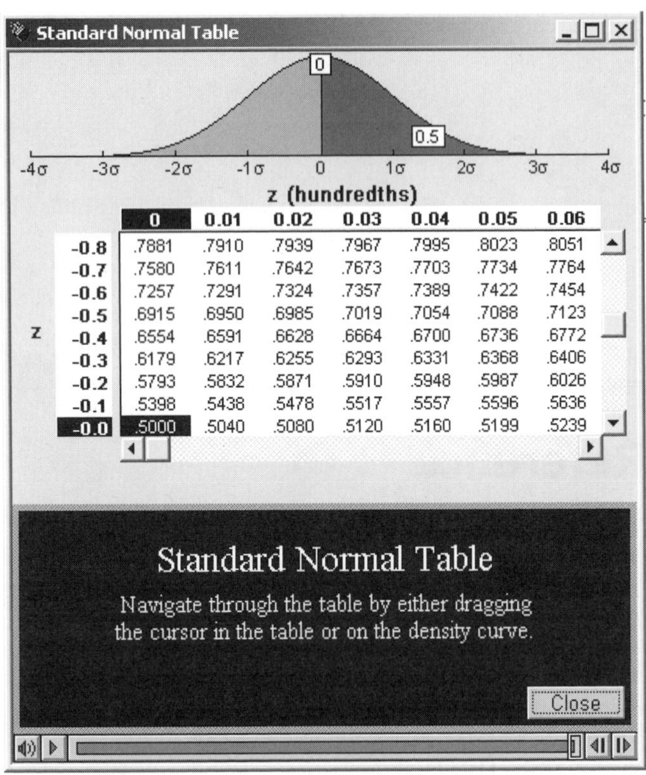

z	0	0.01	0.02	0.03	0.04	0.05	0.06
-0.8	.7881	.7910	.7939	.7967	.7995	.8023	.8051
-0.7	.7580	.7611	.7642	.7673	.7703	.7734	.7764
-0.6	.7257	.7291	.7324	.7357	.7389	.7422	.7454
-0.5	.6915	.6950	.6985	.7019	.7054	.7088	.7123
-0.4	.6554	.6591	.6628	.6664	.6700	.6736	.6772
-0.3	.6179	.6217	.6255	.6293	.6331	.6368	.6406
-0.2	.5793	.5832	.5871	.5910	.5948	.5987	.6026
-0.1	.5398	.5438	.5478	.5517	.5557	.5596	.5636
-0.0	.5000	.5040	.5080	.5120	.5160	.5199	.5239

t Distribution Table

Chi Square Distribution Table

F Distribution Table

16-3 Other Programs for Analyzing Epidemiologic Data

A special student version Data Desk has been integrated into ActivEpi to provide you with a platform to practice applying the concepts you have learned. Several other popular analysis programs are available to analyze epidemiologic data. The names of the program, and the companies that make the programs, are listed below. Web links to each of these companies can be found in the Web page for this lesson. Click on the web icon in the title bar to access the web page.

- **Epi Info**, from the Centers for Disease Control and Prevention. The CDC is located in Atlanta, GA.
- **SAS**, from the SAS Institute. The SAS Institute is one of the oldest data analysis software companies. They are located in Cary, NC.
- **SPSS**, from SPSS Inc.. SPSS, Inc is located in Chicago, IL.
- **Stata**, from the Stata Corporation. Stata Corporation is located in College Station TX.
- **S Plus**, from Insightful. Insightful is located in Seattle, W A.

Index

R

S

T

U

V

W

Z